W9-CDX-280

An Introduction to Analysis and Integration Theory

ESTHER R. PHILLIPS
Professor of Mathematics and Computer Science
Herbert H. Lehman College
of the City University of New York

With a New Historical Introduction
and Notes by the Author

Dover Publications, Inc.
New York

To Lipman Bers

Copyright © 1971 by International Textbook Company.
Copyright © 1984 by Esther R. Phillips.
All rights reserved under Pan American and International Copyright Conventions.

Published in Canada by General Publishing Company, Ltd., 30 Lesmill Road, Don Mills, Toronto, Ontario.
Published in the United Kingdom by Constable and Company, Ltd., 10 Orange Street, London WC2H 7EG.

This Dover edition, first published in 1984, is a corrected and enlarged republication of the work first published by Intext Educational Publishers, Scranton, Pa., in 1971. A new Preface, a Historical Introduction, and new Notes have been specially prepared for this edition by the author, who has also made extensive corrections.

Manufactured in the United States of America
Dover Publications, Inc., 31 East 2nd Street, Mineola, N.Y. 11501

Library of Congress Cataloging in Publication Data

Phillips, Esther R.
An introduction to analysis and integration theory.

"A corrected and enlarged republication of the work first published by Intext Educational Publishers, Scranton, Pa., in 1971"—T.p. verso.
Bibliography: p.
Includes index.
1. Integrals, Generalized. 2. Measure theory.
I. Title.
QA312.P48 1984 515.4 84-6138
ISBN 0-486-64747-1 (pbk.)

Preface to the Dover Edition

In this edition a new Historical Introduction has been added to the original text. Because the subject is obviously too vast and complex to be treated adequately in a few pages, the discussion will be restricted mainly to two topics: the evolution of the concept of the integral and the construction of the real-number system. This choice appeared natural for two reasons: first of all, these subjects occupy the greater part of the text—indeed, in the original Preface, the integral is referred to as the heart of the book. Further, the discussion of the real-number system and the integral includes, in a natural way, the closely related topics of trigonometric and Fourier series, the concept of function, the arithmetization of analysis, and the evolution of set and measure theoretical ideas. The reader will find suggestions for further reading in the Notes and the References (to the Historical Introduction). Admittedly, the References are very incomplete and are not meant for the specialist. Note 32 indicates those sources containing extensive and useful bibliographies.

In addition, five new Notes have been added to the main text (see page 447). These are indicated in the text by a bullet (•) in the outer margin next to each line for which there is a note.

I would like to thank Joseph W. Dauben, Godfrey L. Isaacs and David Rothchild of Herbert H. Lehman College of the City University of New York and Gregory Moore of the University of Toronto for having read a preliminary version of the Historical Introduction and for having made many valuable suggestions. Above all, I thank Morris Kline for his encouragement and for making possible the publication of this edition.

Esther R. Phillips

Preface to the First Edition

The unifying theme of this text is the completion of a space. This idea is first introduced in the construction of the real numbers as the smallest complete (Archimedean) ordered field which contains the rationals. The subsequent extension of this construction (Chapter 2) to general metric and normed linear spaces is scarcely more than a notational exercise. The heart of the book, however, is Part II, where the (real) Lebesgue integrable functions are constructed by completing a space of step functions, upon which has been defined a norm that assigns to each step function the sum of the areas under its steps as the norm of that function. This quantity may be thought of as an integral (Lebesgue or Riemann) of the absolute value of the step function. However, the choice of step functions as the first functions to be integrated, permits a development which is independent of the Riemann integral (or any other) and assumes only the simple fact that the area under each step is the area of a rectangle, and therefore known. By completing this normed linear space the Lebesgue integrable functions are obtained, with the integral and norm extended to the completion by taking limits. In this treatment of integration and measure theory, the integrable functions appear before both the measurable functions and sets, the latter being defined in terms of the measurability of their characteristic functions, whose integrals, if they exist, give the measures of the sets.

The basic outline of this text originated in a graduate course given by Professor Lipman Bers at New York University. These lectures were received as enthusiastically as they were given, and the accompanying lecture notes, although uneven in quality and lacking the original impact of the lectures, were in great demand, both at the time that the course was given and in subsequent years. Professor Bers often spoke of correcting and editing these notes for future publication, but other duties prevented his carrying out this plan.

In teaching this course, both to undergraduates and to graduate students, the material in Part I was expanded into its present form. In order to cope with improperly prepared students (some with little or no advanced calculus) it seemed more realistic to postpone the construction of the integral to the second semester, and to concentrate on the metric and topological properties of some familiar spaces, many of which would appear in the subsequent development. By the time it was agreed that I should write this text, Part I contained not only the construction of the real numbers and the necessary definitions and properties of metric spaces, but a considerable amount of the standard material on complete, connected, and compact spaces. For example, the Baire Category theorem, the theorems on contracting and nonexpanding mappings, the Intermediate Value theorem, theorems about continuous functions defined on connected or compact spaces, and the Stone-Weierstrass theorem were proved, and all of these were accompanied by examples and applications. In the case of a more adequately prepared group of students (who have completed a good advanced calculus course) much of Part I may be condensed or even omitted, and it is not unrealistic in this case to think of beginning the treatment of Lebesgue integration by mid-semester or even earlier.

Part II begins with a discussion of step functions and their integrals, and gives the extension of this integral to the full space of (real) Lebesgue integrable functions. This is followed by the introduction of (Lebesgue) measurable functions and sets. In Chapter 8, other approaches to Integration and Measure theory, in particular one which begins with the notion of exterior measure, are discussed, and the equivalence of these approaches is demonstrated. This section of the book concludes with a chapter on differentiation, which contains the "Fundamental theorem" relating differentiation and Lebesgue integration.

Part III contains a short introduction to Banach and Hilbert spaces and the linear functionals defined on them. Chapter 10 deals primarily with Hilbert spaces and contains a rather long section on Fourier series, which may be entirely omitted from any presentation since no subsequent material depends in any way upon it. In Chapter 11 the Riesz Representation theorem is proved for bounded, linear functionals defined on the L_p spaces, as well as on the space $C[a, b]$. In the latter case, the functionals are represented by Stieltjes integrals. The chapter concludes with a proof of the Hahn-Banach theorem and some applications.

In the fourth and final part of the book, many of the earlier results of the Lebesgue theory are extended to abstract spaces. Again, in order to obtain a theory of integration and measure for abstract spaces, either of the methods discussed in Part I may be extended. Thus, we may begin first with the (Daniell) integral defined on a class of *elementary* functions (replacing the step functions for the real case), complete this space to obtain the class of integrable functions, and then introduce measurable functions and sets. Or, as in the real case, we may begin with the measurable sets, obtaining the measurable and integrable functions later.

In the final chapter, the Lebesgue-Stieltjes measures are introduced and used to answer a question similar to the one posed earlier (Chapter 9) for Lebesgue measurable functions — when can a given (distribution) function be represented by a Lebesgue-Stieltjes integral? The answer — whenever it is absolutely continuous — is the same as in the preceding case. Finally in the Radon-Nikodym theorem, the conditions are given which insure that one measure may be represented as an integral with respect to a second measure.

Integrals over product spaces are discussed briefly in the final section of Chapter 13, and the Fubini theorem which relates the integral over a product space and an iterated integral is proved.

The exercises are an important part of the text. In particular, those which are starred (*) appear later in subsequent discussions or in other exercises.

I should like to thank Professor Marvin Tretkoff of the Stevens Institute of Technology, who read the bulk of the manuscript and offered many valuable criticisms, and Professor Godfrey Isaacs of Lehman College, who taught from an almost final version of the text and was exceedingly generous with both constructive criticisms and encouragement. But above all, I am deeply grateful to Professor Lipman Bers with whom the entire project originated. The structure of the book, in particular the unifying idea of completion, was suggested, if not determined by, the lectures that he gave. Needless to say, any misdirected embellishments of the main theme, as well as any errors that may appear in the final text, are entirely my own responsibility. Finally, let me express my gratitude to Intext, and in particular to their Mathematics Editor, Mr. Charles J. Updegraph, whose patience and cooperation have been greatly appreciated.

Esther R. Phillips

Bronx, New York
March, 1971

Historical Introduction

1 The Integral Before Lebesgue

When the integral was introduced in the seventeenth century as a tool for finding areas and volumes, it was initially conceived as an infinite sum of "indivisibles" or "infinitesimals," suggesting the modern definition of the (Cauchy-Riemann) integral as a limit of approximating sums. However, in creating the infinitesimal calculus, both Isaac Newton and Gottfried W. Leibniz recognized the inverse relation between differentiation and integration (as did some of their predecessors in special cases), and until the early nineteenth century the integral was primarily characterized in applications as the inverse of differentiation. This was due largely to Newton, who used series expansions to obtain antiderivatives of many functions, thus greatly simplifying the computation of what we would call the definite integral. Consideration of the area under the curve $y = ax^{n/m}$ led Newton to show that the derivative of the function representing the area is the ordinate of the corresponding point of the curve. This special case of the *Fundamental Theorem of Calculus* was quickly extended to finite sums of such functions and then (illegally by today's standards) to functions whose antiderivatives could not be found directly, but which could be expanded in a series of fractional powers and integrated term by term. (The most famous example is the binomial expansion derived by Newton for $(1 + x)^{1/2}$ in connection with the quadrature of the circle.)

Although his predecessors had used series in special cases, Newton was the first to regard series expansions as a *universal method of representing functions*.[1] Sometimes term-by-term integration leads to familiar functions (as does the integration of the series for $\sin x$, $\cos x$, e^x, etc.), but this is not always the case: for example, integrating the series corresponding to e^{-x^2} does not produce a series of a familiar transcendental function. Thus the extensive use of series by Newton and his successors (especially Leonhard Euler) greatly enlarged the class of functions studied by mathematicians, and, before long, series had become a separate topic of study.[2]

Throughout most of the eighteenth century few mathematicians questioned the *existence* of an antiderivative for a given function. Since integrals were used mainly in problems of geometry and physics, their existence appeared to be assured by the nature of the geometrical or physical problem. Where an antiderivative was not known or where an integrand could not be expanded into a series to be integrated term by term, mathematicians were often able to evaluate (or approximate) the integrals by calculating what we would call their Cauchy-Riemann sums. However, these sums were not used, before Cauchy's time, to *define* the integral; their frequent use in *evaluating* integrals, together with the general development of techniques for summing series, made Cauchy's contribution possible. Yet the availability of these techniques cannot alone account for Cauchy's attempt to formulate a rigorous theory of the integral, based upon carefully articulated concepts of *limit* and *continuity*. Why, after more than a century of unprecedented progress in mathematics, did Cauchy decide that it was necessary to introduce the concepts of continuity and limit and to *establish the existence* of the integral for continuous functions?

From the time of Newton and Leibniz there had always been critics of the calculus who deplored the vagueness of its concepts. One of the earliest, Bishop George Berkeley, in his famous essay, *The Analyst, or a Discourse Addressed to an Infidel Mathematician* (1734), claimed that it was "an unaccountable paradox that mathematicians should deduce true propositons from false principles," while remaining "squeamish about any point of divinity." Attempts had been made from time to time during the eighteenth century to banish arguments involving the infinite or infinitesimals, but few mathematicians were sufficiently concerned with foundations to question the propriety of concepts and techniques that had led to so many useful and remarkable results. Until the end of the eighteenth century, questions of foundations and rigor were neither the frequent nor the regular concerns of working mathematicians. Those who did take up such issues were primarily interested in metaphysics. However, as applications of the calculus multiplied, apparent paradoxes began to appear. One such "paradox" resulted from the uncritical use of the formula $\int_a^b f'(x)\, dx = f(b) - f(a)$, in particular where $f'(x) = x^{-m}$ $(m > 0)$ and the interval contains the point 0. It was hard to explain what was meant by $\int_{-1}^{1} x^{-1}\, dx = \ln(-1)$; the left side appeared to be a real number, whereas the right side was imaginary.[3] And so, by the end of the eighteenth century, as the limitations of the old techniques became increasingly apparent, serious attention to foundational matters became the concern of *working* mathematicians.

Joseph-Louis Lagrange, like Bishop Berkeley (whose essay he had read), was a strong critic of the vague and nonrigorous use of infinitesimals, fluxions and limits (as conceived at that time), asserting that correct results had often been obtained by a compensation of errors. In 1784, during his tenure as mathematics director at the Berlin Academy, a prize was offered for a lucid, precise theory of the infinite and infinitesimals.[4] Later (1797) Lagrange proposed to found the calculus on the algebra of infinite series. (Eighteenth-century mathematicians saw no difficulty in extending the operations on finite sums to infinite sums.) Assuming that (except possibly at finitely many points) every function could be expressed by its Taylor series, Lagrange *defined* the derivative to be the coefficient of the linear term. In this way, he believed, all limits and infinitesimals and the like would be banished from the calculus. Unfortunately Lagrange's method posited the analyticity of all functions. Doubtless it was with this egregious assumption in mind that Cauchy later gave the following well-known example of an infinitely differentiable function, not expressible by its Taylor series:

$$f(x) = e^{-1/x^2} \text{ if } x \neq 0 \text{ and } f(0) = 0.$$

It should be noted that, in early attempts (such as Lagrange's) to lay the foundations of the calculus, the derivative, rather than the integral, received the greater attention. This is not surprising in light of the prevailing *definition of the integral as an antiderivative.* From this point of view, properties of the integral were derived from the logically prior properties of the derivative.

In the 1760s and 1770s a stimulating and eventually very productive debate arose over the possibility of representing "arbitrary" functions by trigonometric series. The problem originated in the analysis of the vibrating string; the arbitrary function was to represent the initial position of the string, which might, at worst, have "corners." This polemic, whose participants included Jean le Rond d'Alembert, Euler, Daniel Bernoulli, Lagrange and Pierre-Simon Laplace, highlighted the vague and inconsistent working definitions of such concepts as function and continuity; it further emphasized the need for a definition of the integral for functions whose primitives (antiderivatives) could not easily be found. J.-B.-J. Fourier, in a paper on heat diffusion presented in 1807, appeared to give an affirmative answer to the question of whether an "arbitrary," perhaps discontinuous, function could be represented by a trigonometric series. He was able to show, by solving the equation

$$f(x) = a_0 + \sum_{n=1}^{\infty} a_n \cos nx + b_n \sin nx$$

for the infinitely many unknown coefficients, that the latter were expressible in terms of integrals. Using a term-by-term integration (whose validity Fourier did not question) the coefficients were obtained in integral form (see pp. 296 ff.); in particular, a_0 is given by the integral of $f(x)$ over the interval $[-\pi,\pi]$. For some physical problems $f(x)$ may have corners (the plucked string) or be discontinuous. Since such functions are not always representable as the sums of algebraic functions, their antiderivatives could not be found. On the other hand, physical or geometrical considerations make it possible to talk about the *area* under the curve belonging to $f(x)$. In this way Fourier's work, which opened the way to the consideration of a wider class of functions, emphasized the need for a reformulation of the concept of the integral that would include these more general functions.[5]

The integral equality $\int_{-1}^{1} x^{-1}\, dx = \ln(-1)$, cited above as an apparent paradox arising from the uncritical use of antiderivatives in the eighteenth century, was not, of course, an isolated example. In a major paper published in 1814, Augustin-Louis Cauchy launched a study of complex integration containing an analysis of such singular integrals. It is well known that the value of a complex integral taken over a plane curve does not (generally) depend only upon the endpoints of the curve, and that the identification of the integral and antiderivative breaks down. The subsequent effect of this study on real analysis has been described as "nothing less than an attack on the hallowed inversion principle relating differentiation and integration" ([14], p. 35).* Indeed, it has been argued that Cauchy's study of complex integrals may well have spurred his efforts toward a reconsideration of the (real) integral and a careful formulation of the basic concepts of analysis (see [13], pp. 144 ff.).

Henri Lebesgue, on the other hand, believed that Cauchy had initiated his program of rigorization mainly for pedagogical reasons, a view that appears to be supported by the fact that Cauchy's rigorous reformulation of the calculus, including his theory of the integral, is contained mainly in three works—*Cours d'analyse algébrique* (1821), *Résumé des leçons sur le calcul infinitésimal* (1823) and *Leçons sur le calcul différentiel* (1829)—all published from lectures given at the École Polytechnique.[6]

Although pedagogical reasons may well have influenced Cauchy, it is difficult to regard such motives as decisive. Cauchy's efforts at rigorization and reformulation were not unique, although they represented the broadest, most ambitious and most influential reformulation made at that time.[7] The rapid development of mathematics throughout the eighteenth century, the increasing complexity of the problems studied and the uncritical extension of finite techniques and operations to infinite ones, though fruitful, had raised many foundational questions. Further progress in analysis, in particular in such new fields as complex function theory, required a careful formulation of the concepts on which the operations were based.

It is not certain to what extent Cauchy was influenced by Fourier's paper of 1807. (Citing the work of others was rare at that time. Indeed, it has been suggested that some of Cauchy's ideas were taken from works he failed to cite.)[8] In any case, his integral constituted a response to questions raised by Fourier: with the old definition, the existence of the integral as an antiderivative was ensured by the function's analytic representation as a sum of algebraic terms—this could be integrated term by term. With Cauchy's definition, the existence of the integral depended on a *property of the integrand*—its *continuity*—not an analytic representation. This apparent generality permitted the area under a curve belonging to $f(x)$ to be expressed as an integral, even if $f(x)$ could not be represented by an analytic expression whose antiderivative could readily be

*See References below. All reference numbers in square brackets indicate numbered sources in that section.

found. Moreover, once $\int_{-\pi}^{\pi} f(x)\, dx$ was defined, so were the remaining Fourier coefficients; thus Fourier's conjecture that "arbitrary" functions could be represented by trigonometric series appeared to be justified. Precisely what Fourier meant by an arbitrary function was never made entirely clear. Certainly he wished to admit all functions appearing in physical problems (the initial configuration of the plucked string, for example), including many that could not be given by a single equation (or expression).

It was P. G. L. Dirichlet (continuing Fourier's investigations) who gave the modern definition of a function in a paper containing a careful presentation of the first set of sufficient conditions for the convergence of a Fourier series (see p. 305, below).[9] Dirichlet gave, as an example of a function failing to satisfy the conditions, $f(x) = c$ if x is rational, and $f(x) = d \neq c$ if x is irrational (see Example 6.3.5, p. 196, below). He hoped, at some future time, to extend Cauchy's integral to a class of functions having an infinite set of discontinuities (although he did not expect the above-mentioned function to be integrable in the extended sense), thereby widening the class of functions representable by Fourier series.

Although Dirichlet never made such an extension, Georg Friedrich Bernhard Riemann, who had briefly studied with Dirichlet in Berlin and later followed him to Göttingen, did. In 1854 Riemann submitted his Habilitationsschrift (not published until 1868), "On the Representation of Functions by Trigonometric Series." In this work Riemann succeeded in extending Cauchy's integral to a wide class of functions, some of whose singularities are dense on the interval of integration. Notwithstanding Dirichlet's example of an everywhere dense discontinuous function, such functions did not enter significantly into mathematical investigations before Riemann's time. Indeed, their entry into the mainstream of mathematics was long delayed (see Chapter 6, Section 3).

It is beyond the scope of this brief survey to describe and analyze the concepts of this great work. We will simply cite some of the major results, in particular those discussed in the present text, and indicate some aspects of Riemann's theory, the further development of which led ultimately to Lebesgue's contribution.

Riemann's goal was to discover necessary and sufficient conditions for a function to be represented by a Fourier series.[10] The first step was to extend Cauchy's integral to a wide class of functions, thereby enlarging the set of functions whose Fourier coefficients could be defined. Cauchy had first proved the existence of an integral for functions continuous on closed, bounded intervals; later the integral was extended to piecewise continuous functions. Instead of considering further extensions, Riemann set up the approximating sums for *any* function *bounded* on such an interval, defining the integral to be the limit (when it exists) of these sums as the norm of the subdivisions tends to zero (see Chapter 6, Section 3). Thus Riemann set out to obtain the *widest* class of functions integrable in Cauchy's sense. He derived a characteristic property of such functions: if $a = a_0 < a_1 < \cdots < a_n = b$ is a subdivision (or partition) of $[a, b]$, $\delta_i = a_i - a_{i-1}$ and $D_i = [\sup f(x) - \inf f(x)]$ for $a_{i-1} \le x \le a_i$, then $f(x)$ is integrable over $[a, b]$ if and only if the sum, $\delta_1 D_1 + \cdots + \delta_n D_n$, tends to zero as the maximum length of the δ_i tends to zero. With some very minor notational changes, this *integrability condition* is easily seen to be equivalent to Theorem 6.3.4 as well as to Theorem 6.3.8, which asserts that $f(x)$ is integrable over $[a, b]$ if and only if its set of discontinuities has Lebesgue measure 0. Naturally, the latter form of the integrability condition was not given by Riemann or by any of his immediate successors. Indeed, it was only after the concepts of content and measure were articulated that the stage was set for Lebesgue's extension.

Riemann's integrability criterion admitted to the class of integrable functions some that were very discontinuous. Although he conceded that such functions are not found in nature, he argued that their study is nevertheless worthwhile. For one thing, there are number-theoretical applications of Fourier series in which the functions do not always satisfy Dirichlet's conditions (see [4], p. 199).[11]

Riemann was the first to distinguish between trigonometric and Fourier series, giving an example of an integrable function whose Fourier series does not converge anywhere.[12] He also showed that there are convergent trigonometric series whose coefficients are not the Fourier coefficients of an integrable function (see Example 10.4.11).

Some other results in Riemann's Habilitationsschrift appearing in this text are the *Riemann-Lebesgue Lemma* (Theorem 10.4.3): as n tends to ∞, the Fourier coefficients, a_n and b_n, tend to 0; and the *Localization Theorem* (10.4.4): the convergence of a Fourier series at a point depends only on the local behavior of the function.

2 The Real-Number System

Although Cauchy had initiated the rigorization and arithmetization of analysis in his three texts—*Cours* (1821), *Résumé* (1823) and *Leçons* (1829)—he left more than a few details for his successors to complete. In the *Résumé* he had announced his goal as no less than to "reconcile rigor, which I have made a law unto myself in the *Cours d'analyse*, with the simplicity which the direct consideration of infinitely small quantities produces." This required him "to reject the development of functions into infinite series each time that the series obtained is not convergent . . ." and to "demonstrate generally the existence of integrals or primitive functions before making known their diverse properties." Although the latter goal was attained, some of his proofs were marred by a failure to distinguish between (pointwise) continuity on an interval and uniform continuity and between convergence and uniform convergence.[13] However, these proofs are easily repaired, and rarely did Cauchy make errors because he failed to make these distinctions.[14]

The most glaring omission in Cauchy's work (from our point of view) is the absence of a theory of real numbers. In discussing series Cauchy had introduced (in the *Cours*) the sequence of partial sums, and defined convergence as we do today. He had no difficulty proving that if a series converges then its sequence of partial sums $\{s_n\}$ is what we would call today a *Cauchy sequence*, i.e., $|s_n - s_m|$ tends to 0 as n, m tend to ∞. Following this he *asserted* that the condition is also *sufficient* for convergence of the sequence, but said nothing either about the need to prove this assertion or to assume it (see pp. 17–23). Throughout his works Cauchy used—quite correctly—this statement of the *completeness of the real numbers* as well as the various equivalent forms. (See Chapter 1, especially Theorems 1.2.4, 1.3.16, 18, 19, 22 and 27 and Exercise 1.3.27.) The only contemporary of Cauchy who recognized the need for a statement of completeness was Bernhard Bolzano, who, in a paper published in 1817, attempted to analyze some of the basic notions of the calculus. Not only did Bolzano state Cauchy's convergence criterion, but he attempted to prove the sufficiency of the condition. Bolzano's interest in foundations came from his view that mathematics was a part of philosophy, so that he was concerned with the metaphysical status of the objects of the calculus, in particular the real-number system on which the calculus was built. On the other hand Cauchy, like most of his contemporaries, felt no need for such a theory, for confidence in the real-number system had not been shaken by inconsistencies and anomalies such as those arising from the uncritical use of the operations of the calculus. However, Bolzano's work remained little known until the late 1850s and early 1860s, when Karl Weierstrass first introduced a construction of the real-number system in his Berlin lectures and completed much of the rigorization begun by Cauchy and Bolzano.[15]

The two earliest constructions of the real-number system (from the rationals) were given by Richard Dedekind and Karl Weierstrass. In his monograph, *Continuity and Irrational Numbers* (1872), Dedekind explained that when he had given his first lectures on the calculus in 1858, he had felt deeply dissatisfied with the foundations of arithmetic. Although conceding that geometric intuition had its place in a first calculus course, it was pedagogically necessary to replace these intuitive concepts and to construct a satisfactory theory of continuous magnitudes to serve as the foundation of the calculus. He pointed out that a precise characterization of the *continuity* of the number line (its "hanging together") had so far eluded mathematicians. What, Dedekind asked, distinguishes the interval [0,1] from the rational points it contains? Nothing, he wrote, is gained by "vague remarks" about its "unbroken connection." He insisted that the problem was to define a "precise characteristic" from which valid deductions could be made. Weierstrass, independently of Dedekind, began in 1859 to include a construction of the real numbers in his lectures at Berlin. His principal reason for carrying out such a construction was his belief that it was indispensable for a rigorous construction of a theory of (complex) analytic functions (see [4], p. 221; [10], p. 8).

Although Dedekind's notebooks reveal that he had discovered his construction of the irrational numbers in November 1858, he did not attempt to publish these results until he learned that Georg Cantor (1872) had just published a construction (see [7], p. 48; also Section 3 of this introduction). Dedekind's construction differed radically from those given by Weierstrass, Cantor and Charles Méray; the latter used *fundamental* (= Cauchy) sequences (or their equivalent series) of rational numbers to define the irrational numbers, whereas Dedekind turned to Euclid for his inspiration, basing his definition on the Eudoxian theory of the ratios of magnitudes. Dedekind may also have been motivated by recent investigations in analysis. Almost certainly he had read Riemann's "On the Representation of Functions by Trigonometric Series" before beginning his teaching career. Indeed, Riemann and Dedekind were friends, and the latter was responsible for the posthumous publication in 1868 of the monograph on trigonometric series. It will be recalled that in that monograph Riemann had given a completely unexpected example of an integrable function having a dense set of singularities. This and other counterexamples—some found by Weierstrass and others earlier by Bolzano (see Note 15)—further emphasized the need for a theory of the real numbers (see Chapter 4, Section 3, especially Corollary 28 and Exercises 17, 18 and 19). Having constructed the irrational numbers from the rationals, Dedekind promised to present, at a later date, a complete theory of the number system. Here he was influenced less by analysis than by his own investigations in algebra, algebraic number fields and factorization—which led ultimately to his theory of ideals (see [22], pp. 820–24).

There were other developments in mathematics, not directly related to analysis, that also encouraged mathematicians to seek a nongeometric theory of the real numbers. For about two thousand years the soundness or the "truth" of Euclidean geometry had never been questioned. However, the almost simultaneous discovery in the 1820s by János Bolyai, Karl F. Gauss and Nikolai I. Lobachevskii of the non-Euclidean geometries—and the possibility that one of these geometries might provide as good a description of the physical world as did Euclid's geometry—began to challenge the latter's privileged position. Riemann's influential Probevorlesung (1854), "On the Hypotheses Which Underlie Geometry," further weakened the long-held belief that Euclidean geometry was *the* geometry of space. This in turn raised doubts that it could serve as the basis for other branches of mathematics: above all, for some mathematicians Eudoxus' theory of the ratios of magnitudes had long served as a basis for the real numbers.

A series of investigations paralleled the revolution in thought brought about by the non-Euclidean geometries—investigations that may be characterized as the first stages in the development of modern algebra. These began, not surprisingly, with attempts to justify and clarify arithmetic and algebraic operations mathematicians had been using—sometimes incorrectly and often with little or no justification. Around 1800 a satisfactory definition of a complex number as a point in the Euclidean plane was found by several mathematicians, and addition and multiplication of complex numbers were justified by simple geometric arguments. Encouraged by the similarity of operations on the real and complex numbers, several mathematicians directed their efforts toward studying the *universal* and *general* properties of algebra. One of the most prominent of these was George Peacock of Cambridge University, who articulated a principle of the *permanence of form* that allegedly justified the universal validity of a range of operations on numbers of all kinds.

Such efforts were brought to an unsuccessful conclusion by a series of discoveries, as startling and perhaps as shocking as those made in analysis. In the 1830s, William Rowan Hamilton, who first gave the modern definition of a complex number as an ordered pair of real numbers and defined their addition and multiplication as we do today, sought a three-dimensional analogue of the complex numbers. He never succeeded in this quest (no such analogue in R_3 exists), but discovered a four-dimensional extension of the real numbers, the *quaternions*, which shares with the former all its field properties save one—the commutativity of multiplication. Shortly thereafter algebras with more unexpected (and perhaps disturbing) properties were found by Arthur Cayley, Hermann Grassmann, William K. Clifford and others. For example, matrices—introduced when determinants were first used to solve systems of linear equations—appeared to have unusual multiplicative properties. Besides being noncommutative, the product of two nonzero matrices may be the zero matrix. These discoveries obviously brought to an end any hope to develop a universal arithmetic or algebra. Proliferation of algebraic structures sharing some, but not all, of the familiar properties of the real-number system, intensified the need to place arithmetic on a firm foundation. It was not that anyone seriously doubted the validity of arithmetic operations (on the real numbers), but the presence of other (apparently) consistent and useful algebraic systems forced mathematicians to examine closely the familiar ones. Gradually they came to view the real numbers as just one (although, practically speaking, the most important) of the many algebraic structures.[16]

3 Sets, Content and Measure

From the integrability condition (see Section 1, above) Riemann derived the following necessary and sufficient condition for a function to be integrable on an interval $[a,b]$:

> If $\mathcal{S}$ is a subdivision (see p. 193) of $[a,b]$, $\sigma > 0$, and $S(\mathcal{S},\sigma)$ is the sum of those $\delta_i = a_i - a_{i-1}$ for which $[\sup f(x) - \inf f(x)] > \sigma$ (where $a_{i-1} \leq x \leq a_i$), then $f(x)$ is integrable if and only if, for each $\epsilon > 0$, there is a $\delta > 0$ such that whenever $|\mathcal{S}| < \delta$, then $S(\mathcal{S},\sigma) < \epsilon$.

That is, the *sum of the intervals over which* f(x) *varies by more than an arbitrary positive number* σ *can be made less than an arbitrarily prescribed* $\epsilon > 0$, *provided the subdivision is fine enough*. In this version of the integrability criterion the intervals do not cover all the points of discontinuity of $f(x)$, but only those having a jump or oscillation greater than σ. Using the sequence $\epsilon_n = 2^{-n}\epsilon$, it is easily seen that the condition is equivalent to stating that the discontinuities form a set having (Lebesgue) measure zero (Theorem 6.3.8). In 1870 a former student of Riemann, Hermann Hankel, revised the integrability condition in the direction of a measure-theoretic formulation. He introduced, for each $f(x)$, a function (closely related to the modulus of continuity; see pp. 143 ff.) vanishing at points of continuity and giving what we would call the *oscillation* of the function at each point of discontinuity. Letting S_σ be the set on which this function is greater than σ (i.e., at each point of S_σ the oscillation of $f(x)$ is greater than σ), Hankel proved that f is integrable if and only if S_σ can be covered by finitely many intervals whose total length can be made arbitrarily small. Since $\bigcup_{n=1}^{\infty} S_{1/n} = S$ is the set of all discontinuities of $f(x)$, it follows that the integrability condition is equivalent to stating that S can be covered by *countably* many intervals whose total length can be made arbitrarily small; i.e., S *is a null set.* But Hankel did not draw this latter conclusion; indeed, this formulation would not be given until measure-theoretical concepts were articulated for Riemann's integral. In the 1870s no one had a very clear idea of what characterized a set that is "negligible" from the point of view of integration. In fact, Hankel conjectured (as did Dirichlet before him) that the points of discontinuity must be nowhere dense, thus confusing what may be considered a *topologically* negligible set with one that is *measure-theoretically* negligible.[17] The first example of a nowhere dense set that is *not* negligible from the point of view of Riemann integration (its outer content is positive) was given by H. J. S. Smith in 1875. A few years later Vito Volterra and Georg Cantor constructed similar sets (see Example 9.4.2).

It is possible only to sketch some of the events leading to the measure-theoretical formulation of Riemann's integral.[18] As a first step it was necessary for mathematicians to focus greater attention on sets—mainly sets of discontinuity—and to untangle the confusion that was due largely to the paucity of examples of sets having unusual characteristics. Cantor's investigations launching the theory of sets began, innocently enough, with a series of papers (1871–72) in which he extended a theorem proved in 1870 by his colleague Eduard H. Heine: if a trigonometric series converges uniformly to a function $f(x)$, continuous except at finitely many points of a bounded interval, then there is no other such series. Cantor was able to generalize this uniqueness theorem first by removing the condition of uniformity and, more significantly for this discussion, by extending the theorem from finite to *first-species sets*,[19] which share with finite sets the important property that they can be covered by finitely many sets having arbitrarily small total length. Cantor's results initiated a long series of investigations to characterize the so-called *sets of uniqueness*—the sets on which a trigonometric series may fail to converge to the given function or may not converge at all, while remaining the *unique* trigonometric series representing that function.[20] It is not difficult to imagine the confusion surrounding the sets of uniqueness, sets negligible from the point of view of integration, first-species sets, nowhere dense sets and (by the mid-1870s) countable and uncountable sets. By the 1880s many examples and counterexamples of incorrect conjectures had been constructed; in particular, it was finally demonstrated that, although every first-species set is nowhere dense, the converse is false (the Cantor set is an example). This disproved one conjecture—that first-species sets were the ones negligible in integration. Finally, with the introduction of the notion of outer content, the stage was set for a measure-theoretic characterization of the sets on which an integrable function may be discontinuous.

Most decisive in bringing measure-theoretic notions to the attention of French mathematicians (who had either ignored Cantor's investigations up to this point or opposed them) were the efforts of Camille Jordan, whose formulation of the notions of *inner* and *outer content* and their subsequent inclusion in the second edition of his *Cours d'analyse* (1892) brought about the acceptance of set and measure-theoretical ideas in France. (Giuseppe Peano had introduced a similar notion in 1887.) By the turn of the century these ideas bore fruit in the works of Émile Borel, Louis-René Baire and Henri Lebesgue.

Anyone who has ever lectured on the two-dimensional Riemann integral will appreciate Jordan's motivation for introducing the notion of the content of a set. In an earlier (1881) edition of his text, Jordan discussed the integral of $f(x,y)$ over a plane region bounded by a closed curve. In forming the approximating sums, Jordan had assumed that it did not matter if the terms $f(x,y)\Delta A$, which correspond to rectangles of the subdivision meeting the boundary curve, were retained or discarded. This, of course, is equivalent to assuming that the sum of such terms tends to zero as the norm of the subdivision tends to zero, i.e., that the *boundary curve has zero outer* (two-dimensional) *content*. In 1890 Peano had discovered his famous space-filling curve, a continuous curve passing through *all* the points of a rectangle. This may have led Jordan to introduce the condition that the boundary curve of the region of integration must have zero content.

Having introduced content in the 1892 edition of the *Cours,* Jordan was able to give a measure-theoretic definition of the Riemann integral: the subdivision of $[a,b]$ into finitely many *intervals* was replaced by a subdivision into finitely many *measurable sets* (i.e., sets having equal inner and outer content). This gave the same class of integrable functions as did Riemann's definition. As we can see, Jordan was not dissatisfied with Riemann's integral. He wished only to give an alternate definition that could be extended to higher dimensions. Another factor influencing Jordan's reformulation was the introduction in the 1870s of the *upper* and *lower sums* corresponding to a bounded function and a particular subdivision of the interval. Most significant was the observation that, whether or not the function is integrable, the *infimum* and *supremum* of these sums, taken over all possible subdivisions, exist. These numbers, commonly called the *upper integral* ($\overline{\int} f$) and the *lower integral* ($\underline{\int} f$)—or simply the Darboux integrals—were first introduced in 1881 by Volterra.[21] It is easily seen that if f is a bounded function, then $\underline{\int} f \leq \overline{\int} f$; in the case where they are unequal—i.e., where the function is not Riemann integrable—it may be possible to obtain equality if the class of measurable sets over which the supremum and infimum are taken is widened. This may be what motivated Lebesgue to replace *finite* additivity by *countable* additivity.

In 1894 Borel published his thesis on the continuation of complex functions across a curve containing a dense set of singularities. It was in this work (which, incidentally, contains the first explicit statement of the Heine-Borel Theorem) that Borel introduced the notion of a *countably additive measure* (replacing finite by countable covers). This paper exemplifies the increasing use of countably additive processes.[22] However, Borel did not, at this early date, introduce a new integral.

Thus by the mid-1890s all the techniques needed to generalize Riemann's integral were in place—the set and measure-theoretical notions disseminated in France through Jordan's text and the generalization of Riemann's (finitely) additive measure to one that is countably additive. Further, Jordan's 1892 edition of the *Cours* contained a detailed discussion of functions of bounded variation, first introduced in the 1881 edition. These functions would play a central role in the theory of differentiation and in the discussion of the length of curves (see Chapter 9, Section 3).

At the same time it had been shown that Riemann's integral was not sufficiently general, for example, to express the length of all rectifiable curves: arc length had been expressed by an integral of $\sqrt{1 + (f')^2}$, which requires that the discontinuities of f' form a null set. In 1884 L. Scheeffer gave an example of a rectifiable curve whose length could not be expressed by a Riemann integral because f' is too discontinuous.[23]

Another unsatisfactory feature of Riemann's integral was its failure to maintain the inverse relation between integral and derivative: Cauchy's integral was defined only for functions having, at the worst, finitely many jump discontinuities—it is differentiable except at these (finitely many) points of discontinuity, and its derivative is equal to the integrand. This inverse relation, the *Fundamental Theorem of Calculus*, appeared to break down for Riemann's integral because the set of discontinuities of the integrand may be dense. Moreover, in 1881 Volterra had constructed a function having a *bounded* but not (Riemann) integrable derivative.

The systematic study of discontinuous functions, begun by Weierstrass and his students in the 1860s, was being taken up in other countries as well. Using the new set-theoretical concepts, Baire (who had studied with and been influenced by Volterra) carried out the first general study of discontinuous functions. These investigations have been described (in [28], p. 37) as the completion of studies, initiated in the 1870s and 1880s by U. Dini, C. Arzela and G. Ascoli, of the limits of sequences of continuous functions. Among Baire's earliest results (1898–99) was the characterization of the class B_1 (of discontinuous limits of sequences of continuous functions) in terms of the discontinuities of the limit functions. Although the Baire classes are not discussed in the present book, Baire's results were a major factor in convincing mathematicians that the theory of sets and the study of discontinuous functions were central to analysis. In particular, his proof that all derivatives are either in B_0 (the class of continuous functions) or in B_1, together with the fact that most of the functions in B_1 are too discontinuous to be Riemann integrable (a reflection of the incompleteness of the space of Riemann integrable functions; see p. 152), spurred efforts to extend the integral. In the course of classifying discontinuous functions, Baire introduced the notion of *category* (pp. 134–36) and studied the sets $[f \geq \alpha]$, later used by Lebesgue in his discussion of measurable sets and functions (see pp. 91, 233, 217 and 240).[24]

4 The Integral of Lebesgue

Five brief "Notes," published in the *Comptes-Rendus* between 1899 and 1901, formed the basis of Lebesgue's thesis *Integral, Area, Volume* (1902). In defining an integral for a bounded function, Lebesgue altered the Cauchy-Riemann sums by subdividing the *range* of the function instead of the (interval) *domain*. This method (described in Chapter 8, Section 2) requires the measurability of the sets $[y_{k-1} \leq f < y_k]$ which, in general, are neither intervals nor Riemann measurable sets. Since Lebesgue's thesis and his subsequent works have been described at length elsewhere, we will cite here only a few highlights.[25]

A major theorem—called by Charles de la Vallée Poussin the crowning glory of the dissertation—states that if $\{f_n\}$ is a sequence of integrable functions converging almost everywhere (a.e.; see p. 173) to a function f, where $|f| < g$ and g is integrable, then $\lim_{n\to\infty} \int f_n = \int f$ (Theorem 6.2.4). This is precisely the property that distinguishes Lebesgue's integral from that of Riemann.

Describing his goal as the restoration of the inverse relationship between integral and derivative, Lebesgue succeeded for *bounded* derivatives. (A further extension was made in 1912 by Arnaud Denjoy, who restored the *Fundamental Theorem* to unbounded derivatives of continuous functions.) That the integral of f is a.e. differentiable—even if f is everywhere discontinuous—was regarded at the time to be a startling fact.[26]

Lebesgue was able to show that his integral could be used to express the length of any rectifiable curve; he also extended the concept of area.

Trigonometric series were not discussed in the dissertation—Lebesgue did not say (as did Riemann) that he had extended the integral in order to enlarge the class of functions representable by such series. However, shortly after completing his dissertation Lebesgue turned to the subject of Fourier series. He extended Riemann's lemma to his integral (Theorem 10.4.3) in 1903, giving a new set of sufficient conditions for the convergence of a Fourier series, and then used these conditions to prove Parseval's equality for bounded, integrable functions.[27] Lebesgue also proved that a Fourier series (whether or not it converges) may be integrated term by term (Theorem 10.4.10 and Example 10.4.11).

Jordan had already (in 1881) introduced functions of bounded variation, but it was Lebesgue who first proved that they are a.e. differentiable. (F. Riesz called this one of the most important and startling theorems.) Although the concept of *absolute continuity* did not appear in the dissertation, Lebesgue proved a related result: among all the primitives of a given function, the indefinite integral has the minimum variation.[28] In 1904 Lebesgue showed that the indefinite integral is absolutely continuous and stated (without proof here) that every absolutely continuous function is an indefinite integral.

Despite some detractors Lebesgue's theory of measure and the integral was widely accepted and applied to a host of problems. The theory of measure, which had relied so heavily on the new set theory, in turn strengthened and enlarged the latter. A similar statement can be made about the notion of functionality. Indeed, Lebesgue's investigations could not have taken place without the earlier studies of discontinuous functions; conversely, Lebesgue's results encouraged further investigations of functions of bounded variation, absolutely continuous and measurable functions and Baire functions. Some of these investigations led to prolonged and heated controversies concerning the acceptability or meaningfulness of certain classes of functions and sets. For example, Borel objected to using the (complete) class of Lebesgue measurable sets, arguing that it made sense to admit only what we would call the *Borel sets* (see Chapter 7, Section 3). Admittedly, this would hardly restrict the applicability of Lebesgue's theory: Theorem 7.3.3 and its corollaries indicate that it is possible to work only with the Borel sets, since every finitely (Lebesgue) measurable set is "sandwiched" between two such sets. Similarly, Borel never agreed that the full class of measurable functions was meaningful.[29]

5 Other Integrals

In contrast to the detailed and extensive analyses of Lebesgue's theory of measure and the integral, not much beyond a few expository (as opposed to historical) papers and sections of older texts have been written (in English) about extensions of Lebesgue's integral or about the various attempts to redefine it. A similar statement may be made about Stieltjes' integral and its extensions.[30] In this concluding section we will briefly cite a few examples of these other integrals, starting with efforts made by some of Lebesgue's contemporaries.

W. H. Young, independently of Lebesgue, defined an integral (in 1904–5) equivalent to that of Lebesgue [41]. Young's intentions, however, could not have been more different from Lebesgue's, and therein, very likely, lies the reason that Young did not fully appreciate the significance of his integral extension: once Young had defined his integral—which he failed at first to see was equivalent to Lebesgue's—he did not go on to study its properties (as did

Lebesgue) or to develop a complete and coherent theory containing the major theorems on integration (see Chapter 6, Sections 1 and 2). (It is probably for this reason that the entire theory of measure and the integral bears only Lebesgue's name.) As an advocate of the new set theory, Young wished to see how this theory could be used to redefine Riemann's integral, and what would happen to this integral if the subdivision of the interval into finitely many parts was replaced by a subdivision into countably many intervals or measurable sets. After showing that several such variations of the definition led nowhere, he constructed an integral, fully equivalent to that of Lebesgue. Young's construction was more classical, in the sense that he did not attempt anything so radical as the subdivision of the range of the function. Further, his method avoids the introduction of the class of measurable functions. Indeed, Young regarded the introduction of the measurable functions as an unnecessary complication of the theory, preferring instead to use semicontinuous functions and to define the integral by means of Darboux's upper and lower integrals. By utilizing methods and classes of functions more familiar to his contemporaries, Young may well have hastened the general acceptance of Lebesgue's theory after it was fully articulated in 1904 in the first edition of the *Leçons*.[31]

Young was not alone in objecting to the logical priority of measure theory to the integral. Others found the theory of measure quite formidable and suggested variants they believed to be pedagogically preferable. For example, F. Riesz's exposition begins with the notion of a *null set* (measure 0) and is followed by the integration of step functions. The integral is extended to the full class of (Lebesgue) integrable functions by taking the differences of monotone limits of step functions. This method is carried out in the classic text of Riesz and Nagy (see p. 442) and is closely related to (and doubtless a direct ancestor of) the method used in the present text (see Method II on p. 239).

O. Perron also wished to avoid the theory of measure. Using major and minor functions (introduced earlier by de la Vallée Poussin), he defined an integral (in 1914) to be a particular primitive of the given function. It was not his intention to *extend* Lebesgue's integral; indeed, for bounded functions the integrals of Lebesgue and Perron are equivalent. However, it will be recalled that Lebesgue was able to prove only that *bounded* derivatives are integrable, so that he was not entirely successful in restoring the inverse relation between derivative and integral. Lebesgue realized this, and in the *Leçons* he cited the example $F(x) = x^2 \sin(1/x^2)$ if $x \neq 0$ and $F(0) = 0$, whose unbounded derivative is not integrable on $[0,1]$. The function $F(x)$ is, however, the indefinite Perron integral of its derivative.

An extension equivalent to Perron's was given in 1912 by Arnaud Denjoy. The construction of the *total*—as Denjoy named his integral—required a transfinite induction, a procedure not acceptable to some mathematicians. Two years later a descriptive definition of Denjoy's total was given by N. N. Luzin in *The Integral and Trigonometric Series* [26]. Luzin was able to characterize the total by giving a suitable extension of the concept of a function of bounded variation, thereby avoiding the transfinite induction. This characterization, however, does not provide a construction. As the title of Luzin's monograph suggests, the author, strongly influenced by Riemann's Habilitationsschrift (see Section 1, above), was attempting to extend Denjoy's integral to determine the widest possible class of functions representable by Fourier series. Although Luzin did not succeed in finding the extension he sought, his monograph was extremely influential, and, only two years after it appeared, one of his students, A. Khinchin, found such an extension (see [32]). Like Lebesgue's (indefinite) integral, Denjoy's total is a continuous function; but further extensions required the admission of discontinuous primitives. This in turn required the generalization of the concepts of continuity, differentiability, absolute continuity and bounded variation. Khinchin's extension [21], incidentally, was received in 1916 by the editors of the *Comptes-Rendus* within days of a completely equivalent extension made by Denjoy!

Each of the integrals mentioned so far, from Cauchy's to the one defined by Denjoy and Khinchin, was defined on a subset of the class of real-valued functions of a real variable. Starting with Riemann's integral, each is an *extension* of an earlier one, in the sense described on p. 193 (see also Theorem 6.3.6). Further, these integral extensions are closely related either by their applications to such fundamental questions in analysis as the representation of a function by a trigonometric series or by their having originated in a search for a primitive function.

The integral of T. J. Stieltjes lies outside this tradition. Introduced in 1894 in a study of continued fractions, it received little attention in France until the second decade of this century. Lebesgue, who had proved that every Stieltjes integral can be represented as a Lebesgue integral, did not believe it was of any importance. However, Stieltjes' integral received more attention after F. Riesz proved that it can be used to represent a continuous linear functional on $C[a,b]$ (see Chapter 11, Section 4, especially Theorem 11.4.6) and after the publication in 1913 of J. Radon's memoir in which the Lebesgue-Stieltjes integral was introduced (see Chapter 13, Section 2 and [30]).

We regret that limited space prevents us from describing other types of integrals, some of which appear in this text. For example, in 1910 Lebesgue introduced the notion of a *set function* in connection with multiple integration; he also defined differentiation and integration of these functions with respect to other set functions. (He was apparently unaware that in 1897 Peano had defined such functions with even greater generality.) These investigations were continued by Radon, who extended step functions to σ-rings of sets in R_n, gave the corresponding definition of an absolutely continuous function and proved that such functions may be expressed as integrals. In 1915 abstract spaces were introduced by M. Fréchet, and in 1917 P. J. Daniell described a general integral (see p. 441 and Chapter 12). Finally, in 1930 the characterization of absolutely continuous functions received its abstract formulation in the Radon-Nikodym Theorem (Chapter 13, Section 4) [30].

It would be possible—if not of great interest—to fill nearly a page (if not more) citing other integrals introduced since, say, Riemann's time. Many of these have apparently sunk into perhaps a well-deserved oblivion, while others are described elsewhere.[32]

Notes

Numbers in square brackets indicate references (see following section).

1Although Leibniz was the first to use the word *function*, it was not until Euler's work that the concept was central [42].

2[3], Chapter 5; [11], Chapters 7, 8 and 10; [22], Chapter 20; and [39].

3Examples of these inconsistencies and "paradoxes" may be found in [27], pp. 162–66.

4See [13], pp. 40 ff.; [16], p. 101.

5The vibrating-string debate and its consequences are discussed in [4], Chapter 1; [14], [22], [23] and [42].

6Selections from these treatises have been translated and may be found in [20].

7Gauss's investigations of the hypergeometric series may be considered as an early example of rigor (see [22], p. 962). Descriptions of Bolzano's efforts are in [13] and [15]; see also [22], pp. 947 and 964.

8See note 15, below, and [15].

9Lobachevskii, whose work in analysis is not widely known in the West, gave a similar definition (see [42], pp. 75 ff.).

10Riemann's work on trigonometric series is described in [4], pp. 196–213; [17], chapter 1; and [22], pp. 967–72.

11For further discussion of Riemann's extension of Cauchy's integral, see [17], [22], [27] and [31].

12The example given by Riemann is

$$f(x) = (x)/12 + (2x)/2^2 + \cdots + (nx)/n^2 + \cdots,$$

where (x) is the difference between x and the nearest integer; if x is midway between two integers, then $(x) = 0$ (see [22], p. 955).

13Uniform continuity of the integrand is required to prove the existence of the integral. It is likely that Riemann realized that pointwise continuity is not enough, for he did not attempt to prove the existence of the integral of a continuous function.

14In at least one case Cauchy, like his contemporaries, failed to appreciate the diversity and complexity of discontinuous functions. For example, he believed that, if $f(x,y)$ is continuous in both x and y (separately), then it is a continuous function of x and y.

Further, he "proved" that an infinite series of continuous functions is continuous! In 1826, a counterexample was given by N. Abel (see [13], p. 12, and [22], pp. 947 and 965).

15Owing mainly to his isolation from the major mathematical centers—he lived and worked in Prague and his works were published in pamphlets and not widely distributed—Bolzano's ideas were not well known during his lifetime. In 1830, he gave an example of a continuous, nowhere differentiable function to illustrate the pitfalls of founding analysis on a geometric rather than arithmetic basis. This example went unnoticed until the time of Weierstrass. Although it has been argued that Bolzano's ideas were too far ahead of the times to be appreciated by his contemporaries, Grattan-Guinness [15] has argued that Cauchy actually read Bolzano's paper of 1817 (in which the intermediate-value theorem is proved) and that Cauchy's definitions of limit and continuity—indeed, his reformulation of analysis—were inspired by this work.

See also [13], pp. 99–109.

[16]See Chapters 32 and 33 of [22].

Cited in Section 2 are constructions of the irrational numbers from the rationals. These in turn may be constructed from the integers and the latter from the natural numbers (see Chapter 1). In 1889 Peano gave a set of axioms for the natural numbers, and later (1899) David Hilbert provided a set of axioms for the complete (Archimedean) ordered field of real numbers (see pp. 987–92 of [22]).

[17]For an account of the confusion surrounding this question, see [17], pp. 28–42 and [7].

[18]Further discussion appears in [17], [18] and [27].

[19]If S is any set, then the *first derived set* $S^{(1)}$ is the set of accumulation points of S; for $n > 1$, $S^{(n)}$ is the set of accumulation points of $S^{(n-1)}$. The set S is a *first-species set* if, for some integer n, the nth derived set is empty.

[20]Sets of uniqueness are described in [1]. An early and completely unexpected result, obtained in 1916 by D. Menshov, a student of N. Luzin [32], disproved a widely held conjecture that the sets of uniqueness are identical with the null sets.

[21]Volterra, Ascoli and Peano were the first to define Riemann's integral as the common value of the upper and lower integrals (see [17], p. 41; [28], p. 49).

[22]Borel's thesis is described in [17].

In a report commissioned by the Deutsche Mathematiker Vereinigung and published in 1900, Arthur Schoenflies suggested that Borel's extension of the concept of content was useless! The irony, of course, is that no sooner had Schoenflies expressed this opinion than Lebesgue completed his extension of Riemann's integral by using a countably additive process.

[23]See [17], pp. 79–85; [27], p. 231. For a discussion of the area problem, including Peano's definition, see [28], pp. 42–43.

[24]For more details see [27], pp. 228–32; [28], pp. 35–40.

[25]An English translation is in Lebesgue [1], listed on p. 442. See also [12]; [17], Chapters 5–7; [22], pp. 1044–50; [28], Chapter 6; and [31], Chapter 4.

[26]Charles Hermite did not want to accept for publication in the *Comptes-Rendus* Lebesgue's "Note" in which his integral was first described. Indeed, in an earlier letter, Hermite had written, "Je me détourne avec effroi et horreur de cette plaie lamentable des fonctions qui n'ont pas de dérivées" (quoted from [36], which contains a wealth of bibliography). This was not the first rebuff Lebesgue received; his suggestion, in his first "Note" (1899), that discontinuous functions be admitted in a hitherto classical problem of differential geometry "shocked" (scandalisé) Darboux ([6], p. 57).

[27]In 1906 P. Fatou removed the boundedness condition.

[28]If the characteristic function of the Cantor set (p. 167) is added to an indefinite integral, then the sum remains a primitive, but its total variation is increased by 1.

[29]These controversies are described in [29] and in [28], Chapters 3 and 4. See also [3], especially the famous "Cinq lettres."

[30]The following contain descriptions of Stieltjes' work and extensions of Lebesgue's integral: [17], [19], [28], [31] and [40].

[31]The much expanded 1928 edition is listed in the References [25].

[32]For more information about the early history of the calculus, see [5], [11] and [22]; from Euler to Weierstrass, see [4], [9], [11], [12], [13], [14] and [22]; on divergent series and summability methods, [39]; on the Stone-Weierstrass Theorem, [35] and [38]; on abstract spaces, [2].

The following contain additional references: [1], [7], [12], [13], [14], [17], [22], [27], [28], [31], [35], [36], [40] and [41].

References

[1] Bary (Bari), Nina K. *A Treatise on Trigonometric Functions.* 2 vols., New York: Macmillan, 1964.

[2] Bernkopf, Michael. "The Development of Function Spaces with Particular Reference to Their Origins in Integral Equation Theory," *Archive for History of Exact Sciences* 3, (1966), 1–96.

[3] Borel, Émile. *Leçons sur la théorie des fonctions.* 2nd ed. Paris: Gauthier-Villars, 1914 (contains "Les polémiques sur le transfini," including the "Cinq lettres").

[4] Bottazzini, Umberto. *Il calcolo sublimo: storia dell'analisi matematica da Euler a Weierstrass.* Torino: Boringhieri, 1981.

[5] Boyer, Carl B. *The History of the Calculus and Its Conceptual Development.* New York: Dover, 1959.

[6] Burkill, J. C. "Henri Lebesgue," *Journal of the London Mathematical Society* 19 (1944), 56–65.

[7] Dauben, Joseph W. *Georg Cantor: His Mathematics and Philosophy of the Infinite.* Cambridge: Harvard University Press, 1979.

[8] Denjoy, Arnaud. "Une extension de l'intégrale de M. Lebesgue," *Comptes-Rendus . . . Académie des Sciences, Paris* 154, (1912), 859–62.

[9] Dugac, Pierre. "Éléments d'analyse de Karl Weierstrass," *Archive for History of Exact Sciences* 10 (1971) 41–176.

[10] ——. "Weierstrass et Dedekind," *Historia Mathematica* 3 (1976), 5–19.

[11] Edwards, C. H. *The Historical Development of the Calculus.* New York: Springer Verlag, 1980.

[12] *From the Calculus to Set Theory: 1630–1910.* Edited by I. Grattan-Guinness. London: Duckworth, 1980.

[13] Grabiner, Judy V. *The Origins of Cauchy's Calculus.* Cambridge: M.I.T. Press, 1981.

[14] Grattan-Guinness, I. *The Development of the Foundations of Mathematical Analysis from Euler to Riemann.* Cambridge: M.I.T. Press, 1970.

[15] ——. "Bolzano, Cauchy and the 'New Analysis' of the early Nineteenth Century," *Archive for History of Exact Sciences* 6 (1970), 372–400.

[16] ——. "The Emergence of Mathematical Analysis and Its Foundational Progress, 1780–1880," in [12], 94–148.

[17] Hawkins, Thomas W. *Lebesgue's Theory of Integration: Its Origins and Development.* Madison: University of Wisconsin Press, 1970.

[18] ——. "The Origins of Modern Theories of Integration," in [12], 149–80.

[19] Hildebrandt, T. H. "On Integrals Related to and Extensions of the Lebesgue Integral," *Bulletin of the American Mathematical Society* 2 (1917–18), 24, 113–44 and 177–202.

[20] Iacobacci, R. F. "Augustin-Louis Cauchy and the Development of Mathematical Analysis." Ph.D. dissertation, New York University, 1965.

[21] Khintchine (Khinchin), A. "Sur une extension de l'intégrale de M. Denjoy," *Comptes-Rendus . . . Académie des Sciences, Paris* 158 (1916), 287–91.

[22] Kline, Morris. *Mathematical Thought from Ancient to Modern Times.* New York: Oxford University Press, 1972.

[23] Langer, R. E. "Fourier's Series: The Genesis and Evolution of a Theory," *American Mathematical Monthly* 54 (1947), supp. 1–86.

[24] Lebesgue, Henri. "Sur les fonctions représentables analytiquement," *Journal de Mathématiques pures et appliquées* 6 (1905), 1, 139–216.

[25] ——. *Leçons sur l'intégration et la recherche des fonctions primitives.* 2nd ed. 1928. Reprint. New York: Chelsea, 1973.

[26] Luzin, N. N. *Integral i trigonometricheskii ryad* (The Integral and Trigonometric Series). Moscow, 1915. (Also in *Collected Works* and *Matematicheskii Sbornik* 30, no. 1 (1916), 1–242.)

[27] Medvedev, F. A. *Razvitie ponyatiya integrala* (The Development of the Concept of the Integral). Moscow: Nauka, 1974.

[28] ——. *Frantsuzskaya shkola teorii funktsii i mnozhestv, na rubezhe XIX–XX vv.* (The French School of the Theory of Functions and Sets, at the Turn of the Century). Moscow: Nauka, 1976.

[29] Monna, A. F. "The Concept of Function in the 19th and 20th Centuries, in Particular with Regard to the Discussions Between Baire, Borel and Lebesgue," *Archive for History of Exact Sciences* 11 (1972), 57–84.

[30] Nikodym, O. "Sur une généralisation des intégrales de M. Radon," *Fundamenta Mathematica* 15 (1930), 131–79.

[31] Pesin, Ivan N. *Classical and Modern Integration Theories*. Translated from the 1966 Russian edition. New York: Academic Press, 1970.

[32] Phillips, E. R. "Nikolai N. Luzin and the Moscow School of the Theory of Functions," *Historia Mathematica* 5 (1978), 275–305.

[33] Radon, J. "Theorie und Anwendungen der absolut additiven Mengenfunktionen," *Akademie der Wissenschaften, Wien* 122 (1913), 1295–1438.

[34] Riemann, Bernhard. *Ueber die Darstellbarkeit einer Funktion durch eine trigonometrische Reihe*. Read in 1854; published in 1867. Reprint. New York: Dover, 1953.

[35] Rodi, Stephen. "A History of the Stone-Weierstrass Theorem," *Historia Mathematica* (to appear).

[36] Saks, S. *Theory of the Integral*. Translated from the 2nd, revised ed., 1937. Reprint. New York: Dover, 1964.

[37] Stieltjes, T. J. "Recherches sur les fractions continues," *Annales de la Faculté des Sciences de Toulouse* 8 (1894), 1–122.

[38] Stone, Marshall H. "The Generalized Weierstrass Approximation Theorem," *Mathematics Magazine* 21 (1948), 167–83, 237–54.

[39] Tucciarone, John. "The Development of the Theory of Summable Divergent Series from 1880 to 1925," *Archive for History of Exact Sciences* 10 (1973), 1–40.

[40] van Dalen, D., and Monna, A. F. *Sets and Integration: An Outline of the Development*. Amsterdam: Wolters Noordhoff, 1972.

[41] Young, W. H. "On the General Theory of Integration," *Philosophical Transactions, Royal Society of London* 204 (1905), 221–52.

[42] Youschkevitch, A. P. "The Concept of Function up to the Middle of the 19th Century," *Archive for History of Exact Sciences* 16 (1976), 37–85.

Contents

PART I

The Real Numbers; Metric and Normed Vector Spaces

CHAPTER **0**

Preliminaries

For the most part the reader is probably familiar with the basic operations involving sets, as well as the algebraic properties of the fields of real and rational numbers. Therefore, with the exception of Sec. 4, in which the incompleteness of the field of rational numbers is discussed, this chapter is primarily a catalogue of definitions, notation, and some preliminary material, all of which will be used later in the text.

1 The Algebra of Sets and Mappings

Sets will be denoted by capital letters, $S, T, A, B, X, Y, \ldots$, and their respective members by $s, t, a, b, x, y, \ldots$.

Whenever each element of a set S is also an element of a set T, we write $S \subset T$, or $T \supset S$. We say that S is a *subset* of T. If $S \subset T$, but $S \neq T$, we say that S is a *proper* subset of T and write $S \subsetneq T$.

The symbol for the *empty set*, which is the set having no members, is ϕ. It is easy to see that if S is any set, then $S \supset \phi$ (Exercise 1).

Since a set S is completely determined by a "list" of its members, the set S whose elements are the numbers 3, -5, 8, and 2, is denoted by

$$S = \{3, -5, 8, 2\}.$$

Where it is awkward, or perhaps even impossible, to enumerate the members of a set, it may be given descriptively. For example, the set of real numbers that do not exceed 12 is written as

$$\{x\colon x \text{ is a real number and } x \leq 12\}.$$

Wherever convenient, the following abbreviations will be used:

"$\forall$" reads "for all," "for every," ....
"$\exists$" reads "there is," "there exists," "there are," ....
"$\in$" and "$\notin$" read "belongs to" and "does not belong to."
":" reads "such that."

To illustrate the use of these symbols, let S be the subset of the integers Z, that may be written as the squares of other integers. Then

$$S = \{x\colon x \in Z \text{ and } \exists \; y \in Z \text{ such that } y^2 = x\},$$

or equivalently,

$$S = \{x \in Z\colon \exists \; y \in Z \quad \text{and} \quad y^2 = x\}.$$

If S and T are nonempty sets, then the product $S \otimes T$ of these sets is the totality of *ordered pairs* (s,t), where $s \in S$ and $t \in T$. A subset f of $S \otimes T$ is said to be a *function* with *domain* S and *range* T if,

[1] for every s in S, there is one and only one t in T, such that $(s,t) \in f$.

Using set notation, f is the subset

$$\{(s,t); \forall s \in S, \exists t \in T \text{ and } (s,t) \in f;\ (s,t), (s,t') \in f \text{ implies } t = t'\}.$$

In view of [1], the function (or mapping or transformation) f may be thought of as the assignment of an (unique) element t of T to each s in S. In keeping with the "classical" character of this text, we shall eschew the more modern ordered-pair notation, and write simply $f(s) = t$ whenever the pair (s,t) is in f; to emphasize the "mapping" idea, we shall often write

$$f\colon S \to T \quad \text{or} \quad S \xrightarrow{f} T.$$

The following descriptive terms are applied to mappings:

f is *surjective*, or f maps S *onto* T if, whenever $t \in T$, there is (at least one) an element $s \in S$, such that $f(s) = t$. This is frequently denoted by

$$f\colon S \xrightarrow{\text{onto}} T.$$

f is *injective*, or f is a 1–1 mapping of S into T, if

$$f(s) = f(s') \text{ implies } s = s'.$$

A mapping that is both surjective and injective is said to be *bijective*.

Let a mapping $S \xrightarrow{f} T$ be given. Then for every subset A of S, the *image* of A *under* f, denoted by $f(A)$ is the following subset of T:

$$\{t \in T\colon \exists\, s \in A \text{ such that } f(s) = t\}.$$

If f is surjective then $f(S) = T$.

Whenever $S \xrightarrow{f} T$ is bijective, there is a (uniquely determined) mapping $T \xrightarrow{f^{-1}} S$ such that

$$\forall s \in S,\ f^{-1}(f(s)) = s \quad \text{and} \quad \forall t \in T,\ f(f^{-1}(t)) = t.$$

f^{-1} is called the *inverse of* f. Clearly, $f^{-1}(T) = S$.

Although f^{-1} is not defined unless the mapping f is bijective, it is possible to talk of the *inverse image* of a *subset* C *of* T:

$$f^{-1}(C) = \{s \in S : f(s) \in C\}.$$

It is left to the reader (Exercise 4) to translate the preceding definitions into the language of product sets.

Remark: The word "if" as it appears in the definitions given above, is used in a rather imprecise manner. It is always understood that a definition is an "if and only if" statement, and replacing this by the single word "if" is done to avoid some rather awkward sentences. Needless to say, no such liberties will be taken when stating theorems, and the following three cases will be scrupulously differentiated (A and B denote statements):

(i) "A if B," or "B implies A," or "$B \Rightarrow A$";
(ii) "A only if B," or "A implies B," or "$A \Rightarrow B$";
(iii) "A if and only if B," or "A implies B and B implies A," or "$A \Leftrightarrow B$."

1 Definition. A set S is said to be *indexed by a second set* $\mathfrak{A}$ if there is a mapping f of $\mathfrak{A}$ onto S.

Setting $s_\alpha = f(\alpha)$, for each α in $\mathfrak{A}$, we have $S = \{s_\alpha : \exists\, \alpha \in \mathfrak{A}$ and $f(\alpha) = s_\alpha\}$. For convenience, we shall frequently write $S = \{s_\alpha\}$. In such cases, it will be made clear from the preceding discussion that $\{s_\alpha\}$ does *not* denote the set which contains a single element s_α.

2 Examples. (a) The set $S = \{s_1, s_2, s_3, s_4\}$ is indexed by $\mathfrak{A} = \{1,2,3,4\}$ The elements s_i of S may be anything at all: numbers, people, . . . , etc. A nonempty *finite* set is one that may be indexed by the set $\{1, 2, \ldots, n\}$, for some positive integer n. An *infinite* set is one that cannot be indexed by such a set.

(b) Let $\mathfrak{A}$ be the set of the natural numbers, $N = \{1, 2, 3, \ldots\}$. The collection $S = \{I_n\}$, where each I_n is the subset of N defined by

$$I_n = \{m \in N : m \leq n\} = \{1, 2, \ldots, n\},$$

is indexed by N.

A *countable set* is one which may be indexed by N.

(c) Let $\mathfrak{A} = R$, the set of all real numbers. The family (collection) of *intervals*

$$(-\infty, r] = \{s \in R : s \leq r]$$

is indexed by R.

Remark. The familiar properties of the real numbers will be used freely in the examples, although we have not yet formally defined R.

3 Definition. Let $\{S_\alpha\}$ be a family (collection) of sets indexed by a set $\mathfrak{A}$. The *union* and *intersection* of these sets are defined as follows:

(i) $\bigcup_{\alpha \in \mathfrak{A}} S_\alpha = \{s : \exists \alpha \in \mathfrak{A} \text{ such that } s \in S_\alpha\}$. That is, the set S is said to be the union of the S_α, if each element s in S is in at least one of the S_α.

(ii) $\bigcap_{\alpha \in \mathfrak{A}} S_\alpha = \{s : \forall \alpha \in \mathfrak{A}, s \in S_\alpha\}$. In this case, s is said to be in the intersection of the S_α if s is in every S_α. (We shall write simply $\bigcup S_\alpha$ and $\bigcap S_\alpha$ for the union and intersection respectively, whenever the index set is unambiguously determined by the context of the discussion.)

4 Definition. For any pair of sets S and T, the *complement* of S *in* T is the set

$$T - S = \{t : t \in T \text{ and } t \notin S\}.$$

The inclusion $T \supset S$ is not required for $T - S$ to be defined. For, although the set R of real numbers contains the set Z of integers, $Z - R$ is defined. It is, of course, the empty set.

The proofs of the properties stated in Propositions 5 and 6 are left to the reader in the exercises, as well as their generalizations, when meaningful, to arbitrarily indexed families of sets.

5 Proposition. If P, S, and T are sets, then

(i) $S \cup T = T \cup S$, and $S \cap T = T \cap S$ (commutative laws).

(ii) $S \cup (T \cup P) = (S \cup T) \cup P$ and $S \cap (T \cap P) = (S \cap T) \cap P$ (associative laws).

(iii) $S \cap (T \cup P) = (S \cap T) \cup (S \cap P)$, and
$S \cup (T \cap P) = (S \cup T) \cap (S \cup P)$ (distributive laws).

(iv) If $S \subset T$ and $T \subset P$, then $S \subset P$ (transitivity).

(v) $S \subset S \cup T$; also, $S = S \cup T$ if and only if $T \subset S$.

6 Proposition. (DeMorgan's laws.) Let $\{S_\alpha\}$ be a family of sets and let X be any set. Then

(i) $X - \bigcup S_\alpha = \bigcap (X - S_\alpha)$, and

(ii) $X - \bigcap S_\alpha = \bigcup (X - S_\alpha)$.

Exercises

***1.** Prove that the empty set is contained in every set. (*Hint*: The statement "*A* implies *B*" is true whenever "*A*" is false.)

***2.** Show that the empty set is unique.

3. Show that $\emptyset = \{x : x \neq x\}$.

4. Using the language of product spaces, redefine surjective, injective, bijective, and inverse mappings.

5. Show that a finite set is countable.

***6.** Prove Proposition 5.

***7.** State generalizations of parts (i), (ii), and (iii) of Proposition 5 and prove them. (For example, (iii) becomes $S \cap (\bigcup T_\alpha) = \bigcup(S \cap T_\alpha)$, where $\{T_\alpha\}$ is any indexed family of sets.)

***8.** Prove Proposition 6, DeMorgan's laws.

9. Prove that if $T \supset S$, then $T = S \cup (T - S)$. How must this be modified if T does not contain S?

10. Verify each of the following equalities:
(a) $S \cap (T - S) = \emptyset$.
(b) $S - (S - T) = S \cap T$.
(c) $S \cup (T - S) = S \cup T$.
(d) $(S - P) \cup (T - P) = (S \cup T) - P$.
(e) $(S - T) \cup (T - S) = (S \cup T) - (S \cap T)$.

11. Show that the set of even (or odd) natural numbers is countable. Show also that the integers are countable.

***12.** Let $f: S \to T$ be a mapping. Show that the following statements are true for any pair of subsets A and B of T.
(a) $f^{-1}(A \cup B) = f^{-1}(A) \cup f^{-1}(B)$.
(b) $f^{-1}(A \cap B) = f^{-1}(A) \cap f^{-1}(B)$.
(c) $f^{-1}(A - B) = f^{-1}(A) - f^{-1}(B)$.
(d) $f(f^{-1}(B)) \subset B$; And $f(f^{-1}(B)) = B$ if and only if the mapping f is surjective.

***13.** Let $f: S \to T$ be a mapping. Show that the following statements are true for all pairs of subsets A and B of S.
(a) $f(A \cup B) = f(A) \cup f(B)$.
(b) $f(A \cap B) \subset f(A) \cap f(B)$. Equality holds for all pairs A and B if and only if f is injective.
(c) $A \subset f^{-1}(f(A))$. Equality holds for all A if and only if f is injective.

*Exercises which request the proof of a statement made in the text, as well as those referred to in the text are starred.

2 Equivalence Relations and Partial Orderings

1 Definition. A *relation* $\mathcal{R}$ defined on a set X is (determined by) a subset $\mathcal{Y}$ of $X \otimes X$. We say that x *is related to* y, written $x \mathcal{R} y$, if the ordered pair (x,y) is in $\mathcal{Y}$.

2 Examples. (a) Let $X = \{1,2,3\}$ and $\mathcal{Y} = \{(1,2),(1,3),(2,3)\}$. We see that $\mathcal{Y}$ defines the relation "<," since $(x,y) \in \mathcal{Y}$ if and only if $x < y$.

(b) Let X be a fixed but arbitrary set and let $\mathcal{F}$ be the family of all subsets of X. The inclusion relation is defined by the set $\mathcal{Y} = \{(S,T): S \subset T \subset X\}$.

3 Definition. A relation $\mathcal{R}$ defined on a set X is said to be
(i) *reflexive*, if for all x in X, $x \mathcal{R} x$;
(ii) *symmetric*, if $x \mathcal{R} y \iff y \mathcal{R} x$;
(iii) *antisymmetric*, if $x \mathcal{R} y$ and $y \mathcal{R} x$ imply $x = y$;
(iv) *transitive*, if $x \mathcal{R} y$ and $y \mathcal{R} z$ imply $x \mathcal{R} z$.

Two relations that appear frequently in this text are *equivalence relations* and *partial orderings*, the first satisfying properties (i), (ii), and (iv), and the second (i), (iii), and (iv). An equivalence relation will be denoted by $x \equiv y$ (read "x is equivalent to y"), and a partial ordering on X is written: $x \prec y$, (read "x is less than y" or "x comes before y.")[1]

Some examples follow, several of which are undoubtedly familiar to the reader. Others may be found in the exercises.

4 Examples. (a) The most obvious equivalence relation that may be defined on any set is the identity relation. Here

$$\mathcal{Y} = \{(x, x) : x \in X\}.$$

That is, $x \,\mathcal{R}\, y$, or $x \equiv y$, if and only if $x = y$.

(b) Let p be a fixed positive integer. The relation $x \equiv y \pmod p$ is an equivalence relation on the set Z of integers. In this case,

$$\mathcal{Y} = \{(x, y) : x, y \in Z \text{ and } (x - y) \text{ is a multiple of } p\}.$$

Verification is left to the reader.

(c) Let X be the set of all real valued functions whose domain is the interval $[0,1]$. The subset

$$\mathcal{Y} = \{(f(x), g(x)) : f(x) = g(x) \text{ except on a finite set}\}$$

of $X \otimes X$ defines an equivalence relation.

(d) By varying (c) slightly, we obtain an example which occurs in the study of integration. Let X be the set of real functions whose Riemann integrals over the interval $[0,1]$ exist, and let $\mathcal{Y}$ be defined as in (c). That is, $f \equiv g$ means that $f(x) = g(x)$ except possibly at finitely many points. We know that two functions which differ at finitely many points, although different as *functions*, are "indistinguishable" from the point of view of Riemann integration.[2] Indeed, the integral $\int_0^1 f(x)\,dx$ remains unchanged if f is replaced by an equivalent function. We therefore group together equivalent functions into classes. That is, the equivalence class $[f]$ of a function f in X is the set

$$\{g \in X : f \equiv g\}.$$

It is easily seen that the given set X can be written as the union $\bigcup_{f \in X} [f]$, since each function in X is in some equivalence class. We show next that

[1] When convenient, we shall write $y \succ x$ for $x \prec y$ and say that "y is greater than x" or "y comes after x."

[2] Later (Sec. 3 of Chapter 6) we shall prove that two Riemann integrable functions may differ on certain infinite sets, and yet be "indistinguishable" from the point of view of Riemann integration.

these classes are disjoint. Let h be an element of X which is in both $[f]$ and $[g]$. Then $h \equiv f$ and $h \equiv g$. The transitivity of the relation implies that $f \equiv g$, which in turn yields $f \in [g]$ and $g \in [f]$. This implies that $[f] = [g]$. For, if $f' \in [f]$, then $f' \equiv f$. Taken together with $f \equiv g$, we have $f' \equiv g$, or, $f' \in [g]$. Thus $[f] \subset [g]$. A similar argument yields the opposite inclusion. Thus, two classes are disjoint or identical.

Of course, *any* equivalence relation will partition a set into disjoint equivalence classes in the manner described in the preceding paragraph (Exercise 3).

5 Examples. (a) The inclusion relation [Example 2(b)] is a partial ordering on $\mathfrak{F}$, the family of subsets of a given set X, since

(i) $\forall S, S \subset S$ (reflexivity);

(ii) if $S \subset T$ and $T \subset S$, then $S = T$ (antisymmetry);

(iii) if $S \subset T$ and $T \subset Q$, then $S \subset Q$ (transitivity).

(b) The relation "less than or equal to," written $\leq$, is a partial ordering of Z, the set of all integers.

It should be observed, in part (a), that if X contains more than one element, then there will be subsets S and T of X for which both $S \subset T$ and $T \subset S$ are false; that is, neither the pair (S,T) nor (T,S) is in the subset $\mathfrak{Y}$ of $\mathfrak{F} \otimes \mathfrak{F}$ which defines the inclusion relation. On the other hand, if n and m are integers, then either $n \leq m$ or $m \leq n$, so that in this case either (n,m) or (m,n) is in $\mathfrak{Y}$. The relation "$\leq$" is an example of a *total ordering*:

6 Definition. X is said to be *totally ordered* by a relation $\prec$ if

(i) $\prec$ partially orders X, and

(ii) if $x,y \in X$, then either $x \prec y$ or $y \prec x$.

Remark. Property (ii) includes the case that *both* $x \prec y$ and $y \prec x$. Here the antisymmetry of $\prec$ implies that $x = y$.

Hereafter, a set X which is ordered (partially or totally) by the relation $\prec$ will be called an *ordered set*, and written as an ordered pair $(X,\prec)$, or simply as X when it is quite clear from the context of the discussion how $\prec$ is defined.

Since some pairs of elements in a partially ordered set may not be related by $\prec$, it does not generally make sense to speak of an element in X which is greater than all other elements of X [Example 8(c).] There may be, however, elements of X which are not smaller than any other elements of X:

7 Definition. An element μ of a partially ordered set $(X,\prec)$ is said to be *maximal* if

$$\mu \prec x \quad \text{implies} \quad \mu = x.$$

This means that if $x \in X$, then one of the following statements is true: (i) $x = \mu$, (ii) $x < \mu$, or (iii) x and μ are not related by $<$.

8 Examples. (a) The ordered set $(Z, \leq)$ contains no maximal elements. For, given any number n whatsoever, the number $n + 1 \geq n$, and $n + 1 \neq n$. (*Query:* What can you say about the subset of maximal elements of a totally ordered set? See Exercise 5.)

(b) X is a maximal element of $(\mathfrak{F}, \subset)$, since every member of $\mathfrak{F}$ is a subset of X. This says also that X is the only maximal element.

(c) Removing the set X from the family $\mathfrak{F}$, we obtain $\mathfrak{F}' = \mathfrak{F} - \{X\}$, the family of proper subsets of X. Each subset of X obtained by removing a single point from X, is maximal, and it is easily seen that there are no other maximal elements.

Some existence theorems may be stated in the language of partially ordered sets. Here the object whose existence we wish to assert is a maximal element. A typical example follows.

9 Example. Let X be a real vector space. A finite set of vectors $\{x_1, x_2, \ldots, x_n\}$ in X is said to be *linearly independent* if, whenever $\Sigma c_i x_i = \bar{0}$, where $c_i \in R$ and $\bar{0}$ is the additive identity of X, then $c_i = 0$, $i = 1, 2, \ldots, n$. An infinite subset (countable or uncountable) Y of X is called *linearly independent* if every finite subset of Y is linearly independent in the sense just described. Y is said to be a *maximal* set of linearly independent vectors if for any x in X, the set $Y \cup \{x\}$ is *not* linearly independent. This implies that if Y is maximal, then every element x in X may be written as a finite linear combination of vectors in Y. A maximal set is therefore a *basis* for the vectors in X. The collection $\mathfrak{F}$ of linearly independent subsets of X is partially ordered by the inclusion relation $\subset$. We have just shown that a subset of X is a basis if and only if the subset is a maximal element of the partially ordered set $(\mathfrak{F}, \subset)$.

We turn our attention now to the problem of verifying the existence of a maximal element. For Example 9, the proof rests upon what is called *Zorn's Lemma*, an inaccurate nomenclature in this case since it is taken here as an axiom. It is a question of taste which of the dozen (perhaps even more) equivalent statements, which include the Axiom of Choice, the Principle of Transfinite Induction, and the Well Ordering Principle, is assumed. For some, all are equally distasteful. For a more complete discussion, the reader is urged to consult Kelley [1].

10 Zorn's Lemma. Let $(X, <)$ be a partially ordered set which possesses the following property: To each totally ordered subset S of X there is an element μ in X such that $\mu > s$ for all s in S. (μ is called an *upper bound* for S.) Then the partially ordered set X contains a maximal element.

If in Example 9, $\mathcal{S}$ is a totally ordered subset of $\mathfrak{F}$, then

$$Y, Y' \in \mathcal{S} \text{ implies } Y \subset Y' \text{ or } Y' \subset Y.$$

It is not difficult to show that the union of the sets in $\mathcal{S}$ is in $\mathfrak{F}$ and is an upper bound for $\mathcal{S}$. Thus Zorn's lemma may be used to obtain a maximal element, which, as we have already seen, is a basis for X. The details are left as an exercise.

Although Zorn's lemma and the Axiom of Choice are equivalent, the latter is often more readily accepted by students. In fact, The Axiom of Choice is frequently used in textbook proofs without any formal statement of its use. On the other hand, it is rare that Zorn's Lemma will be used (in textbooks) without some sort of formal announcement. We state next the Axiom of Choice and give a very simple example of its use. Later, in Sec. 4 of Chapter 7 it appears in the proof of the existence of nonmeasurable sets.

11 Axiom of Choice. Let $\{X_\alpha\}$ be a family of nonempty sets indexed by a set $\mathfrak{A}$. Then there exists a set $X = \{x_\alpha\}$ also indexed by $\mathfrak{A}$, and for each α in $\mathfrak{A}$, $x_\alpha \in X_\alpha$.

Roughly speaking, the Axiom of Choice asserts the existence of a set X which is obtained by *choosing* an element x_α from each X_α. For a finite index set $\mathfrak{A}$ there is certainly no need for a special axiom. No one would question the existence of a set formed by choosing one element from each of a finite number of sets. It is the infinite "simultaneous" choice that is unacceptable to some mathematicians.

12 Example. *Every infinite set contains a countable subset.* An infinite set is one that cannot be put in a 1–1 correspondence with the set $\{1,2,\ldots,n\}$, for any value of n. The most natural way to demonstrate this assertion is simply to select elements from the infinite set X, naming them $x_1, x_2, \ldots$. Since X is infinite, the supply is never exhausted. More formally, we "choose" x_1 from the set $X_1 = X$, then x_2 from $X_2 = X_1 - \{x_1\}, \ldots, x_n$ from $X_n = X_{n-1} - \{x_{n-1}\}$, thus obtaining the countable set.

Exercises

1. Give examples of relations that are
  (a) reflexive but neither transitive nor symmetric,
  (b) symmetric but neither transitive nor reflexive,
  (c) transitive but neither symmetric nor reflexive,
  (d) reflexive and transitive but not symmetric, etc.

2. (a) Prove that "congruence mod p" is an equivalence relation. [Example 4(b).] Describe the partition of Z into equivalence classes.
  (b) Let $R[x]$ be the totality of polynomials with real coefficients, and let $p(x)$ be a fixed but arbitrary polynomial. Prove that the following relation is an equivalence relation:

$$f \equiv g \qquad \text{if and only if } f - g \text{ is a multiple of } p,$$

that is, of the form $m(x) \cdot p(x)$, where $m(x) \in R[x]$. Discuss the corresponding partition of $R[x]$.

*3. Prove that in general an equivalence relation on a set X partitions X into disjoint equivalence classes. (See Example 4.)

4. For each of the following, decide which of the properties of relations that are stated in Definition 3 are satisfied.
   (a) Example 2(a).
   (b) Example 2(b).
   (c) Let X be the family of real-valued functions defined on the interval $[0,1]$, and let the relation be defined by the subset

$$\mathcal{Y} = \{(f,g) : f(1/2) = g(1/2)\} \qquad \text{of the product } X \otimes X.$$

   (d) Let X be the set described in (c), and let

$$\mathcal{Y} = \{(f,g) : f(0) < g(1)\}.$$

   (e) For the same X, let

$$\mathcal{Y} = \{(f,g) : [f(1) - g(1)] \in Z\}.$$

5. Prove that a totally ordered set possesses at most one maximal element.

6. Prove that Zorn's lemma implies the Axiom of Choice. (*Outline of proof:* Let $\{X_\alpha\}$ be a family of disjoint sets indexed by $\mathcal{A}$, and let $\mathfrak{F}$ be the totality of sets of the following type: $S \in \mathfrak{F}$ if

$$s, t \in S \quad \text{and} \quad s \neq t \rightarrow \exists\, \alpha, \beta \in \mathcal{A},\ \alpha \neq \beta, \quad \text{and} \quad s \in X_\alpha,\ t \in X_\beta.$$

$\mathfrak{F}$ is partially ordered by inclusion. *A* set X that contains one and only one element from each X_α is clearly a maximal element of $\mathfrak{F}$, and conversely. Zorn's lemma will yield the existence of a maximal element if it can first be verified that every totally ordered subset of $\mathfrak{F}$ has an upper bound in $\mathfrak{F}$. It is left to the reader to show that the union of sets in any totally ordered subset of $\mathfrak{F}$ is an upper bound for this subset.)

7. Prove that the Axiom of Choice implies Zorn's lemma.

*Exercises which request the proof of a statement made in the text, as well as those referred to in the text, are starred.

3 The Rational Numbers As an Ordered Field

The construction of the complete ordered field of real numbers falls naturally into two parts. The first is entirely algebraic; that is, the concept of a limit does not appear, and all operations are finite. The standard procedure is to begin with the Peano Axioms (see Landau [1]) for the natural numbers, and to extend this set first to the integers and then to the rational numbers, defining the binary operations of addition and multiplication, and an ordering relation. Since this construction is rather lengthy and not really a part of Analysis, we shall begin by assuming that the *ordered field* (Definitions 1 and 2) Q of *rational numbers* is given, and proceed to the construction of the real numbers. All the algebraic material omitted, as well as a slightly different but equivalent way to define the real number can be found in Landau [1] or Isaacs [1].

In this section we shall list the basic properties of the rational numbers and of sequences of rational numbers. For the most part, proofs will be assigned as exercises, and it is generally assumed that these properties, if not their proofs, are familiar to the reader. Further discussion of the algebraic ideas may be found in Birkhoff and MacLane [1], Herstein [1], McCoy [1], and almost any other standard algebra text. Supplementary material about sequences is easily obtained from any advanced calculus text.

1 Definition. A *field* is a triple $(X; +, \cdot)$, consisting of a nonempty set X and two *binary operations* "+" and "·" (addition and multiplication respectively) defined on X. Each ordered pair (a,b) of elements in X is mapped into its sum $a + b$, and product $a \cdot b$, both of which are elements of X. The following properties are satisfied:

(i) *Addition and multiplication are associative:*

$$\forall a,b,c \text{ in } X, \quad a + (b + c) = (a + b) + c, \quad a \cdot (b \cdot c) = (a \cdot b) \cdot c.$$

(ii) *Addition and multiplication are commutative:*

$$\forall a,b \text{ in } X, \quad a + b = b + a, \quad a \cdot b = b \cdot a.$$

(iii) *The Distributive Law holds:*

$$\forall a,b,c \text{ in } X, \quad a \cdot (b + c) = a \cdot b + a \cdot c.$$

(iv) *X contains an additive and multiplicative identity:* $\exists$ elements $\bar{0}$, e, in X such that if a is any element of X, then $a + \bar{0} = a$, and $a \cdot e = a$.

(v) Every member of X has an additive inverse, and every $a \neq \bar{0}$ has a multiplicative inverse:

$$\forall a \in X, \exists -a \in X, \text{ such that } a + (-a) = \bar{0};$$
$$\forall a \neq \bar{0}, \exists a^{-1} \in X \text{ such that } a \cdot a^{-1} = e.$$

For convenience, we shall simply write X in place of $(X; +, \cdot)$, and when completely unambiguous, "ab" in place of "$a \cdot b$."

2 Definition. An *ordered field* is a pair $(X,\prec)$, where X is a field and $\prec$ is a relation defined on X that possesses the following properties:

(i) For each x in X, exactly one of the following is true: $x \prec \bar{0}$, $\bar{0} \prec x$, or $x = \bar{0}$. (x is said to be *positive* if $\bar{0} \prec x$, and ***negative*** if $x \prec \bar{0}$.)

(ii) $\bar{0} \prec x$ and $\bar{0} \prec y$ imply $\bar{0} \prec x + y$ and $\bar{0} \prec xy$.

(The sum and product of positive elements are positive.)

(iii) $x < y$ if and only if $\bar{0} < y - x$.
(The standard substitution of "$y - x$" for "$y + (-x)$" is made here.)

An equivalent set of defining properties of an ordered field is obtained by substituting (i′) for (i):

(i′) For any pair of elements x,y in X exactly one of the following statements is true: $x < y$, $y < x$, or $x = y$.

Verification of the equivalence is left to the reader.

Remark. Property (i) of an ordered field implies that the relation $<$ • is neither reflexive nor antisymmetric. Thus $<$ is not an example of a partial ordering. To illustrate the distinction between the order relations of a partially ordered set and of an ordered field, consider the pairs $(Q,\leq)$ and $(Q,<)$, Q being the rational numbers. The relation "less than or equal to" is a partial ordering of Q, and the relation "$<$" makes Q into an ordered field.

3 Proposition. If $(X,<)$ is an ordered field, then

(i) $<$ is a transitive relation on X;
(ii) for all x in X, $x < \bar{0}$ if and only if $\bar{0} < -x$;
(iii) for all z in X, $x < y$ if and only if $x + z < y + z$.
(iv) if $x < \bar{0}$ and $y < \bar{0}$, then $\bar{0} < xy$,
if $x < \bar{0}$ and $\bar{0} < y$, then $xy < \bar{0}$,
if $x \neq \bar{0}$, then $\bar{0} < xx = x^2$.

(Again we adopt the conventional notation and write "x^n" for the n-fold product of x with itself.)

The proof of Proposition 3 is left to the reader.

4 Definition. The *absolute value* of an element x of an ordered field X is given by

$$|x| = \begin{cases} x & \text{whenever } \bar{0} < x \text{ or } x = \bar{0}. \\ -x & \text{whenever } x < \bar{0}. \end{cases}$$

The function $|x|$ maps X into the subset of non-negative elements of X, and possesses the usual familiar properties, which are stated below and in the exercises. Again the proofs are left to the reader.

5 Proposition.

(i) $x^2 = |x|^2$.
(ii) $|xy| = |x||y|$.
(iii) If $\bar{0} < x$, then $x^n = |x|^n$, and if $x < \bar{0}$ then $x^n = (-1)^n|x|^n$. (n is an integer)

6 Definition. An ordered field $(X,<)$ is said to be *Archimedean* if, for any pair of positive elements x and y, there is an integer n such that $y < nx$. (Here, $nx = \underbrace{x + x + \cdots + x}_{n \text{ times}}$.)

•See special note on page 447.

It is easily seen that $(Q,<)$ is an Archimedean ordered field. For if x and y are positive rational numbers, then there are positive integers a, b, c, d such that $x = a/b$ and $y = c/d$. We seek an integer n satisfying the inequality $y < nx$. This is equivalent to the integral inequality $cb - nad < 0$. Thus the given problem is equivalent to asking: for given positive integers cb and ad, is there an integer n such that $cb - nad < 0$? An affirmative answer is given by the *Well Ordering Principle* for the natural numbers, which asserts that *every nonempty subset of the natural numbers (positive integers) contains a least element*. If the desired inequality is satisfied for no value of n, then the set $\{cb - nad\}$ is a set of non-negative integers and must therefore have a least element, which we denote by $cb - Nad$. But then a contradiction is obtained by setting $n = N + 1$, for in this case we have $cb - nad < cb - Nad$, which cannot be if $cb - Nad$ is the smallest element in the set.

Remark. The Well Ordering Principle or the equivalent statement of the Principle of Mathematical Induction, is one of the axioms for the system of real numbers (Exercise 9).

7 Example of a Non-Archimedean Ordered Field. Let $X = Q(x)$, the field of quotients of polynomials whose coefficients are rational (see Birkhoff and MacLane [1], Chapter XIII). $Q(x)$ may be made into an ordered field in the following way. Let $p(x)$ and $q(x)$ be distinct elements of $Q(x)$ that are written in reduced form; that is, $p(x) = f(x)/g(x)$, f and g having no common factors other than constants. Similarly for $q(x)$. It is not difficult to show that there is a positive number x_0 such that one of the following statements is true:

(a) If $x > x_0$ then $p(x) > q(x)$, or
(b) If $x > x_0$ then $p(x) < q(x)$.

(See Exercise 10.) If (a) is true we say that $p \succ q$, and if (b) holds, $p \prec q$. In Exercise 11 the reader is asked to show that $(Q(x),\prec)$ is an ordered field that does not possess the Archimedean property.

Although Q is an Archimedean ordered field, these properties cannot be used to define Q since there are many Archimedean ordered fields (the real numbers, for example). What distinguishes Q from the other Archimedean ordered fields is that $(Q,<)$ is the *smallest* ordered field in the following sense: If $(X,\prec)$ is an ordered field, then X contains a subfield $\hat{Q}$ which is (field) isomorphic to Q. This means that there is a bijective map $\Phi: Q \to \hat{Q}$ which preserves the field operations; that is, $\Phi(a + b) = \Phi(a) \hat{+} \Phi(b)$ and $\Phi(a \cdot b) = \Phi(a) \hat{\cdot} \Phi(b)$. Furthermore Φ preserves the order relation: If $a \prec b$ then $\Phi(a) \prec \Phi(b)$. The assertion that Q does not contain a proper subfield follows from the easily verified fact that if a field has *characteristic zero* (if $x \neq 0$, then $nx \neq 0$ for all $n = 1,2,\ldots$) then it contains a subfield isomorphic to Q. This follows directly from Definition 1, (i), (ii), and (iv). The preservation of the order relation is also not difficult to prove. (See Birkhoff and MacLane[1], Chapter II, Secs. 4–6.)

Exercises

*1. Prove that (i) and (iii) of Definition 2 are equivalent to (i′) and (iii).

*2. Prove Proposition 3.

*3. Prove Proposition 5.

In exercises 4-7 it is assumed that all elements are in an ordered field $(X,\prec)$.

4. Show that there is no x in X which satisfies $x^2 + 1 = \bar{0}$.

5. Show that $\bar{0} \prec x \prec y$ implies $\bar{0} \prec y^{-1} \prec x^{-1}$.

*6. Show that if x is an element of X for which $x \prec y$ whenever y is positive, then either $x = \bar{0}$ or $x \prec \bar{0}$.

*7. Prove that $|x + y| \preceq |x| + |y|$, $|x - y| \preceq |x| + |y|$, and $||x| - |y|| \preceq |x - y|$ ("$\preceq$" reads "less than or equal to"). Using the Principle of Mathematical Induction, prove that

$$(x/y)^n = x^n/y^n.$$

$(1/y = y^{-1}$ and $1/y^n = (y^{-1})^n = y^{-n})$.

8. Complete the proof that $(Q,<)$ is an Archimedean ordered field.

9. The Principle of (finite) Mathematical Induction states that if S is a subset of the natural numbers N which satisfies (i) $1 \in S$, and (ii) if $n \in S$ then $n + 1 \in S$, then $S = N$. Prove that this is equivalent to the Well Ordering Principle.

10. Prove that if $f(x)$ and $g(x)$ are distinct polynomials with rational coefficients, then there is a positive number x_0 such that either (a) $x > x_0$ implies $f(x) > g(x)$, or (b) $x > x_0$ implies $g(x) > f(x)$. Use this result to show that the same is true for quotients (See Example 7).

11. Verify that $Q(x)$ (Example 7) is an ordered field. Using the polynomials $p(x) = 1$ and $q(x) = x$, show that the Archimedean property is not satisfied.

12. Show that a field of characteristic zero contains a subfield which is isomorphic to Q.

*13. Show that if x is a rational number, then there is a uniquely determined integer n such that $n \leq x < n + 1$. Show also that there is a uniquely determined integer m such that $m < x \leq m + 1$.

*14. Using the Principle of Mathematical Induction, prove that if α and β are rational numbers, then for all positive integers n.

$$(\alpha + \beta)^n = \alpha^n + n\alpha^{n-1}\beta + \frac{n(n-1)}{2}\alpha^{n-2}\beta^2 + \cdots + \binom{n}{k}\alpha^{n-k}\beta^k + \cdots + n\alpha\beta^{n-1} + \beta^n,$$

where

$$\binom{n}{k} = \frac{n(n-1)\cdots(n-k-1)}{k(k-1)\cdots 2\cdot 1},$$

$1 \leq k \leq n$, are the *Binomial Coefficients*.

15. Show that for all positive integers n,

$$1 + 2 + \cdots + n = (n + 1)n/2.$$

*Exercises which request the proof of a statement made in the text, as well as those referred to in the text, are starred.

4 Incompleteness of the Rational Numbers

In this section another property of ordered fields, *completeness*, is introduced. Unlike the field and order properties, which are algebraic, completeness is a concept of analysis and is defined by means of sequences or in equivalent ways that are discussed later. Before giving a formal definition of completeness and proving that Q is incomplete, we shall give a geometric description of the incompleteness of the rational numbers: Suppose that to each rational number β a unique point on a "calibrated" line is assigned. This point is located $|\beta|$ units from some fixed point to which the number 0 has been attached, and lies to the left of 0 if $\beta < 0$, and to the right if $\beta > 0$. Thus an injection of Q into the points of a line is defined. As we shall see later, this mapping is not surjective, so that although every interval on the line contains "rational points," there are also gaps on the line to which no (rational) number has been assigned. This is what is meant by the incompleteness of Q. By creating (or defining) new numbers that correspond to these gaps, the *complete, ordered field of real numbers* is obtained.

We begin by listing some of the familiar properties of sequences of rational numbers.

1 Definition. A *sequence* of rational numbers, denoted by $\{a_n\}$, is a mapping of the natural numbers (positive integers) into the rational numbers. The symbol a_n denotes the image of the positive integer n under this mapping.

A *sequence* $\{a_n\}$ of rational numbers is said to *converge* to the rational number a, written

$$\lim_{n\to\infty} a_n = a, \quad \lim a_n = a, \quad \text{or} \quad a_n \to a,$$

if for any rational number $\epsilon > 0$, there is an integer N such that

$$n > N \quad \text{implies} \quad |a - a_n| < \epsilon.$$

An equivalent formulation which eliminates the explicit use of the integer N is

$$\lim a_n = a \Leftrightarrow \forall \epsilon > 0,\ |a - a_n| < \epsilon,$$

for all but finitely many integral values of n, or as we shall frequently say, "for almost all values of n."

For reference purposes, the following list of properties of convergent sequences is included.

2 Proposition. Let $\{a_n\}$ and $\{b_n\}$ be sequences of rational numbers which converge to the rational numbers a and b respectively.

(i) There is a positive rational number M such that $|a_n| < M$, for all n. (A convergent sequence is bounded.)

(ii) $\lim (a_n \pm b_n) = \lim a_n \pm \lim b_n = a \pm b$.
(iii) $\lim (a_n b_n) = (\lim a_n)(\lim b_n) = ab$.
(iv) If $\{a_n\}$ converges to a', then $a' = a$ (uniqueness of the limit).
(v) If $b \neq 0$, then $\lim (a_n/b_n) = \lim a_n / \lim b_n = a/b$.

Remark. Although the assumption $b \neq 0$ does not imply that *every* $b_n \neq 0$, it is easily shown that all but finitely many $b_n \neq 0$. This result is obtained by combining Propositions 4(v) and 5, which are proved below. Then the quotient sequence $\{a_n/b_n\}$ is understood to be the sequence obtained by discarding those (finitely many) terms for which $b_n = 0$.

(vi) If, for all n, $a_n \leq b_n$, then a $\leq b$.

Let $n_1 < n_2 < \cdots < n_k < n_{k+1} < \cdots$ be an increasing sequence of natural numbers. The subset $\{a_{n_k}\}$ of a given sequence $\{a_n\}$ is called a *subsequence*. Or, using the function definition (See Definition 1), a subsequence $\{a_{n_k}\}$ of a sequence $\{a_n\}$ is a mapping of the natural numbers into a subset of the collection of distinct a_n. That is, the image a_{n_k} of k is one of the elements of the original sequence. Although it is not necessary, it is assumed here that for all values of k, a_{n_k} appears before $a_{n_{k+1}}$ in the original sequence. It is easily verified that

$$\lim a_n = a \qquad \text{implies} \qquad \lim a_{n_k} = a,$$

if $\{a_{n_k}\}$ is a subsequence of $\{a_n\}$.

3 Definition. $\{a_n\}$ is called a *Cauchy sequence* if, for each rational $\epsilon > 0$, there is an integer N such that

$$n, m > N \qquad \text{implies} \qquad |a_n - a_m| < \epsilon;$$

or, if $|a_n - a_m| < \epsilon$ for almost all n and m.

4 Proposition. Let $\{a_n\}$ and $\{b_n\}$ be Cauchy sequences of rational numbers. Then

(i) $\{a_n\}$ is bounded;
(ii) $\{a_n \pm b_n\}$ and $\{a_n b_n\}$ are Cauchy sequences;
(iii) Every subsequence of $\{a_n\}$ is a Cauchy sequence;
(iv) If a subsequence of $\{a_n\}$ converges to a rational number a, then the sequence $\{a_n\}$ also converges to a.
(v) If $\{b_n\}$ does not converge to zero, then only finitely many of the $b_n = 0$, and discarding those which vanish, $\{a_n/b_n\}$ is a Cauchy sequence.

Verification of (i)–(iv) is left to the reader.

Proof of (v). We begin by showing that there is a positive (rational) number δ such that $|b_n| > \delta$ for almost all n: $\lim b_n \neq 0$ implies that there is an $\epsilon > 0$, such that either,

(1) infinitely many $b_n \geq \epsilon > 0$, or
(2) infinitely many $b_n \leq -\epsilon < 0$.

Both (1) and (2) can not be simultaneously true; for if they were there would be arbitrarily large values of n and m for which $|b_n - b_m| > \epsilon$, contradicting the assumption that $\{b_n\}$ is a Cauchy sequence. Let us suppose that (1) is true. (The argument is the same for (2).) Since $\{b_n\}$ is a Cauchy sequence, $|b_n - b_m| < \epsilon/2$ for all n and m that exceed some integer N. Choose an index $m' > N$ for which (1) is satisfied. Then

$$n > N \qquad \text{implies} \qquad b_n > b_{m'} - \epsilon/2 \geq \epsilon/2 > 0.$$

Thus δ may be taken as $\epsilon/2$.

Having shown that $\{b_n\}$ is bounded away from zero for almost all values of n, it follows that

$$\left|\frac{a_n}{b_n} - \frac{a_m}{b_m}\right| = \left|\frac{b_m a_n - b_n a_m}{b_n b_m}\right|$$

$$= \left|\frac{(b_m a_n - b_m a_m) + (b_m a_m - b_n a_m)}{b_n b_m}\right|.$$

Proposition 4(i) implies the existence of a (common) positive upper bound M for the numbers $|a_n|$ and $|b_n|$, and for sufficiently large values of n and m, $|b_n|, |b_m| > \delta > 0$. Thus, for almost all n and m,

$$\left|\frac{a_n}{b_n} - \frac{a_m}{b_m}\right| \leq \frac{M}{\delta^2} \, |(a_n - a_m) + (b_m - b_n)|$$

$$\leq \frac{M}{\delta^2}[\,|a_n - a_m| + |b_m - b_n|\,].$$

The convergence of the two sequences implies that the term on the right tends to zero as n and m tend to infinity, thus proving that $\{a_n/b_n\}$ is a Cauchy sequence. ∎

5 Proposition. A convergent sequence is a Cauchy sequence.

Proof. If $\lim a_n = a$, then the triangle inequality yields

$$|a_n - a_m| \leq |a_n - a| + |a - a_m|.$$

Both terms on the right-hand side tend to zero as n and m tend to infinity. ∎

In the ordered field $(Q,<)$, the converse of Proposition 5 is false. Indeed, there exist Cauchy sequences of rational numbers that do not converge (to rational numbers). This is what is meant by the *incompleteness* of Q. To establish the existence of such sequences, we introduce first the *decimal representation* of a rational number. Although the introduction of decimals to establish the incompleteness of Q is by no means necessary, the decimal representation is of sufficient interest in itself to justify its introduction. Besides lending themselves to the interpretation of a real number as a point on a line, they may in fact serve as a *definition* of a real number. Moreover, decimals may be used to prove that the real numbers are not countable (see Kamke [1], Sec. 4 of Chapter I).

6 Definition. A (nonnegative) *decimal*, written

[1] $$a_0 . a_1 a_2 \ldots a_n \ldots$$

is a sequence of nonnegative integers a_i, $i = 0, 1, 2, \ldots$, for which $0 \leq a_i \leq 9$ whenever $1 \leq i$.

The decimal [1] is said to be a *repeating decimal* if there are integers N and k such that

$$j > N \qquad \text{implies} \qquad a_j = a_{N+p} \qquad \text{for some integer } p,\ 1 \leq p \leq k,$$

and $j - N = p(\text{mod } k)$. This simply means that the block of digits "$a_{N+1}a_{N+2}\ldots a_{N+k}$" is repeated indefinitely in the sequence.

For example, 8.71678783456456456... is a repeating decimal with $N = 8$ and $k = 3$. The block of digits $a_9 a_{10} a_{11} = 456$ is repeated.

To each decimal there is associated a sequence of rational numbers called the nth *truncations of the decimal*:

[2] $$x_1 = a_0 + \frac{a_1}{10}, \quad x_2 = x_1 + \frac{a_2}{10^2}, \ldots, x_n = x_{n-1} + \frac{a_n}{10^n}, \ldots$$

7 Proposition. The sequence $\{x_n\}$ of truncations of a decimal [1] is a Cauchy sequence.

Proof. If $n > m$,

$$0 \leq x_n - x_m = \frac{a_{m+1}}{10^{m+1}} + \cdots + \frac{a_n}{10^n} \leq \frac{9}{10^{m+1}}\left[1 + \frac{1}{10} + \cdots + \frac{1}{10^{n-m-1}}\right],$$

since $0 \leq a_i \leq 9$ whenever $i \geq 1$. The term in brackets is a partial sum of a geometric series, and therefore

$$0 \leq x_n - x_m \leq \frac{9}{10^{m+1}}\left[\frac{1 - (1/10)^{n-m}}{1 - 1/10}\right] = (1/10)^m(1 - (1/10^{n-m})),$$

from which it follows that $\lim_{n,m\to\infty} (x_n - x_m) = 0$, proving that $\{x_n\}$ is a Cauchy sequence. ∎

Remark. The final step of the proof of Proposition 7 relies on the convergence to zero of the sequence $\{10^{-m}\}$. This follows from the inequality $10^{-m} < 1/m$, which is true for all $m \geq 1$, and from the convergence of the sequence $\{1/m\}$ to zero, which the reader is asked to verify in Exercise 9 using the Archimedean property.

Since there are decimals of both the repeating and non-repeating types, (.112123123412345... is an example of the latter). and *all* decimals give rise to Cauchy sequences in Q, the incompleteness of Q will be demonstrated once we have proved the following theorem:

8 Theorem. The Cauchy sequence of truncations [2] of a given decimal $a_0 . a_1 a_2 \ldots a_n \ldots$ converges to a rational number if and only if the decimal is of the repeating type.

Proof (Necessity). Let a repeating decimal be given. It may be assumed without loss of generality that $a_0 = N = 0$ (Why?) Then the decimal is of the following type:

[3] $$0 . a_1 a_2 \ldots a_k a_1 a_2 \ldots a_k a_1 \ldots$$

The sequence of truncations will converge in Q if it contains a convergent subsequence. [Proposition 4(iv).] Let us look at the subsequence of truncations that corresponds to the indices k, $2k$, $3k, \ldots mk, \ldots$. The identity $(1 - \beta^{n+1})/(1 - \beta) = 1 + \beta + \beta^2 + \cdots + \beta^n$ yields

$$x_{mk} = \left[\frac{a_1}{10} + \cdots + \frac{a_k}{10^k}\right] + \frac{1}{10^k}\left[\frac{a_1}{10} + \cdots + \frac{a_k}{10^k}\right]$$

$$+ \cdots + \frac{1}{10^{(m-1)k}}\left[\frac{a_1}{10} + \cdots + \frac{a_k}{10^k}\right]$$

$$= x_k\left[1 + \frac{1}{10^k} + \cdots + \left(\frac{1}{10^k}\right)^{m-1}\right] = x_k\left[\frac{1 - (1/10)^{km}}{1 - (1/10)^k}\right],$$

proving that

$$\lim_{m \to \infty} x_{mk} = \frac{x_k}{1 - (1/10)^k},$$

which is an element of Q.

(Sufficiency.) Suppose now that a rational number n/m is given. We may assume that $m > n > 0$, for the more general case follows easily. Using the Division Algorithm, a unique integer a_1, $0 \leq a_1 \leq 9$, is obtained that satisfies

$$0 \leq n/m - a_1/10 = \left[\frac{10n - a_1 m}{m}\right]\frac{1}{10} < \frac{1}{10}.$$

This implies that $0 \leq 10n - a_1 m = r_1 < m$. A second application of the division algorithm yields a number a_2, bounded by 0 and 9, which satisfies

$$0 \leq n/m - a_1/10 - a_2/10^2 = (r_2/m)(1/10^2) < 1/10^2.$$

Again, a_2 is uniquely determined and the remainder term r_2 is nonnegative and less than m. Continuing in this manner, one of two things must happen: if at some point a remainder $r_j = 0$, then for all $i > j$, $a_i = 0$, and the decimal is repeating. If no $r_j = 0$, then there must be integers

N and k such that $r_N = r_{N+k}$. This occurs because the total number of distinct remainders does not exceed m, the denominator of the given rational number. This yields

$$\frac{n}{m} - \frac{a_1}{10} - \cdots - \frac{a_N}{10^N} = \frac{r_N}{m}\left(\frac{1}{10^N}\right)$$
$$= \frac{1}{10^N}\left[\frac{a_{N+1}}{10} + \cdots + \frac{a_{N+k}}{10^k}\right] + \frac{r_{N+k}}{m}\left(\frac{1}{10^{N+k}}\right).$$

Since $r_{N+k} = r_N$, continued application of the Division Algorithm yields

$$\frac{n}{m} - \frac{a_1}{10} - \cdots - \frac{a_N}{10^N} = \left(\frac{1}{10^N}\right)A + \left(\frac{1}{10^{N+k}}\right)A + \cdots,$$

where

$$A = \left[\frac{a_{N+1}}{10} + \cdots + \frac{a_{N+k}}{10^k}\right].$$

It follows as in the necessity proof, that the sequence

$$x_{jk} = A\left[\frac{1}{10^N} + \frac{1}{10^{N+k}} + \cdots + \frac{1}{10^{N+k+j}}\right], \quad j = 1, 2, \ldots,$$

converges as j tends to infinity to the number

$$\frac{n}{m} - \frac{a_1}{10} - \cdots - \frac{a_N}{10^N}.$$

We have just proved that to each nonnegative rational number which is bounded by 1, there is a repeating decimal whose associated sequence of truncations converges to the given number. However it still remains to be proved that there cannot *also* be a nonrepeating decimal whose sequence of truncations converges to this rational number. For there *do* exist pairs of distinct decimals whose truncations converge to the same number; for example, the decimals 1.00000... and 0.9999... have truncation sequences that converge to the same number. Although it can be shown that only such pairs exist (Exercise 12), it suffices for our purposes to verify the following statement:

* If $b_1 b_2 \ldots b_n \ldots$ is *not* a repeating decimal, then the Cauchy sequence $y_n = y_{n-1} + b_n/10^n$, of truncations does not converge to a rational number.

Let us suppose that the y_n converge to a rational number n/m. Construct, as above, the repeating decimal $0.a_1a_2\ldots a_k a_1\ldots$ whose truncations converge also to n/m. Since $.b_1b_2\ldots$ is not a repeating decimal, there is a smallest integer j for which $b_j \neq a_j$. It is not difficult to show that for $n > j$, this implies $|y_n - x_n| > 1/10^{j+1}$, from which it follows that the sequences $\{x_n\}$ and $\{y_n\}$ cannot both converge to the same number. The details are left to the reader (Exercise 11). ∎

All definitions and properties stated for sequences in Q are equally applicable to any ordered field, $(X,\prec)$. In particular, every convergent sequence in an ordered field is a Cauchy sequence. The converse, as we have just seen is not always true.

9 Definition. An ordered field is said to be *complete* if every Cauchy sequence $\{x_n\}$ in X converges to a point x in X.

Thus the ordered field $(Q,<)$ is not a complete ordered field. In the next chapter it will be shown that $(Q,<)$ is "contained" in a complete ordered field. The smallest complete ordered field containing Q is the field of real numbers.

Exercises

***1.** Prove that if a sequence $\{a_n\}$ of rational numbers converges to a rational number a, then every subsequence of $\{a_n\}$ also converges to a.

***2.** Prove Proposition 4, (i)–(iv).

3. Give examples of bounded divergent sequences.

***4.** Show that $\lim a_n = a$, $\lim b_n = b$, and $a_n < b_n$ for all values of n, do not suffice to conclude that $a < b$. What additional assumption must be made to draw this conclusion?

***5.** Prove that $\lim a_n = 0$ and $|b_n| \leq |a_n|$ imply that $\lim b_n = 0$.

***6.** Prove that if $\{b_n\}$ is bounded and if $\lim a_n = 0$, then $\lim a_n b_n = 0$.

***7.** Show that $\lim a_n = a$ implies that $\lim |a_n| = |a|$.

***8.** Show that if $\{a_n\}$ is a Cauchy sequence, then $\{|a_n|\}$ is also a Cauchy sequence.

***9.** Prove that $\lim 1/n = 0$. (*Hint:* Use the Archimedean property.)

10. By imitating the proof of Theorem 8, part 2, show that the decimal associated with the rational number 5/7 may be obtained by the elementary school algorithm

$$\begin{array}{r} .71428571\ldots \\ 7\,\overline{)\,5.00000000\ldots} \end{array}$$

What are the a_i, r_i, N, and k?

***11.** Complete the proof of Theorem 9.

12. It is possible to define the real numbers as the totality of equivalence classes of decimals; two decimals are said to be equivalent if the difference of their truncation sequences converges to zero. Prove that if two different decimals are equivalent, then one must have repeating 9's and the other repeating 0's.

The *sum* of two decimals $a_0 . a_1 a_2 \ldots$ and $b_0 . b_1 b_2 \ldots$ is defined by adding their truncation sequences termwise. Let $\{x_n\}$ and $\{y_n\}$ denote their respective sequences. Form the sequence $z_n = x_n + y_n = (a_0 + b_0) + 1/10(a_1 + b_1) + \cdots + 1/10^n(a_n + b_n)$. This sequence determines a decimal $c_0 . c_1 c_2 \ldots$ as follows: If all the sums $a_n + b_n < 9$, then the c_n are taken to be these sums. Otherwise there is the problem of "carrying." The reader is asked to resolve this problem and to show that the addition operation that he defines is commutative and associative. Similarly for multiplication.

The order relation is not difficult to define, but again it is rather tedious and lengthy to verify everything. Try.

(This exercise was included to illustrate the difficulties that arise even in defining the real numbers by means of decimals, which is perhaps the most intuitive way.)

*Exercises which request the proof of a statement made in the text, as well as those referred to in the text, are starred.

CHAPTER 1

The Real Number System

1 Construction of the Field R

Roughly speaking, R is obtained by adding to Q points that correspond to the Cauchy sequences of rational numbers that do not converge in Q. *Every* Cauchy sequence of rational numbers determines a real number in the following way: If $\lim a_n = a \in Q$, then a is called a "rational" real number. If $\{a_n\}$ does not converge in Q, then it determines an "irrational" real number. One difficulty that arises here is that a given real number is determined by many Cauchy sequences. For example, both $\{1/n\}$ and $\{0\}$, the latter being the sequence each of whose terms is 0, converge to 0. Similarly if r is any rational number, then both $\{r + 1/n\}$ and the constant sequence $\{r]$ converge to r. Thus more than one Cauchy sequence may be associated with each number. For this reason, a real number will be defined as an *equivalence class* of Cauchy sequences of rational numbers.

Let A denote the collection of all Cauchy sequences of rational numbers.

1 Definition. Two Cauchy sequences $\{a_n\}$ and $\{b_n\}$ (of rational numbers) are called *equivalent*, written $\{a_n\} \sim \{b_n\}$, if $\lim (a_n - b_n) = 0$, or as we shall say, if $\{a_n - b_n\}$ is a *null sequence.*

2 Examples. (a) If r is in Q, then

$$\{r\} \sim \{r + 1/n\}.$$

(b) If $\{a_{n_k}\}$ is a subsequence of $\{a_n\}$, a Cauchy sequence, then

$$\{a_{n_k}\} \sim \{a_k\}$$

(Exercise 1).

3 Proposition. If $\{a_n\} \sim \{b_n\}$ and $\lim a_n = a$, $(a \in Q)$, then $\lim b_n = a$.

Proof: $\{a_n\} \sim \{b_n\} \Leftrightarrow \lim (b_n - a_n) = 0$

$$\Rightarrow a = a + 0 = \lim a_n + \lim (b_n - a_n) = \lim b_n. \quad \blacksquare$$

4 **Proposition.** $\lim a_n = \lim b_n = a$ implies $\{a_n\} \sim \{b_n\}$.
The proof is left to the reader.

5 **Proposition.** "$\sim$" is an equivalence relation on A.

Proof: Reflexivity: $\{a_n\} \sim \{a_n\} \Leftrightarrow \{a_n - a_n\} = \{0\}$ is a null sequence.
Symmetry: $\{a_n\} \sim \{b_n\} \Leftrightarrow \lim (a_n - b_n) = 0$

$$\Leftrightarrow \lim (b_n - a_n) = 0 \Leftrightarrow \{b_n\} \sim \{a_n\}.$$

Transitivity: $\{a_n\} \sim \{b_n\}$ and $\{b_n\} \sim \{c_n\} \Leftrightarrow$

$$\lim (a_n - b_n) = \lim (b_n - c_n) = 0 \Leftrightarrow \lim (a_n - c_n)$$
$$= \lim (a_n - b_n) + \lim (b_n - c_n) = 0 \Leftrightarrow \{a_n\} \sim \{c_n\}. \quad \blacksquare$$

Thus "$\sim$" partitions A into disjoint equivalence classes of Cauchy sequences of rational numbers. The class determined by the sequence $\{a_n\}$, written $[\{a_n\}]$, is the collection of Cauchy sequences each of which is equivalent to $\{a_n\}$. Let R denote the collection of these equivalence classes. The elements of R fall into two categories:

(i) If a class $[\{a_n\}]$ contains one sequence which converges to a rational number a, then every sequence in this class converges also to a (Proposition 3). Conversely, every sequence that converges to a is in this class (Proposition 4), in particular the constant sequence $\{a\}$. This establishes a 1–1 correspondence between Q and a subset $\bar{Q}$ of R, which is given by the mapping, $\phi: Q \longrightarrow \bar{Q}$, $\phi(a) = [\{a\}]$.

(ii) The class $R - \bar{Q} \neq \phi$, since there are Cauchy sequences (in Q) which do not converge to points in Q. Each class $[\{a_n\}]$ in $R - \bar{Q}$ is determined by a nonconvergent Cauchy sequence of rational numbers, and no class in $R - \bar{Q}$ may contain a convergent sequence.

6 **Definition.** The class R, described above is the set of *real numbers*. That is, a *real number* is an *equivalence class of Cauchy sequences* of *rational numbers*; two sequences are said to be equivalent, written $\{a_n\} \sim \{b_n\}$, if $\lim (a_n - b_n) = 0$.

$\bar{Q}$ is called the subset of "rational" real numbers, and $R - \bar{Q}$ is the set of "irrational" real numbers. In this section and the next, the following properties of R will be verified:

PI: R can be made into a field (proved in Sec. 1).

PII: If $\oplus$ and $\odot$ are the field operations of R, then $(\bar{Q}; \oplus, \odot)$ is a subfield which is isomorphic to $(Q; +, \cdot\,)$ (proved in Sec. 1).

PIII: A relation $\otimes$ may be defined on R in such a way that $(R, \otimes)$ is an ordered field and $(\bar{Q}, \otimes)$ is a subfield which is (order) isomorphic to $(Q, <)$ (proved in Sec. 2).

PIV: R is complete (proved in Sec. 2).

PV: If $R \supset \hat{R} \supset \bar{Q}$, and $\hat{R}$ satisfies PI, PII, PIII, and PIV, then $R = \hat{R}$. That is, R is the *smallest* complete, ordered, field which contains $\bar{Q}$ (proved in Sec. 2).

Let us begin by defining the field operations on R.

7 **Definition.** The *sum* $\oplus$ and the *product* $\odot$ of two real numbers $\bar{a} = [\{a_n\}]$ and $\bar{b} = [\{b_n\}]$ are, respectively,

(i) $\bar{a} \oplus \bar{b} = [\{a_n + b_n\}]$, and (ii) $\bar{a} \odot \bar{b} = [\{a_n b_n\}]$.

From Proposition 0.4.4 it follows that $\{a_n + b_n\}$ and $\{a_n b_n\}$ are Cauchy sequences. However, it remains to be proved that the operations $\oplus$ and $\odot$ are *well defined*, which means that they do not depend upon the particular sequences chosen from each class. Thus we must show that if $\{a_n\} \sim \{a_n'\}$ and $\{b_n\} \sim \{b_n'\}$ then $\{a_n + b_n\} \sim \{a_n' + b_n'\}$ and $\{a_n b_n\} \sim \{a_n' b_n'\}$. This follows easily from Definition 1 and Propositions 0.4.2 and 0.4.4: $\{a_n\} \sim \{a_n'\}$ and $\{b_n\} \sim \{b_n'\} \Leftrightarrow \lim (a_n - a_n') = \lim (b_n - b_n') = 0 \Rightarrow 0 = \lim [(a_n - a_n') + (b_n - b_n')] = \lim [(a_n + b_n) - (a_n' + b_n')] \Leftrightarrow \{a_n + b_n\} \sim \{a_n' + b_n'\}$. A similar argument may be used for the product. ∎

8 **Theorem.** $(R; \oplus, \odot)$ is a field.

Proof: (The field properties to be verified are listed in Defintion 0.3.1.)

(i) The associativity of addition in $(Q; +, \cdot)$ and Definition 7 yield

$$(\bar{a} \oplus \bar{b}) \oplus \bar{c} = [\{a_n + b_n\}] \oplus [\{c_n\}] = [\{(a_n + b_n) + c_n\}]$$

$$= [\{a_n + (b_n + c_n)\}] = [\{a_n\}] \oplus [\{b_n + c_n\}] = \bar{a} \oplus (\bar{b} \oplus \bar{c}).$$

Similarly for multiplication, and for (ii) (commutative law), and (iii) (distributive law).

(iv) The additive identity is $\bar{0} = [\{0\}]$, the class containing the constant sequence each of whose terms is 0. The multiplicative identity is $\bar{1} = [\{1\}]$. We shall verify only the former: If $\bar{a} \in R$, then $\bar{0} \oplus \bar{a} = [\{0 + a_n\}] = [\{a_n\}] = \bar{a}$.

(v) If $\bar{a}$ is any real number, then its additive inverse is $-\bar{a} = [\{-a_n\}]$.

If $\bar{a} \neq \bar{0}$, then $\lim a_n \neq 0$. Assuming then that no $a_n = 0$, or discarding those that are equal to zero, the sequence $\{a_n^{-1}\}$ is itself a Cauchy sequence. [Proposition 0.4.4(v)]. Then

$$[\{a_n\}] \odot [\{a_n^{-1}\}] = [\{1\}] = \bar{1},$$

proving that $\bar{a}^{-1} = [\{a_n^{-1}\}]$.

This completes the verification of the field properties of R, (PI). ❙

Proof of PII: It has already been shown that $\Phi(a) = [\{a\}] = \bar{a}$ is a bijective mapping of Q into $\bar{Q}$. For typographical purposes, let us denote the rational real number $[\{a\}]$ by $\dot{a}$. To differentiate $\dot{a}$ from a, the former will be called a "dotted" rational. It remains to be shown that $(Q;\oplus,\odot)$ is a subfield, and that

$$[1] \qquad \Phi(a + b) = \dot{a} \oplus \dot{b}, \quad \text{and} \quad \Phi(ab) = \dot{a} \odot \dot{b}.$$

It suffices to prove [1], for this implies that the image $\Phi(Q) = \bar{Q}$ is also a field. This is left to the reader.

Thus the field R may be considered as an "extension" of the field Q, in the sense that the elements of Q and $\bar{Q}$ are paired in a natural way, and that this correspondence is preserved under the field operations $+,\cdot$, and $\oplus,\odot$, in Q and $\bar{Q}$ respectively. For this reason it is no longer necessary to distinguish between the two sets of field operations, and henceforth "$+$" and "$\cdot$" will be used to denote operations in both R as well as Q (or $\bar{Q}$).

Exercises

*1. Prove that a Cauchy sequence of rational numbers is equivalent to all of its subsequences (Example 2).

*2. Prove that if two (rational) sequences have the same limit, then they are equivalent (Proposition 4).

*3. Prove that the product $\bar{a} \odot \bar{b}$ (Definition 7) is well defined. Show also that the product is associative and commutative.

*4. Prove that the distributive law holds in $(R;\oplus,\odot)$.

*5. Complete the proof of PII by verifying [1], and show that this implies $\bar{Q}$ is a subfield.

6. Show directly, without using the already proved existence of a multiplicative inverse for nonzero real numbers, that the equation $\bar{b}\bar{x} = \bar{a}$ has a unique solution in R.

*7. Show that if $\{a_n\}$, $\{b_n\}$ are Cauchy sequences of rational numbers contained in the equivalence classes (real numbers) $\bar{a}$ and $\bar{b}$ respectively, and if $a_n < b_n$ for all n, then there is a real number $\bar{\delta}$ which contains a rational Cauchy sequence $\{\delta_n\}$, each of whose terms δ_n is positive, and $\bar{b} = \bar{a} + \bar{\delta}$. Show by example that $\delta = \dot{0}$ is possible.

2 Ordering and Completeness of *R*

In this section, the order relation "$<$" on Q is extended to R in such a way that R becomes an ordered field. Following this is the demonstration of the completeness of the ordered field R.

*Exercises which request the proof of statement made in the text, as well as those referred to in the text, are starred.

1 Definition. The real number $\bar{a} = [\{a_n\}]$ is said to be *greater than* $\bar{b} = [\{b_n\}]$, or $\bar{b}$ is *less than* $\bar{a}$, written

$$\bar{a} \circledgt \bar{b}, \qquad \text{or} \qquad \bar{b} \circledlt \bar{a},$$

if there is positive rational number δ such that for almost all n,

$$a_n > b_n + \delta,$$

for any pair of sequences $\{a_n\}$ and $\{b_n\}$ in $\bar{a}$ and $\bar{b}$ respectively. (See Exercises 0.4.4 and 1.1.7.)

"$a \geq b$" will appear in place of the longer expression "$\bar{a} \circledgt \bar{b}$ or $\bar{a} = b$."

Again, the relation $\circledgt$ is neither reflexive nor symmetric, and therefore $\circledgt$ is not a partial ordering of R. On the other hand we shall see that $\circledgt$ totally orders R.

In verifying the field properties for R, it was necessary to show first that $\oplus$ and $\odot$ were "well defined," that is, independent of the particular sequences $\{a_n\}$ and $\{b_n\}$ that were chosen to represent the numbers $\bar{a}$ and $\bar{b}$ (see Definition 1.1.7). Here a similar argument is required to show that $\circledgt$ is well defined, and this is contained in Proposition 2:

2 Proposition. If $\{a_n\} \sim \{a'_n\}$ and $\{b_n\} \sim \{b'_n\}$, then the following two statements imply each other:

(i) There is a positive rational number δ, and $a_n > b_n + \delta$ for almost all n.

(ii) There is a positive rational number δ', and $a'_n > b'_n + \delta'$ for almost all n.

(This means that the conditions stated in Definition 1 hold for every pair $\{a_n\}$ and $\{b_n\}$ in $\bar{a}$ and $\bar{b}$ respectively, or for no pair.)

Proof: Assume (i). The equivalence of the two pairs of sequences implies that both $|a_n - a'_n|$ and $|b_n - b'_n|$ are less than $\delta/4$ for almost all n. Therefore

$$a'_n > a_n - \delta/4 > b_n + 3\delta/4 > b'_n + \delta/2,$$

proving that δ' [of statement (ii)] may be taken to be $\delta/2$. A similar argument yields "(ii) implies (i)." ▎

It is readily seen that the relation $\circledgt$ restricted to $\bar{Q}$ coincides with $>$ on Q: for, if $\dot{a} = [\{a\}] > \dot{b} = [\{b\}]$, then Definition 1 implies the existence of a positive rational number δ, such that $a > b + \delta$. But this is exactly what it means to assert that a is greater than b, or $a > b$. Thus, $\circledgt$ may be considered as an extension to R of the relation $>$ on Q, and henceforth we shall denote the ordering relation on R by "$>$."

Propositions 3, 4, and 5, which follow, yield PIV, that $(R, <)$ is an ordered field.

3 Proposition. If $\bar{a}$ is a real number, then exactly one of the following statements is true:

$$\bar{a} < \dot{0}, \quad \bar{a} > \dot{0}, \quad \text{or} \quad \bar{a} = \dot{0}.$$

Proof: If $\bar{a} \neq \dot{0}$, and $\{a_n\}$ is any sequence in $\bar{a}$, then there is a positive rational number δ such that either $a_n > \delta > 0$ for almost all n, or $a_n < -\delta < 0$ for almost all n; but both can not be simultaneously true. [See proof of Proposition 0.4.4(v).] This means that if $\bar{a} \neq \dot{0}$, then either $\bar{a} > \dot{0}$, or $\bar{a} < \dot{0}$, but not both.

If both $\bar{a} < \dot{0}$ and $\bar{a} > \dot{0}$ are false, it follows from Definition 1 that to each rational $\delta > 0$, there must be arbitrarily large values of n for which $-\delta < a_n < \delta$. Since $\{a_n\}$ is a Cauchy sequence, this must be true for almost all n. Thus $\lim \bar{a}_n = \dot{0}$ and $\bar{a} = \dot{0}$ follows.

Finally, if $\bar{a} > \dot{0}$ (or $\bar{a} < \dot{0}$) is known to be false, then the preceding argument yields:

Either $\bar{a} = \dot{0}$ or $\bar{a} < \dot{0}$ ($\bar{a} > \dot{0}$) is true, but not both. ∎

4 Proposition. $\bar{a}, \bar{b} > \dot{0}$ implies $\bar{a} + \bar{b} > \dot{0}$ and $\overline{ab} > \dot{0}$.

Proof: If $\{a_n\} \in \bar{a}$ and $\{b_n\} \in \bar{b}$, then there is a positive rational number δ, such that $a_n, b_n > \delta$ for almost all n. Then $\{a_n + b_n\}$ is a Cauchy sequence in the class of $\bar{a} + \bar{b}$ and, $a_n + b_n > 2\delta$ for almost all n implies that $\bar{a} + \bar{b} > \dot{0}$.

A similar argument holds for the product $\overline{ab}$. ∎

5 Proposition. $\bar{a} > \bar{b}$ if and only if $\bar{a} - \bar{b} > \dot{0}$.

The proof is left to the reader.

It is also left to the reader to prove that "$\leq$" is a total ordering of R (Exercise 3), and that "$<$" possesses also the following properties:

(1) $\bar{a} < \bar{b}$ if and only if $\bar{a} + \bar{c} < \bar{b} + \bar{c}$, for all $\bar{c}$ in R.

(2) $\bar{a} < \bar{b}$ and $\bar{c} > \dot{0}$ imply $\overline{ac} < \overline{bc}$.

(3) $\bar{a} < \bar{b}$ and $\bar{c} < \dot{0}$ imply $\overline{ac} > \overline{bc}$.

(See Exercise 4.)

We have seen that both the field operations on R and the order properties are verified by using the corresponding field operations and order properties of the rational numbers, which are assumed to be known. Also "inherited" is the Archimedean property, for which we first prove Lemma 6.

6 Lemma. If $\bar{a} > \dot{0}$ there is a positive rational number γ such that $\bar{a} > \dot{\gamma} > \dot{0}$.

Proof: Definition 1 implies that there is a $\delta > 0$ such that $a_n > \delta > 0$ for almost all values of n. ($\{a_n\}$ is in the class of $\bar{a}$.) Let $\gamma = \delta/2$. Then,

$$a_n > \delta/2 + \delta/2 = \delta/2 + \gamma, \quad \text{for almost all } n,$$

proving that $\bar{a} > \dot{\gamma}$. ∎

7 Corollary. If $\bar{a} > \bar{b}$ then there is a rational number γ such that $\bar{a} > \dot{\gamma} > \bar{b}$.

Proof: Use Lemma 6 and Property (iii) of an ordered field. ∎

8 Theorem. $(R,<)$ is an Archimedean ordered field.

Proof: Let $\bar{b} > \bar{a} > \dot{0}$ be given. An integer n is required for which $\dot{n}\bar{a} > \bar{b}$. Lemma 6 implies that there is a positive rational number γ, and $\bar{a} > \dot{\gamma} > \dot{0}$. Let q be a rational number such that $\dot{q} > \bar{b}$ (Exercise 6). Since Q possesses the Archimedean property, there is an integer n satisfying $n\gamma > q$. This yields $\dot{n}\bar{a} > \dot{n}\dot{\gamma} > \dot{q} > \bar{b}$. The transitivity of the order relation implies $\dot{n}\bar{a} > \bar{b}$. ∎

We turn next to PIV, the completeness property. This requires defining the absolute value of a real number, convergent sequences, and Cauchy sequences. The procedure is relatively straightforward but rather awkward, for until the completeness property is verified we dwell in a notational nightmare of sequences of equivalence classes of Cauchy sequences of rational numbers! For that is what is meant by a sequence of real numbers. The calculations become somewhat repetitious, and for this reason only those that illustrate new techniques or ideas will be carried out. It is strongly recommended that the reader supply the omitted proofs, at least until the difficulties are transformed into boredom. This exercise will pay off when the completion of abstract spaces is discussed (Chapter 2, Sec. 4) and again in the construction of the Lebesgue integrable functions.

If $\{a_n\}$ is a Cauchy sequence in Q, then the inequality

$$\big|\,|a_n| - |a_m|\,\big| \le |a_n - a_m|$$

implies that $\{\,|a_n|\,\}$ is also a Cauchy sequence.

9 Definition. The *absolute value* of a real number $\bar{a} = [\{a_n\}]$ is the number $|\bar{a}| = [\{\,|a_n|\,\}]$.

10 Proposition. (i) $|\bar{a}|$ is well defined. (This means that if $\{a_n\} \sim \{a'_n\}$ then $\{\,|a_n|\,\} \sim \{\,|a'_n|\,\}$.)

(ii) $$|\bar{a}| = \begin{cases} \bar{a} & \text{whenever } \bar{a} \ge \dot{0}. \\ -\bar{a} & \text{whenever } \bar{a} < \dot{0}. \end{cases}$$

(iii) $|\overline{ab}| = |\bar{a}|\,|\bar{b}|$, in particular $a^2 = |a|^2$.

(iv) $|\bar{a} + \bar{b}| \le |\bar{a}| + |\bar{b}|$, and $|\bar{a} + \bar{b}| = |\bar{a}| + |\bar{b}|$ if and only if either both $\bar{a}$ and $\bar{b}$ are positive, or both are negative.

To avoid using a double subscript, a sequence of real numbers will be denoted by $\{\bar{a}^m\}$; each $\bar{a}^m$ is a real number determined by the Cauchy sequence

$$\{a_n^m\} = \{a_1^m, a_2^m, \ldots, a_n^m \ldots\}$$

of rational numbers. A sequence of rational real numbers will be denoted by $\{\dot{a}^m\}$, where $\dot{a}^m$ contains the constant sequence $\{a^m\} = \{a^m, a^m, \ldots\}$.

11 Definition. Let $\{\bar{a}^m\}$ and $\{\bar{b}^m\}$ be sequences of real numbers and let $\bar{\beta}$ be a real number. The sum and product of the sequences, the product of a sequence and a real number, and the quotient of two sequences are given by

(i) $\{\bar{a}^m + \bar{b}^m\} = \{\bar{c}^m\}$; $\bar{c}^m = [\{a_n^m + b_n^m\}]$.

(ii) $\{\bar{a}^m \bar{b}^m\} = \{\bar{c}^m\}$; $\bar{c}^m = [\{a_n^m b_n^m\}]$.

(iii) $\{\bar{\beta}\bar{a}^m\} = \{\bar{d}^m\}$; $\bar{d}^m = [\{\beta_n a_n^m\}]$ and $\bar{\beta} = [\{\beta_n\}]$.

(iv) If $\bar{b}^m \neq \dot{0}$ for any value of m, then $\{\bar{a}^m/\bar{b}^m\} = \{\bar{c}^m\}$; $\bar{c}^m = [\{a_n^m/b_n^m\}]$. (Again it has been assumed that if some $b_n^m = 0$, they have been discarded. The subsequence obtained by discarding the zero elements is equivalent to the original one.)

It is left to the reader to show that the sequences $\{\bar{c}^m\}$ and $\{\bar{d}^m\}$ of Definition 11 are well defined.

12 Definition. $\lim \bar{a}^m = \bar{a}$ if, for all $\bar{\epsilon} > \dot{0}$, $|\bar{a}^m - \bar{a}| < \bar{\epsilon}$ for almost all m.

To make better use of the properties of rational numbers, it is convenient to recast Definition 12 into the language of sequences of rational numbers: From Lemma 6 it is readily seen that the real number $\bar{\epsilon}$ of Definition 12 may be replaced by a rational number. For, if $\bar{\epsilon} > \dot{0}$ is given, there is a rational number $\epsilon > 0$ satisfying $\bar{\epsilon} > \dot{\epsilon} > \dot{0}$, and if $|\bar{a}^m - \bar{a}| < \dot{\epsilon}$ for almost all m, then $|\bar{a}^m - \bar{a}| < \bar{\epsilon}$ for almost all m. Definition 1 implies further that the inequality on the real numbers, $|\bar{a}^m - \bar{a}| < \dot{\epsilon}$, which holds for almost all m, is equivalent to: To each rational positive number ϵ, there is a $\delta > 0$ such that $|a_n^m - a_n| < \epsilon - \delta$ for almost all m and n. (Remember, $\bar{a}^m = [\{a_n^m\}]$ and $\bar{a} = [\{a_n\}]$.) Thus the convergence of a sequence of real numbers is equivalent to stating that the doubly indexed rational sequence $\{|a_n^m - a_n|\}$ tends to zero as n and m tend to infinity.

If $\{\bar{a}^m\}$ is a Cauchy sequence of real numbers, then $\lim_{j,k\to\infty} |\bar{a}^j - \bar{a}^k| = \dot{0}$. This too may be translated into a statement about rational numbers: Setting $\bar{a}^j = [\{\bar{a}_n^j\}]$, $\bar{a}^k = [\{a_n^k\}]$, we get $\lim_{j,k,n\to\infty} |a_n^j - a_n^k| = 0$ as the equivalent statement about rational numbers.

By similar translations, Propositions 0.4.2, 0.4.4, and 0.4.5 may be proved for real numbers. We offer one example:

If $\lim \bar{a}^m = \bar{a}$ and $\lim \bar{b}^m = \bar{b}$, then $\lim_{n,m\to\infty} |a_n^m - a_n| = 0$ and $\lim_{n,m\to\infty} |b_n^m - b_n| = 0$. This implies that $|(a_n^m + b_n^m) - (a_n + b_n)|$ tends also to zero as n and m tend to infinity. But this is equivalent to

$$\lim_{n,m\to\infty} |(\bar{a}^m + \bar{b}^m) - (\bar{a} + \bar{b})| = \dot{0},$$

thus proving that $\lim (\bar{a}^m + \bar{b}^m) = \lim \bar{a}^m + \lim \bar{b}^m$.

The completeness of R rests on the following Lemma:

13 Lemma. If $\bar{a} \in R$, there is a sequence $\{\dot{a}^m\}$ of rational real numbers which converges to $\bar{a}$.

Proof: Let $\{\dot{a}_m\}$ be a Cauchy sequence in the class of $\bar{a}$, and let $\{\dot{a}^m\}$ be the corresponding sequence of "dotted" rationals (in $\bar{Q}$). Then, since $\{a_m\}$ is a Cauchy sequence of rational numbers, $\lim_{n,m\to\infty} |a_m - a_n| = 0$. But this is equivalent to $\lim_{m\to\infty} |\dot{a}_m - \bar{a}| = \dot{0}$, or $\lim_{m\to\infty} \dot{a}_m = \bar{a}$. ∎

14 Theorem. (Cauchy) Every Cauchy sequence of real numbers converges to a real number.

Proof: Let $\{\bar{a}^m\}$ be a Cauchy sequence in R. To each m there is a dotted rational $\dot{b}^m$ and

$$|\bar{a}^m - \dot{b}^m| < (1/\dot{m})$$

(Lemma 13).

The triangle inequality yields

$$|\dot{b}^n - \dot{b}^m| \leq |\dot{b}^n - \bar{a}^n| + |\bar{a}^n - \bar{a}^m| + |\bar{a}^m - \dot{b}^m|$$
$$< (1/\dot{n}) + |\bar{a}^n - \bar{a}^m| + (1/\dot{m}).$$

Thus $\{\dot{b}^m\}$ is a Cauchy sequence in $\bar{Q}$. But since each $\dot{b}^m$ contains the constant sequence $\{b^m, b^m, \ldots\}$, it follows that $\lim_{n,m\to\infty} |b^n - b^m| = 0$, proving that $\{b^m\}$ is a Cauchy sequence of rational numbers. (Yes, this is not much more than a notational exercise.)

Now let $\bar{b} = [\{b^m\}]$. Clearly, $\bar{b} = \lim \dot{b}^m$, from which it follows that

$$|\bar{b} - \bar{a}^m| \leq |\bar{b} - \dot{b}^m| + |\dot{b}^m - \bar{a}^m| \to \dot{0} \text{ as } m \to \infty,$$

thus proving that the given Cauchy sequence $\{\bar{a}^m\}$ converges to the real number $\bar{b}$. ∎

Only PV remains to be verified. Suppose that the complete ordered field R contains a proper subfield $\hat{R}$ which in turn contains Q (or $\bar{Q}$).

Since $R - \hat{R} \neq \phi$, there is real number $\bar{b}$ which is not in $\hat{R}$. Let $\{b^m\}$ be a sequence of dotted rationals which converges to $\bar{b}$ (Lemma 13). Since $\hat{R} \supset \bar{Q}$, this shows that $\hat{R}$ contains a Cauchy sequence which does not converge in $\hat{R}$, thus proving that any proper subfield of R is incomplete.

To summarize, R is a *complete* (*Archimedean*) *ordered field* that *contains* the (*Archimedean*) *ordered field* Q of *rational numbers* (more precisely, a field $\bar{Q}$ which is isomorphic to Q). Moreover, each real number is the limit of a sequence of rationals.

In many texts, R is *defined* to be a complete, Archimedean ordered field. To justify this as a definition, two things must be proved:

(i) There exists at least one complete Archimedean ordered field.

(ii) There is no more than one complete Archimedean ordered field.

By constructing R, we have verified (i). The uniqueness of R (ii) is proved in Birkhoff and MacLane [1], McShane and Botts [1], and Isaacs [1]. Hereafter, it will be unnecessary to return to the sequence definition of a real number. PI–PV may be thought of as the defining properties of R; that is, R is the smallest, complete (Archimedean) ordered field that contains Q.

Exercises

***1.** Let $\{a_n\}$ be a Cauchy sequence of rational numbers. Prove that if to every $\delta > 0$, there are arbitrarily large values of n for which $-\delta < a_n < \delta$, then $\lim a_n = 0$. (See the proof of Proposition 3.)

***2.** Prove Proposition 5.

***3.** Prove that "$\leq$" totally orders R.

***4.** Verify Properties (1), (2), and (3) that follow the statement of Proposition 5.

***5.** State and prove the theorems about real numbers which correspond to Propositions 0.4.2, 0.4.4, and 0.4.5.

***6.** Show that if $\bar{b} \in R$, then there is a rational number q such that $\dot{q} > \bar{b}$. (See proof of Theorem 8.)

7. Using the results of this section as well as Sec. 4 of Chapter 0, including Exercise 0.4.12, show that the decimals may be used to define the real numbers.

*Exercises which request the proof of a statement made in the text, as well as those referred to in the text, are starred.

3 Theorems about Real Numbers

At the beginning of Sec. 4 of Chapter 0, a geometric description of the incompleteness of Q was given. Each rational number was assigned a point on a calibrated line, and the claim was made, (although not proved at that time) that there were "gaps" on the line to which no rational number was attached. These gaps, of course, are the "irrational" points. A short description, using decimals, of the correspondence between R

and the points on a line follows. Naturally, no subsequent proofs will depend upon "pictures." However, the geometry of the real-number system is of great heuristic value, and in all cases it will be relatively easy to translate a geometric argument to a rigorous analytic one.

We begin with a line on which a point called the *origin* has been chosen. The number 0 is attached to this point. Taking an arbitrary, but hereafter fixed line segment as a unit, the integer ν is attached to the point on the line whose *directed* distance from the origin is ν. By repeatedly subdividing into ten equal parts the segments that lie between the "points" ν and $\nu + 1$, we locate those points whose directed distance from the origin can be written as

$$[1] \qquad \pm\left\{a_0 + \frac{a_1}{10} + \frac{a_2}{10^2} + \cdots \quad \cdots + \frac{a_\nu}{10^\nu}\right\};$$

the a_i are nonnegative integers, and for $i \geq 1$, $0 \leq a_i \leq 9$. These are the points that correspond to the terminating decimals $\pm a_0 \cdot a_1 a_2 \ldots$ $\ldots a_\nu 0000 \ldots$; the sequence of truncations of such a decimal converges to the directed distance of this point from the origin. To each point P whose distance from the origin can not be represented by a finite sum [1], a nonterminating decimal is assigned as follows: a_0 is chosen so that the distance of P from 0 lies between a_0 and $a_0 + 1$; a_1 is chosen so that this distance is between $a_0 + \dfrac{a_1}{10}$ and $a_0 + \dfrac{a_1 + 1}{10} \ldots$; in general, a_n is chosen so that the distance from P to 0 is between

$$\sum_0^n a_i\, 10^{-i} \qquad \text{and} \qquad \sum_0^{n-1} a_i\, 10^{-i} + (a_n + 1)\, 10^{-n}.$$

The details are left to the reader. In this way every point that does not correspond to a finite decimal determines uniquely a nonterminating decimal. Note that repeating 9's cannot appear. Why?

Conversely, it is intuitively clear that each decimal determines uniquely a point P on the line. This point corresponds to the real number which is the limit of the sequence of truncations of the given decimal. Geometrically, P can be described as follows: The points corresponding to truncations of the decimal lie to the left of (or coincide with) P. If Q is a point lying to the left of P, then for sufficiently large n, the nth truncation point will lie to the right of Q. The existence of the point P is taken as a geometric axiom which corresponds to the completeness of R.

To avoid unnecessary formalism, the words "point" and "real number" will be interchangeable. In all cases, a geometric argument or geometric language will be used only if it is obviously and easily translatable into an analytic statement.

This section deals with several properties of real numbers which are equivalent to the completeness property. All of them are most easily described as geometric properties.

We begin with a brief discussion of *open* and *closed* sets of real numbers.

1 Definition. If a and b are real numbers and if $a < b$, then the set

$$(a,b) = \{x \in R : a < x < b\}$$

is called the *open bounded interval* with endpoints a and b.

The sets $(a,\infty) = \{x \in R : a < x\}$ and $(-\infty,a) = \{x \in R : x < a\}$ are the *open unbounded intervals*.

The *closed bounded interval* whose endpoints are a and b is denoted by $[a,b] = \{x \in R : a \leq x \leq b\}$. Similarly for the *closed unbounded intervals* $[a,\infty)$ and $(-\infty,a]$, and the *half open* or *half closed intervals* $[a,b)$ and $(a,b]$.

Remark. If $(R,<)$ is replaced by any ordered field $(X,\prec)$, then Definition 1 as well as much of the subsequent discussion remains meaningful (see Exercise 27).

2 Definition. A subset G of the real numbers is called *open* if every point ζ in G is contained in a "symmetric" interval $(\zeta - \delta, \zeta + \delta)$, $\delta > 0$, that lies entirely in G. This interval is called a *δ-neighborhood* of the *point* ζ.

3 Examples. (a) The open intervals, both bounded and unbounded, are open sets (Exercise 1).

(b) The empty set $\emptyset$ is open (Exercise 2).

(c) R is open (Exercise 2).

(d) The intervals $[a,\infty)$ and $[a,b)$ are not open, because in both cases every δ-neighborhood of a contains points that lie outside the interval.

Although there are nonempty open sets which are not intervals [see Proposition 4 (i)], it can be shown that every open subset of the real numbers may be written as the disjoint countable union of open intervals (see either Exercise 9 or Proposition 7.3.1).

4 Proposition.

(i) If $\{G_\alpha\}$ is a family of open sets (not necessarily countable), then $\bigcup G_\alpha$ is an open set.

(ii) If $\{G_1,G_2,\ldots,G_n\}$ is a finite collection of open sets, then their intersection

$$\bigcap_{\nu=1}^{n} G_\nu$$

is an open set.

The proof is left to the reader.

5 Example. For $n \geq 1$, the intervals $I_n = (-1/n, 1/n)$ are open. Their intersection however, is the set $\{0\}$ which is not open, thus proving that Proposition 4 (ii) is not true for infinite collections of open sets.

6 Definition. A subset F of R is said to be *closed* if its complement in R, $R - F$, is an open set.

7 Examples. (a) The complements of the open sets ϕ and R are the closed sets R and ϕ respectively. (Are there other subsets of R which are both open and closed? You should be able to prove that these are the only such sets before finishing this section. Otherwise see Chapter 3, Sec. 3).

(b) The set $G = (-\infty, a) \cup (b, \infty)$ is open (Proposition 4). Therefore its complement $R - G = [a,b]$ is closed. Similarly the interval $[a, \infty)$ is closed. (It is assumed that $a < b$; otherwise the complement is the empty set, which is of course also closed.)

(c) The finite set $S = \{a_1, a_2, \ldots, a_n\}$ is closed. (Show first that a set consisting of a single point is closed.)

(d) The set $[a,b)$ is neither open nor closed.

(e) $S = \{1/2, 1/3, \ldots, 1/n \ldots\}$ is not closed because its complement contains the point 0, all of whose neighborhoods $(-\delta, \delta)$ intersect S. However, the set $S' = S \cup \{0\}$ is closed (Exercise 6).

8 Definition. ζ is called a *cluster point* of a *set* S, or a *point* of *accumulation* of S, if every δ-neighborhood of ζ contains a point of S other than ζ itself.

9 Proposition. ζ is a cluster point of S if and only if every δ-neighborhood of ζ contains infinitely many points of S.

The proof is left to the reader.

10. Examples. (a) 0 is a cluster point of both $S = \{1, 1/2, \ldots\}$ and $S' = S \cup \{0\}$, proving that a cluster point of a set may or may not be in the set.

(b) Every point of an open set is a cluster point of that set. (In verifying this, be sure to include the open set ϕ.)

11 Proposition. A set S is closed if and only if it contains all its cluster points.

Proof. Let ζ be a cluster point of a closed set F. If $\zeta \in R - F$, an open set, there is a δ-neighborhood of ζ which is contained entirely in $R - F$, contradicting the assumption that ζ is a cluster point of F.

Conversely, if F contains all its cluster points, then no point of $R - F$ can be a cluster point of F. This means that if $\theta \in R - F$, then there is a δ-neighborhood of θ which contains no points of F; that is, it lies entirely in $R - F$, proving that $R - F$ is open. ▌

12 Proposition. (i) If $\{F_\alpha\}$ is a family of closed sets, then their intersection is closed. (ii) If $\{F_1, F_2, \ldots, F_n\}$ is a finite collection of closed sets, then their union is also closed.

Proof. Use Proposition 4 and the De Morgan laws (Proposition 0.1.6). ▌

13 Definition. A set S of real numbers is said to be *bounded*, if there is a number $M > 0$ such that $|a| \leq M$ for all a in S; or equivalently, if S is contained in a bounded interval $[-M,M]$.

The set S is *bounded from above* if there is a number $M > 0$ and $a \leq M$. If there is a number m such that $m \leq a$ for all a in S, then S is *bounded from below*. The numbers M and m are called *upper* and *lower bounds* of S.

14 Definition. A number λ is a *least upper bound* of a nonempty set S, or the *supremum* of S, written "sup S," if

(i) λ is an upper bound for S, and

(ii) λ' is an upper bound for S implies $\lambda' \geq \lambda$.

15 Example. The number b is a least upper bound of both (a,b) and $[a,b]$.

16 Theorem (Principle of the Least Upper Bound; abbreviated "LUB"). A nonempty set of numbers that is bounded from above has a unique least upper bound.

Proof. If λ and λ' are least upper bounds of the same set S, then Definition 14(ii) implies that they are equal.

Let S be a nonempty set which is bounded from above by the number M. We shall construct a Cauchy sequence $\{\lambda_n\}$ which converges to the least upper bound of S. Set $M = \lambda_1$. Choose any point β in S. If the midpoint $\dfrac{\beta + M}{2} = \dfrac{\beta + \lambda_1}{2}$ is an upper bound for S, call it λ_2. If not, let $\lambda_2 = \lambda_1$. If $\lambda_2 = \dfrac{\beta + \lambda_1}{2}$ (see Figure 1), bisect $[\beta,\lambda_2]$, setting $\lambda_3 =$

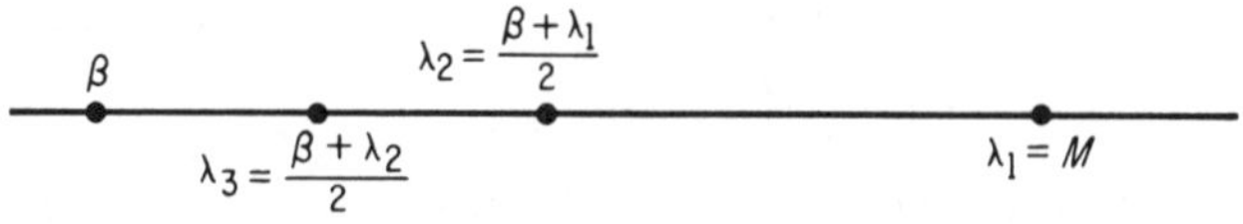

FIGURE 1

$\dfrac{\beta + \lambda_2}{2}$ if this number is an upper bound for S. Otherwise set $\lambda_3 = \lambda_2$. If however, we had set $\lambda_2 = \lambda_1$, then $[(\beta + M)/2, \lambda_2]$ must be bisected, and λ_3 is chosen as described above (see Figure 2). Continuing in this manner, a nonincreasing sequence $\{\lambda_n\}$, $(\lambda_n \geq \lambda_{n+1})$ of upper bounds of S is obtained. If ν is any integer, then $n,m > \nu$ implies $|\lambda_m - \lambda_n| < \dfrac{(M - \beta)}{2^\nu}$, proving that $\{\lambda_n\}$ is a Cauchy sequence in R. Since R is complete, the limit λ exists.

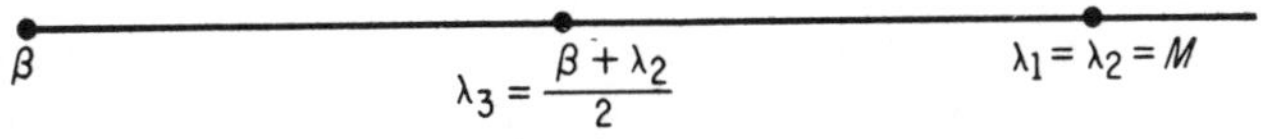

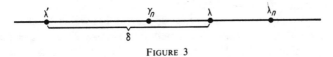

FIGURE 2

To complete the proof, it must be shown that $\lambda = \sup S$. If $\alpha \in S$, then for all values of n, $\alpha \leq \lambda_n$, from which it follows that λ is an upper bound for S (Exercise 1.2.5). If, however, λ' is also an upper bound and if $\lambda - \lambda' = \delta > 0$, then for sufficiently large values of n, there will be points γ_n in S such that

$$\lambda - \gamma_n < \frac{M - \beta}{2^{n-2}} < \frac{\delta}{2}.$$

This implies that $\gamma_n > \lambda'$, from which it follows that λ' can not be an upper bound for S. (See Figure 3.) ▮

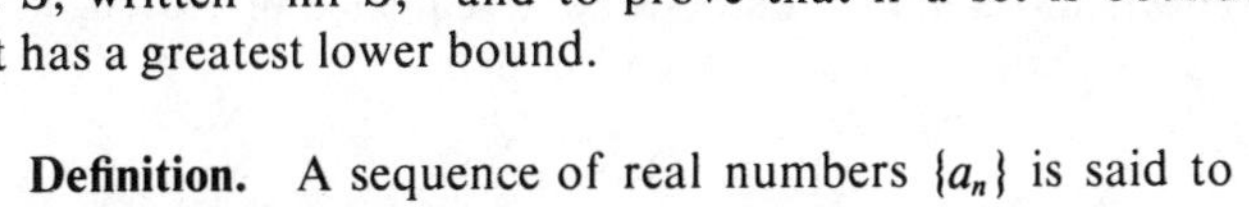

FIGURE 3

It is left to the reader to define the *greatest lower bound*, or *infimum*, of a set S, written "inf S," and to prove that if a set is bounded from below it has a greatest lower bound.

17 Definition. A sequence of real numbers $\{a_n\}$ is said to be *nondecreasing*, denoted by $a_n \uparrow$, if $a_n \leq a_{n+1}$ for all n. The sequence is *strictly increasing*, or *increasing* if $a_n < a_{n+1}$.

A similar definition may be given for nonincreasing sequences, $a_n \downarrow$.

A *monotone* sequence is either nondecreasing or nonincreasing.

18 Theorem (Principle of Monotone Convergence; abbreviated "MONO"). A monotone bounded sequence converges. (This includes both nondecreasing sequences which are bounded from above, and nonincreasing sequences which are bounded from below.)

Proof. If $a_n \uparrow$ and $a_n \leq M$, then Theorem 16 implies that the set of sequence elements has a least upper bound λ. We shall show that $\lim a_n = \lambda$:

If $\epsilon > 0$, then $\lambda - \epsilon$ is not an upper bound for the a_n, and therefore some member a_ν of the sequence must lie in the interval $(\lambda - \epsilon, \lambda]$. However, the monotonicity of the sequence and the inequality $a_n \leq \lambda$, for all n, imply that $a_n \in (\lambda - \epsilon, \lambda]$ whenever $n \geq \nu$. Thus almost all the a_n lie in an arbitrary neighborhood of λ, proving that $\lim a_n = \lambda$. ∎

It is not at all difficult to give examples demonstrating the falseness of both the Principles of the Least Upper Bound and Monotone Convergence in Q: Let $\{x_n\}$ be the sequence of truncations of a nonrepeating decimal. Surely $\{x_n\}$ is nondecreasing and bounded from above by a rational number. But its limit is not in Q (Theorem 0.4.8).

The falsity of both MONO and LUB in Q is a result of its incompleteness. It can be shown, in the sense described below, that these three properties are equivalent, and that therefore any one of them could be used to define the real numbers.

The definitions of open and closed sets, cluster points, and upper and lower bounds are meaningful in any ordered field $(X, <)$. It is not difficult to reverse the order of proving these theorems, and to show that $(X, <)$ is a complete Archimedean ordered field if and only if

(i) $(X, <)$ is an ordered field in which every subset which is bounded from above has a least upper bound (LUB), or

(ii) $(X, <)$ is an ordered field in which all bounded monotone sequences converge (MONO).

Additional equivalent properties appear in Theorems 19, 22, and 27, and their corollaries. These are listed in Exercise 27 and are accompanied by a diagram which suggests the order of their appearance in an equivalence proof.

19 Theorem (Cantor). Let $\{S_n\}$ be a sequence of closed, bounded, nonempty subsets of R satisfying also, for all n,

$$S_n \supset S_{n+1}.$$

Then the intersection $\bigcap_1^\infty S_n \neq \phi$.

Proof. Since the S_n are bounded and nonempty, there are numbers

$$a_n = \inf S_n \leq \sup S_n = b_n.$$

Clearly, the closed intervals $[a_n, b_n]$ contain the sets S_n. The closed sets S_n contain all their cluster points (Proposition 11), from which it follows that $a_n, b_n \in S_n$. Moreover, the inclusion $S_n \supset S_{n+1}$ implies also that

$$a_1 \leq a_2 \leq \cdots \leq a_n \leq a_{n+1} \leq \cdots \leq b_{n+1} \leq b_n \leq \cdots \leq b_2 \leq b_1 .$$

Thus both $\{a_n\}$ and $\{b_n\}$ are bounded monotone sequences. Theorem 18 implies that they converge. We shall show that their respective limits α and β, are in the intersection of the S_n.

Case 1. If there are only finitely many distinct a_n, then the monotonicity property implies that there is an integer ν such that $a_n = a_\nu$ whenever $n \geq \nu$. This says that $\lim a_n = a_\nu = \alpha$. The inclusion $S_n \supset S_{n+1}$ and $a_n \in S_n$ yield

$$\text{If } n \geq \nu \text{, then } \alpha = a_n \in S_n \subset S_{n-1} \subset \cdots \subset S_\nu \subset \cdots \subset S_1,$$

proving that α is in all the S_n. A similar argument holds for $\beta = \lim b_n$.

Case 2. If there are infinitely many distinct a_n, then $\alpha = \lim a_n$ is a cluster point of the a_n. Let ν be a fixed but arbitrary integer. Then if $n \geq \nu$, $S_\nu \supset S_{\nu+1} \supset \cdots \supset S_n$, and therefore $a_n \in S_\nu$. Thus the point α is also a point of accumulation of the closed set S_ν, and therefore $\alpha \in S_\nu$. Since this argument holds for all values of ν, it follows that α is in the intersection. A similar argument holds for β. ∎

20 Corollary. Let S_n, a_n, b_n be as described in the preceding theorem. Then the intersection consists of a single point if and only if $\lim (a_n - b_n) = 0$.

Proof. If $\gamma < \alpha$ or $\gamma > \beta$, then γ cannot be in all the S_n. (Why?) Thus the intersection consists of a single point if and only if $0 = \beta - \alpha = \lim (a_n - b_n)$. ∎

The following examples show that none of the assumptions made in Theorem 19 and its corollary can be completely discarded.

21 Examples. (a) The "nested" sets $S_n = (0,1/n]$ are bounded, nonempty, but not closed. Their intersection is the null set.

(b) The nested sets $S_n = [n, \infty)$ are nonempty, closed, but not bounded. Again, the intersection is empty.

(c) If some $S_\nu = \emptyset$, then the intersection is $\emptyset$.

22 Theorem (Bolzano-Weierstrass Property; abbreviated "BW"). A bounded infinite set has at least one cluster point.

Proof: Let $M > 0$ be chosen so that the bounded infinite set $S \subset [-M,M]$. Since S contains infinitely many points, either $S \cap [-M,0]$ or $S \cap [0,M]$ contains infinitely many points. (It is possible that both do.) Set $S_1 = [-M,M]$, and choose S_2 to be one of the half-intervals which contains infinitely many points of S. Bisecting S_2, either the left or right closed half-interval must contain infinitely point of S. Call such an interval S_3. Continuing in this way (see the *Remark* following this proof), a nested sequence, $S_n \supset S_{n+1}$, of closed, bounded, nonempty intervals is obtained whose lengths, $M/2^{n-2}$, converge to zero. Let ζ be the unique point which lies in each of the S_n (Corollary 20). In order that ζ be a cluster point of S, it must contain infinitely many points of S in each of its δ-neighborhoods. By choosing an integer n for which the inequality $M/2^{n-2} < \delta$ is true, the corresponding set $S_n \subset (\zeta - \delta, \zeta + \delta)$. Since each S_n contains infinitely many points of S, the conclusion follows. ∎

Remark. Whenever both halves of a bisected interval contained infinitely many points of S, we "chose" one of them to be an S_n. This was an example of the "hidden" use of the Axiom of Choice (see 0.2.11). In this particular case however, it could be avoided by giving more explicit instructions for choosing the S_n. For example, we could say that whenever both parts of a bisected interval contained infinitely many points of S, then the left-hand interval was to be chosen.

23 Corollary. Every bounded sequence of real numbers contains a convergent subsequence.

Proof: If there are integers $n_1 < n_2 < \cdots < n_k < n_{k+1} \cdots$ for which the corresponding a_{n_k} are all equal to a single number α, then $\lim a_{n_k} = \alpha$.

If no constant subsequence exists, then the set of distinct sequence elements, called the *trace* of $\{a_n\}$ and denoted by Tr $\{a_n\}$, is infinite. The Bolzano-Weierstrass Property (BW) implies that Tr $\{a_n\}$ has at least one cluster point. If ζ is a cluster point of this set, then to each integer k, there is a sequence element $\{a_{n_k}\}$ satisfying

(i) $a_{n_k} \in (\zeta - 1/k, \zeta + 1/k)$, and

(ii) $n_1 < n_2 < \cdots < n_k < n_{k+1} < \cdots$.

The existence of an element a_{n_k}, for each positive integer k, is assured by the fact that the intersection Tr $\{a_n\} \cap (\zeta - 1/k, \zeta + 1/k)$ is infinite. (The Axiom of Choice gives the subsequence.) Clearly $\lim_{k\to\infty} a_{n_k} = \zeta$. ∎

24 Definition. A *covering* of a set S is a collection $\{S_\alpha\}$ of sets whose union contains S. If the index set $\mathfrak{A}$ ($\alpha \in \mathfrak{A}$) is finite, then the covering is said to be *finite*.

Let $\mathfrak{B}$ be a subset of the index set $\mathfrak{A}$. Then the subcollection $\{S_\beta\}$ indexed by $\mathfrak{B}$ is said to be a *subcovering* if its union also contains S.

A covering of a set of real numbers is called *open* if the S_α are open sets.

25 Definition. A set of real numbers is *compact* if every open covering contains a finite subcovering.

26 Examples of Noncompact Sets. (a) The set of all real numbers is not compact. Indeed, the family of open intervals $S_n = (-n,n)$, $n = 1,2,\ldots$ is an open covering of R which has no finite subcovering.

(b) The open sets $S_n = (1/n, 3/2)$, $n = 1,2,\ldots$ cover the set $(0,1]$. Again there is no finite subcovering.

We should like to give also examples of compact sets. But at this point the verification of compactness requires testing *all* open coverings if we wish to use a direct argument. The Heine-Borel theorem, which follows, provides a set of simpler criteria for compactness. By assuming that a set is *not* compact, we arrive at the conclusion that either the set is unbounded or it is not closed [Examples 26(a) and (b)]:

27 Theorem (Heine-Borel; abbreviated "HB"). A subset of the real numbers is compact if and only if it is both closed and bounded.

Proof: If a set is unbounded, then the open covering of Example 26(a) can have no finite subcovering, thus proving that no unbounded set is compact.

If a set is not closed, then one of its accumulation points ζ, is not in the set. Letting ζ correspond to the point 0 in Example 26(b), it is not difficult to describe an open covering which has no finite subcovering (Exercise 21).

Suppose now that S is a closed, bounded set which is contained in the interval $[-M,M]$. If there is an open covering $\{S_\alpha\}$ which possesses no finite subcovering, then $\{S_\alpha\}$ is also an open covering of both $S \cap [-M,0]$ and $S \cap [0,M]$, and there cannot be a finite subcovering of *both* these sets without there being a finite subcovering of the original set S. Assume that there is no finite subcovering for $S \cap [0,M]$. (Compare this with the proof of Theorem 22.) Set $S = T_1$ and $S \cap [0,M] = T_2$. Bisecting $[0,M]$, T_3 is taken to be either $S \cap [0,M/2]$ or $S \cap [M/2,M]$, requiring that the set T_3 be one for which no finite subcover of $\{S_\alpha\}$ exists. Continuing in this manner, a sequence of bounded, closed, non-empty sets $T_1 \supset T_2 \supset \cdots$ is obtained, each of which is contained in a closed interval whose length is $M/2^{n-2}$. Theorem 19 assures the existence of a point ζ in the intersection of the T_n. The inclusion $T_n \subset S$, for all n, implies that $\zeta \in S$, and must therefore be in (at least) one of the open sets S_α of the covering of S. From the openness of S_α it follows that some δ-neighborhood of ζ is contained in S_α. By choosing n so that $M/2^{n-2} < \delta$, we have

$$T_n \subset (z - \delta, \zeta + \delta) \subset S_\alpha.$$

However, in selecting the T_n it was assumed that each T_n was a subset of S for which no finite subcover exists. Here we have shown that a T_n is contained in a single S_α, proving that a finite subcovering must exist if S is both closed and bounded. ∎

We conclude this section with a short discussion of the *limit superior* ($\overline{\lim}$) and the *limit inferior* ($\underline{\lim}$) of a sequence.

28 Definition. If $\{a_n\}$ is a sequence which is unbounded from above, then the *limit superior* of $\{a_n\}$, $\overline{\lim}\ a_n = +\infty$.

If $\{a_n\}$ is bounded from above, then

$$\lambda_n = \sup\,[a_n, a_{n+1}, \ldots]$$

is a nonincreasing sequence which either tends to $-\infty$, or if bounded from below to a number $\bar{a}$. The *limit superior* of the sequence is

$$\overline{\lim}\ a_n = \lim \lambda_n = \bar{a} \text{ or } -\infty.$$

Similarly, if the sequence is not bounded from below, then the *limit inferior*, $\underline{\lim}\ a_n = -\infty$.

If $\{a_n\}$ is bounded from below, then $\gamma_n = \inf\,[a_n, a_{n+1}, \ldots]$ is nondecreasing and

$$\underline{\lim}\ a_n = \lim \gamma_n = \underline{a} \text{ or } +\infty.$$

29 Proposition. (i) $\underline{\lim}\ a_n \leq \overline{\lim}\ a_n$. (ii) The sequence $\{a_n\}$ converges to a number a if and only if $a = \lim a_n = \underline{\lim}\ a_n = \overline{\lim}\ a_n$.

Proof of (i). For all n, $\gamma_n \leq \lambda_n$. The generalization of Proposition 0.4.2(vi) to the real numbers (Exercise 1.2.5) yields the desired inequality.

Proof of (ii). For all n, $\gamma_n \leq a_n \leq \lambda_n$. If $\lim \lambda_n = \lim \gamma_n = a$, then

$$|a - a_n| \leq |a - \gamma_n| + |\gamma_n - \lambda_n| + |\lambda_n - a| \to 0$$

as $n \to \infty$.

Conversely, if $\lim a_n = a$, and $\epsilon > 0$, then $a - \epsilon < a_n < a + \epsilon$ for almost all n, that is, for n greater than some integer ν. This implies that if $n > \nu$ then

$$\gamma_n = \inf\,[a_n, a_{n+1}, \ldots] \geq a - \epsilon, \text{ and } \lambda_n = \sup\,[a_n, a_{n+1}, \ldots] \leq a + \epsilon,$$

from which it follows that $\lim \lambda_n = \lim \gamma_n = a$. ∎

30 Examples. (a) If $a_n = \cos n\pi$, then $\underline{\lim}\ a_n = -1$ and $\overline{\lim}\ a_n = 1$.

(b) If $a_n = (-1)^n n/(n+1)$, then $\underline{\lim}\ a_n = -1$ and $\overline{\lim}\ a_n = 1$.

(c) If $a_n = n \cdot \cos \dfrac{n\pi}{2}$, then $\underline{\lim}\ a_n = -\infty$, and $\overline{\lim}\ a_n = +\infty$.

Alternate definitions of $\underline{\lim}$ and $\overline{\lim}$ are provided by:

31 Proposition. $\overline{\lim}\ a_n = \bar{a} < \infty$ if and only if, for all $\epsilon > 0$,

(i) almost all $a_n < \bar{a} + \epsilon$, and

(ii) if ν is any integer, then there is an integer $n > \nu$ such that $a_n > \bar{a} - \epsilon$.

(The corresponding statement for $\underline{\lim}$, and its proof, is left to the reader.)

Proof: Suppose that $\overline{\lim}\, a_n = \bar{a}$ and that ϵ is a positive number. Then

$$\sup\,[a_k, a_{k+1}, \ldots] = \lambda_k \downarrow \bar{a},$$

from which it follows that $0 \leq \lambda_k - \bar{a} < \epsilon$ for almost all k. Since $n > k$ implies $a_n \leq \lambda_k$, we have $a_n < \bar{a} + \epsilon$ for almost all n.

To prove (ii), let an integer ν be given. From the definition of λ_ν it follows that there must be integers $n > \nu$ for which $a_n > \lambda_\nu - \epsilon \geq \bar{a} - \epsilon$.

Conversely, if $\bar{a}$ is a number for which (i) and (ii) are satisfied, then $\{a_n\}$ is a bounded sequence and λ_n is finite for all values of n, and tends monotonically (down) to a finite value $\bar{a}'$, which is the limit superior of the sequence $\{a_n\}$. If $\bar{a}' - \bar{a} = \delta > 0$, then by taking $\epsilon = \delta/2$, a contradiction is arrived at in the following way (see also Figure 4): From the first part of the proof, $\bar{a}'$ satisfies (ii), and therefore infinitely many of the a_n are in the interval $(\bar{a}' - \epsilon, \bar{a}')$. However, $\bar{a}$ also satisfies both (i) and (ii), so that almost all the a_n are in the interval $(\bar{a}, \bar{a} + \epsilon) = (\bar{a}, \bar{a}' - \epsilon)$, and consequently only a finite number may be in $(\bar{a}' - \epsilon, \bar{a}')$, which contradicts the preceeding sentence. A similar contradiction is arrived at if it is assumed that $\bar{a} - \bar{a}' > 0$. ∎

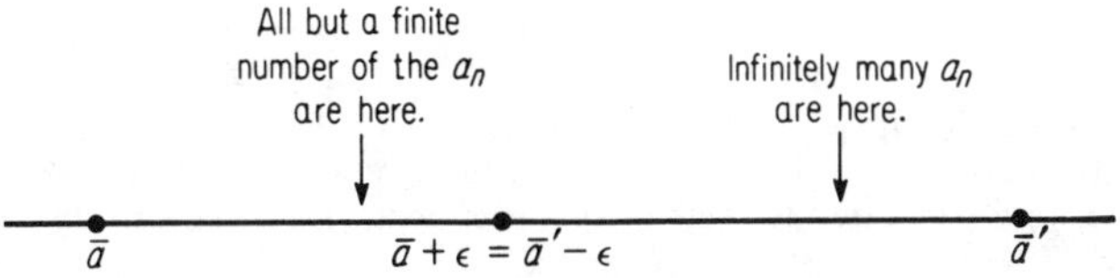

FIGURE 4

We shall postpone the statements and proofs of many of the standard theorems about continuous functions until Sec. 2 of Chapter 3 where they appear in a more general setting. Meanwhile, it is assumed that the reader is familiar with the basic properties of continuous functions, whose domain and range are the real numbers, and they will be used freely in the examples and in Sec. 1 of Chapter 2. In the latter case, the continuity of exponential functions is needed, and the necessary theorems are outlined in the Exercises.

Exercises

***1.** Show that the open intervals are open sets (Example 3).
***2.** Show that both R and $\emptyset$ are open.
***3.** Show that a set which consists of a single point is closed. Show that a finite set is closed.
***4.** Prove that no finite set is open. (The empty set is excluded.)
***5.** Prove Proposition 4.

*Exercises which request the proof of a statement made in the text, as well as those referred to in the text, are starred.

***6.** Show that the set $\{0, 1/2, 1/3, \ldots, 1/n, \ldots\}$ is closed.

***7.** Prove that if a set of real numbers is both both open and closed then it is either empty or the entire set of real numbers. [*Hint:* Arrive at a contradiction by showing that a proper subset of R will either contain points each of whose neighborhoods has points outside the given set (thus the set is not open) or there is at least one cluster point of the set which is not contained in the set (thus the set is not closed).]

8. Describe the set of accumulation points of the set

$$S = \{x \in R: \exists\, n,m \in Z, \text{and } x = n/2^m\}.$$

Does S contain any nonempty open subsets? Does its set of accumulation points contain any open subsets?

9. Show that every nonempty open set can be written as the countable, disjoint, union of open intervals. [*Hint:* Show first that if x is in the open set S, then there is a largest open interval in the set which contains x: $I_x = (a_x, b_x)$, where $a_x = \inf [a]$ for which $x \in (a,b) \subset S$. Similarly for b_x.]

***10.** Define $\inf S$, and prove that if S is bounded from below, then $\inf S < \infty$. [*Hint:* Define the *reflection* of S to be the set $\{x : -x \in S\}$. Then use Theorem 16.]

***11.** Prove that if $\lambda = \sup S$ or $\inf S$, then either $\lambda \in S$ or λ is a cluster point of S.

12. What is $\sup S$ and $\inf S$ in each of the following cases?

$$S = \{x : x^2 < 4\}, \quad \{x : x^2 \leq 4\}, \quad \{x : 2 < x^2 \leq 4\},$$

$$\left\{x : x = \frac{\cos n\pi}{n}, n = \pm 1, \pm 2, \ldots\right\}.$$

13. If $\sup S$ and $\sup T$ are finite, what is $\sup (S \cup T)$? Generalize this to $\sup (\cup S_\alpha)$ and $\inf (\cup S_\alpha)$. What can you say about intersections?

14. Let S and T be bounded sets of real numbers. Describe sup and inf of the set $S \cdot T = \{x : x = st, s \in S \text{ and } t \in T\}$.

***15.** Prove that if β is a real number, there are sequences $\{a_n\}$ and $\{b_n\}$ of rational numbers such that

$$a_n \uparrow \beta \qquad \text{and} \qquad b_n \downarrow \beta.$$

(*Hint:* Use the method for assigning to each point on a line a decimal. This produces *both* desired sequences.)

16. Let $f(x) = 0$ whenever x is an irrational number. If x is rational and written in lowest terms as n/m, $m > 0$, then set $f(x) = 1/m$. Show that if y is irrational, then to every $\epsilon > 0$ there is a δ-neighborhood of y in which $0 < f(x) < \epsilon$. (This says that $f(x)$ is continuous at irrational points. What happens at rational points?)

17. Show that the theorem on nested sets (Theorem 19), remains true if the condition that all the S_n are bounded is replaced by the condition that a single S_n is bounded.

***18.** Show that $\lim a_n = a$ implies that Tr $\{a_n\}$ has at *most* one point of accumulation. Must there be at *least* one point of accumulation if the sequence converges? Is convergence implied by "Tr $\{a_n\}$ has one and only one point of accumulation"?

19. Using the binomial theorem (Exercise 0.3.15) and MONO, show that the sequence $a_n = (1 + 1/n)^n$ converges. This limit is of course the number e.

*Exercises which request the proof of a statement made in the text, as well as those referred to in the text, are starred.

20. Find $\lim (1 + 2/n)^n$. [*Hint:* Factor $(1 + 2/n) = (1 + 1/n)(1 + 1/n+1)$.]

***21.** Complete the proof that every compact set is closed.

***22.** Show that the rational numbers are countable. (See Kamke [1].)

23. Show, using decimals that the real numbers are not countable. (See Kamke [1].)

24. Show that an uncountable subset of R must have at least one cluster point. Does every countably infinite subset of R have cluster points?

***25.** State and prove Proposition 31 for $\underline{\lim}\, a_n < \infty$, and then for the infinite case.

***26.** Prove that if $\{a_n\}$ and $\{b_n\}$ are bounded sequences, then

$$\text{(i)}\quad \underline{\lim}\, a_n + \underline{\lim}\, b_n \le \underline{\lim}\,(a_n + b_n), \text{ and}$$

$$\text{(ii)}\quad \overline{\lim}\,(a_n + b_n) \le \overline{\lim}\, a_n + \overline{\lim}\, b_n.$$

Give examples for which the strict inequalities hold.

27. It was stated in the text that some of the proven properties of real numbers (MONO, LUB, HB, BW) could be taken instead as *defining* properties of the real numbers, replacing the completeness property (All Cauchy sequences converge). This is stated formally below. The reader will observe that the definitions and properties of sequences, intervals, bounds, etc. are applicable to any ordered field. It can then be shown that the following statements are equivalent:

(a) $(X,<)$ is an Archimedean ordered field in which every Cauchy sequence converges.

(b) $(X,<)$ is an Archimedean ordered field in which every nested sequence of closed, bounded, nonempty intervals has a nonempty intersection.

(c) $(X,<)$ is an ordered field in which every subset which is bounded from above has a least upper bound.

(d) $(X,<)$ is an ordered field in which bounded monotone sequences converge.

(e) $(X,<)$ is an ordered field in which every bounded sequence contains a convergent subsequence.

(f) $(X,<)$ is an ordered field in which every covering of a closed bounded interval by open intervals contains a finite subcovering.

Only in statements (a) and (b) was the Archimedean property added. In the remaining cases, it can be *proved.* The reader may verify that statements (a)–(f) are equivalent. A possible sequence of steps that will result in such an equivalence is shown in Figure 5.

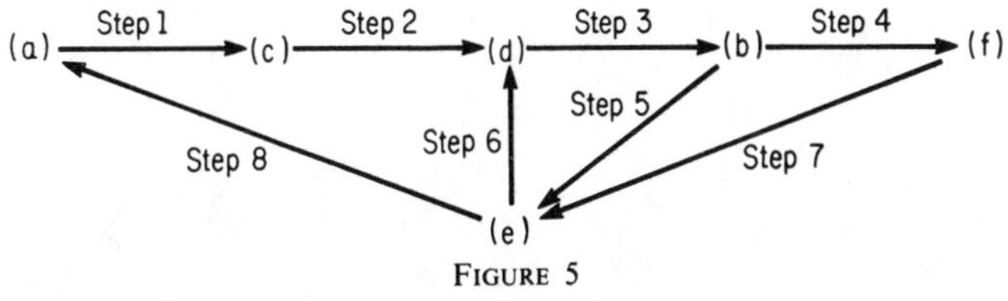

FIGURE 5

28. Show directly that Q violates each of the "completeness" properties stated in Exercise 27. In each case, how is a real number defined? (For example, define first the notion of equivalent nested sequences of closed intervals whose endpoints are rational and whose lengths tend to zero. A real number may be defined as such an equivalence class.)

*Exercises which request the proof of a statement made in the text, as well as those referred to in the text, are starred.

29. *Dedekind's Theorem.* Let R (the real numbers) be the union of two non-empty sets S and T, having the property that if $s \in S$ and $t \in T$, then $s < t$. Show that there is a unique real number ρ, which is in either S or T both not both, such that:

$$x < \rho \text{ implies } x \in S; \text{and } x > \rho \text{ implies } x \in T.$$

30. Show that Dedekind's theorem is false in Q. Show also that if $(X,\prec)$ is any ordered field in which Dedekind's theorem (restated for arbitrary fields) is true, then that field is simply R. This means that Dedekind's theorem is equivalent to all of the statements of Exercise 27.

*Exercises which request the proof of a statement made in the text, as well as those referred to in the text, are starred.

CHAPTER **2**

Metric and Normed Linear Spaces

After deriving some inequalities in Sec. 1, some examples will be given of metric and normed linear spaces, many of which reappear in the discussions of the Lebesgue integral and of uniform approximation of functions.

Convergence of sequences is then defined, and the concept of completeness is extended to these more general spaces. With scarcely more than some notational variation, the construction of R as the completion of Q may be imitated to obtain the abstract completion of a metric space.

1 Inequalities

This section could be read casually, or even omitted if the reader desires. It may instead be used only for reference purposes.

If a and b are positive numbers, then $\sqrt{ab}$ is called the *geometric mean of a and b*. If a and b are the lengths of the sides of a rectangle, then $\sqrt{ab}$ is the length of an edge of a square whose area is equal to that of the given rectangle. Keeping the sum $a + b$ fixed, the inequality

[1] $$\left(\frac{a+b}{2}\right)^2 \geq ab$$

reflects the geometrical fact that the greatest area is attained when $a = b$. Algebraically, $(a + b)^2 \geq 4ab \Leftrightarrow a^2 + 2ab + b^2 \geq 4ab \Leftrightarrow (a - b)^2 \geq 0$. The corresponding inequality for n nonnegative numbers is stated and proved next. Its geometrical interpretation in R^n, the space of real n-tuples, is similar.

1 Theorem. The geometric mean of n nonnegative numbers does not exceed their arithmetic mean; that is,

[2] $$(a_1 a_2 \cdots a_n)^{1/n} \leq \frac{1}{n}(a_1 + a_2 + \cdots + a_n).$$

Proof: We prove first that if the inequality [2] is satisfied for $n = k$, then it is satisfied for $n = 2k$.

Let $a_1, a_2, \ldots, a_k, b_1, b_2, \ldots, b_k$ be $2k$ nonnegative numbers. Setting $a = (a_1 \cdots a_k)^{1/k}$ and $b = (b_1 \cdots b_k)^{1/k}$, [1] and the assumption that [2] holds for $n = k$, yield

$$(a_1 \cdots a_k b_1 \cdots b_k)^{1/2k} = [(a_1 \cdots a_k)^{1/k}(b_1 \cdots b_k)^{1/k}]^{1/2}$$

$$\leq \frac{1}{2}\,[(a_1 \cdots a_k)^{1/k} + (b_1 \cdots b_k)^{1/k}] \leq \frac{1}{2k}\sum_{1}^{k}(a_i + b_i).$$

We have already shown that the inequality is satisfied for $k = 2$. Hence [2] holds whenever $n = 2^m$.

Now let n be an integer which is not a power of 2. Choose m so that $n < 2^m$ and let $a = \frac{1}{n}\sum_{1}^{n} a_i$. Since [2] is satisfied for 2^m terms, we have

$$(a_1 \cdots a_n a^{2^m - n})^{1/2^m} \leq \frac{\sum_{1}^{n} a_i + (2^m - n)a}{2^m} = \frac{na + (2^m - n)a}{2^m} = a,$$

which implies $(a_1 \cdots a_n a^{2^m-n}) \leq a^{2^m}$. Dividing by a^{2^m-n} yields

$$(a_1 \cdots a_n) \leq a^n = \left[\frac{1}{n}\sum_{1}^{n} a_i\right]^n. \quad \blacksquare$$

2 Theorem. If $\lambda_i \geq 0$ and $b_i > 0$, $i = 1, 2, \ldots, n$, then

[3] $$\left[\sum_{1}^{n} \lambda_i\right]^{\sum_{1}^{n}\lambda_i} \prod_{1}^{n} b_i^{\lambda_i} \leq \left[\sum_{1}^{n} \lambda_i b_i\right]^{\sum_{1}^{n}\lambda_i}.$$

Proof: The inequality [3] is satisfied if $\lambda_i = 0$, $i = 1, 2, \ldots, n$. If some but not all of the λ_i vanish, then those terms corresponding to $\lambda_i = 0$ may simply be omitted, since such terms add only $0b_i$ to the right side, and contribute multiplicative terms $(b_i)^0 = 1$ and additive terms of 0 to the left. Thus nothing is lost by assuming that each $\lambda_i > 0$.

Case 1. If the λ_i are positive integers, then Theorem 1 yields

$$\Pi b_i^{\lambda_i} = \overbrace{b_1 \cdots b_1}^{\lambda_1 \text{ times}} b_2 \cdots \quad \cdots \overbrace{b_n \cdots b_n}^{\lambda_n \text{ times}} \leq \left(\frac{\Sigma \lambda_i b_i}{\Sigma \lambda_i}\right)^{\Sigma \lambda_i},$$

which is the desired result.

Case 2. If the λ_i are rational then $\lambda_i = \mu_i/\mu$, where the μ_i are positive integers and μ is the least common denominator of the λ_i. Case 1 yields

$$(\Pi b_i^{\mu_i})^{1/\mu} \leq \left(\frac{\Sigma \mu_i b_i}{\Sigma \mu_i}\right)^{\Sigma \mu_i/\mu} = \left(\frac{\Sigma(\mu_i/\mu) b_i}{\Sigma(\mu_i/\mu)}\right)^{\Sigma(\mu_i/\mu)} = \left(\frac{\Sigma \lambda_i b_i}{\Sigma \lambda_i}\right)^{\Sigma \lambda_i}.$$

Case 3. To each irrational λ_i there is a nondecreasing sequence $\{\lambda_{i,j}\}$ of rational numbers and $\lim_{j\to\infty} \lambda_{i,j} = \lambda_i$. By showing that the functions

$b^x, b_1^{x_1} \cdot b_2^{x_2} \cdot \cdots \cdot b_n^{x_n}$, and $\left(\sum_1^n b_i x_i\right)^{\sum_1^n x_i}$ are continuous functions of one, n, and n variables respectively, [3] is verified for irrational values of λ_i. [See Exercises 3(f), (g) and (h) for details.]

The preceding argument will be referred to hereafter as a "proof by continuity."

3 Theorem (Hölder's Inequality). If $p,q > 1$, $1/p + 1/q = 1$, and $a_i, b_i \in R$, $i = 1, 2, \ldots, n$, then

[4] $$|\Sigma a_i b_i| \leq (\Sigma |a_i|^p)^{1/p} (\Sigma |b_i|^q)^{1/q}.$$

Proof: The triangle inequality for real numbers implies that

[5] $$|\Sigma a_i b_i| \leq \Sigma |a_i b_i|.$$

Thus it suffices to prove that the right-hand side of [5] does not exceed the right-hand side of [4], or equivalently, to assume that the a_i and b_i are nonnegative. It may also be assumed that for some value of i, $a_i b_i > 0$. For if every product $a_i b_i$ vanishes, then [4] is trivially satisfied.

Theorem 2 implies

$$\frac{\Sigma a_i b_i}{(\Sigma a_i^p)^{1/p} (\Sigma b_i^q)^{1/q}} = \sum \left(\frac{a_i^p}{\Sigma a_j^p}\right)^{1/p} \left(\frac{b_i^q}{\Sigma b_j^q}\right)^{1/q}$$

$$\leq (1/p + 1/q) \sum_{i=1}^n \left[\left(\frac{a_i^p}{\sum_{j=1}^n a_j^p}\right) + \left(\frac{b_i^q}{\sum_{j=1}^n b_j^q}\right)\right].$$

Summing over i, it follows from the identity $1/p + 1/q = 1$ that the right-hand side is equal to 1. What remains is clearly equivalent to [4]. ∎

For the special case $p = q = 2$, and $n = 2$ or 3, Hölder's inequality expresses the fact that if ψ is an angle, then $|\cos \psi| \leq 1$. Indeed, if ψ is the acute angle formed by the vectors (a_1, a_2, a_3) and (b_1, b_2, b_3) then

$$\cos \psi = \frac{|a_1 b_1 + a_2 b_2 + a_3 b_3|}{\sqrt{a_1^2 + a_2^2 + a_3^2}\sqrt{b_1^2 + b_2^2 + b_3^2}} \leq 1.$$

In this special case as well as for the more general situation described in Theorem 3, [4] is an equality if and only if there is a real number t such that for $i = 1, 2, \ldots, n$, $ta_i = b_i$ (or $tb_i = a_i$). (See either Exercise 4 or Kazarinoff [1].)

4 Theorem (Minkowski's Inequality). If a_i, b_i, $i = 1, 2, \ldots, n$, are real numbers, and if $p \geq 1$, then

[6] $$(\Sigma \mid a_i + b_i \mid^p)^{1/p} \leq (\Sigma \mid a_i^p \mid)^{1/p} + (\Sigma \mid b_i^p \mid)^{1/p}.$$

Proof: As in Theorem 3 it is only necessary to verify [6] for non-negative a_i and b_i.

When $p = 1$ equality is attained.

If $p > 1$, the number $q = p/(p - 1)$ is *conjugate* to p, $(1/p + 1/q = 1)$. Hölder's Inequality [4] yields

$$\begin{aligned}\Sigma(a_i + b_i)^p &= \Sigma a_i(a_i + b_i)^{p-1} + \Sigma b_i(a_i + b_i)^{p-1} \\ &\leq (\Sigma a_i^p)^{1/p}[\Sigma(a_i + b_i)^p]^{1/q} + (\Sigma b_i^p)^{1/p}[\Sigma(a_i + b_i)^p]^{1/q} \\ &= [\Sigma(a_i + b_i)^p]^{1/q}[(\Sigma a_i^p)^{1/p} + (\Sigma b_i^p)^{1/p}].\end{aligned}$$

If this inequality is divided by the first term on the right, then, remembering that $1 - 1/q = 1/p$, we obtain Minkowski's Inequality. ∎

We show next that both Hölder's and Minkowski's Inequalities remain valid for infinite sums.

5 Theorem. If $\sum_1^\infty \mid a_i \mid^p$ and $\sum_1^\infty \mid b_i \mid^q$ are convergent series, and if p and q are conjugate indices, then $\sum_1^\infty \mid a_i b_i \mid$ converges and

$$\Sigma \mid a_i b_i \mid \leq (\Sigma \mid a_i \mid^p)^{1/p}(\Sigma \mid b_i \mid^q)^{1/q}.$$

Proof: Let

$$S_n = \sum_1^n \mid a_i b_i \mid, \quad A_n = \left(\sum_1^n \mid a_i \mid^p\right)^{1/p}, \quad B_n = \left(\sum_1^n \mid b_i \mid^q\right)^{1/q}.$$

Hölder's Inequality for finite sums implies that

$$S_n \leq A_n B_n \leq \left(\sum_1^\infty \mid a_i \mid^p\right)^{1/p}\left(\sum_1^\infty \mid b_i \mid^q\right)^{1/q} = AB.$$

Therefore $\{S_n\}$ is a bounded monotone sequence of real numbers converging to the sum $\sum_1^\infty \mid a_i b_i \mid$, which remains bounded by AB (Theorem 1.3.18 and Proposition 0.4.2). ∎

6 Theorem. If $p \geq 1$ and $\sum_1^\infty |a_i|^p, \sum_1^\infty |b_i|^p < \infty$, then $\left(\sum_1^\infty |a_i + b_i|^p\right)^{1/p}$ converges and is bounded by $\left(\sum_1^\infty |a_i|^p\right)^{1/p} + \left(\sum_1^\infty |b_i|^p\right)^{1/p}$.

The proof is left to the reader.

Since the Riemann integral $\int_a^b f(x)\,dx$ is defined as the "limit" of *approximating sums* $\sum_1^n f(\zeta_i)(x_i - x_{i-1})$, $(x_{i-1} \leq \zeta_i \leq x_i,\ a = x_0 < x_1 < \cdots < x_n = b)$ as $\max_{1 \leq i \leq n} (x_i - x_{i-1})$ tends to zero, "continuous analogues" of Hölder's and Minkowski's Inequalities may be proved for integrals. The idea is to apply these inequalities to the sums and then to take limits. Later, simpler proofs will be given for Lebesgue integrable functions. Here, these results are stated for continuous functions and Riemann integrals.

7 Theorem. If $f(x)$ and $g(x)$ are continuous functions defined on the interval $[a,b]$, then

$$\left|\int_a^b fg\,dx\right| \leq \int_a^b |fg|\,dx \leq \left(\int_a^b |f|^p\,dx\right)^{1/p}\left(\int_a^b |g|^q\,dx\right)^{1/q}.$$

where p and q are conjugate indices.

8 Theorem. If $f(x)$ and $g(x)$ are continuous functions defined on the interval $[a,b]$ and if $p \geq 1$, then

$$\left[\int_a^b |f + g|^p\,dx\right]^{1/p} \leq \left(\int_a^b |f|^p\,dx\right)^{1/p} + \left(\int_a^b |g|^p\,dx\right)^{1/p}.$$

Exercises

1. Use Theorem 1 to prove if $a_1, a_2, \ldots, a_n$ are positive real numbers whose sum $a_1 + a_2 + \cdots + a_n = D$ is given, then the product $a_1 a_2 \cdots a_n$ is greatest when $a_1 = a_2 = \cdots = a_n = D/n$.
2. Again let $a_1, a_2, \ldots, a_n$ be positive numbers. Prove that if their product is given, then their sum is least when $a_1 = a_2 = \cdots = a_n$.

*3. Let b be a positive real number and let n be a positive integer. We define b^n to be the product of b with itself n times.
   (a) Prove that if $a,b > 0$ and if n and m are positive integers, then (i) $b^n b^m = b^{n+m}$; (ii) $(b^n a^n) = (ba)^n$; (iii) if $n > m$, then $b^n/b^m = b^{n-m}$; (iv) $(b^n)^m = b^{nm}$.
   (b) Writing b^{-n} for $1/b^n$, prove that (i)–(iv) of part (a) remain true for negative integers. (The condition $n > m$ of (iii) is hereby removed.)
   (c) Define the nth root of a positive number b, written $b^{1/n}$, to be the number r that satisfies $r^n = b$. Prove that $b^{1/n}$ is well defined; that is, show that

there is such a number r and only one such number. (*Hint:* Use one of the completeness properties of R.)

(d) Write $b^{-n/m}$ for $(b^{-n})^{1/m}$. Show that this is equal to $(b^{1/m})^{-n}$ and that (i)–(iv) continue to be satisfied for fractional exponents.

(e) Prove that for rational values of q, b^q is an increasing function of q if $b > 1$, and decreasing if $b < 1$.

(f) If $b > 1$ and r is a real number, define

$$b^r = \sup_{\substack{q \in Q \\ q < r}} b^q.$$

Show that b^r is a well defined increasing function of r.

(g) Define $b_1^{x_1} \cdots b_n^{x_n}$ for $b_i > 0$. Show that it is a continuous function of its n exponents.

(h) Show that $\left(\sum_{i=1}^{n} b_i x_i \right)^{\sum_1^n x_i}$ is a continuous function of the n variables x_i. It is assumed that the x_i are nonnegative and the b_i are positive. Use this result and parts (f) and (g) to prove case (iii) of Theorem 2.

***4.** Prove that equality is attained in [4] if and only if one vector is a multiple of the other.

5. $f(x)$ is said to be *convex* on the interval $[a,b]$ if whenever $x,y \in [a,b]$,

[7] $$f\left(\frac{x+y}{2}\right) \leq \frac{f(x) + f(y)}{2}.$$

Interpret this geometrically. Prove that if $f(x)$ is also continuous then for $x_1, x_2, \ldots, x_n \in [a,b]$,

[8] $$f\left(\frac{\Sigma x_i}{n}\right) \leq \frac{1}{n} \Sigma f(x_i).$$

Show also that if [7] and the continuity are replaced by

$$f(\lambda x + (1 - \lambda)y) \leq \lambda f(x) + (1 - \lambda) f(y); \quad 0 \leq \lambda \leq 1,$$

then [8] follows.

***6.** Prove Theorem 6.

7. Prove Hölder's and Minkowski's Inequalities for integrals as stated in Theorems 7 and 8.

*Exercises which request the proof of a statement made in the text, as well as those referred to in the text, are starred.

2 Metric and Normed Linear Spaces: Definitions and Examples

As its name suggests, a metric space is one in which it is possible to measure distances between pairs of points. Upon reading Definition 1, the reader will observe that the "defining" properties are suggested by some familiar properties of Euclidean distances.

1 Definition. The pair (X,d) consisting of a nonempty set X and a nonnegative function $d(x,y)$ defined on the product $X \otimes X$ is said to be a *metric space* if:

(i) $d(x,y) = d(y,x)$ for all $x,y \in X$, (symmetry);
(ii) $d(x,y) = 0$ if and only if $x = y$, (positivity);
(iii) $d(x,y) \leq d(x,z) + d(z,y)$ for all $x,y,z \in X$, (triangle inequality).

The function $d(x,y)$ is called a *distance function* or a *metric*.

2 Examples of Metric Spaces.

(a) $X = R$, $d(x,y) = |x - y|$.

(b) $X = R^n$, the space of real n-tuples. If x and y are points in R^n whose coordinates are respectively x_i and y_i, then the following functions are metrics defined on R^n:

(i) $d_1(x,y) = \sqrt{\Sigma(x_i - y_i)^2}$, the *Euclidean metric*. (For this metric as well as the one that follows, the summation is taken over the coordinates; that is, from $i = 1$ to $i = n$).

Properties (i) and (ii) of Definition 1 are readily verified, and the triangle inequality (iii) follows from Minkowski's inequality (Theorem 2.1.4) for $p = 2$.

(ii) More generally, if $p \geq 1$, then Minkowski's inequality yields also the triangle inequality for

$$d_2(x,y) = (\Sigma\,|x_i - y_i|^p)^{1/p}.$$

This metric space is denoted by l_p^n. [See (c).]

(iii) $d_3(x,y) = \max\limits_{1 \leq i \leq n} |x_i - y_i|$. Verification of the metric properties is left to the reader.

(c) For $p \geq 1$, the set l_p of sequences $\{a_n\}$ of real numbers for which $\sum_1^\infty |a_n|^p < \infty$, can be made into a metric space by

$$d(a,b) = (\Sigma\,|a_n - b_n|^p)^{1/p}; \qquad [1]$$

a and b denote the sequences $\{a_n\}$ and $\{b_n\}$ respectively. Again Minkowski's inequality, this time for infinite sums (Theorem 2.1.6), assures the convergence of [1] and gives the triangle inequality.

3 Example. Let $X = C[a,b]$ denote the set of continuous real-valued functions defined on the interval $[a,b]$.

(a) Since a continuous function defined on a closed bounded interval is itself bounded,

$$d(f,g) = \sup_{a \leq x \leq b} |f(x) - g(x)|$$

is defined for all pairs f and g in $C[a,b]$. Furthermore, each such function attains the value of its least upper bound at some point of the interval. Thus we may write

$$d(f,g) = |f(\xi) - g(\xi)|,$$

where ξ is a point at which the maximum is attained (see Theorem 3.2.14). If $h(x)$ is some other function in X, then the triangle inequality for real numbers yields

$$\begin{aligned} d(f,g) = |f(\xi) - g(\xi)| \leq |f(\xi) - h(\xi)| + |h(\xi) - g(\xi)| \\ \leq d(f,h) + d(h,g). \end{aligned}$$

The symmetry and positivity of the distance function are easily verified, thus proving that $(C[a,b],d)$ is a metric space.

(b) $C[a,b]$ may also be made into a metric space by defining

$$d'(f,g) = \int_a^b |f - g|\, dx.$$

Certainly d' is nonnegative and symmetric. The triangle inequality follows from $|f(x) - g(x)| \leq |f(x) - h(x)| + |h(x) - g(x)|$, which is satisfied for all functions f,g,h in $C[a,b]$ and for all x in $[a,b]$, and from the following property of the Riemann integral:

$$F \leq G \quad \text{implies} \quad \int_a^b F\, dx \leq \int_a^b G\, dx.$$

To prove (ii) we need another property of continuous functions: If at some point ξ in $[a,b]$, $|f(\xi) - g(\xi)| = \epsilon > 0$, then there is a δ-neighborhood of ξ in which $|f(x) - g(x)| > \epsilon/2$. It follows that

$$d'(f,g) \geq \int_{\xi-\delta}^{\xi+\delta} \epsilon/2\, dx = \delta\epsilon > 0.$$

(Later, many of the properties of continuous functions that were used in the examples will be proved in a more general setting.)

4 Example. A metric d is said to be *bounded* if there is a positive number M such that $d(x,y) \leq M$ for all $x,y \in X$. None of the metrics in either Examples 2 or 3 are bounded. However, if δ is any metric whatsoever on X, then a bounded metric is defined by

$$d(x,y) = \frac{\delta(x,y)}{1 + \delta(x,y)}.$$

Properties (i) and (ii) of Definition 1 are satisfied and $d(x,y) < 1$. To verify the triangle inequality, it suffices to show that if α, β, and γ are nonnegative numbers and if $\alpha \leq \beta + \gamma$, then

$$\frac{\alpha}{1+\alpha} \leq \frac{\beta}{1+\beta} + \frac{\gamma}{1+\gamma}.$$

5 Definition. Let $a = a_0 < a_1 < \cdots < a_n = b$ be a finite subdivision of the interval $[a,b]$ into subintervals $[a_{i-1},a_i]$, $i = 1,2,\ldots,n$. If $\alpha_1, \alpha_2, \ldots, \alpha_n, \lambda_0, \lambda_1, \ldots, \lambda_n$ are real numbers, then

$$\sigma(x) = \begin{cases} 0 & \text{if} \quad x \in R - [a,b] \\ \alpha_i & \text{if} \quad a_{i-1} < x < a_i, \quad i = 1,2,\ldots,n \\ \lambda_i & \text{if} \quad x = a_i, \quad i = 0,1,\ldots,n \end{cases}$$

is called a *real step function*, or simply a step function. Thus σ is a function which vanishes outside a closed, bounded interval, and which takes on only finitely many values, each of these on either an open interval or at one of its end points.

Let τ be a second step function determined by the subdivision $c = c_0 < c_1 < \cdots < c_m = d$ of the interval $[c,d]$, and constants $\gamma_1, \gamma_2, \ldots, \gamma_m, \nu_0, \nu_1, \ldots, \nu_m$. Then the sum $\rho(x) = \sigma(x) + \tau(x)$ vanishes outside the closed interval $[s,t]$, where $s = \min\{a,c\}$ and $t = \max\{b,d\}$. This is the smallest closed interval which contains both $[a,b]$ and $[c,d]$. We wish to show that ρ is a step function which vanishes outside $[s,t]$. A subdivision of $[s,t]$ is determined by ordering the distinct points in the union $\{a_i\} \cup \{c_i\}$. Let us call these points s_j, $j = 0,1,\ldots,k$, where $s_0 = s$ and $s_k = t$. Thus each s_j is either one of the a_i or c_i (or possibly both), and each a_i and c_i appears as an s_j. Each open interval $I_j = (s_{j-1},s_j)$ is of one of the following types: (i) I_j may be one of the (a_{i-1},a_i) or (c_{i-1},c_i); (ii) I_j may be an intersection $(a_{i-1},a_i) \cap (c_{l-1},c_l)$, for some pair of integers i and l; (iii) if $[a,b] \cap [c,d] = \phi$, then one of the (s_{j-1},s_j) will be equal to either (b,c) or (d,a). (Draw a picture.)

It is easily seen that $\rho = \sigma + \tau$ is a step function which is constant on the interior of each interval (s_{j-1},s_j). In case (i), $\rho(x) = \alpha_i$ or γ_i on I_j; for (ii), $\rho(x) = \alpha_i + \gamma_l$; and, for (iii), $\rho(x) \equiv 0$ on (b,c) or (d,a).

It is left to the reader to show that the product of two step functions, and the product of a step function and a real number, are step functions.

We define next a function on the product of the set of step functions, Σ, with itself. This function is almost, but not quite, a metric on Σ. The positivity property is *not* satisfied. Using the notation of the preceding paragraph, the step function $|\sigma(x) - \tau(x)| = |\alpha_i - \gamma_l| = \beta_j$ on the open subintervals $(s_{j-1},s_j) = (a_{i-1},a_i) \cap (c_{l-1},c_l)$, and vanishes elsewhere. •
Let

[2] $$\delta(\sigma,\tau) = \sum_1^k \beta_j(s_i - s_{j-1}).$$

•See special note on page 447.

It is easily seen that δ is symmetric and nonnegative. The triangle inequality for δ is a simple consequence of the triangle inequality for real numbers and its verification is left to the reader. Since [2] is independent of the values that either σ or τ takes on at the end points s_j, it is easy to see that if σ and τ are a pair of step functions which differ only at finitely many points, then $\delta(\sigma,\tau) = 0$. Conversely, if $\delta(\sigma,\tau) = 0$, then σ and τ must be equal except at finitely many points. We forego a formal proof and urge the reader instead to look at Figure 6 which depicts two

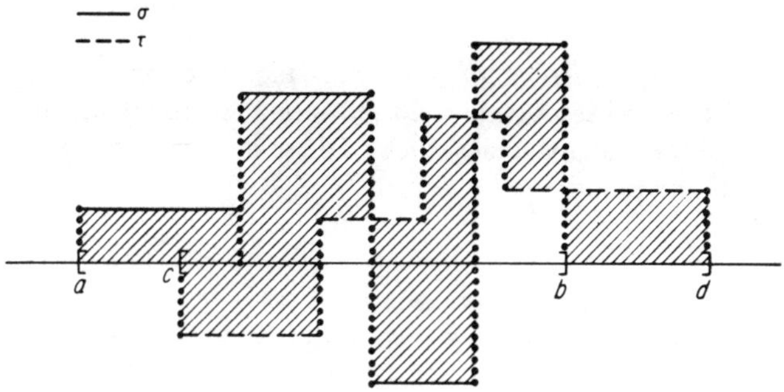

FIGURE 6

step functions σ and τ. The quantity $\delta(\sigma,\tau)$ is simply the (positive) shaded area that lies between the graphs of $y = \sigma(x)$ and $y = \tau(x)$, which are denoted respectively by an unbroken and a slashed line.

Thus, [2] satisfies all but (ii) of the properties of Definition 1. This happens because the function δ does not recognize the distinction between step functions which are identical except for finitely many points. This suggests partitioning the set Σ of step functions into disjoint classes in such a manner that δ treats all functions in a given class in the same way. This is accomplished by defining on Σ, the following equivalence relation:

6 Definition. Two *step functions* σ and σ' are said to be *equivalent*, written $\sigma \sim \sigma'$, if $\delta(\sigma,\sigma') = 0$.

7 Proposition. $\sigma \sim \sigma'$ if and only if $\sigma(x) = \sigma'(x)$ except at finitely many points.

The argument appears in the preceding paragraphs.

8 Proposition. $\sigma \sim \sigma'$ is an equivalence relation.

The proof is left to the reader.

Let $\bar{\Sigma}$ denote the collection of disjoint equivalence classes of step functions. (See Sec. 2 of Chapter 0.) In very much the same way that field operations were defined in R (Definition 1.1.7) the sum of two equivalence classes of step functions and the product of such a class with a real number may be defined so that $\bar{\Sigma}$ is a real vector space. If $\bar{\sigma}$ and $\bar{\tau}$ are classes of step functions which contain the step functions σ and τ respectively, then we define $\bar{\sigma} + \bar{\tau}$ to be $(\overline{\sigma + \tau})$, which is the class that contains the step function $\sigma + \tau$. Similarly, if r is a real number, then $r \cdot \bar{\sigma}$, or simply $r\bar{\sigma}$, is the equivalence class containing $r\sigma$. It must be shown that if σ and τ are replaced by equivalent step functions, σ' and τ', then $\sigma + \tau \sim \sigma' + \tau'$, and $r\sigma \sim r\sigma'$, thus proving that the operations on $\bar{\Sigma}$ are well defined. The details and the verification of the vector space axioms are left to the reader, as well as the proof that the absolute value of a class, $|\bar{\sigma}| = \overline{|\sigma|}$ is also well defined.

In view of Definition 6 and Proposition 7, the function

$$[3] \qquad d(\bar{\sigma},\bar{\tau}) = \delta(\sigma,\tau); \qquad \sigma \in \bar{\sigma} \text{ and } \tau \in \bar{\tau},$$

is well defined on the product of $\bar{\Sigma}$ with itself, is nonnegative, vanishes if and only if $\bar{\sigma} = \bar{\tau}$, (that is, when $\sigma(x) = \tau(x)$ except at finitely many points), and is symmetric. The triangle inequality is "inherited" from δ.

Thus $(\bar{\Sigma},d)$ is a metric space.

It is however inconvenient to talk about classes of functions. For this reason, we take the liberty of saying that Σ itself is a metric space whose distance function is d. It must be kept in mind that in doing this, two step functions are considered to be the "same" if they differ only at finitely many points. For the reader who objects to such casual language, a more formal procedure is outlined that also does away with the language of equivalence classes. This involves choosing from each class a "special" step function as a *representative* of that class. The special step function σ_0 chosen from the class $\bar{\sigma}$ takes on the following values at the end points of the open intervals over which σ_0 is constant: $\sigma_0(a_i) = (\alpha_i + \alpha_{i+1})/2$ the average value, when $i = 1,2,\ldots,n - 1$. $\sigma_0(a) = \sigma_0(a_0) = \alpha_1/2$ and $\sigma_0(b) = \sigma_0(a_n) = \alpha_n/2$. It is readily seen that there is a 1–1 correspondence between the classes in $\bar{\Sigma}$ and the set Σ_0 of special step functions. The vector space[1] operations and taking of the absolute value are preserved under this correspondence. (The special step functions are a real vector space under the usual addition of functions and multiplication by a real number. See Exercise 13.) Again the reader is warned that we shall frequently fail to distinguish between step functions, classes of step functions, and special step functions as elements of a metric space whose distance function is given by [2] or [3].

[1]The definition of a vector space is given in Exercise 12.

In each of the examples described above, the spaces possessed additional algebraic properties. For example, a pair of vectors (in R^n) may be added coordinatewise, and multiplication of a vector by a real number could be defined so that R^n is a vector space. (Similar statements can be made for $C[a,b]$ and Σ.) There do however exist metric spaces which do not possess such algebraic properties. Indeed, any nonempty set X, whose elements are denoted by $x,y,\ldots$, may be made into a metric space by setting

$$d(x,y) = \begin{cases} 0 & \text{whenever} \quad x = y \\ 1 & \text{whenever} \quad x \neq y. \end{cases}$$

The set X need satisfy no other conditions. This metric is called the *discrete metric*. We shall however often need a little *more* than a metric. The most important such example, for the purposes of this book, is given next.

9 Definition. A real vector space[2] X is said to be a *normed*, linear (or vector) space if there is defined on X a nonnegative function, $\|x\|$, which has the following properties:

(i) $\|x\| = 0$ if and only if $x = \bar{0}$.

(ii) If r is a real number and x is in X, then $\|rx\| = |r|\|x\|$.

(iii) If x and y are in X, then $\|x + y\| \leq \|x\| + \|y\|$.

The function $\|x\|$ is called the *norm of* x. The normed linear space will be denoted by $(X, \| \ \|)$.

It follows immediately from the definition, that if $(X, \| \ \|)$ is a normed linear space, then $d(x,y) = \|x - y\|$ is a metric defined on X. Most of the metrics defined earlier are easily seen to arise from norms. Most typical are:

Example 2(b): $X = R^n$, $\|x\| = \sqrt{x_1^2 + \cdots + x_n^2}$.

Example 3(a): $X = C[a,b]$, $\|f\| = \max_{a \leq x \leq b} |f(x)|$.

However, not all of the metric spaces discussed earlier are also normed spaces. It is not difficult to show that a bounded metric cannot arise from a norm. For if X is a normed linear space which satisfies $\|x - y\| = d(x,y) < M$, for some number M and all x,y in X, a contradiction is obtained as follows: Let x be any nonzero vector in X. Then $0 < \delta = \|x\|$. Definition 9 and the boundedness of the distance function imply that

$$M > d\left(\frac{2M}{\delta}x, \bar{0}\right) = \left\|\frac{2M}{\delta}x\right\| = \frac{2M}{\delta}\|x\| = 2M > M,$$

which cannot be.

[2] See footnote 1.

The metric ([2] and [3]) defined on the space Σ (or Σ_0) of step functions arises from the norm

[4] $$\| \sigma \| = \sum_1^n | \alpha_i | (a_i - a_{i-1}).$$

This quantity is the sum of the (nonnegative) areas under the "steps," and can be written as a Riemann integral,

[5] $$\| \sigma \| = \int_a^b | \sigma(x) | \, dx.$$

The reader may have guessed by now that the normed space Σ will play an important role in the theory of integration. It is in fact the starting point of this theory, the step functions being the first functions for which an integral is defined.

Exercises

***1.** Prove that if X is any nonempty set, then

$$d(x,y) = \begin{cases} 1 & \text{if } x \neq y, \\ 0 & \text{if } x = y. \end{cases}$$

is a metric. (X,d) is called a *discrete* metric space.

***2.** Let X be the set of real valued bounded functions defined on $[a,b]$. Show that

$$d(f,g) = \sup_{a \leq x \leq b} | f(x) - g(x) |$$

is a metric.

3. Let X be the space of all real sequences $\{a_n\}$. Show that

$$d(a,b) = \sum_1^\infty \frac{1}{2^n} \left(\frac{| a_n - b_n |}{1 + | a_n - b_n |} \right)$$

is a bounded metric.

4. Let X' be the space of all bounded sequences; that is, if $a = \{a_n\} \in X'$, then there is a positive number M such that all $| a_n | < M$. Show that both $d(a,b)$, as defined in Exercise 3, and $d'(a,b) = \sup_n \{ | a_n - b_n | \}$ are metrics.

***5.** Let $S[a,b]$ be the space of functions which vanish outside $[a,b]$ and which are continuous on $[a,b]$ with the exception of finitely many points. At these points, which we shall denote by $a_1, a_2, \ldots, a_n$, both left- and right-hand limits exist. Such functions are called *sectionally continuous.* Their Riemann integrals exist, as do the integrals of their absolute values. Show that

$$d'(f,g) = \int_a^b | f(x) - g(x) | \, dx$$

possesses the properties of a metric with the exception of the positivity property. Construct a metric space by using equivalence classes.

***6.** If X is any nonempty set and if $\delta(x,y)$ is a nonnegative function that satisfies only (i) and (iii) of Definition 1, then the pair (X,δ) is called a *pseudo-metric space.*

(a) Show that the relation, $x \sim y$ if and only if $\delta(x,y) = 0$, is an equivalence relation on X.

(b) Let $\overline{X}$ be the set of equivalence classes of elements of X. Show that if $\overline{x}$ and $\overline{y}$ are in $\overline{X}$, then the function

$$d(\overline{x},\overline{y}) = \delta(x,y)$$

where $x \in \overline{x}$ and $y \in \overline{y}$ is a metric on $\overline{X}$.

(c) Interpret the preceding results for Exercise 5.

7. Show that none of the metrics of Examples 2 and 3 are bounded metrics.

***8.** Prove that if $a,b,c \geq 0$, then

$$a \leq b + c \quad \text{implies} \quad \frac{a}{1+a} \leq \frac{b}{1+b} + \frac{c}{1+c}.$$

(See Example 4.)

9. Two metrics d and d', defined on the same set of points X, are said to be *topologically equivalent* if the following pair of symmetric conditions is satisfied:

(i) For each real positive number δ and point ξ in X, there is a second positive number δ' and

$$d'(\xi,x) < \delta' \quad \text{implies} \quad d(\xi,x) < \delta.$$

(ii) For each real positive number δ' and point ξ in X, there is a positive number δ and

$$d(\xi,x) < \delta \quad \text{implies} \quad d'(\xi,x) < \delta'.$$

(a) Prove that if d is any metric, then the metric $d' = d/(1 + d)$ (the bounded metric) is equivalent to d.

(b) Show that the three metrics of Example 2(b) are topologically equivalent. For $n = 2$, describe the sets

$$S_i(\xi) = \{x \in R^2 : d_i(\xi,x) < 1\}, \quad i = 1,2,3.$$

(c) Show that the metrics defined on $C[a,b]$ (Example 3) are not equivalent.

(d) Are the metrics described in Exercise 4 equivalent?

***10.** Verify the triangle inequality for the pseudo-metric [2].

11. Show that $\sim$ of Proposition 8 is an equivalence relation.

***12.** A *real vector space* X is a set on which there is defined a binary operation called "addition" which is both commutative and associative. (See Definition 0.3.1.) The set X contains an additive identity and additive inverses of every element. There is also defined the product rx, of a real number r with an element x in X which satisfies: (i) $\forall\ r,r' \in R,\ x \in X,\ r(r'x) = (rr')x$, (ii) $1x = x$, (iii) $r(x + y) = rx + ry$ and $(r + r')x = rx + r'x$. Complete the proof that Σ is a real vector space.
Show also that if σ is a step function then $|\overline{\sigma}| = \overline{|\sigma|}$ is well defined.

***13.** Show that Σ_0, the set of special step functions, is also a real vector space and is isomorphic to the vector space of equivalence classes of step functions.

***14.** Prove that if $(X, \|\ \|)$ is a normed vector space, then $d(x,y) = \|x - y\|$ is a metric on X.

*Exercises which request the proof of a statement made in the text, as well as those referred to in the text, are starred.

3 Topological Properties and Sequences

Many of the properties of real numbers may be restated for metric spaces. For example, the definitions of neighborhoods, open and closed sets, convergent and Cauchy sequences in R may be translated into metric space language simply by substituting $d(x,y)$ for $|x - y|$. Generally we shall omit any metric space proofs that are simply "translations" into metric space language of statements that have been proved for real numbers. On the other hand, we shall include the verification of some metric properties that appeared earlier without proof or in the exercises on the real numbers.

1 Definition. If ξ is a point of a metric space (X,d) and if ϵ is a positive number, then the set

$$S_\epsilon(\xi) = \{x \in X : d(\xi,x) < \epsilon\}$$

is called an *ϵ-neighborhood of ξ*.

2 Definition. A subset G of a metric space is said to be *open* if every point of G has an ϵ-neighborhood which is contained entirely in G.

3 Proposition. For each point ξ in X and real positive ϵ, $S_\epsilon(\xi)$ is an open set.

Proof: We must show that if $x \in S_\epsilon(\xi)$ then there is a $\delta > 0$ such that $S_\delta(x) \subset S_\epsilon(\xi)$. If x is such a point, then $d(x,\xi) = \rho < \epsilon$. Set $\delta = (\epsilon - \rho)/2$. Then if $y \in S_\delta(x)$, the triangle inequality yields

$$d(y,\xi) \leq d(y,x) + d(x,\xi) < \delta + \rho = \epsilon/2 + \rho/2 < \epsilon,$$

thus proving that a δ-neighborhood of x lies in the given ϵ-neighborhood of ξ. ∎

4 Definition. A subset F of a metric space X is *closed* if its complement in X, $X - F$, is open.

5 Definition. A point ζ is said to be a *point* of *accumulation* or a *cluster point* of a set S in X, if every ϵ-neighborhood of ζ contains points of S other than ζ itself.

By choosing an appropriate sequence $\{\epsilon_n\}$ of positive numbers which converges monotonically to zero, and considering the corresponding sequence of neighborhoods, $\{S_{\epsilon_n}(\zeta)\}$, it is easily seen that Definition 5 is equivalent to

5′ Definition. ζ is a *cluster point* of S if every ϵ-neighborhood of ζ contains infinitely many points of S.

6 Proposition. F is closed if and only if F contains all its points of accumulation.

See the proof of Proposition 1.3.11.

7 Proposition. (i) If $\{G_\alpha\}$ is a family of open subsets of a metric space X, then $\bigcup G_\alpha$ is an open set.

(ii) If $G_1, G_2, \ldots, G_n$ are open sets, then $\bigcap_1^n G_i$ is open.

(iii) If $\{F_\alpha\}$ is a collection of closed sets, then $\bigcap F_\alpha$ is closed.

(iv) If $F_1, F_2, \ldots, F_n$ are closed sets then $\bigcup_1^n F_i$ is closed.

Proof of (i): $x \in \bigcup G_\alpha$ implies that for some α, $x \in G_\alpha$. Since G_α is open some ϵ-neighborhood of x is contained in G_α, which in turn is contained in the union.

Proof of (ii): $x \in \bigcap G_i$ implies that there are positive numbers ϵ_i, $i = 1,2,\ldots,n$, such that $S_{\epsilon_i}(x) \subset G_i$. Let $\epsilon = \min\{\epsilon_1, \ldots, \epsilon_n\} > 0$. Then $S_\epsilon(x) \subset \bigcap S_{\epsilon_i}(x) \subset \bigcap G_i$, implying that the intersection is open.

(iii) and (iv) follow from (i) and (ii) by taking complements and using DeMorgan's laws (Proposition 0.1.6). ∎

8 Proposition. (Hausdorff Separation Property) If x and y are distinct points of a metric space, then there are disjoint open sets G_x and G_y which contain x and y respectively.

Proof: $x \neq y$ implies that $d(x,y) = \delta > 0$. Let $G_x = S_{\delta/4}(x)$ and $G_y = S_{\delta/4}(y)$. These sets are open (Proposition 3). If the intersection contains a point z, then the triangle inequality yields $\delta = d(x,y) \leq d(x,z) + d(z,y) < \delta/4 + \delta/4 = \delta/2 < \delta$, which is a contradiction. ∎

9 Definition. The intersection of all closed sets which contain a given set S is called the *closure* of S and is denoted by $\overline{S}$.

From Proposition 7(iii) it follows that $\overline{S}$ is a closed set. If F is any closed set which contains S, then $F \supset \overline{S} \supset S$. Thus $\overline{S}$ may be described as the *smallest* closed set which contains S.

10 Theorem. Let S be a subset of a metric space (X,d).

(i) S is closed if and only if $S = \overline{S}$.

(ii) If S_c denotes the set of cluster points of a set S, then $\overline{S} = S \cup S_c$.

(iii) S is closed if and only if, $x \in X - S$ implies that

$$\inf_{y \in S} d(x,y) = d(x,S) > 0.$$

($d(x,S)$ is called the *distance between the point x and the set S.*)

(iv) $d(x,S) = 0$ if and only if $x \in \overline{S}$.

Proof: (i) follows from the preceding remarks which characterize $\overline{S}$ as the smallest closed set which contains S.

Proof of (ii): Since $\overline{S}$ is closed it contains all its points of accumulation (Proposition 6), and therefore all the points of accumulation of S. Thus $\overline{S} \supset S \cup S_c$. To reverse the inclusion, it suffices to show that the set $S \cup S_c$ is a closed set. Let ζ be a cluster point of $S \cup S_c$. Then every ϵ-neighborhood of ζ contains a point $x \neq \zeta$ which is in $S \cup S_c$. If x is not in S, then $x \in S_c$ follows, and since $S_\epsilon(\zeta)$ is an open set, there is a δ-neighborhood of x which is contained in $S_\epsilon(\zeta)$ (see Figure 7). Defini-

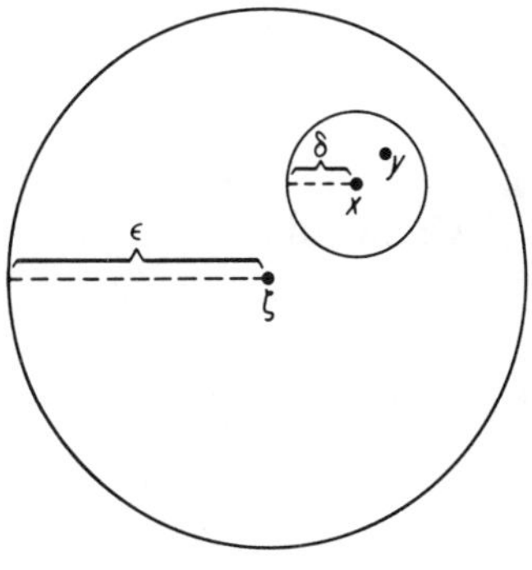

FIGURE 7

tion 5 assures the existence of a point y in $S \cap S_\delta(x)$, and $y \neq x$. By taking δ to be sufficiently small, we can be sure that $y \neq \zeta$. This proves that every accumulation point of $S \cup S_c$ is in S_c. Therefore $S \cup S_c$ is closed and must contain $\overline{S}$.

Proof of (iii): If S is closed then $X - S$ is open. Thus every point x in $X - S$ has an ϵ-neighborhood which is contained entirely in $X - S$. It follows that

$$d(x,S) = \inf_{y \in S} d(x,y) \geq \epsilon > 0.$$

Conversely, if $d(x,S) = \delta > 0$ for a point x in $X - S$, then $S_{\delta/2}(x) \subset X - S$, proving that $X - S$ is open and therefore S is closed.

The proof of (iv) is left as an exercise. ∎

Remark. The properties so far discussed in this chapter are said to be *topological*. Although it goes beyond the scope of this book to discuss general topologies, we shall say that the topological properties of a metric space (X,d) and of its subsets are those which remain invariant if d is replaced by a topologically equivalent metric d'. In Exercise 2.2.9, topological equivalence is defined, and it follows immediately from the definition that the open (and therefore closed) sets of (X,d) and (X,d') are identical whenever d and d' are topologically equivalent. After defining convergent sequences (Definition 11) the reader may verify that convergence is also a topological property, in the sense that every sequence in X will converge in *both* (X,d) and (X,d') or diverge in both of these spaces if the metrics are topologically equivalent. This is not true of Cauchy sequences. (See Exercises 4 and 5.)

11 Definition. A sequence of points $\{x_n\}$ in a metric space (X,d) *converges* to a point x in X, written $\lim_{n\to\infty} x_n = x$, or $x_n \to x$, if the sequence of real numbers $d(x,x_n)$ converges to zero.

In a normed linear space, $\lim_{n\to\infty} x_n = x$ means $\|x - x_n\| \to 0$. Sometimes we write simply $\lim x_n = x$.

12 Definition. $\{x_n\}$ is a *Cauchy sequence* in (X,d) if

$$\lim_{n,m\to\infty} d(x_n,x_m) = 0.$$

In norm notation, this becomes

$$\lim_{n,m\to\infty} \|x_n - x_m\| = 0.$$

13 Definition. A subset S of a metric space X is *bounded*, if every point ζ in X has a neighborhood which contains S.

The triangle inequality yields the following equivalent definition of boundedness:

13′ Definition. S is a *bounded* subset of (X,d) if there is a (single) point ζ in X which possesses a neighborhood that contains S.

Thus in a normed linear space, boundedness may be described using neighborhoods of $\bar{0}$:

14 Proposition. S is a bounded subset of a normed linear space if and only if there is a positive number M such that

$$x \in S \qquad \text{implies} \qquad \|x\| < M.$$

The proof is left to the reader.

15 Definition. The *diameter* of a subset S of a metric space is

$$\operatorname{diam} S = \sup_{x,y\in S} d(x,y).$$

$\operatorname{diam} S = +\infty$ is permitted.

16 Proposition. S is bounded if and only if diam $S < \infty$.

Proof: If S is bounded, then $S \subset S_M(\zeta)$, for some point ζ in X and real number M. Then, for all $x, y \in S$,

$$d(x,y) \leq d(x,\zeta) + d(\zeta,y) < 2M,$$

from which it follows that diam $S \leq 2M$. The proof of the converse is left to the reader. ∎

We state next some properties of convergent and Cauchy sequences in metric and normed linear spaces. Of course, any metric space property is automatically a normed linear space property. The converse is obviously false. For example, it makes no sense to talk about the sum of two sequences in a metric space.

17 Theorem. (a) Properties of sequences in a metric space (X,d):

(i) The limit of a convergent sequence is unique.

(ii) If $\{x_n\}$ is a convergent sequence, then Tr $\{x_n\}$ (the set of distinct sequence elements) is bounded.

(iii) Every convergent sequence is a Cauchy sequence.

(iv) If $\lim x_n = x$, then every subsequence converges also to x.

(v) Every subsequence of a Cauchy sequence is a Cauchy sequence.

(vi) If $\{x_n\}$ is a Cauchy sequence which contains a subsequence converging to x, then $\{x_n\}$ converges to x.

(b) If $(X, \| \ \|)$ is a normed vector space, then it has the following additional properties:

(vii) If $\{x_n\}$ and $\{y_n\}$ converge to x and y respectively, and if c is a real number, then $\{x_n + y_n\}$ and $\{cx_n\}$ converge to $x + y$ and cx respectively.

(viii) If $\{x_n\}$ and $\{y_n\}$ are Cauchy sequences, and if c is a real number, then $\{x_n + y_n\}$ and $\{cx_n\}$ are Cauchy sequences.

(ix) If $\{x_n\}$ converges or is a Cauchy sequence, then the sequence $\{\|x_n\|\}$ of real numbers converges in R.

The proofs are left to the reader.

18 Definition. (X,d) is a *complete metric space* if every Cauchy sequence in X converges to an element of X.

Many illustrations of the concepts introduced in this section are furnished by the complete normed space of real numbers. Others are given in Example 19 and in the exercises.

19 Examples. (a) Let $X = R^2$. The reader was asked to show in Exercise 2.2.9(b) that the metrics d_1, d_2, and d_3 of Example 2.2.2(b) were topologically equivalent. In the Euclidean metric d_1, the ϵ-neighborhood

$S_\epsilon^1(\zeta)$ is disclike; on the other hand the neighborhood $S_\epsilon^3(\zeta)$ is a square whose vertices are given by $(\zeta_1 + \epsilon, \zeta_2 + \epsilon)$, $(\zeta_1 + \epsilon, \zeta_2 - \epsilon)$, $(\zeta_1 - \epsilon, \zeta_2 - \epsilon)$, and $(\zeta_1 - \epsilon, \zeta_2 + \epsilon)$. The topological equivalence of d_1 and d_3 simply means that if a δ-neighborhood of ζ in the d_1-metric is given (which is a disc whose radius is δ), a square with edge equal to $\sqrt{2}\delta$ can be inscribed in the disc. This square is of course a $\sqrt{\delta/2}$-neighborhood in the d_3 metric. Conversely, a circular neighborhood of size δ' can be inscribed in every square neighborhood of size δ'. It follows immediately from this that (R^2,d_1) and (R^2,d_3) have the same open sets. Definition 11 implies further that a sequence in R^2 converges in the d_1 metric if and only if it converges in the d_3-metric. Although it is true here that a sequence is a Cauchy sequence in both or neither metric, it is not true in general that Cauchy sequences are preserved under topological mappings (Exercise 5). Finally, the completeness of R implies that R^2 is complete in both metrics (Exercise 17).

(Everything in the preceding paragraph is equally applicable to the d_2-metric. Here the neighborhoods are "ovals.")

(b) Let Λ_p be the totality of real sequences each of which has only finitely many nonvanishing terms. Using Minkowski's inequality, it can be shown that for $p \geq 1$,

$$\|x\| = \left(\sum_1^\infty |x_m|^p\right)^{1/p}$$

is a norm on Λ_p. It is not difficult to show that Λ_p is incomplete in this norm. Indeed the sequence $x^1 = (1,0,0,\ldots)$, $x^2 = (1,1/2,0,\ldots),\ldots,$ $x^m = (1,1/2,\ldots,(1/2)^{m-1}, 0,0,\ldots),\ldots$ is a Cauchy sequence that does not converge to an element of Λ_p (Exercise 18). In Sec. 4, we shall show that the "completion" of Λ_p is the space l_p discussed in Example 2.2.2(c).

(c) (i) If $X = C[a,b]$ and $d(f,g) = \max\limits_{a \leq x \leq b} |f(x) - g(x)|$, then $S_\epsilon(f)$ is the set of those continuous functions whose graphs lie between the curves $y = f(x) + \epsilon$ and $y = f(x) - \epsilon$ (see Figure 8). The convergence

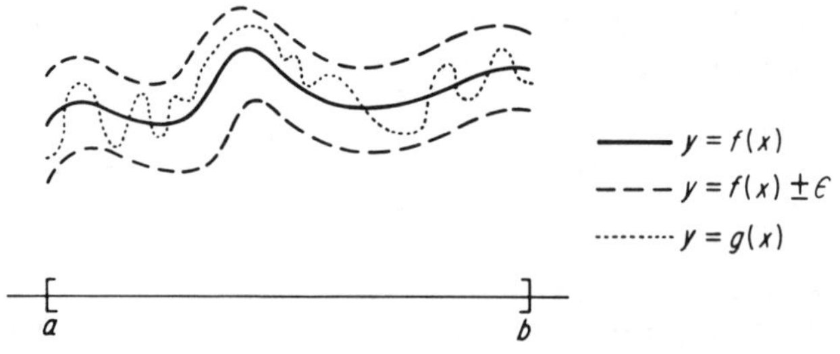

FIGURE 8

of a sequence of functions in this norm is uniform convergence. It is left to the reader to show that the space $(C[a,b],d)$ is complete.

(ii) The neighborhoods in the metric $d'(f,g) = \int_a^b |f - g| \, dx$ are not easy to describe; convergence is not uniform, or even pointwise, as we shall see below. Let us first compare neighborhoods in the d and d'-metrics:

If $g \in S_\epsilon(f) = \{h \in C[a,b] : \max_{a \le x \le b} |f(x) - h(x)| < \epsilon\}$, then $d'(f,g) < \epsilon(b - a)$, proving that $g \in S'_{\epsilon(b-a)}(f)$. Thus every ϵ-neighborhood of f (in the d-metric) is contained in an ϵ'-neighborhood (in the d'-metric). It follows that

$$\lim d(f,f_n) = 0 \quad \text{implies} \quad \lim d'(f,f_n) = 0.$$

The converse, however, is false. It can be shown that an ϵ'-neighborhood of a function f contains a sequence of functions g_n for which $d(f,g_n) = \max_{a \le x \le b} |f(x) - g_n(x)| \ge n$. This means that a given ϵ'-neighborhood is not contained in any $S_n(f)$, no matter how large n is taken to be. We illustrate this by setting $f(x) = \overline{0}$ and taking $g_n(x)$ to be the *polygonal* or *piecewise linear* functions shown in Figure 9 by the slashed line. Then

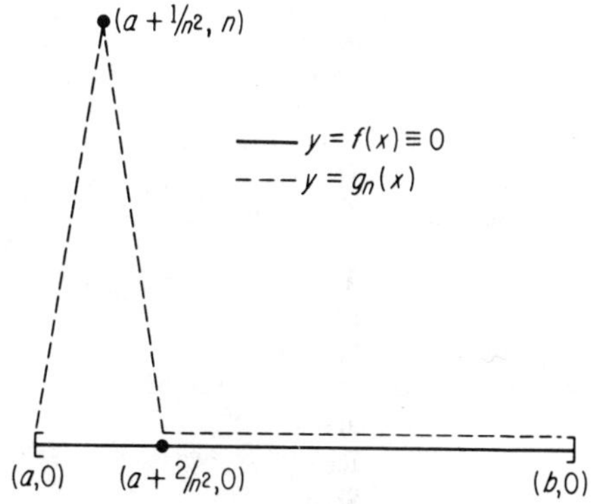

FIGURE 9

$d'(\overline{0},g_n) = 1/n$, but $d(\overline{0},g_n) = n$, proving that in the d'-metric, $\lim g_n = \overline{0}$. However, $\{g_n\}$ diverges in the d-metric. Later it will be proved that $(C[a,b],d')$ is not complete.

(d) Let $X = S[a,b]$ be the space of sectionally continuous functions defined on the interval $[a,b]$ (Exercise 2.2.5). It is easy to see that $S[a,b]$ contains $C[a,b]$, as well as $\Sigma[a,b]$, the collection of step functions that vanish outside $[a,b]$. Again both d and d' of part (c) define metrics on this space, or more precisely on the space of equivalence classes. (The reader will recall that f and g are equivalent if they differ only on a finite set.) It is left as an exercise to show that

$$d(f,g) = 0 \Leftrightarrow d'(f,g) = 0 \Leftrightarrow f(x) = g(x)$$

except at finitely many points.

Like $C[a,b]$, neither $S[a,b]$ nor $\Sigma[a,b]$ is complete in the d'-metric. Although this is not difficult to visualize, the proof is postponed until Sec. 2 of Chapter 5.

Exercises

1. Describe the ϵ-neighborhoods for each of the following metrics defined on R^2. The points are denoted by $x = (x_1,x_2)$ and $y = (y_1,y_2)$.

$$d_1(x,y) = \sqrt{(x_1 - y_1)^2 + (x_2 - y_2)^2},$$

$$d_2(x,y) = |x_1 - y_1| + |x_2 - y_2|,$$

$$d_3(x,y) = \max\{|x_1 - y_1|, |x_2 - y_2|\}, \qquad d_4(x,y) = \begin{cases} 0 & \text{if } x = y \\ 1 & \text{if } x \neq y \end{cases}.$$

Prove that d_1, d_2, and d_3 are topologically equivalent metrics.

***2.** Prove that two metrics defined on the same set are equivalent if and only if every sequence in the space either converges in both metrics or in neither.

3. Describe convergent and Cauchy sequences in the d_4-metric of Exercise 1. Describe open and closed sets and points of accumulation.

***4.** Two metric spaces (X,d) and (X',d') are called *topologically equivalent* if there is a bijection $\psi: X \to X'$ satisfying

(i) If G is open in X, then $\psi(G)$ is open in X', and

(ii) if G' is open in X' then $\psi^{-1}(G')$ is open in X.

The topological equivalence of two metrics defined on the same set is a special case of this. Here $X = X'$ and $\psi(x) = x$ for all $x \in X$.

Show, as in Exercise 2, that two metric spaces are equivalent if and only if the sequences $\{x_n\}$ and $\{\psi(x_n)\} = \{x'_n\}$ either both converge or both diverge.

5. Let (R,d) denote the metric space of real numbers with $d(x,y) = |x - y|$. If d is restricted to the subspace R_+ of positive numbers, then (R_+,d) is obviously also a metric space. Setting $R_+ = X = X'$, show that $\psi(x) = 1/x$ satisfies the conditions stated in Exercise 4. Then it follows that $\lim x_n$ exists if and only if $\lim 1/x_n$ exists, which is not very surprising. However, the image of a Cauchy sequence under ψ may not be a Cauchy sequence. Show by example that this is the case.

***6.** Show that Definitions 13 and 13′ are equivalent.

***7.** Prove that diam $S < \infty$ implies that S is bounded (Proposition 16).

***8.** Prove that $d(x,S) = 0$ if and only if $x \in \bar{S}$ [Theorem 10(iv)].

9. Prove that if S and T are disjoint closed subsets of a metric space (X,d), then there are disjoint open sets G_S and G_T which contain S and T respectively. [*Hint:* From Exercise 8 it follows that $d(s,T) = \delta_s > 0$ for all $s \in S$. Take $G_S = \bigcup S_{\delta_s/3}(s)$, the union being taken over all s in S. Similarly for G_T.

10. Define the distance between two sets S and T to be

$$d(S,T) = \inf d(s,t),$$

where $s \in S$ and $t \in T$. Prove the following statements:

(a) $S \cap T \neq \phi$ implies $d(S,T) = 0$.

(b) If either S or T is a bounded subset of R^2, then $d(S,T) = 0$ implies that $\bar{S} \cap \bar{T} \neq \phi$. Give an example which shows that for unbounded sets this is not in general true.

***11.** Prove that $d(x,S) = d(x,\bar{S})$ for all subsets S and points x of X.

12. Prove that diam S = diam $\bar{S}$.

13. Give an example of a set S for which $d(x,y) <$ diam S for all pairs $x,y \in X$.

14. If S and T are subsets of a metric space X, show that diam $(S \cup T) \leq$ diam S + diam $T + d(S,T)$. Give examples for which the strict inequality holds, and for which there is equality.

***15.** Prove that every point of an open subset of a normed linear space is a point of accumulation of that set. Show that this is not true for closed sets. A closed set each of whose points is a cluster point is called a *perfect* set. Give examples of perfect sets.

***16.** Let d' denote any of the three metrics d_1, d_2, d_3 of Exercise 1. If $\{x^m\} = \{(x_1^m, x_2^m)\}$ is a sequence in R^2, show that $\lim x^m = x = (x_1, x_2)$ in the d'-metric, if and only if both $\lim x_1^m = x_1$ and $\lim x_2^m = x_2$. This is the familiar statement that convergence in R^2, with respect to any of the usual metrics, is equivalent to *coordinatewise convergence.* A similar statement holds for all R^n, as well as for the metric

$$d_5(x,y) = (|x_1 - y_1|^p + |x_2 - y_2|^p)^{1/p}, p \geq 1.$$

17. Prove that (R^2, d_i), $i = 1,2,3,4$ and 5, is a complete metric space.

18. Show that Λ_p of Example 19(b) is incomplete.

***19.** Let $\Pi[a,b]$ be the set of real functions that vanish outside $[a,b]$ and are equal to polynomials on $[a,b]$. $\Pi[a,b]$ is a subset of $C[a,b]$ and may therefore be made into a metric space by setting $d(f,g) = \max_{a \leq x \leq b} |f(x) - g(x)|$. Show, using Taylor's theorem, that this space is incomplete.

20. Show that a discrete metric space is complete. ($d(x,y) = 1$ if $x \neq y$.)

21. Let X be a metric space. By the *closed ball* with radius δ and center at ζ, we mean the set

$$B_\delta(\zeta) = \{x \in X : d(x,\zeta) \leq \delta\}.$$

Show that the closed ball is a closed set. Show by means of an example that the closed ball $B_\delta(\zeta)$ is not always equal to the closure of the δ-neighborhood of ζ, that is to $\overline{S_\delta(\zeta)}$. (*Hint:* Use the space of Exercise 20.)

***22.** Prove that if X is a normed linear space, then $\overline{S_\delta(\zeta)} = B_\delta(\zeta)$.

***23.** Let S and T be subsets of a metric space X. Show that

$$\overline{S \cup T} = \bar{S} \cup \bar{T}.$$

Discuss the relation between $\overline{S \cap T}$ and $\bar{S} \cap \bar{T}$.

***24.** Let S be any subset of a metric space X. A point of S is said to be in the *interior* of S, denoted by $I(S)$, if S contains a neighborhood of that point. If x is a point in $X - S$ which has a neighborhood that lies entirely in $X - S$, then x is said to be in the *exterior*, $E(S)$, of S. The set $B(S) = X -$

$(I(S) \cup E(S))$ is called the *boundary* of S. (Draw some pictures of planar sets.)

Prove that $E(S)$ and $I(S)$ are open subsets of X and hence $B(S)$ is closed.

Prove that if $x \in B(S)$ then every neighborhood of x must contain points of both S and $X - S$.

Show that S is closed if and only if $S = I(S) \cup B(S)$.

Show that S is open if and only if $S = I(S)$.

Give examples of sets that satisfy (i) $S = B(S)$, (ii) $B(S) = \phi$, (iii) $I(S) = \phi$ (one example each of $E(S) = \phi$ and $\neq \phi$), (iv) $E(S) = \phi$.

*Exercises which request the proof of a statement made in the text, as well as those referred to in the text, are starred.

4 Completion of Metric and Normed Linear Spaces

The construction of the abstract completion of a given metric space is patterned after the construction of the complete field R as the collection of equivalence classes of Cauchy sequences of rational numbers. For the most part, it requires little more than to translate that construction into metric space language. As a consequence, we shall do little more than outline the steps. We begin by defining an *isometry* between two metric spaces, and then describe what is meant by the completion of a metric space.

1 Definition. Two metric spaces (X,d) and (X',d') are *isometric* if there is a bijection $X \xrightarrow{\psi} X'$ that preserves distances. This means that if $x,y \in X$ then $d(x,y) = d'(\psi(x),\psi(y))$. The mapping ψ is called an *isometry*.

If X and X' are normed vector spaces with $\| \ \|$ and $\| \ \|'$ as their respective norms, and if ψ is an isometry which preserves the vector space operations (ψ is a vector space isomorphism), then we say that $(X, \| \ \|)$ and $(X', \| \ \|')$ are *isonormal*.

Roughly speaking, isometric (isonormal) spaces are indistinguishable as metric (normed linear) spaces.

2 Examples. (a) The spaces Q (rational numbers) and $\overline{Q}$ ("rational" real numbers) are isonormal. (See Sec. 1 of Chapter 1.) The mapping is given by $\psi(a) = [\{a\}] = \dot{a}$; a is a rational number and $\dot{a}$ is the equivalence class of Cauchy sequences of rational numbers that contains the constant sequence $\{a\}$. Definition 1.2.9 yields $\|\dot{a}\|' = |\dot{a}| = |a| = \|a\|$. (The primed norm is in $\overline{Q}$.)

(b) No two of the metric spaces (R^2,d_i), $i = 1,2,3,4$, of Exercise 2.3.1 are isometric.

3 Definition. The *completion* of a metric space (X,d) is the "smallest" complete metric space that contains a subspace isometric to

(X,d). That is, $(\bar{X},\bar{d})$ is a *completion* of (X,d) if

(i) $\bar{X}$ is complete,

(ii) $\bar{X}$ contains a subspace X' which is isometric to X, and

(iii) if $\bar{X} \supset Y \supset X'$ and Y is complete, then $\bar{X} = Y$.

A similar definition may be given for normed spaces.

Let (X,d) be a metric space. We shall outline the procedure for completing the space, omitting details that are similar to the completion of Q. Included is a uniqueness proof. This asserts that if there is a second metric space $(\tilde{X},\tilde{d})$ that satisfies the conditions of Definition 3, then $(\tilde{X},\tilde{d})$ is isometric to the particular completion that is constructed in the "existence" part of the proof.

4 Definition. Two Cauchy sequences $\{x_n\}$ and $\{x'_n\}$ in (X,d) are called *equivalent*, written $\{x_n\} \sim \{x'_n\}$, if $\lim d(x_n,x'_n) = 0$.

It is readily seen that "$\sim$" defines an equivalence relation on the family of Cauchy sequences in X. This relation partitions this family of sequences into disjoint equivalence classes. The collection of these classes $\bar{x}$ is denoted by $\bar{X}$. The next step is to define a metric $\bar{d}$ on $\bar{X}$, and then to show that $(\bar{X},\bar{d})$ is the desired completion.

5 Proposition. If $\{x_n\}$ and $\{y_n\}$ are Cauchy sequences in (X,d), then $\lim d(x_n,y_n)$ exists. Furthermore, if $\{x_n\} \sim \{x'_n\}$ and $\{y_n\} \sim \{y'_n\}$ then $\lim d(x_n,y_n) = \lim d(x'_n,y'_n)$.

Proof: The sequence of real numbers $d_n = d(x_n,y_n)$ converges if and only if it is a Cauchy sequence. Let m and n be a pair of nonnegative integers. If $d_n \geq d_m$ then the triangle inequality yields

$$d_n - d_m = d(x_n,y_n) - d(x_m,y_m) \leq d(x_n,x_m) + d(x_m,y_m)$$
$$+ d(y_m,y_n) - d(x_m,y_m) = d(x_n,x_m) + d(y_n,y_m).$$

By reversing the roles of n and m (assume that $d_m \geq d_n$) and combining the two inequalities, we obtain

$$|d_n - d_m| \leq d(x_n,x_m) + d(y_n,y_m),$$

which tends to zero as n,m tend to infinity, thus proving that $\{d_n\}$ is a Cauchy sequence of real numbers and must therefore converge to a real number d.

If $\{x_n\} \sim \{x'_n\}$ and $\{y_n\} \sim \{y'_n\}$, then we must show that $\lim d(x'_n,y'_n) = d' = d$. A double application of the triangle inequality for $d_n \geq d'_n$ and $d'_n \geq d_n$ yields

$$|d_n - d'_n| = |d(x_n,y_n) - d(x'_n,y'_n)| \leq |d(x_n,x'_n) + d(x'_n,y'_n)$$
$$+ d(y'_n,y_n) - d(x'_n,y'_n)| = d(x_n,x'_n) + d(y'_n,y_n) \rightarrow 0$$

as $n \rightarrow \infty$. ∎

6 Definition. If $\overline{x} = [\{x_n\}], \overline{y} = [\{y_n\}] \in \overline{X}$, then

$$\bar{d}(\overline{x},\overline{y}) = \lim d(x_n,y_n)$$

is a metric on $\overline{X}$.

Proposition 5 implies that $\bar{d}$ is well defined, which in this case means that $\bar{d}(\overline{x},\overline{y})$ does not depend upon the sequences $\{x_n\}$ and $\{y_n\}$ in $\overline{x}$ and $\overline{y}$ respectively that are chosen. The verification of the metric properties of $\bar{d}$ is left to the reader.

7 Theorem. $(\overline{X},\bar{d})$ is a completion of (X,d) and is unique up to an isometry.

Proof: We begin with the verification of property (ii) of Definition 3: Let X' be the set of classes $\dot{x} = [\{x\}]$ that contain constant sequences. Clearly, if $\dot{x}, \dot{y} \in X' \subset X$, then

$$d(\dot{x},\dot{y}) = \lim d(x_n,y_n) = d(x,y),$$

$\{x_n\} \sim \{x\}$ and $\{y_n\} \sim \{y\}$, which proves that X' is isometric to X.

Proof of property (i): This is exactly the same as the proof given for the completeness of R. First we prove that every $\overline{x}$ in $\overline{X}$ may be approximated by elements in X'. (See Lemma 1.2.13.) Then if $\{\overline{x}^n\}$ is a Cauchy sequence in $\overline{X}$, we choose $\{\dot{y}^n\}$ in X' so that $\bar{d}(\overline{x}^n,\dot{y}^n) < 1/n$, and prove next that $\{\dot{y}^n\}$ converges to an element $\overline{x}$ in $\overline{X}$. The final step of the proof is to show that $\{\overline{x}^n\}$ converges also to $\overline{x}$. (See the proof of Theorem 1.2.14.)

Property (iii) is also proved in the same way as in the case of the real numbers.

We turn next to proving that there is no other completion. Suppose that $(\widetilde{X},\tilde{d})$ is a second completion of (X,d). Since $(\widetilde{X},\tilde{d})$ must satisfy the three conditions of Definition 3, it contains a subspace X'' which is isometric to X and therefore also to X'. Furthermore, every Cauchy sequence $\{x_n''\}$ of elements in X'' converges to a point $\tilde{x}$ in $\widetilde{X}$, and equivalent Cauchy sequences in X'' converge to the same element. The isometry $\overline{X} \xrightarrow{\psi} \widetilde{X}$ is defined as follows: For each $x' \in X'$, there is an element x in X which corresponds to x' under the isometry between X' and X. Let x'' be the corresponding element in the subspace X'' of X. We define the isometry to be given by $\psi(x') = x''$. To extend the isometry from the subspaces X' and X'' to $\overline{X}$ and $\widetilde{X}$, we make use of the Cauchy sequences which determine the elements of $\overline{X}$: If $\overline{x} = [\{x_n\}] \in \overline{X}$, then let $\{x_n''\}$ be the Cauchy sequence in X'' which corresponds to $\{x_n\}$. Set $\psi(\overline{x}) = \tilde{x} = \lim x_n''$. It is not difficult to prove that this mapping is well defined (depends only upon the *class* of $\{x_n\}$) and is injective. If it were not also surjective, then $\widetilde{X}$ would not be the *smallest* complete space containing X (more precisely, X''); that is, property (iii) of Definition 3 would be violated. ∎

Theorem 7 may be extended to normed linear spaces as follows: Every normed linear space $(X, \| \ \|)$ is automatically a metric space and therefore possesses a unique completion $(\overline{X},\overline{d})$, as a metric space, and $\overline{d}(\overline{x},\overline{y}) = \lim d(x_n,y_n) = \lim \|x_n - y_n\|$, where $\overline{x} = [\{x_n\}]$ and $\overline{y} = [\{y_n\}]$. To make $(\overline{X},\overline{d})$ into a normed linear space, we must define the vector space operations on $\overline{X}$: Let

$$\overline{x} + \overline{y} = [\{x_n + y_n\}], \qquad \overline{cx} = [\{cx_n\}];$$

$\{x_n\}$ and $\{y_n\}$ are Cauchy sequences in X and c is a real number. The reader may verify that these operations are well defined and that $\overline{X}$ is thereby made into a vector space. The norm is also easily extended by setting

$$\|\overline{x}\| = \lim \|x_n\|.$$

The details are left to the reader.

8 Definition. A complete normed linear space is called a *Banach space.*

Before giving some examples of metric spaces and their completions, we shall derive a very useful characterization of the completeness property. This property, which is equivalent to completeness in a normed linear space, is the generalization of the following well known fact about series of real numbers: If a series $\sum_1^\infty a_n$ of real numbers converges absolutely $\left(\text{that is, } \sum_1^\infty |a_n| < \infty\right)$, then $\sum_1^\infty a_n < \infty$.

9 Definition. Let X be a normed linear space and let $\{x_n\}$ be a sequence in X. The (formal) series $\sum_1^\infty x_n$ is said to *converge* or be *summable* to *an element x in X* if

$$\lim_{n\to\infty} \|x - \sum_1^n x_i\| = 0.$$

The series is said to be *absolutely summable* if the series $\Sigma \|x_n\|$ of real numbers converges.

10 Theorem. A normed linear space is complete if and only if every absolutely summable series is summable to an element in the space.

Proof: Suppose that X is a complete normed linear space and that $\Sigma \|x_n\| < \infty$. Then if $\epsilon > 0$ is given, there is an integer N for which $\sum_N^\infty \|x_n\| < \epsilon$. Let $s_n = \sum_1^n x_i$. If $n > m \geq N$, then

$$\| s_n - s_m \| = \| \sum_{j=m+1}^{n} x_j \| \leq \sum_{j=m+1}^{n} \| x_j \| < \epsilon,$$

proving that $\{s_n\}$ is a Cauchy sequence in X. The completeness of X assures the existence of an element x in X to which the sequence of partial sums converges. Thus

$$x = \lim_{n\to\infty} s_n = \lim_{n\to\infty} \sum_1^n x_j = \sum_1^\infty x_j.$$

Let us assume now that X is a normed linear space in which every absolutely summable series is summable. To show that a given Cauchy sequence $\{s_n\}$ in X converges, it suffices to show that $\{s_n\}$ contains a convergent subsequence [Theorem 2.3.17 (vi)]. Such a subsequence is obtained by choosing integers $n_1 < n_2 < \cdots < n_k < n_{k+1} < \cdots$ for which the inequality

$$\| s_{n_k} - s_{n_{k+1}} \| < 1/2^{k-1}$$

is satisfied. The existence of this sequence is assured by the assumption that $\{s_n\}$ is a Cauchy sequence. Setting $x_k = s_{n_k} - s_{n_{k-1}}$, the inequality implies that Σx_k is absolutely summable. It follows that the series is summable to an element x in X. Thus

$$\| x - s_{n_k} \| = \| x - \sum_1^k x_j \| \to 0 \text{ as } k \to \infty,$$

proving that the subsequence converges. ∎

11 Example. Let Λ_p be the set of sequences of real numbers, $x = \{x_n\}$, whose terms are almost all equal to zero (Example 2.3.19b) and Exercise 2.3.18). If N_x is the largest index of the nonvanishing terms of an element x in this space, then

$$\| x \|_p = \left(\sum_{i=1}^{N_x} | x_i |^p \right)^{1/p}$$

defines a norm on Λ_p. Since $(\Lambda_p, \|\|_p)$ is incomplete, we may form its abstract completion $\bar{\Lambda}_p$, which is the collection of equivalence classes of Cauchy sequences in Λ_p. The extended norm is given by

$$\| \bar{x} \|_p = \lim_{n\to\infty} \| x^n \|_p,$$

where

$$\bar{x} = [\{x^n\}] \in \bar{\Lambda}_p \quad \text{and} \quad x^n = (x_1^n, x_2^n, \ldots, x_j^n, 0, 0, \ldots) \in \Lambda_p.$$

This is a rather clumsy way to describe the completion of Λ_p, and it should come as no great surprise that $\bar{\Lambda}_p$ is isonormal to l_p, the space of sequences $x = \{x_n\}$ for which

$$\sum_{n=1}^{\infty} |x_n|^p < \infty.$$

The norm in l_p is defined by

[1] $$\|x\|_p^* = \left(\sum_{i=1}^{\infty} |x_i|^p\right)^{1/p}.$$

After proving the isometry, we may omit "$*$," and simply consider [1] to be an extension of the norm defined on Λ_p.

We begin by defining a mapping $\psi: l_p \to \overline{\Lambda}_p$: Let

$$x = (x_1, x_2, \ldots, x_n, \ldots) \in l_p.$$

It is easily seen that the "truncations" of this sequence,

[2] $$x^n = (x_1, x_2, \ldots, x_n, 0, 0, \ldots)$$

form a Cauchy sequence in Λ_p. For, if $n > m$,

$$\|x^n - x^m\|_p = \left(\sum_{j=m+1}^{n} |x_j|^p\right)^{1/p} \to 0 \text{ as } m \to \infty.$$

Therefore $\{x^n\}$ determines an element of $\overline{\Lambda}_p$. We define

[3] $$\psi(x) = [\{x^n\}] = \bar{x}.$$

This mapping is clearly injective. For, if x and y are unequal elements of l_p, then their sequences of truncations $\{x^n\}$ and $\{y^n\}$ defined by [2] are not equivalent and therefore $[\{x^n\}]$ and $[\{y^n\}]$ are distinct elements of $\overline{\Lambda}_p$. Furthermore, the norm is preserved under ψ:

$$\|\bar{x}\|_p = \lim_{n\to\infty} \|x^n\|_p = \lim_{n\to\infty}\left(\sum_{1}^{n} |x_j|^p\right)^{1/p} = \|x\|_p^*.$$

Since ψ is an injection which preserves the norm, l_p is isonormal to a subspace $\bar{l}_p$ of $\overline{\Lambda}_p$. To show that $\bar{l}_p = \overline{\Lambda}_p$, it suffices to demonstrate the completeness of l_p, for $\overline{\Lambda}_p$ is the *smallest* complete space which contains Λ_p. Let $\xi_n = (x_{n1}, x_{n2}, \ldots)$ be a Cauchy sequence in l_p. Then, given $\epsilon > 0$, there is an integer N such that

$$n,m > N \quad \text{implies} \quad \|\xi_n - \xi_m\|_p^p = \sum_{i=1}^{\infty} |x_{ni} - x_{mi}|^p < \epsilon^p.$$

It follows that for all values of i, and $m,n > N$,

$$|x_{ni} - x_{mi}| < \epsilon,$$

proving that each coordinate sequence, (here the ith coordinate)

$$x_{1i}, x_{2i}, x_{3i}, \ldots, x_{ni}, \ldots$$

is a Cauchy sequence of real numbers. The completeness of R yields

[4] $$\lim_{n \to \infty} x_{ni} = x_i \in R, \qquad \text{for } i = 1,2,\ldots.$$

It remains to be shown that

[5a] $x = (x_1, x_2, \ldots, x_i \ldots) \in l_p$ (x_i is defined by [4]), and

[5b] $\lim \xi_n = x$; that is, $\lim \| x - \xi_n \| = 0$.

The boundedness of the Cauchy sequence $\{\xi_n\}$ implies that $\sum_{i=1}^{\infty} |x_i|^p < \infty$, thus proving that $x \in l_p$ (Exercise 4).

Turning now to [5b], we observe that in "norm" language this means that

$$\| x - \xi_n \|_p^p = \sum_{i=1}^{\infty} | x_i - x_{ni} |^p \to 0 \qquad \text{as } n \to \infty.$$

For the finite case (l_p^n or R^n), this is assured by the equivalence between convergence in the norm and coordinatewise convergence. (See Exercise 2.3.16.) This is not a valid argument in the infinite-dimensional case, for it is possible to give an example of a sequence $\{\xi_n\}$ in l_p such that $\| \xi_n \| = 1$ for all values of n, but the coordinate sequences $\{x_{ni}\}$ each converge to 0 (Exercise 6). In this case, however, the sequence is not a Cauchy sequence. It is left to the reader to show that if the sequence is a Cauchy sequence, then coordinatewise convergence and convergence in the norm are the same.

Exercises

***1.** Verify that $\bar{d}(\bar{x},\bar{y}) = \lim d(x_n, y_n)$ is a metric on $\bar{X}$ (Definition 6).

***2.** Complete the uniqueness proof of Theorem 7 by showing that the mapping ψ is well defined.

***3.** Complete the proof that a normed vector space has a completion by showing that the extended vector space operations and the norm are well defined.

***4.** Prove that the sequence that appears in [5a] is in l_p by showing that $\sum_{i=1}^{\infty} |x_i|^p < \infty$.

***5.** Complete the proof of [5b].

6. Construct a sequence in l_p which possesses the following properties:

(i) $$\| \xi_n \|_p = \left(\sum_{i=1}^{\infty} |x_{ni}|^p \right)^{1/p} = 1 \text{ for all } n, \text{ and}$$

(ii) $\lim_{n \to \infty} x_{ni} = x_i = 0$ for all i.

Show that $\{\xi_n\}$ is not a Cauchy sequence.

7. Let X be a metric space, and let $\mathfrak{F}$ be the family of all nonempty, closed, bounded subsets of X. We define first a nonnegative function on the *ordered* pairs in $\mathfrak{F}$. If S and T are in $\mathfrak{F}$, let $\delta(S,T) = \sup_{s \in S} d(s,T)$. (See Theorem 2.3.10.) The boundedness of the sets assures the finiteness of $\delta(S,T)$. Show that

$$d(S,T) = \max[\delta(S,T), \delta(T,S)]$$

is a distance function on $\mathfrak{F}$. Describe a Cauchy sequence in this space and show that $\mathfrak{F}$ is complete if X is complete.

8. Let l_∞ denote the collection of bounded sequences of real numbers. Show that if $\xi = \{x_n\} \in l_\infty$, then

$$\|\xi\|_\infty = \sup_n \{x_n\}$$

is a norm defined on the real vector space l_∞, where addition is defined to be coordinatewise addition and $c\xi = \{cx_n\}$. Show also that $(l_\infty, \| \ \|)$ is complete.

*Exercises which request the proof of a statement made in the text, as well as those that are referred to in the text, are starred.

CHAPTER **3**

Compactness, Continuity, and Connectedness

1 Compact Spaces

In view of the uniqueness (up to isometry) of the completion of a metric or normed linear space, we are justified now in *defining* R, the set of real numbers, to be *the* completion of Q. That is, R is the *smallest, complete, ordered field* that contains Q, the ordered field of rational numbers. Although a complete metric space has been defined as one in which every Cauchy sequence converges, we have seen that in the case of the real numbers, the convergence of all Cauchy sequences is equivalent to several other properties. (See Theorems 1.3.16, 1.3.18, 1.3.19, 1.3.22, and 1.3.27, and Exercise 1.3.27.) In a general metric space not all of these properties are equivalent. In fact, two of them are meaningful only in an ordered field: that bounded monotone sequences converge, and that a set which is bounded from above has a least upper bound. We shall restate those properties that remain meaningful in a metric space, and discuss the relation between them. Each property is prefaced by an abbreviation that will be used throughout the chapter.

C.S. (Cauchy sequences): A metric space is said to be *complete*, or possess the property C.S. if every Cauchy sequence in the space has a limit.

C.N.S. (Cantor nested sets): A metric space (X,d) has the *C.N.S. property* if every sequence of nested, $(F_n \supset F_{n+1})$, closed, nonempty subsets of X satisfying

$$\lim_{n\to\infty} \operatorname{diam} F_n = 0$$

has one and only one point in its intersection.

Remark. Unlike the metric space of real numbers, the omission of the condition that diam $F_n \to 0$ may result in an empty intersection (Example 3).

B.W. (Bolzano-Weierstrass): A metric space (X,d) possesses the *B.W. property* if every infinite subset of X has a point of accumulation in X.

H.B. (Heine-Borel): A metric space (X,d) is *compact*, or possesses the H.B. property if every open covering of X has a finite subcovering.

Several variations of B.W. and H.B., such as the *Finite Intersection Property* and *Sequential Compactness* will also appear later.

Before setting about to discover the relations between these properties, a few remarks about subsets of a metric space are in order. If S is a nonempty subset of a metric space (X,d), then by restricting the distance function d to $S \otimes S$, it is easily seen that (S,d) is itself a metric space. It inherits from X all the necessary properties. Thus a *subset* S of a metric space X may be said to be compact. This simply means that if $\{G'_\alpha\}$ is a covering of S by sets which are open in the metric space (S,d), then a finite number of these sets will also cover S. It is, however frequently convenient to use coverings of S by sets which are open in X. It turns out that if S is a compact subset in the sense described above, then coverings by sets that are open in X also possess finite subcoverings. Thus the compactness of a subset S of X may be described *either* by open coverings in S or in X. This is assured by the following simple correspondence between the open sets of a metric space and those of a subset: If G' is an open subset of (S,d), where $S \subset X$, then there is an open subset G of X and $G' = G \cap S$. And conversely, if G is open in X then $G' = G \cap S$ is open in S. To prove the first statement, suppose that G' is open in S. If $\zeta \in G'$, then there is a neighborhood $S'_\epsilon(\zeta) = \{x \in S : d(\zeta,x) < \epsilon\}$ contained entirely in G'. But this neighborhood of ζ in the metric space S, may also be written as $\{x \in X : d(\zeta,x) < \epsilon\} \cap S = S_\epsilon(\zeta) \cap S$, where $S_\epsilon(\zeta)$ is a neighborhood in the metric space X. Since each $S_\epsilon(\zeta)$ is an open set of X for every ζ in G', it follows that the union of such neighborhoods, $\bigcup_{\zeta \in G'} S_{\epsilon(\zeta)}(\zeta) = G$ is an open set of X, and $G \cap S = G'$. Conversely, if G is open in X, then $G' = G \cap S$ is open in S. The proof is similar to the one just given and is left to the reader.

From these remarks, it follows easily that the compactness of the subset S of X may be described by using sets which are either open in S or in X:

1 Proposition. If S is a nonempty subset of a metric space (X,d), then (S,d) is a compact metric space if and only if every covering of S by sets which are open in X has a finite subcovering.

The details are left to the reader.

We prove next that C.S. $\Leftrightarrow$ C.N.S. and B.W. $\Leftrightarrow$ H.B., and that the latter two imply the former.

2 Theorem (C.S. $\Leftrightarrow$ C.N.S.). (X,d) is a metric space in which every Cauchy sequence converges $\Leftrightarrow$ Every nonempty, closed, nested sequence of sets $\{F_n\}$ in X, whose diameters tend to zero, has a nonempty intersection.

Proof: Let $\{F_n\}$ be a sequence of nonempty, closed, nested subsets of a complete metric space whose diameters tend to zero. To show that $\bigcap F_n \neq \phi$, we construct a Cauchy sequence whose limit is in the intersection of the F_n. Since the F_n are not empty, we may choose a point x_n in each F_n (Axiom of Choice). If $\epsilon > 0$, there is an integer ν such that

$$n > \nu \qquad \text{implies} \qquad \operatorname{diam} F_n < \epsilon.$$

Then $m > n > \nu$ implies $F_m \subset F_n$, and therefore $x_n, x_m \in F_n$. It follows that $d(x_n,x_m) \leq \operatorname{diam} F_n < \epsilon$, proving that $\{x_n\}$ is a Cauchy sequence in the complete space (X,d), and it therefore converges in X to some point x. If x is not in the intersection of the F_n, then let ν be an integer for which $x \in F_\nu$ is false. Then it must follow that if $n > \nu$, $x \neq x_n$, for otherwise we have $x = x_n \in F_n \subset F_\nu$, which is a contradiction. In this case, the set of distinct elements of the sequence $\{x_n\}$, called the *trace* of $\{x_n\}$, and denoted by $\operatorname{Tr}\{x_1, x_2, \ldots\}$, is infinite, and x is an accumulation point of that set. It follows that for all n the sets $F_n \cap \operatorname{Tr}\{x_1, x_2, \ldots\}$ are infinite subsets of the closed sets F_n, each having x as a point of accumulation, from which it follows that x is in each F_n.

Now let (X,d) be a metric space which possesses the C.N.S. property. If $\{x_n\}$ is a Cauchy sequence in X, we define a sequence of closed, nonempty sets by

$$T_1 = \overline{\operatorname{Tr}\{x_1, x_2, \ldots\}},\ T_2 = \overline{\operatorname{Tr}\{x_2, x_3, \ldots\}}, \ldots,\ T_n = \overline{\operatorname{Tr}\{x_n, x_{n+1}, \ldots\}};$$

that is, each T_n is the closure of the trace of the sequence $\{x_{n+i}\}$, $i = 0, 1, 2, \ldots$. Since $\operatorname{diam} \overline{S} = \operatorname{diam} S$ for all subsets of a metric space, it follows that as ν tends to infinity,

$$\operatorname{diam} T_\nu = \sup_{m,n \geq \nu} d(x_n,x_m) \to 0.$$

The C.N.S. property yields the existence of a point x in the intersection of the T_n. It is easily seen that $x = \lim x_n$, which proves that (X,d) is complete. ▎

If $X = R$ and $d(x,y) = |x - y|$ then the substitution of the boundedness of the F_n for the condition $\lim \operatorname{diam} F_n = 0$ results only in the loss of the uniqueness of the point which is common to all the F_n. The following example shows that in a general metric space, this may lead to an empty intersection.

3 Example. Let $X = R$, and let $\delta(x,y) = |x - y|/(1 + |x - y|)$. ($\delta$ is a bounded metric.) The sets $F_n = [n, \infty)$ are closed, nested, bounded, and nonempty. However, for each value of n, diam $F_n = 1$. It is obvious that no point in R can be in every F_n.

To explain this perhaps unexpected behavior, it should be noted that if δ is replaced by $d(x,y) = |x - y|$, then the sets F_n remain (in the new metric), closed and nonempty. However, here each F_n is unbounded, thus violating one of the conditions of Theorem 1.3.19. The reader might also note that these two metrics on R are topologically equivalent (Exercise 2.2.9), from which it follows that boundedness is not a topological property.

4 Definition. A subset S of a metric space is said to be *sequentially compact* if every sequence in S contains a subsequence which converges to a point of S.

5 Theorem. A subset S of a metric space X possesses the B.W. property if and only if it is sequentially compact.

Proof: Let $\{x_n\}$ be a sequence of points in a subspace S which has the B.W. property. If $\mathrm{Tr}\{x_n\}$ is a finite set, then there are infinitely many indices $n_1 < n_2 < \cdots < n_k \cdots$, and a point x in S such that $x_{n_k} = x$ for all values of k. This sequence of course converges to x. If $\mathrm{Tr}\{x_n\}$ is infinite, then B.W. yields a point of accumulation of $\mathrm{Tr}\{x_n\}$ which is a member of S. This point x contains infinitely many of the x_n in each of its neighborhoods, so that it is possible to select a subsequence $\{x_{n_k}\}$ which satisfies $d(x,x_{n_k}) < 1/k$.

Conversely, if S is sequentially compact and if T is an infinite subset of S, then we may select from T a sequence of distinct points x_n (Axiom of Choice). The sequential compactness of S assures a convergent subsequence $\{x_{n_k}\}$. Clearly, the limit of this subsequence is a point of accumulation of $\mathrm{Tr}\{x_{n_k}\}$ and therefore of T. By assumption, this limit is in S. ∎

6 Definition. A nonempty family of sets $\mathfrak{F}$ is said to have the *finite intersection property* if every (nonempty) finite subfamily has a nonempty intersection.

It is left to the reader (Exercise 5) to demonstrate the equivalence of the following characterizations of compact spaces:

7 Theorem. (a) (X,d) is compact $\Leftrightarrow$ every family of closed subsets of X which has the finite intersection property has a nonempty intersection.

(b) (X,d) is compact $\Leftrightarrow$ every family of closed subsets of X whose intersection is empty, contains a finite subcollection whose intersection is also empty.

8 Theorem. (H.B. $\Leftrightarrow$ B.W.) A subset of a metric space is compact if and only if it possesses the B.W. property.

Proof: Let T be an infinite subset of a compact space (S,d) that has no points of accumulation in S. Since no point of S can be a point of accumulation of points in T, it follows that every $s \in S$ has a neighborhood which contains no points of T other than possibly s itself. The collection $S_{\delta_s}(s)$ of these neighborhoods is an open covering of S. The compactness of S implies that there is a finite subcover. However, since each of these neighborhoods contains at most one point of T (its center), this implies that T is a finite set, which is a contradiction.

We assume now that every infinite subset of S has a point of accumulation in S. We wish to show that if $\{G_\alpha\}$ is an open covering of S, it has a finite subcovering. If $s \in S$, there is a G_α (at least one) and a positive number ϵ which depends on both s and α such that $S_\epsilon(s) \subset G_\alpha$. Let

$$\rho(s) = \sup\{\epsilon : S_\epsilon(s) \subset G_\alpha, \text{ for some } \alpha\}.$$

Although this does not mean that $S_{\rho(s)}(s)$ itself is contained in one of the G_α, it implies that if $0 < \epsilon < \rho(s)$, then there is at least one G_α which contains $S_\epsilon(s)$. Now let $\delta = \inf_{s \in S} [\rho(s)]$. Clearly δ is nonnegative since each $\rho(s) > 0$. We prove next that $\delta > 0$. If $\delta = 0$, then there are points s of S for which $\rho(s)$ can be made arbitrarily small. In particular, there is a sequence of distinct (why?) points s_n in S such that $\rho(s_n) < 1/n$. Theorem 5 implies that there is a subsequence $\{s_{n_k}\}$ which converges to a point s in S. Choose G_α so that it contains the neighborhood $S_\gamma(s)$ where $\gamma = (2/3)\,\rho(s)$. Since $\lim s_{n_k} = s$, it follows that for sufficiently large n_k, we have

$$s_{n_k} \in S_{\gamma/2}(s), \quad \text{and} \tag{1}$$

$$\rho(s_{n_k}) < 1/n_k < \gamma/2. \tag{2}$$

(See Figure 10.) From [1] it follows that

$$S_{\gamma/2}(s) \subset S_\gamma(s) \subset G_\alpha.$$

But [2] implies that $S_{\gamma/2}(s_{n_k})$ cannot be contained in any G_α, since $\gamma/2 = \rho(s)/3 > \rho(s_{n_k}) = \sup\{\epsilon : S_\epsilon(s_{n_k}) \subset G_\alpha, \text{ for some } \alpha\}$.

Thus $\delta > 0$. This number is called the *Lebesgue number* of the covering $\{G_\alpha\}$. Since $\rho(s) \geq \delta$ for all s in S, it follows that to each s in S, there is a G_α which contains $S_{\hat{\delta}}(s)$ whenever $\hat{\delta} < \delta$. If we can show that the covering of S by the open neighborhoods $S_{\hat{\delta}}(s)$ has a finite subcovering then we are through. For suppose that $S_{\hat{\delta}}(s_i)$, $i = 1, 2, \ldots, n$, covers S. Choose indices $\alpha(i)$ so that $S_{\hat{\delta}}(s_i) \subset G_{\alpha(i)}$. Then clearly the finite set $\{G_{\alpha(i)}\}$ also covers S.

Again, we use an indirect proof. Suppose that there is no finite subcovering of $\{S_{\hat{\delta}}(s)\}$. Then we can select a sequence of points from S in the

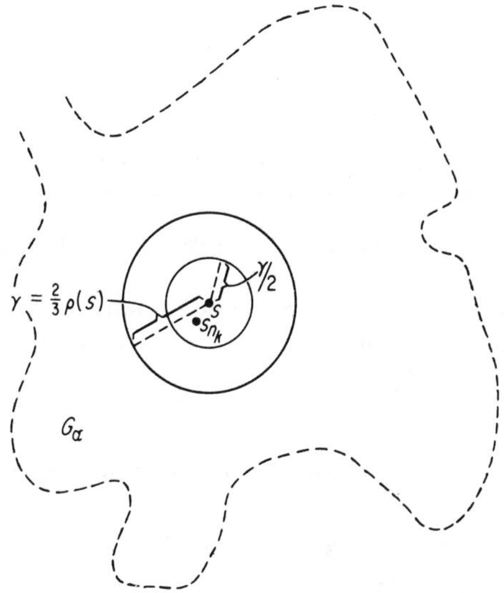

FIGURE 10

following way: Let s_1 be any point in S. Since it is assumed that no finite subcollection of the $\hat{\delta}$-neighborhoods covers S, the set $S - S_{\hat{\delta}}(s_1) \neq \phi$. Let $s_2 \in S - S_{\hat{\delta}}(s_1)$. Continuing in this way, we choose a sequence $s_n \in S - \bigcup_{i=1}^{n-1} S_{\hat{\delta}}(s_i)$. Since no s_n is in any $S_{\hat{\delta}}(s_i)$ for $i = 1,2,\ldots,n-1$, it follows that $d(s_n,s_m) > \hat{\delta}$ whenever $n \neq m$. The B.W. property implies that the infinite set Tr $\{s_n\}$ has an accumulation point in S. Let $\{s_{n_k}\}$ be a subsequence which converges to this accumulation point s. Theorem 2.3.17 implies that $\{s_{n_k}\}$ is a Cauchy sequence, contradicting $d(s_n,s_m) > \hat{\delta}$. ∎

We have already seen that the family of compact subsets of R, $(d(x,y) = |x - y|)$, is the collection of closed and bounded subsets of R (Theorem 1.3.27). The same proof implies that a compact subset of a general metric space is closed and bounded. The following example demonstrates that the converse is false.

9 Example. Let $X = C[0,1]$, $d(f,g) = \max_{0 \leq x \leq 1} |f(x) - g(x)|$. The subset

$$S = \overline{S_1(\bar{0})} = \{f \in C[0,1] : \max |f(x)| \leq 1\}$$

is closed and bounded. However the open covering of S by the neighborhoods $S_{1/4}(f)$, $f \in S$, does not have a finite subcovering. For if there were functions $f_1, f_2, \ldots, f_n$, such that

$$\bigcup_{i=1}^{n} S_{1/4}(f_i) \supset S,$$

then it would follow that to each function g in S, that is if $|g(x)|$ does not exceed 1, there is an f_i such that $\max_{0 \le x \le 1} |g(x) - f_i(x)| < 1/4$. The reader should convince himself that this cannot be the case.

To obtain necessary and sufficient conditions for a subset of a metric space to be compact, we must replace the "boundedness" condition by a slightly stronger property which is stated below.

10 Definition. A subset S of a metric space is said to be *totally bounded* if to each $\epsilon > 0$, there are finitely many points $s_1, s_2, \ldots, s_n$ in S such that

$$S \subset \bigcup_{i=1}^{n} S_\epsilon(s_i).$$

11 Proposition. If S is a totally bounded subset of a metric space (X,d) then S is bounded.

Proof: It suffices to prove that diam $S < \infty$ (Proposition 2.3.16). The total boundedness of S implies that there are points, $s_1, \ldots, s_n$, such that $\bigcup_{i=1}^{n} S_1(s_i) \supset S$. Let $\delta = \max[d(s_i,s_j)]$, the maximum being taken over all pairs s_i,s_j; $i,j, = 1,2,\ldots,n$. If s and t are any two points in S, then there are indices i and j (possibly the same) such that $s \in S_1(s_i)$ and $t \in S_1(s_j)$. The triangle inequality yields

$$d(s,t) \le d(s,s_i) + d(s_i,s_j) + d(s_j,t) < 1 + \delta + 1,$$

implying that diam $S < 2 + \delta$. ∎

An illustration of a bounded set which is not totally bounded was given in Example 9. See also Exercise 11.

12 Proposition. A compact subset of a metric space is totally bounded.

Proof: Let $\epsilon > 0$ be given. Since $\{S_\epsilon(s)\}$, for $s \in S$, is an open covering of S, the compactness yields a finite subcovering. ∎

13 Proposition. A compact subset S of a metric space X is complete. (This means of course that (S,d) is a complete metric space.)

Proof: Let $\{s_n\}$ be a Cauchy sequence in S. If Tr$\{s_n\}$ is finite, then a subsequence converges and therefore $\{s_n\}$ itself converges. If however,

Tr $\{s_n\}$ is infinite, then S contains a point of accumulation of the sequence (Theorem 8). Since a Cauchy sequence is bounded and cannot have more than one point of accumulation, it follows that lim s_n exists and is equal to that point of accumulation. ∎

Propositions 12 and 13 imply that if S is compact, then S is totally bounded and complete. The completeness of S implies in turn that S is a closed subset of X. We shall state as a theorem these necessary conditions for compactness.

14 Theorem. A compact subset of a metric space X is closed and totally bounded.

To give sufficient conditions for compactness, the completeness of X must be added, for if X is not complete, there will be closed, totally bounded subsets which are not compact (Exercise 9).

15 Theorem. A closed, totally bounded subset of a complete metric space is compact.

Proof: In view of Theorem 8, it suffices to prove that if S is a closed, totally bounded subset of a complete space X, then every infinite subset T of S has an accumulation point in S. The total boundedness of S implies that there are finitely many points $s_{11}, s_{12}, \ldots, s_{1,n(1)}$, in S such that $\bigcup_{i=1}^{n(1)} S_1(s_{1i}) \supset S$. If T is infinite, then at least one of these neighborhoods, say $S_1(s_{11})$, must contain infinitely many points of T. Since every subset of a totally bounded set is itself totally bounded (Exercise 6), it follows that there are finitely many points $s_{21}, s_{22}, \ldots, s_{2,n(2)}$, in $S_1(s_{11})$ such that $\bigcup_{i=1}^{n(2)} S_{1/2}(s_{2i}) \supset S_1(s_{11})$. Again, one of these neighborhoods, say $S_{1/2}(s_{21})$, contains infinitely many points of T. Cover this neighborhood with neighborhoods whose diameter is $1/3$. Continuing in this way, we obtain a sequence of sets $S_{1/m}(s_{m1}) = \tilde{S}_m$ such that $\tilde{S}_m \supset \tilde{S}_{m+1}$ and each $\tilde{S}_m \cap T$ is infinite. It is therefore possible to select a distinct sequence $t_m \in T \cap \tilde{S}_m$. It is easily seen that if $n,m > \nu$, then $d(t_n,t_m) < 2/\nu$, proving that $\{t_m\}$ is a Cauchy sequence. The completeness of X insures the existence of a limit $t \in X$. However, S is closed, from which it follows that $t \in S$. And finally, Tr$\{t_m\}$ is infinite implies that t is an accumulation point of T. ∎

Exercises

***1.** Let S be a subspace of a metric space (X,d). Prove that if G is an open set in X, then $G' = G \cap S$ is an open subset of S. Show also that F is closed in X implies that $X \cap F = F'$ is a closed subset of S.

***2.** Prove Proposition 1.

***3.** Show that under the conditions stated in Theorem 2, we may conclude that the nonempty intersection of the F_n contains exactly one point.

***4.** Prove that if ξ is a point of accumulation of T, then there is a sequence of distinct points $\{t_n\}$ in T such that $d(\xi, t_n) < 1/n$.

***5.** Prove Theorem 7.

***6.** Prove that every subset of a totally bounded set is totally bounded.

***7.** Show that a compact set is closed and bounded.

***8.** Show that a complete subset of a metric space X is closed in X. Show that a closed subset of a compact space is compact. Show that a closed subset of a complete space is complete.

9. Give an example of an incomplete metric space X which contains a subspace S that is closed and totally bounded but incomplete. (This justifies the assumption made in Theorem 15 that X is complete.)

10. Show that if (X,d) is a discrete metric space, then a subset S is compact if and only if it contains finitely many elements.

11. The complete metric space l_2 of sequences $x = \{x_n\}$, where $\Sigma x_i^2 < \infty$, contains closed, bounded subsets which are not compact. One of these is the closed ball

$$B_1(\bar{0}) = \{x : \| x \| \leq 1\}, \qquad \| x \| = (\Sigma x_i^2)^{1/2}.$$

Use Theorem 8 to prove that it is not compact. Show directly that it is not totally bounded.

12. Use Theorem 8 to give a second proof that the closed ball of radius 1 of Example 9 is not compact.

13. Show that the following subset of l_2 is compact:

$$I = \{x : \forall n, \ |x_n| \leq 1/2^n\}.$$

***14.** Prove that if $S_1, S_2, \ldots, S_n$, are compact subsets of X, then $\bigcap_{i=1}^{n} S_i$ and $\bigcup_{i=1}^{n} S_i$ are compact. Does this remain true for infinite intersections and unions?

15. Let S and T be disjoint, nonempty subsets of a complete metric space X. Show that if S is closed and T is compact, then

$$d(S,T) = \inf_{\substack{s \in S \\ t \in T}} d(s,t) > 0.$$

Give an example in R^2 which demonstrates that this is false if it is not assumed that one of the sets is compact. In particular, find a pair of closed, unbounded disjoint sets S and T of R^2 having the property that for all values of $\nu > 0$, there are points $s_\nu \in S$ and $t_\nu \in T$ such that $d(t_\nu, s_\nu) < 1/\nu$. (Use the Euclidean metric.)

*Exercises which request the proof of a statement made in the text, as well as those referred to in the text, are starred.

2 Continuous Functions and Compactness

For many of the examples that appeared earlier in the text, it was assumed that the reader was familiar with certain theorems about functions in $C[a,b]$. Although these results could easily have been obtained in Chapter 1, we have chosen to postpone the proofs of these theorems in order that they may appear in the more general setting of a metric space.

We begin by defining continuous functions (or mappings) from one metric space to another, in the familiar ϵ,δ manner. Ultimately, two additional equivalent definitions are given, the first using open sets and the second convergent sequences. Following these definitions, are some important theorems about continuous functions defined on compact sets.

1 Definition. Let (X,d) and $(\hat{X},\hat{d})$ be two metric spaces. A mapping

$$(X,d) \xrightarrow{f} (\hat{X},\hat{d})$$

is said to be *continuous* at a *point* ζ of X, if to each $\epsilon > 0$, there is a $\delta > 0$, which depends on both ϵ and ζ, such that

$$f(S_\delta(\zeta)) \subset \hat{S}_\epsilon(f(\zeta)), \tag{1}$$

or equivalently, if

$$d(x,\zeta) < \delta \qquad \text{implies} \qquad \hat{d}(f(x),f(\zeta)) < \epsilon.$$

The mapping f is said to be *continuous on* X if f is continuous at every point of X.

If the number δ depends only upon ϵ, that is if [1] is true for all $\zeta \in X$, then f is *uniformly continuous* on X.

2 Examples. (a) If $X = \hat{X} = R$, with the standard metric $d(x,y) = |x - y|$, then Definition 1 becomes: For each $\epsilon > 0$, and $\zeta \in R$, there is a $\delta > 0$ such that

$$|x - \zeta| < \delta \qquad \text{implies} \qquad |f(x) - f(\zeta)| < \epsilon.$$

(b) If $X = R^2$, equipped with the Euclidean metric, and $\hat{X} = R$, then $f: R^2 \to R$ is continuous at a point $\zeta \in R^2$ means that for any ϵ-neighborhood of $f(\zeta)$ (this is an interval $(f(\zeta) - \epsilon,\ f(\zeta) + \epsilon)$) there is a δ-neighborhood of ζ in R^2 (a disc with center at ζ and radius equal to δ), whose image under f is contained in the given ϵ-neighborhood. (See Exercises 8, 9, and 10.)

3 Proposition. The mapping $(X,d) \xrightarrow{f} (\hat{X},\hat{d})$ is continuous on X if and only if $f^{-1}(\hat{G})$ is open in X whenever $\hat{G}$ is open in $\hat{X}$.

Remark. The mapping need not be surjective to define $f^{-1}(\hat{G})$. See Sec. 1 of Chapter 0.

Proof: Assume that f is continuous and that $\hat{G}$ is an open subset of $\hat{X}$. We must show that if $\zeta \in f^{-1}(\hat{G})$, there is a neighborhood of ζ which is contained in $f^{-1}(\hat{G})$: The point $f(\zeta)$ is in the open set $\hat{G}$. Hence there is a neighborhood $\hat{S}_\epsilon(f(\zeta))$ which is contained entirely in $\hat{G}$. Definition 1 assures the existence of a positive number δ such that

$$f(S_\delta(\zeta)) \subset \hat{S}_\epsilon(f(\zeta)) \subset \hat{G}.$$

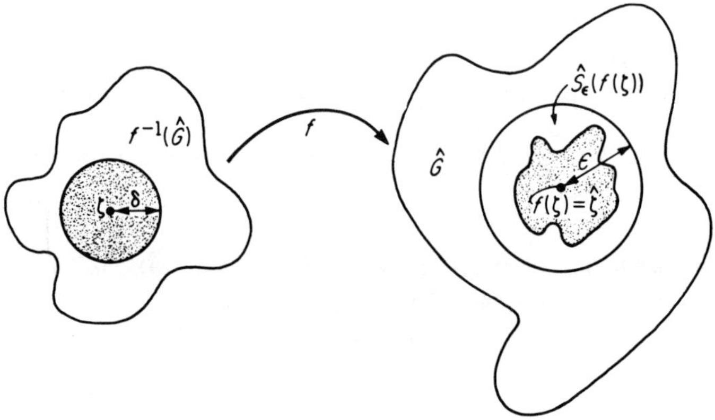

FIGURE 11

(See the shaded portions of Figure 11.) It follows from the definition of the inverse image of a set that $S_\delta(\zeta) \subset f^{-1}(\hat{G})$, thus proving that $f^{-1}(\hat{G})$ is open in X.

Conversely, if the preimage of every open subset of $\hat{X}$ is open in X, then if $\hat{\zeta} = f(\zeta)$, it follows that for every $\epsilon > 0$, $f^{-1}(\hat{S}_\epsilon(\hat{\zeta}))$ is an open set in X, and therefore contains a neighborhood $S_\delta(\zeta)$. It is easily seen that $f(S_\delta(\zeta)) \subset \hat{S}_\epsilon(\hat{\zeta})$. ∎

The following example demonstrates that a continuous mapping does not generally map open sets into open sets. A surjective mapping which does this need not itself be continuous, but its inverse is a continuous mapping.

(The reader will recall that if S is a metric subspace of X, then the open subsets of S are precisely those which can be written as the intersection of S and an open set of X. See the remarks that precede Proposition 3.1.1.)

4 Example. Let $S = [0,2\pi)$ and let d be the standard metric on R. S is mapped onto the unit circle $\hat{S} = \{(x,y): x^2 + y^2 = 1\} \subset R^2$ by the continuous real functions $x = \cos t$ and $y = \sin t$. (The reader should verify that this is a continuous mapping.) The open subsets of S are simply the (relatively) open arcs and their countable unions. The interval $[0,1/4) = S \cap (-1,1/4)$ is open in S but its image in $\hat{S}$ is an arc which is "closed" on one end, and therefore not an open subset of $\hat{S}$.

5 Proposition. If f is a continuous mapping of a metric space X into the real numbers, then for all $\alpha \in R$, the sets

$$[f > \alpha] = \{x \in X : f(x) > \alpha\}$$

are open in X.

Proof: $[f > \alpha] = f^{-1}((\alpha, \infty))$ and (α, ∞) is an open subset of R. ∎

6 Corollary. If f is a continuous mapping of X into the real numbers, then for all real values of $\alpha < \beta$,

(i) $[f < \alpha]$ is open; (ii) $[f \leq \alpha]$ and $[f \geq \alpha]$ are closed; (iii) $[\alpha < f < \beta]$ is open; (iv) $[\alpha \leq f \leq \beta]$ is closed.

The proof is left to the reader.

7 Corollary. If f is a continuous mapping of a subset S of the real numbers into R, and if for some $\zeta \in S$, $f(\zeta) > 0$, then there is $\delta < 0$ such that,

$$|x - \zeta| < \delta \qquad \text{implies} \qquad f(x) > 0.$$

The proof is left to the reader.

A third characterization of continuity is given by its "convergence preserving" property. This definition is most useful in proving that a given function is discontinuous at a point (Example 10).

8 Definition. A mapping $f:(X,d) \longrightarrow (\hat{X},\hat{d})$ is called *sequentially continuous* or *convergence preserving* at a point ζ in X, if

$$\lim x_n = \zeta \text{ (in } X) \qquad \text{implies} \qquad \lim f(x_n) = f(\zeta) \text{ (in } \hat{X}).$$

9 Proposition. A mapping f of X into $\hat{X}$ is continuous at ζ if and only if it is sequentially continuous at ζ.

Proof: Assume that f is continuous at ζ. Then to each $\hat{S}_\epsilon(f(\zeta))$, there is a $S_\delta(\zeta)$ whose image under f is in $\hat{S}_\epsilon(f(\zeta))$.

If $\lim x_n = \zeta$, then almost all the x_n will lie in any δ-neighborhood of ζ. To prove that $\lim f(x_n) = f(\zeta)$, we must show that almost all the points $f(x_n)$ lie in a given ϵ-neighborhood of $f(\zeta)$. Let $S_\delta(\zeta)$ be a neighborhood of ζ in X which is mapped into the given ϵ-neighborhood in X (continuity). Since almost all the x_n are in $S_\delta(\zeta)$, it follows that almost all the $f(x_n) \in f(S_\delta(\zeta))$, which is contained in $\hat{S}_\epsilon(f(\zeta))$.

Suppose now that f is convergence preserving but not continuous at ζ. This implies that there must be a positive number ϵ for which it is not possible to find a δ of continuity. Thus for every positive integer n, there is a point $x_n \in X$ for which $d(\zeta,x_n) < 1/n$ but $d(f(\zeta),f(x_n)) \geq \epsilon$. This means that there is a sequence $\{x_n\}$ which converges to ζ, but $\{f(x_n)\}$

does not converge to $f(\zeta)$, which contradicts the assumption that f is converging preserving. ∎

Obviously the convergence preserving property cannot be used to give a direct proof of the continuity of a function at a point, since it requires testing *all* sequences. However it can be used in many cases to show that a given function is discontinuous at a point ζ. This requires exhibiting a sequence $\{x_n\}$ which converges to ζ but whose images $f(x_n)$ do not converge to $f(\zeta)$.

10 Example. The real-valued function

$$f(x,y) = \begin{cases} \dfrac{x^2 - y^2}{x^2 + y^2} & \text{whenever } x^2 + y^2 \neq 0 \\ 0 & \text{when } (x,y) = (0,0) \end{cases}$$

is discontinuous at (0,0): if μ is any real number, then the sequence $(1/n, \mu/n)$ converges to (0,0) (in the Euclidean metric) and

$$\lim f(1/n, \mu/n) = \frac{1 - \mu^2}{1 + \mu^2}.$$

For the remainder of this section we investigate the properties of continuous functions defined on compact sets.

11 Theorem. Let $(X,d) \xrightarrow{f} (\hat{X},\hat{d})$ be continuous. If X is compact, then $f(X)$ is a compact subset of $\hat{X}$.

Proof: Let $\{\hat{G}_\alpha\}$ be a covering of $f(X)$ by sets that are open in $\hat{X}$. The continuity of f implies that each $f^{-1}(\hat{G}_\alpha) = G_\alpha$ is open in X. Moreover,

$$f^{-1}(\bigcup \hat{G}_\alpha) = f^{-1}(\hat{X}) = X,$$

proving that $\{G_\alpha\}$ is an open covering of X. The compactness of X yields a finite subcovering, $G_1, G_2, \ldots, G_n$. It is easily seen that the sets $\hat{G}_i = f(G_i)$, $i = 1,2,\ldots,n$, cover $f(X)$. ∎

By taking X to be the space of real numbers, Theorem 11 and Theorem 1.3.27 (Heine-Borel), yield:

12 Corollary. A continuous mapping of a compact space into the real numbers is bounded.

Although any real-valued, bounded function defined on a metric space X has both a least upper bound λ and a greatest lower bound μ, it is not in general true that the function "attains" these values. That is, there may not be points ξ and ζ at which the function takes on the values λ and μ respectively. This is illustrated by the following examples:

13 Examples. In all cases $X \subset R$ and $d(x,y) = |x - y|$.

(a) Let $X = (0,1)$, $f(x) = x$, $\lambda = \sup_{0<x<1} f(x) = 1$, and $\mu = \inf_{0<x<1} f(x) = 0$. For all x in X, $0 < f(x) < 1$. Thus f attains neither the value of its least upper bound nor of its greatest lower bound.

(b) Let $X = R$ and $f(x) = 1/(1 + x^2)$. Here, $\lambda = 1$ and $\mu = 0$. In this case $f(0) = 1$, but for all values of x, $f(x) > 0$.

(c) $X = [0,1]$, and

$$f(\mathrm{x}) = \begin{cases} 1/2^n & \text{whenever } x = 1/2^n \\ 1 & \text{whenever } x \in X - \{1, 1/2, 1/2^2, \ldots, 1/2^n, \ldots\}. \end{cases}$$

$f(x)$ attains the value of its least upper bound $\lambda = 1$, at the point $x = 1 = 1/2^0$ and at all points $x \neq 1/2^n$. It does not however, take on the value of its greatest lower bound $\mu = 0$ at any point of X.

The functions defined in parts (a) and (b) are continuous, and it is clear that their failure to attain both the values λ and μ is due to the nature of the sets on which they are defined. In (a) this occurs because X is not closed, and it is easily seen that on the closure $[0,1]$ of X, $f(x)$ does take on both the values $\lambda = 1$ and $\mu = 0$. In (b), it is because X is unbounded. This leads to the (correct) conjecture that if X is a closed, bounded subset of R, then a continuous real valued function defined on X attains both a maximum and minimum value. In a general metric space, it is not enough to assume that X is closed and bounded (Exercise 18). Here it must be assumed that X is compact. For sets of real numbers, compactness is the same as being both closed and bounded.

14 Theorem. Let f denote a continuous mapping of (X,d) into R. If X is compact, then f attains the values of its least upper bound and greatest lower bounds at points in X.

Proof: From Corollary 12, the least upper bound λ exists (is finite). If $f(x)$ does not attain the value λ at any point of X, then the image $f(X)$ is not a closed subset of R, since λ is a point of accumulation of $f(X)$ which lies in its complement. Hence $f(X)$ is not compact (Theorem 1.3.27), which contradicts Theorem 11. A similar argument holds for the greatest lower bound. ∎

Although there exist continuous, bounded functions defined on non-compact spaces (the constant functions, for example), it is possible to characterize compact sets in the following way:

15 Theorem. A metric space X is compact if and only if every continuous mapping of X into the real numbers is bounded.

The proof is left to the reader.

16 Theorem. Let f be a continuous mapping of a metric space (X,d) into a metric space $(\hat{X},\hat{d})$. Then f is uniformly continuous on every compact subset of X.

Proof: Let $\epsilon > 0$ be given. We seek a positive number $\hat{\delta}$ such that for x,y in a compact subset C of X,

$$[2] \qquad d(x,y) < \hat{\delta} \quad \text{implies} \quad \hat{d}(f(x), f(y)) < \epsilon.$$

The continuity of f implies that to each $x \in C$ there is a $\delta_x > 0$ and

$$[3] \qquad f(S_{\delta_x}(x)) \subset \hat{S}_{\epsilon/2}(f(x)).$$

The collection of neighborhoods $\{S_{\delta_x}(x)\}$ is an open covering of the compact set C and therefore has a Lebesgue number δ (See the proof of Theorem 3.1.8). This means that if $0 < \hat{\delta} < \delta$, then whenever $x \in C$, there is a neighborhood $S_{\delta_z}(z)$ which contains $S_{\hat{\delta}}(x)$. We shall show that for any such $\hat{\delta}$, [2] holds:

If $d(x,y) < \hat{\delta}$, then $y \in S_{\hat{\delta}}(x) \subset S_{\delta_z}(z)$. From [3] it follows that $\hat{d}(f(x),f(z))$ and $\hat{d}(f(z),f(y))$ do not exceed $\epsilon/2$. The triangle inequality therefore yields

$$\hat{d}(f(x), f(y)) < \epsilon/2 + \epsilon/2 = \epsilon,$$

which is equivalent to [2]. ▌

Again a "converse" exists:

17 Theorem. X is compact if and only if every continuous real valued function on X is uniformly continuous.

See Exercise 21.

Exercises

1.
$$f(x) = \begin{cases} \sin 1/x & \text{if } x \neq 0 \\ 1 & \text{if } x = 0 \end{cases}; \qquad g(x) = \begin{cases} x \sin 1/x & \text{if } x \neq 0 \\ 0 & \text{if } x = 0 \end{cases}$$

Show, using Proposition 9, that $f(x)$ is discontinuous at $x = 0$. Show that $g(x)$ is continuous at $x = 0$.

2. State where, if anywhere,

$$f(x) = \begin{cases} x & \text{if } x \text{ is irrational} \\ 1 - x & \text{if } x \text{ is rational} \end{cases}$$

is a continuous function. Justify your answer.

***3.** (a) Let X, Y, and Z be metric spaces, and let $f: X \to Y$ and $g: Y \to Z$ be continuous mappings. Show that the mapping $f \circ g = h: X \to Z$, which is defined by

$$(f \circ g)(x) = g(f(x)),$$

is continuous.

(b) Assume that it has already been proved that sums, differences, products, and quotients (where the divisor does not vanish) of continuous functions from X into R are also continuous. (X may be any metric space.) Show that if f and g are continuous on X, then $|f|$, f^+, f^-, $\max(f,g)$, $\min(f,g)$ are continuous. The positive part of f, f^+, is equal to f whenever $f \geq 0$, and vanishes otherwise. The function $f^- = -f$ whenever $f \leq 0$ and vanishes otherwise.

4. (a) Do Exercise 1.3.16.
(b) Discuss also the continuity of

$$f(x) = \begin{cases} x & \text{when } x \text{ is irrational} \\ n \sin 1/m & \text{when } x = n/m \text{ in lowest terms, } m > 0. \end{cases}$$

5. A mapping $f: X \to \hat{X}$ is said to be *open* if

$$G \text{ is an open set in } X \qquad \text{implies} \qquad f(G) \text{ is open in } \hat{X}.$$

Show that if $f: R \to R$ is a continuous open mapping, then f is a *monotone* function; that is, either $x < y$ implies $f(x) \leq f(y)$, or $x < y$ implies $f(x) \geq f(y)$.

***6.** Let f be a mapping of R into itself. The number λ is said to be the *left-hand limit* of $f(x)$ at ζ, written $\lambda = \lim_{x \to -\zeta} f(x)$, if for every $\epsilon > 0$ there is a $\delta > 0$, such that $0 < \zeta - x < \delta$ implies that $|\lambda - f(x)| < \epsilon$. A similar definition may be given for a right-hand limit ρ. We say that $f(x)$ has a *limit* L at ζ, written $\lim_{x \to \zeta} f(x) = L$, if $\lambda = \rho = L$. Show that $f(x)$ is continuous at ζ if and only if the limit exists and is equal to $f(\zeta)$.

***7.** Let $f(x)$ be a real-valued function defined on an interval (a,b). Show that if f has right- and left-hand limits at every point, then the set of discontinuities of f is countable.

8. If

$$f(x,y) = \frac{\sin(x^2 + y^2)}{x^2 + y^2}$$

whenever $(x,y) \neq (0,0)$, how must $f(0,0)$ be defined so that $f(x,y)$ is continuous at that point?

9. Show that

$$f(x,y) = \begin{cases} (\sin xy)/x & \text{when } x \neq 0 \\ y & \text{when } x = 0 \end{cases}$$

is everywhere continuous.

10. Show that

$$f(x,y) = \begin{cases} xy/(x^2 + y^2) & \text{when } (x,y) \neq (0,0) \\ 0 & \text{when } (x,y) = (0,0) \end{cases}$$

is discontinuous at (0,0). Show also that the function of one variable $f(x,0)$ [as well as $f(0,y)$] is continuous on R.

11. Let $\xi \in l_2$ (Exercise 3.1.11). Then the infinite sum

$$\Sigma x_i \xi_i$$

defines a mapping l_2 into the real numbers. Here the x_i and ξ_i are terms of the sequences x and ξ respectively. Show, using Hölder's inequality (Theorem 2.1.5), that this is a continuous function of x.

***12.** Prove Corollaries 6 and 7.

13. Show that $f(x) = 1/x$, $1 \leq x < \infty$, is uniformly continuous. This illustrates the fact that a continuous function on a noncompact set *may be* uniformly continuous.

*14. Let S be a bounded subset of R. Show that if $f: S \to R$ is uniformly continuous, then it is bounded. Show, by example, that the boundedness of S is necessary.

*15. Prove that if X is a discrete metric space, then every mapping of X into a second space is continuous. If on the other hand, $f(X)$ is a discrete space, and the mapping is injective, when will f be continuous?

*16. Prove that compactness is a topological property. (See Sec. 3 of Chapter 2.)

17. (a) Prove Theorem 14 by using the Heine-Borel theorem directly. [*Hint:* If λ is not attained, then $G_n = [f < \lambda - 1/n]$, $n = 1, 2, \ldots$, is an open covering of X which has no finite subcover.]

(b) Prove Theorem 14 by using Bolzano-Weierstrass property directly. [*Hint:* Assume that for all x, $f(x) < \lambda$. Select a distinct sequence $\{x_n\}$ in X such that $f(x_n) > \lambda - 1/n$. Show that this leads to a contradiction.]

18. We have seen that the closed ball in l_2 is not compact (Exercise 3.1.11). Construct a real valued continuous function on this closed bounded set which is bounded but does not attain the value of its least upper bound.

19. Show that if f is a continuous injective mapping of a compact space into a second space, then the inverse mapping f^{-1}, defined on $f(X)$ is continuous. (*Hint:* Show first, using Proposition 3 and taking complements, that a mapping is continuous if and only if the inverse image of a close set is closed. Then use Theorem 11.)

20. Show that a metric space X is compact if and only if every real valued continuous function defined on X is bounded. (Try it first for subsets of R. Use the metric to obtain the desired function.)

21. Show that a metric space is compact if and only if every real valued continuous function defined on this space is uniformly continuous.

22. (Urysohn's lemma). Let S and T be nonempty, disjoint, closed subsets of a metric space (X,d). Show that if α and β are any pair of real numbers, there is a real-valued, continuous function $f(x)$ defined on X which has the following properties: If $x \in S$ then $f(x) = \alpha$ and if $x \in T$ then $f(x) = \beta$. Moreover $f(x)$ can be chosen so that for all $x \in X$, $|f(x)| \leq \max[|\alpha|, |\beta|]$. (*Hint:* Show that

$$g(x) = \frac{d(x,S)}{d(x,S) + d(x,T)}$$

is such a function for the case $\alpha = 0$ and $\beta = 1$. Use $g(x)$ to obtain the more general function $f(x)$.)

*Exercises which request the proof of a statement made in the text, as well as those referred to in the text, are starred.

3 Connected Spaces; The Intermediate Value Theorem

Theorem 3.2.14 asserts that if $f \in C[0,1]$, then there are points of $[0,1]$ at which $f(x)$ attains the values of its least upper bound λ and its greatest lower bound μ. This follows from the continuity of $f(x)$ and the compactness of $[0,1]$. In this section we shall show that $f(x)$ attains (at least once) every value between μ and λ. The following heuristic argument makes this conclusion plausible, and suggests the property which is responsible for it.

Let $f(x)$, λ, and μ be as in the preceding paragraph. Suppose that ξ and ζ are points of $[0,1]$ at which $f(x)$ attains the values μ and λ respectively. Imagine a point moving along this curve from the point (ξ,μ) (the lowest point) to the point (ζ,λ) (the highest point). Since the curve has no breaks, (continuity), it must cross the line $y = A$ at least once, if $\mu < A < \lambda$. (See Figure 12.) However, if the set $[0,1]$ is replaced by the

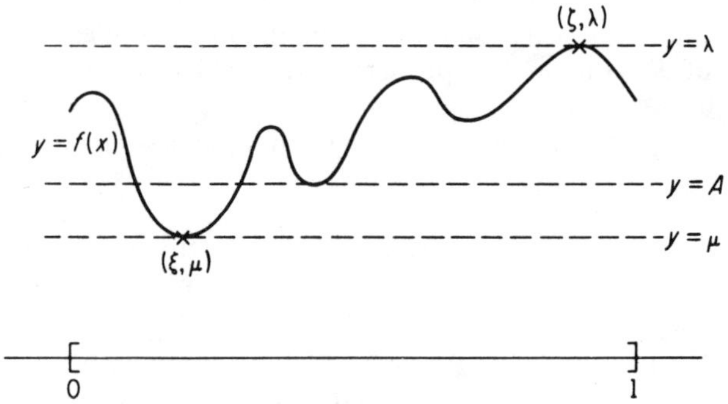

FIGURE 12

set $X = [0,1] \cup [2,3]$, then the function

$$g(x) = \begin{cases} 1 & \text{whenever } 0 \leq x \leq 1 \\ 3 & \text{whenever } 2 \leq x \leq 3 \end{cases}$$

is continuous on X but does not attain the value 2, for example. Thus the continuity of a real function is not enough to insure that it possesses this "intermediate value" property. (Moreover, these two examples show that the compactness of X has no bearing on the problem.) Clearly, the crucial factor is that in the first case the domain of the function is "of one piece" (an interval). It is intuitively clear that its image under a continuous mapping cannot be composed of "disconnected" pieces. In the second case, the domain consists of a disjoint pair of intervals, and it should come as no surprise that the continuous image of such a domain *may* be composed of "disconnected" pieces.

In this section we shall give a general definition of connectedness. Although it may at first seem strange, the reader should bear in mind that this definition allows us to *characterize* a connected metric space X as one for which every function in $C(X)$, the family of real valued continuous functions defined on X, has the Intermediate Value Property. Indeed, this property of $C(X)$ may be used as a definition of a connected space. (See Theorem 7 and Exercises 3.2.20 and 3.2.21.)

One way to describe the disconnectedness of $X = [0,1] \cup [2,3]$ is to say that X is the union of two sets which may be "separated" by a pair of nonintersecting open sets, say $G_1 = (-1,5/4)$ and $G_2 = (7/4,4)$. On the other hand it is intuitively clear that if we were to write $[0,1]$ as a disjoint union $S \cup T$ of nonempty sets, then it would not be possible to "separate" S and T by such a pair of nonintersecting open sets.

Returning to the disconnected set X, the reader will observe also that the sets $G_1 \cap X = [0,1]$ and $G_2 \cap X = [2,3]$ are open, as well as closed, sets in the metric space X. On the other hand $[0,1]$, considered as a metric subspace of R, contains no proper subsets which are both open and closed in $[0,1]$.

These properties are used to define connectedness in two equivalent ways:

1 Definition. A metric space (X,d) is said to be *disconnected* (or not connected) if there are a pair of nonempty, disjoint, open sets S and T of X such that $X = S \cup T$.

A metric space for which no such pair exists is said to be *connected.*

The set $X = [0,1] \cup [2,3]$ is disconnected; S and T may be taken to be the open subsets $[0,1]$ and $[2,3]$ of X.

Like compactness, connectedness may be described using sets that are open in some larger space. By this we mean that if Y is a subset of X, then the following two statements are equivalent:

(i) (Y,d) is a disconnected metric space. Here Definition 1 is used. The sets S and T are open in Y.

(ii) Y is a disconnected subset of (X,d). This means that there are nonempty, open subsets S and T of X such that

$$Y \subset S \cup T,\ Y \cap S \neq \emptyset,\ Y \cap T \neq \emptyset,$$

and

$$Y \cap S \cap T = \emptyset.$$

Verification of the equivalence of (i) and (ii) is left to the reader.

A second characterization of connectedness is suggested by the examples:

2 Proposition. X is connected if and only if no subset of X, other than X itself and $\emptyset$, is both open and closed in X.

Proof: If $X = S \cup T$, where S and T are disjoint, nonempty, open subsets of X, then their complements, $T = X - S \neq X$, and $S = X - T$, are closed nonempty subsets of X.

Conversely, if X is connected and S is a proper subset of X that is both open and closed in X, then $X - S = T$ is a nonempty open subset of X, $S \cap T = \emptyset$, and $X = S \cup T$, again a contradiction. ∎

We show next that the nonempty connected subsets of R are the intervals and the sets which contain a single point. First we state a lemma which characterizes intervals. (See Definition 1.3.1.)

3 Lemma. A nonempty subset X of R which contains at least two points, is an interval if and only if, whenever $c,d \in X$ and $c < d$, then,

$$c < x < d \qquad \text{implies} \qquad x \in X.$$

The proof is left to the reader.

4 Theorem. A nonempty subset X of R is connected if and only if X is an interval or if $X = \{a\}$, the set containing the single element a.

Proof: It is left to the reader to show that $\{a\}$ is a connected set.

Let us assume that X is an interval. If X is disconnected, then there are disjoint, nonempty, open subsets of X, S and T, whose union is equal to X. If $c \in S$ and $d \in T$, then since X is an interval, $c < d$ implies $[c,d] \subset X$. Since S and T are open in X, there is a positive number ϵ such that $[c,c + \epsilon)$ and $(d - \epsilon,d]$ are contained in S and T respectively. Set

$$\delta = \sup \{\epsilon : [c,c + \epsilon) \subset S\}.$$

Clearly, $c < c + \delta < d$ (see Figure 13). The point $c + \delta = \gamma$ is in the interval X and must therefore be in either S or T. If $\gamma \in S$, then a

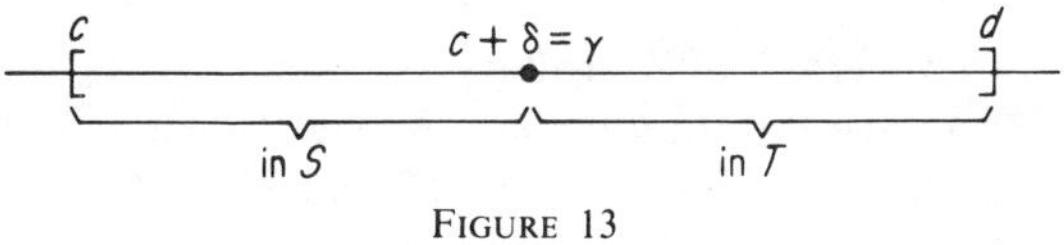

FIGURE 13

neighborhood $(\gamma - \epsilon, \gamma + \epsilon)$ must lie in S. (S is open.) This yields

$$[c,c + \delta) = [c,\gamma) \subset [c,\gamma + \epsilon) = [c,c + \delta + \epsilon] \subset S,$$

which contradicts the definition of δ.

On the other hand if $\gamma \in T$, then there is a positive number ρ such that $(\gamma - \rho,\gamma + \rho) \subset T$. This implies that if $[c,c + \epsilon) \subset S$, then $\epsilon < \delta - \rho$, proving that δ is not a *least* upper bound of those numbers ϵ for which $[c,c + \epsilon) \subset S$. Thus γ is in neither S nor T, and hence not in the interval X, which contradicts Lemma 3.

Now let us assume that X is a nonempty, connected subset of R which contains at least two points, $c < d$. If X is not an interval, then there must be a point ζ which lies between c and d which is not in X. Then the sets

$$S = X \cap (-\infty,\zeta) \qquad \text{and} \qquad T = X \cap (\zeta,\infty)$$

are open in X, nonempty, disjoint, and their union covers X, proving that X is disconnected. ▮

Like compactness, connectedness is preserved under continuous mappings:

5 Theorem. Let f be a continuous mapping of a connected space X into an arbitrary space $\hat{X}$. Then $f(X)$ is a connected subset of $\hat{X}$.

Proof: If $f(X)$ is disconnected, there are disjoint, nonempty, open sets S and T in $f(X)$ whose union $S \cup T = f(X)$. The continuity of f implies that the sets $f^{-1}(S)$ and $f^{-1}(T)$ are open in X. Clearly, they are nonempty (otherwise S or T would be empty) and disjoint (otherwise $S \cap T \neq \emptyset$) and the union is equal to X, thus implying that X is disconnected. ▮

6 Corollary. (Intermediate Value Theorem) If f is a continuous mapping of a connected space X into R, and if $f(x) = A < B = f(y)$ for some pair of points $x,y \in X$, then for any number C lying between A and B, there is a point $\zeta \in X$ such that $f(\zeta) = C$.

Proof: The connectedness of X and the continuity of f imply that $f(X)$ is connected. From Theorem 4 it follows that $f(X)$ is either a point or an interval. The hypothesis $f(x) < f(y)$ for some pair of points in X implies the latter. Lemma 3 implies that $C \in [A,B]$ which means that there is a point ζ of X for which $f(\zeta) = C$. ▮

There is a "converse" of Corollary 6 which asserts that if *all* the functions in $C(X)$ have the Intermediate Value Property, then X is connected. (Compare this with Exercises 3.2.20 and 3.2.21.)

7 Theorem. A metric space X is connected if and only if every member of the family $C(X)$ of real valued functions on X has the Intermediate Value Property.

Proof: The "only if" part is simply Corollary 6.

We show now that if X is not connected, then there is a function in $C(X)$ which does not possess the Intermediate Value Property. Set $X = S \cup T$, where S and T are disjoint, nonempty, open sets. The function

$$f(x) = \begin{cases} 0 & \text{whenever } x \in S \\ 1 & \text{whenever } x \in T \end{cases}$$

does not take on the intermediate value $1/2$ at any point of X. However, $f(x)$ is continuous. This follows from the fact that $f(X)$ consists of but two points, 0 and 1. Therefore a complete list all the open subsets of $f(X)$ is given by ϕ, $\{0,1\}$, $\{0\}$, $\{1\}$. The inverse images of these sets are ϕ, X, S, T respectively, all of which are open, thus proving that there is a continuous function on the disconnected set X which does not take on all intermediate values. ▮

8 Definition. Let x_0 and x_1 be points of a metric space X. A *path* from x_0 to x_1, or an *arc* beginning at x_0 and terminating at x_1 is a continuous mapping $f: [0,1] \to X$, with $f(0) = x_0$ and $f(1) = x_1$.

9 Definition. A metric space is said to be *arcwise connected* if for any pair of points x_0 and x_1 in X, there is a path that begins at x_0 and terminates at x_1.

10 Theorem. If X is arcwise connected then it is connected.

Proof: If X were not connected it would contain a proper subspace S which was both open and closed in X. Let x_0 and x_1 be points in S and $X - S$ respectively. Since X is arcwise connected, there is an arc in X which begins at x_0 and terminates at x_1. Since the path f is a continuous mapping of the connected set $[0,1]$ into X, the image $f([0,1]) = C$ is a connected subspace of X. But then $S \cap C$ and $(X - S) \cap C$ are nonempty sets that are open in C and whose union covers C, proving that C is disconnected. ▮

11 Example. Consider the following subsets of R^2:

$$A = \{(x,y): x = 0 \quad \text{and} \quad -1 \le y \le 1\},$$
$$B = \{(x,y): 0 < x \le 1 \quad \text{and} \quad y = \sin 1/x\}.$$

The set $X = A \cup B$ is not arcwise connected since it is impossible to find an arc f that begins at a point in A and terminates at a point in B, and for which $f([0,1]) \subset X$. Although this is intuitively clear, it requires a rigorous analytic proof. This is left to the reader in the exercises. It may also be found in Mendelson [1].

The space X is, however, connected. If it were not, then X could be written as the disjoint union $S \cup T$, where S and T are disjoint, nonempty, open subsets of X. Since both A and B are connected, it follows that there can not be points of A (or B) in both sets S and T. (Why?) Thus $A \subset S$ or $A \subset T$, and similarly for B. If S and T are to be nonempty, then it must follow that $S = A$ and $T = B$, or vice versa. Although the set $B = X \cap \{(x,y): x > 0\}$ is open in X, the subset A is obviously not open in X. For if it were, there would be an open subset G of R^2 with $A = X \cap G$. Clearly, any such set would contain points of B. (See Figure 14.) Thus X is connected.

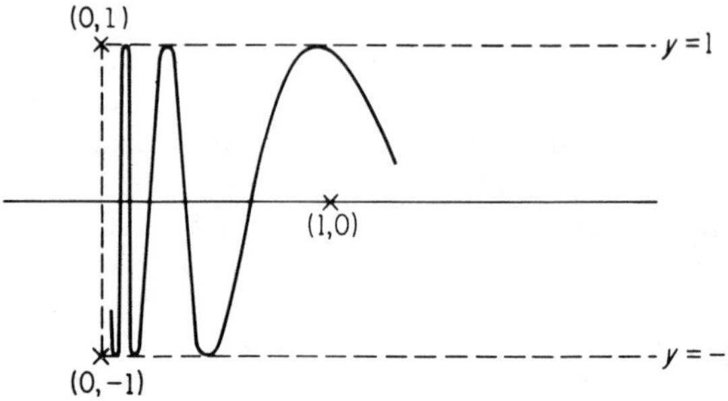

FIGURE 14

The reader is asked to prove in the exercises that an *open*, connected, subset of R^n is also arcwise connected.

Exercises

*1. Prove that the connectedness of a subspace Y of X may be defined using either sets which are open in Y or in X. [See (i) and (ii) that follow Definition 1.]

2. Prove that X is connected if and only if every pair of points x and y in X, is contained in a connected subset of X. (*Hint:* Look at the proof of Theorem 10.)

*3. Prove that Q, the set of rational numbers, is a disconnected subset of R.

*4. Prove that the only connected subsets of a discrete metric space are those which contain a single point.

5. Show that X is disconnected if and only if there are nonempty disjoint subsets A and B of X satisfying:

$$\text{(i) } X = A \cup B, \quad \text{(ii) } \bar{A} \cap B = \phi, \quad \text{and} \quad \text{(iii) } A \cap \bar{B} = \phi.$$

*6. Prove that if f is a real valued continuous function defined on the interval $[a,b]$, then $f(a) \cdot f(b) < 0$ implies that there is at least one point c in $[a,b]$ at which $f(x)$ vanishes. (*Hint:* Use the Intermediate Value Theorem.)

*7. Let f be a continuous mapping of $[0,1]$ into itself. Prove that there is a point ζ in $[0,1]$ at which $f(\zeta) = \zeta$. (Geometrically, this says that the graphs of $y = f(x)$ and $y = x$ must intersect.)

8. Let S be the set of points on the unit circle in R^2. That is, $(x,y) \in S$ if and only if $x^2 + y^2 = 1$. To each point on S there may be assigned a real number $0 \leq \psi < 2\pi$, by requiring that $x = \cos \psi$ and $y = \sin \psi$. The mapping P: $[0,2\pi) \to S$ is continuous. Denoting a point on S parametrically by $P(\psi)$, two points are said to be *antipodal* or *diametrically opposed* if they may be represented by $P(\psi)$ and $P(\psi + \pi)$, for some angle ψ. Using the Intermediate Value Theorem it can be shown that every continuous mapping of S into R sends some pair of diametrically opposed points into the same real number. The parametrization of S allows us to think of this as a mapping of $[0,2\pi)$ into R. The reader is asked to prove that if this mapping is continuous, then there is a point ψ in $[0,2\pi)$ for which $f(\psi) = f(\psi + \pi)$. This may be done by showing that the function $g(x) = f(x) - f(x + \pi)$ vanishes at some point of the interval.

9. A similar type of problem is what is frequently called the "pancake" or "ham sandwich" problem. In the first problem, two bounded regions in R^2 are given. It is assumed that it is possible to assign to each of them an area, say by using a multiple Riemann integral. Is there a line which simultaneously bisects both regions? The answer is yes. You may assume here that if a line is rotating, the area on one side is a continuous function of the angle between this line and some fixed line. For the second problem, one region is given. Find a pair of perpendicular lines which cuts this region into four equal (area) parts. (See Mendelson [1]; Chinn and Steenrod [1].)

10. Let S be a connected set. Then if $S \subset T \subset \bar{S}$, show that T is also connected.

11. Prove that if S is connected and contains at least two points, then S is uncountable. (*Hint:* Use the uncountability of R.)

12. Let X be the union of the sets A and B:

$$A = \{(x,y) : 0 \leq x \leq 1 \quad \text{and} \quad y = x/n, \quad n = 1, 2, \ldots\}$$
$$B = \{(x,y) : y = 0 \quad \text{and} \quad 1/2 \leq x \leq 1\}.$$

Prove that X is connected but not arcwise connected.

13. Let $A = \{(x,y) : x,y \in Q\}$. Show that $R^2 - A$ is connected. Show that this remains true if A is any countable set.

14. Give an example of a compact subset C of R^2 for which $R^2 - C$ is disconnected. Generalize this to R^n, $n > 1$. What about R?

15. Show that if S and T are connected subsets of X and if $S \cap T \neq \phi$, then $S \cup T$ is connected.

State and prove the more general result for a family $\{S_\alpha\}$ of connected subsets of a space X.

16. Show by example that each of the following statements is false:
(a) S, T are connected imply $S \cup T$ is connected.
(b) S, T are connected imply $S \cap T$ is connected.

17. State and prove the Theorem for arcwise connected sets which corresponds to Exercise 15.

18. Prove that every real linear vector space is connected.

A metric space X is said to be *locally connected* if every point of X has a connected neighborhood. This means that if $\zeta \in X$, then for some $\epsilon > 0$, the set $S_\epsilon(\zeta)$ is connected.

19. Each of the following is to be considered as a subspace of R, which is equipped with the usual metric. Which are locally connected?
(a) Z, the set of integers;
(b) $\{1, 1/2, 1/3, \ldots, 1/n, \ldots\} \cup \{0\}$;
(c) $\{1, 1/2, 1/3, \ldots, 1/n\}$;
(d) $\bigcup_{-\infty}^{\infty} I_n$, where $I_n = [n - 1/4, n + 1/4]$.

20. Show that a discrete metric space is locally connected.

21. Prove that an open subset of a locally connected space is locally connected.

22. Use Example 11 to prove that connectedness does not imply local connectedness. Show also that the converse is false.

23. A set is totally disconnected if its only connected subsets are those containing a single point. Give examples of infinite totally disconnected sets.

24. Prove that an open, connected subset of R^n is arcwise connected.

*Exercises which request the proof of a statement in the text, as well as those referred to in the text, are starred.

4 The Space $C(X,\hat{X})$; Uniform Convergence; Equicontinuity and Arzela's (Ascoli's) Theorem

If (X,d) and $(\hat{X},\hat{d})$ are a pair of metric spaces, then the set of continuous mappings of X into $\hat{X}$ is denoted by $C(X,\hat{X})$. For many applications, $\hat{X} = R$, and here as earlier, we shall write $C(X)$ for $C(X,R)$. Using as a model the Banach space $C[a,b]$ with norm defined by $\|f - g\| = \max_{a \le x \le b} |f(x) - g(x)|$, we shall show that if X is compact, as is $[a,b]$, then $C(X,\hat{X})$ can be made into a metric space; if, also, $\hat{X}$ is a normed linear space, as is R, then the function space can be made into a normed linear space.

The distance function d^* on the space $C(X,\hat{X})$ may be defined in very much the same way that the distance function on $C[a,b]$ is defined: If $f,g \in C(X,\hat{X})$, then

$$d^*(f,g) = \sup_{x \in X} \hat{d}(f(x),g(x)).$$

The existence of the supremum follows from the compactness of the spaces $f(X)$ and $g(X)$, which are continuous images of the compact space X. It follows that their union is compact, and that the supremum is bounded from above by diam $(f(X) \cup g(X))$. It is not difficult to show, using the continuity of the distance function, that the value of the supremum is attained on X; thus "sup" may be replaced by "max." However, since this result is not needed throughout the remainder of this section, its verification is left to the reader. The verification of the metric space axioms is also left to the reader.

Let us suppose now that X is compact and $\hat{X}$ is a normed vector space. Then the vector space operations may be defined in $C(X,\hat{X})$ in the usual way: If $f,g \in C(X,\hat{X})$, then for all $x \in X$, $f(x)$, $g(x)$, $f(x) + g(x)$, and $cf(x)$, for any real number c, are in $\hat{X}$. Thus we may define addition and multiplication by a real number in $C(X,\hat{X})$ by

$$(f + g)(x) = f(x) + g(x), \quad (cf)(x) = cf(x).$$

The norm must agree with the metric already defined; thus

$$\|f\|^* = \sup_{x \in X} \|f(x)\| = \sup_{x \in X} \hat{d}(f(x),\hat{0}) = d^*(f,\bar{0}),$$

where $\hat{0}$ is the additive identity in $\hat{X}$ and $\bar{0}$ denotes the function defined on X which takes on the constant value $\hat{0}$, (in $\hat{X}$). The details are left to the reader.

1 Definition. A sequence of mappings $\{f_n\}$ in $C(X,\hat{X})$ is said to *converge pointwise* on X to a mapping $f: X \to \hat{X}$, (f may not itself be in $C(X,\hat{X})$.), if for every positive number ϵ, and x in X, there is an integer

ν which depends upon both ϵ and x, such that

$$n > \nu \qquad \text{implies} \qquad \hat{d}(f_n(x), f(x)) < \epsilon.$$

If ν does not depend upon x, that is, if to each $\epsilon > 0$, there is a single integer ν for which

$$[1] \qquad n > \nu \qquad \text{implies} \qquad \hat{d}(f_n(x), f(x)) < \epsilon \quad \text{for all } x \in X,$$

then we say that f_n *converges uniformly* on X to f.

We observe that if $f \in C(X,\hat{X})$, then [1] is equivalent to

$$n > \nu \qquad \text{implies} \qquad d^*(f_n, f) < \epsilon,$$

which says that the sequence converges to f in the metric d^*. Theorem 2 asserts that the uniform limit of a sequence of functions in $C(X,\hat{X})$ is itself in that space, thus proving that *uniform convergence of functions in* $C(X,\hat{X})$ *is the same as convergence in the metric* (or *norm*) *of that space.*

2 Theorem. Let X be a compact metric space and let $\hat{X}$ be any metric space. If $\{f_n\}$ is a sequence in $C(X,\hat{X})$ which converges uniformly on X, then the limit function f is in $C(X,\hat{X})$.

Proof: Let ϵ be a positive number and let x be a point of X. We seek a positive number δ that satisfies:

$$[2] \qquad d(x,y) < \delta \qquad \text{implies} \qquad \hat{d}(f(x), f(y)) < \epsilon.$$

The triangle inequality yields

$$[3] \quad \hat{d}(f(x), f(y)) \le \hat{d}(f(x), f_n(x)) + \hat{d}(f_n(x), f_n(y)) + \hat{d}(f_n(y), f(y)).$$

The uniform convergence of $\{f_n\}$ to f implies that for sufficiently large n, the first and third terms of [3] will be less than $\epsilon/3$. Choose some integer n for which this is true. Since f_n is continuous, there is a $\delta > 0$ which depends on x, such that

$$[4] \qquad d(x,y) < \delta \qquad \text{implies} \qquad \hat{d}(f_n(x), f_n(y)) < \epsilon/3.$$

Combining [3] and [4] and the uniform convergence, we obtain [2]. ∎

To emphasize the importance of uniform convergence, the following examples of nonuniform convergence are included:

3 Examples. (a) Let $X = [0,1]$, $\hat{X} = R$. Then the sequence of continuous functions

$$f_n(x) = nxe^{-nx}, \qquad 0 \le x \le 1$$

converges pointwise to 0. However, for all n, $f_n(1/n) = e^{-1}$, proving that if $\epsilon < e^{-1}$ is given, it is not possible to find a single integer ν for which

$$n > \nu \qquad \text{implies} \qquad f_n(x) < \epsilon \qquad \text{for all } 0 \le x \le 1.$$

Here the limit function is $f(x) \equiv 0$, which is of course continuous; hence it follows that a nonuniform limit of continuous functions may be continuous. (See Figure 15.)

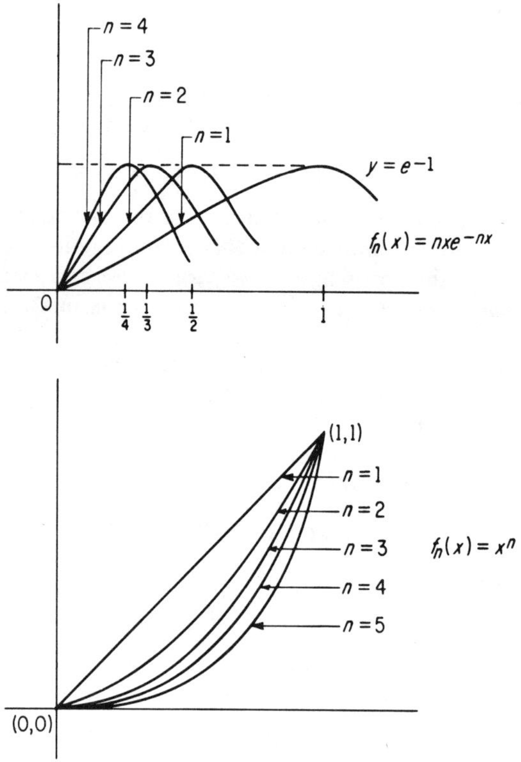

FIGURE 15

(b) On the other hand, nonuniform convergence may also lead to discontinuous functions. A very simple example of this is

$$f_n(x) = x^n, \qquad 0 \leq x \leq 1.$$

Here the limit function

$$f(x) = \begin{cases} 1 & \text{if } x = 1 \\ 0 & \text{if } 0 \leq x < 1, \end{cases}$$

is discontinuous at $x = 1$.

Thus pointwise convergence of continuous functions yields no positive information about the continuity of the limit function.

Uniform convergence also plays an important role in determining the validity of interchanging two limits. For example, we ask under what conditions

$$\lim \frac{d}{dx} f_n(x) = \frac{d}{dx} \lim f_n(x), \quad \text{or}$$

$$\lim \int_a^b f_n(x)\,dx = \int_a^b \lim f_n(x)\,dx$$

(See Exercises 4 and 5.)

It is not always easy to determine whether or not a sequence of functions converges uniformly. Example 3(a) reveals that even if X is compact and the limit function is continuous, it cannot be concluded that the convergence is uniform. Dini's theorem gives sufficient conditions for convergence to be uniform:

4 Theorem (Dini I). If X is a compact metric space, and if $\{f_n\}$ is a sequence of non-negative functions in $C(X)$ which converge monotonically to zero for all $x \in X$, then $\{f_n\}$ converges uniformly to zero.

Proof: Let $\epsilon > 0$ be given. The continuity of the f_n imply that for each positive integer n, $G_n = [f_n < \epsilon]$ is an open subset of X (Proposition 3.2.5 and Corollary 3.2.6). From $f_n(x) \downarrow 0$ at each point of X, it follows that $X = \bigcup_1^\infty G_n$. The compactness of X yields a finite subcover $G_{i_1}, G_{i_2}, \ldots, G_{i_\nu}$. Assuming that $i_1 < i_2 < \cdots < i_\nu$, the monotonicity of the f_n implies that for $k = 1,2,\ldots,\nu - 1$, $G_{i_k} \subset G_{i_{k+1}}$, proving that $X = G_{i_\nu} = [f_{i_\nu} < \epsilon]$. Thus for all $n > \nu$, and all $x \in X$, the monotonicity of the f_n yields $f_n(x) < \epsilon$. ∎

Later we shall need a form of Dini's theorem for step functions. However, the proof of Theorem 4 requires the use of the openness of the G_n, which for step functions is not generally the case. For example, if

$$\sigma(x) = \begin{cases} 0 & \text{if} \quad |x| \geq 1 \\ 1 & \text{if} \quad -1 < x < 0 \\ 4 & \text{if} \quad 0 < x < 1 \\ 2 & \text{if} \quad x = 0, \end{cases}$$

then the set $[\sigma < 3] = R - (0,1)$ is not open. But if $\sigma(x)$ were to be replaced by the equivalent step function $\sigma'(x)$, (Definition 2.2.6 and Propositions 2.2.7 and 2.2.8), where

$$\sigma'(x) = \begin{cases} \sigma(x) & \text{if } x \neq 0, -1, \text{ or } 1 \\ 4 & \text{if } x = 0 \\ 1 & \text{if } x = -1 \\ 4 & \text{if } x = 1 \end{cases}$$

then the sets $[\sigma' < A]$, for all real values of A, are open. This is true because the values of $\sigma'(x)$ at the common end points of two intervals over which $\sigma'(x)$ is constant, are taken to be the values of the height of the larger of the two adjacent steps. To carry through the proof for step functions, we must be sure that the σ_n' vanish outside a common compact set. This is assured by $\sigma_1'(x) \geq \sigma_n'(x)$ for all n and $x \in X$, and the fact that the step function $\sigma_1'(x)$ vanishes outside a closed bounded interval. Thus Dini's theorem holds for this special class of step functions. If, however, $\{\sigma_n\}$ is *any* sequence of step functions which converge monotonically (down) to zero on X, then the corresponding sequence of equivalent step functions $\{\sigma_n'\}$ are equal to the $\{\sigma_n\}$ except at the end points. If $\epsilon > 0$ is given, an integer N_1 is first chosen so that $n > N_1$ implies $\sigma_n'(x) < \epsilon$ for all x. There are at most finitely many points at which $\sigma_n(x) \geq \epsilon$. Therefore $N_2 \geq N_1$ may be selected so that $n > N_2$ implies $\sigma_n(x) < \epsilon$ everywhere. We state this as a theorem:

5 Theorem (Dini II). If $\{\sigma_n\}$ is a sequence of nonnegative step functions which converge monotonically to zero at each point x of R, then $\{\sigma_n\}$ converges uniformly to zero.

It is readily seen that the preceding argument holds equally well for a sequence of functions satisfying the following properties: (i) $\{f_n\}$ is a sequence of nonnegative functions which converges monotonically to zero at each point of a compact set X; (ii) for all real numbers A, the sets $[f_n < A]$ are open.

6 Definition. A real valued function defined on a metric space X is said to be *upper semicontinuous* if for every real number A, the set $[f < A]$ is an open subset of X.

This leads to a generalization of Dini's theorem:

7 Theorem (Dini III). If f_n is a sequence of nonnegative, real valued, upper semicontinuous functions defined on a metric space, which converge monotonically to zero, then $\{f_n\}$ converges uniformly.

In the exercises the reader will find additional examples of familiar theorems that can be extended to the class of upper semicontinuous functions.

For the remainder of this section we shall discuss conditions that insure the compactness of a subset of $C(X)$. Several examples will be given which illustrate the role played by these compact subsets of $C(X)$ in proving the existence of solutions to differential and integral equations. Some of the most interesting examples appear in complex analysis, among them a proof of the Riemann Mapping Theorem (Nehari [1]; See also Hille [1]).

We have proved that a subspace of a complete metric space is compact if and only if it is closed and totally bounded. Since total boundedness is difficult to verify in many applications, it will be replaced by the weaker condition of boundedness, and a compensating condition which is described in Definition 8.

If X is compact, then it is easily seen that the normed linear space $C(X)$ is complete. The proof is almost the same as the one given for $C[a,b]$.

8 Definition. Let X be compact. A family $\mathfrak{F}$ of functions in $C(X)$ is said to be *equicontinuous*, if to every positive number ϵ, there is a positive δ such that

$$d(x,y) < \delta \qquad \text{implies} \qquad |f(x) - f(y)| < \epsilon, \quad \text{for all } f \in \mathfrak{F}.$$

If f is any function in $C(X)$ whatsoever, there is a $\delta > 0$, such that $|f(x) - f(y)| < \epsilon$ whenever $|x - y| < \delta$. This follows from the uniform continuity of every continuous function on the compact set X. In this case δ depends on both ϵ and f. For an equicontinuous family, a single δ (depending only upon ϵ) works for all f in $\mathfrak{F}$.

9 Example. Let $\mathfrak{F}_M$ be the subfamily of $C[a,b]$ of continuously differentiable functions whose derivatives are uniformly bounded by M for all x in $[a,b]$. This family is equicontinuous, for if $\epsilon > 0$ is given, then

$$|f(x) - f(y)| = \left| \int_x^y \frac{df}{dt}\, dt \right| \le \int_x^y \left| \frac{df}{dt} \right| dt < M|x - y| < \epsilon,$$

provided that $|x - y| < \delta = \epsilon/M$. This δ is of course independent of the function f in $\mathfrak{F}_M$.

The equicontinuous family $\mathfrak{F}_M$ appears in the proof of the existence of a local solution of the initial value problem:

$$\frac{dy}{dx} = F(x,y), \quad y(x_0) = y_0.$$

(See Birkhoff-Rota [1].)

10 Example. Let

$$X = \{x = (x_1, x_2) \in R^2 \text{ such that } 0 \le x_1, x_2 \le 1\},$$

be the unit square in R^2. If $\rho(t,s)$ is a function in $C(X)$, and if A is a bounded subset of $C[0,1]$, then for $f \in A$,

$$[5] \qquad F(f) = \int_0^1 \rho(x,s)\, f(s)\, ds = g(x) \in C[0,1].$$

It is easily seen that $F(A)$ is a bounded subset of $C[0,1]$. For if M is a bound for all the functions in A, and if $\|\rho\| = \sup_{x \in X} |\rho(x)| = L$, then $\|g\| = \max_{0 \leq x \leq 1} |g(x)| < LM$. We show next that $F(A)$ is equicontinuous: If $g \in F(A)$, then there is a function f in A and $g = F(f)$. Then [5] yields

$$|g(x) - g(y)| = \int_0^1 [\rho(x,s) - \rho(y,s)] f(s)\, ds.$$

Since $\rho(t,s)$ is uniformly continuous on X, there is a $\delta > 0$ such that

$$|x - y| < \delta \qquad \text{implies} \qquad |\rho(x,s) - \rho(y,s)| < \epsilon/M.$$

It follows that $|g(x) - g(y)| < (\epsilon/M) M = \epsilon$ whenever $|x - y| < \delta$ and $g \in F(A)$.

The equicontinuous family $F(A)$ appears in the proof of the existence of a solution to the integral equation

$$f(x) = \int_0^1 \rho(x,s)\, f(s)\, ds;$$

$\rho(t,s)$ is a given function that is continuous in the unit square. (See Courant and Hilbert [1] or Taylor [2].)

Remark. An existence proof for the initial value problem of Example 9 will be given in Chapter 4, Sec. 2. There the completeness of the space X is used.

We are about to prove (Theorem 11) that the compact subsets of $C(X)$ are the bounded equicontinuous subsets. It was remarked earlier that for function spaces it is generally easier to verify equicontinuity than total boundedness. The reader may well ask why the compactness of families of functions in $C(X)$, such as those of Examples 9 and 10, is desirable. Theorems 3.1.5 and 3.1.8 assert the possibility of selecting a convergent sequence from any infinite subset of a compact space. The solutions of the differential and integral equations in the preceding examples are obtained as the limits of such sequences.

11 Theorem (Arzela-Ascoli). If X is a compact metric space, then a closed subset $\mathfrak{F} \subset C(X)$ is compact if and only if $\mathfrak{F}$ is bounded and equicontinuous.

Proof: If $\mathfrak{F}$ is compact then it is bounded. In fact, since it is totally bounded, there are finitely many functions $f_1, f_2, \ldots, f_n$ in $\mathfrak{F}$ corresponding to each $\epsilon > 0$, such that $\bigcup_1^n S^*_{\epsilon/3}(f_i) \supset \mathfrak{F}$. Each f_i is uniformly continuous on the compact set X. Hence there are positive numbers $\delta_1, \ldots, \delta_n$, such that

$$d(x,y) < \delta_i \qquad \text{implies} \qquad |f_i(x) - f_i(y)| < \epsilon/3, \qquad i = 1, \ldots, n.$$

We claim that $\delta = \min\,[\delta_1, \ldots, \delta_n]$ is a δ of equicontinuity corresponding to the given ϵ. Indeed, if $f \in \mathfrak{F}$ then for some $i = 1, \ldots, n$, $f \in S^*_{\epsilon/3}(f_i)$; that is $\|f - f_i\| = \max_{x \in X} |f(x) - f_i(x)| < \epsilon/3$. Thus if $d(x,y) < \delta$,

$$\begin{aligned} |f(x) - f(y)| &\leq |f(x) - f_i(x)| + |f_i(x) - f_i(y)| + |f_i(y) - f(y)| \\ &< \epsilon/3 + \epsilon/3 + \epsilon/3 = \epsilon, \end{aligned}$$

proving that $\mathfrak{F}$ is equicontinuous.

Suppose now that $\mathfrak{F}$ is closed, bounded and equicontinuous. Since $C(X)$ is complete, the closed subspace $\mathfrak{F}$ is complete. The compactness of $\mathfrak{F}$ is demonstrated by showing that every sequence $\{f_n\}$ in $\mathfrak{F}$ contains a Cauchy subsequence. The completeness of $\mathfrak{F}$ will assure the convergence of this sequence, thus proving that $\mathfrak{F}$ satisfies the Bolzano-Weierstrass property and is therefore compact.

It is not difficult to show that the compact set X contains a countable set $\{x_1, x_2, \ldots\}$ whose closure is X (Exercise 3). This means that if $x \in X$, and $\delta > 0$ is given, there is an x_i in this set such that $d(x,x_i) < \delta$.

Let $S = \{f_n\}$ be a sequence in $\mathfrak{F}$ and let M be a positive number that serves as a uniform bound for all the functions in $\mathfrak{F}$. The set of real numbers $\{f_n(x_1)\}$ is bounded from above and below by $-M$ and M respectively. Since $[-M,M]$ is a compact subset of R, it follows that the sequence $\{f_n(x_1)\}$ of real numbers contains a convergent subsequence $\{f_{1i}(x_1)\}$, $i = 1, 2, \ldots$. Repeating the argument, the sequence of functions $\{f_{1i}\}$ contains a subsequence $\{f_{2i}\}$ chosen so that the sequence $\{f_{2i}(x_2)\}$ of real numbers converges. Continuing in this manner, we select a sequence of subsequences:

$$\{f_{1i}\} \supset \{f_{2i}\} \supset \cdots \supset \{f_{ni}\} \supset \{f_{n+1,i}\} \supset \cdots,$$

requiring that for all values of n, $\lim_{i \to \infty} f_{ni}(x_n)$ exists. Moreover, if $p > 0$,

$$\lim_{i \to \infty} f_{ni}(x_n) = \lim_{i \to \infty} f_{n+p,i}(x_n), \tag{6}$$

since $f_{n+p,i}(x_n)$ is a subsequence of the convergent sequence $f_{ni}(x_n)$. We shall show that the "diagonal" subsequence, $F_i = f_{ii}$, is the desired Cauchy subsequence of the given sequence $\{f_n\}$. From [6] it follows that for fixed but arbitrary n,

$$\{F_i(x_n)\} = \{f_{ii}(x_n)\}$$

is a convergent sequence of real numbers, and therefore a Cauchy sequence of real numbers. It must still be shown that $\{F_n\}$ is a Cauchy sequence in the norm of $C(X)$. The equicontinuity implies that there is a $\delta > 0$ for which

$$d(x,y) < \delta \quad \text{implies} \quad |F_n(x) - F_n(y)| < \epsilon/3, \quad \text{for } n = 1, 2, \ldots.$$

The denseness of the sequence $\{x_i\}$ in X implies further that the sequence of neighborhoods $\{S_\delta(x_i)\}$ covers X. Since X is compact,

$$\bigcup_{i=1}^{N} S_\delta(x_i) = X,$$

for some integer N. The convergence of the real sequences $\{F_i(x_n)\}$, for each x_n, implies that there are integers ν_n that satisfy:

[7] $\qquad j,k > \nu_n \qquad \text{implies} \qquad |F_j(x_n) - F_k(x_n)| < \epsilon/3.$

Choose $\nu = \max[\nu_1, \nu_2, \ldots, \nu_N]$. If $x \in X$, then $x \in S_\delta(x_n)$ for some value of $n = 1, 2, \ldots, N$. It follows therefore that if $j,k > \nu$, then,

$$|F_j(x) - F_k(x)| \leq |F_j(x) - F_j(x_n)| + |F_j(x_n) - F_k(x_n)| \\ + |F_k(x_n) - F_k(x)| < \epsilon/3 + \epsilon/3 + \epsilon/3 = \epsilon.$$

The inequalities on the first and third terms result from the equicontinuity of the family, and the second follows from [7]. ∎

Exercises

***1.** Verify that if X is compact, then $d^*(f,g) = \sup_{x \in X} d(f(x),g(x))$ is a metric on $C(X,\hat{X})$. Show also that there is a point ξ in X at which

$$d^*(f,g) = \hat{d}(f(\xi),g(\xi)).$$

***2.** Show that if X is compact and if $\hat{X}$ is a normed vector space, then $C(X,\hat{X})$ is a normed vector space. The operations are defined at the beginning of the section.

Prove that if $\hat{X}$ is a Banach space, then $C(X,\hat{X})$ is also a Banach space.

***3.** Prove that every totally bounded metric space X contains a sequence $\{x_n\}$ whose closure is X itself. A space that contains such a sequence is said to be *separable*. (*Hint:* To each $\delta = 1/n$, let $x_{n1}, x_{n2}, \ldots, x_{n,j(n)}$ be a finite set of points in X, for which $\bigcup_{i=1}^{j(n)} S_{1/n}(x_{ni}) = X$. Show that the countable set $\{x_{ni}\}$, $n = 1, 2, \ldots$, $i = 1, 2, \ldots, j(n)$, has the desired property.)

4. Prove that the sequence

$$f_n(x) = \frac{x}{1 + nx^2} \qquad -1 \leq x \leq 1,$$

of differentiable functions converges uniformly on $[-1,1]$ to a differentiable function. However, it is not true that

$$\frac{d}{dx}(\lim f_n(x)) = \lim \frac{d}{dx} f_n(x).$$

Why does this happen? Under what conditions may differentiation and the taking of a limit be reversed?

5. Let

$$f_n(x) = \begin{cases} n^2x & \text{whenever } 0 \leq x \leq 1/n \\ -n^2x + 2n & \text{whenever } 1/n < x \leq 2/n \\ 0 & \text{whenever } 2/n < x \leq 1. \end{cases}$$

Why are

$$\lim \int_0^1 f_n(x)\,dx \quad \text{and} \quad \int_0^1 \lim f_n(x)\,dx$$

unequal?

6. Prove that if X is compact and if $\{f_n\}$ is a monotone sequence in $C(X)$ which converges pointwise to a function f in $C(X)$, then the convergence is uniform.

If f is a real-valued function defined on a metric space X, then the *limit superior of f at the point* ξ, written $\overline{\lim}_{x\to\xi} f(x)$, is the quantity $\inf_{\epsilon>0} \sup [f(x) : d(x,\xi) < \epsilon]$. Similarly, the limit inferior of f at ξ is given by

$$\underline{\lim}_{x\to\xi} f(x) = \sup_{\epsilon>0} \inf [f(x) : d(x,\xi) < \epsilon].$$

The limits superior and inferior may of course be infinite.

7. Show that a function f is upper semicontinuous on a metric space X if and only if at each point $\xi \in X$,

$$f(\xi) \geq \overline{\lim}_{x\to\xi} f(x).$$

8. Let f be upper semicontinuous on a compact space X. Prove that f is bounded from above and that it attains the value of its least upper bound at some point of X.
9. Give an example of a bounded upper semicontinuous function which does not attain the value of its greatest lower bound.
10. Show that $f(x)$ is continuous if and only if both f and $-f$ are upper semicontinuous.
11. Show that if $\{f_n\}$ is a sequence of upper semicontinuous functions that converge uniformly to a function f, then f is upper semicontinuous.

A real-valued function defined on a metric space X is *lower semicontinuous* if its negative is upper semicontinuous.

12. State and prove the analogues of Exercises 7–11 for lower semicontinuous functions.
13. Prove that a function is continuous if and only if it is both upper and lower semicontinuous.
14. If f_1, f_2 are upper semicontinuous and g_1, g_2 are lower semicontinuous on X, what can be said about the following functions?
    (i) $f_1 - g_1$ (ii) $af_1 + bf_2$ $a,b \in R$ (iii) $af_1 + bf_2$, $a,b > 0$
    (iv) $f_1 f_2$ (v) If $g_1 > 0$, $g_1 f_1$
15. Show that the family of functions $f_n(x) = \sin nx$, $0 \leq x \leq \pi$, is not equicontinuous.
16. Let $\mathfrak{F}$ be a subset of $C(X)$; X is compact.
    (a) Show that if every sequence $\{f_n\}$ in $\mathfrak{F}$ contains a convergent subsequence, then $\mathfrak{F}$ is an equicontinuous and bounded subset of $C(X)$.
    (b) Let $\mathfrak{F}$ be an equicontinuous subset of $C(X)$. Show that if to each x in X there is a positive number M_x, such that

    $$f \in \mathfrak{F} \quad \text{implies} \quad |f(x)| < M_x, \quad \text{for all } f \in \mathfrak{F},$$

    then there is a positive number M, and $\|f\|^* = \sup_{x\in X} |f(x)| < M$.
17. (Tietze's Extension Theorem for metric spaces). Let f be a bounded, real-valued function defined on a closed subset C of a metric space X. Show that there is a continuous bounded *extension* g of f to the entire space X. That

is, we seek a continuous, real-valued, bounded function g defined on X such that for $x \in C$, $f(x) = g(x)$. [*Hint:* The desired extension g will be the uniform limit of a sequence of continuous functions G_n which are defined below. Let $\gamma_1 = \sup_{x \in C} |f(x)|$, and let $S_1 = [f \leq -\gamma_1/3]$, $T_1 = [f \geq \gamma_1/3]$. There is a continuous function g_1, bounded in absolute value by $\gamma_1/3$, and which takes on the values $-\gamma_1/3$ and $\gamma_1/3$ on S_1 and T_1 respectively (See Exercise 3.2.22). The function $f_2 = f - g_1$ is defined on C. If $\gamma_2 = \sup_{x \in C} |f_2(x)|$, let S_2 and T_2 be the pair of disjoint, closed subsets of X on which f_2 is respectively less than or equal to $-\gamma_2/3$ and greater than or equal to $\gamma_2/3$. Let g_2 be the bounded continus function that takes on the values $-\gamma_2/3$ and $\gamma_2/3$ on S_2 and T_2 respectively, etc. Let $G_n = \sum_{i=1}^{n} g_i$. Show that $\{G_n\}$ converges uniformly to the desired function.]

*Exercises which request the proof of a statement made in the text, as well as those referred to in the text, are starred.

CHAPTER **4**

Metric and Normed Linear Spaces; Special Topics

1 Uniform Approximation of Functions; The Stone-Weierstrass Theorem

A major theme of analysis, which appears in a student's first course in calculus, is the idea of the approximation of certain mathematical objects by means of "simpler" ones: A real number may be approximated by rational numbers, or put another way, an infinite decimal may be approximated by finite decimals (see Sec. 4 of Chapter 0 and Lemma 1.2.13.); the slope of the tangent line at a point of a curve is approximated by the slopes of certain chords (that is, the derivative is approximated by difference quotients); the area under a curve is approximated by a finite sum of areas of rectangles or trapezoids; an infinite sequence in l_2 may bë approximated by the finite sequences in Λ_2, and so on. With the exception of the example of the derivative, all of these objects appear in this text as elements of normed vector spaces. In each case the approximating elements form a subspace whose closure (with respect to the norm) is the space of elements that may be approximated.

In this section, we turn our attention to the problem of approximating the functions of $C(X)$ by elements of a subclass. For example, a special case of the Stone-Weierstrass theorem (4.1.13) asserts that a function in $C[a,b]$ may be uniformly approximated by polynomials defined on $[a,b]$. The functions in $C[a,b]$ may also be approximated by the polygonal or piecewise linear functions (Exercise 2), or by step functions. Since the step functions are discontinuous and therefore not in $C[a,b]$, the problem may be restated so that both $C[a,b]$ and $\Sigma[a,b]$ are subspaces of $S[a,b]$, the normed vector space of (equivalence classes of) sectionally continuous functions on $[a,b]$. In Example 2.3.19(d) and Exercise 2.2.5, $S[a,b]$ was made into a metric space using integrals. Here we define a norm on $S[a,b]$ so that it agrees with the max-norm on $C[a,b]$: Considering any pair of sectionally continuous functions to be the same if they differ only at finitely many points, we define

$$\|f\| = \operatorname*{ess\,sup}_{a \le x \le b} |f(x)|$$

which reads, "essential supremum of f on $[a,b]$." The essential supremum is a number λ which has the following properties:

(i) $\lambda \ge |f(x)|$ for all but finitely many x in $[a,b]$,

(ii) If $\lambda' \ge |f(x)|$ for all but finitely many x in $[a,b]$, then $\lambda' \ge \lambda$.

Observe that for continuous functions the essential supremum is just the maximum. In the exercises the reader is asked to prove that $S[a,b]$ (or the space of equivalence classes) is a normed vector space with this norm. $S[a,b]$ contains as subspaces, $C[a,b]$, $\Sigma[a,b]$, $P[a,b]$, (the piecewise linear functions) and $\Pi[a,b]$, (the polynomial functions).

1 Example. If $f \in C[a,b]$ and $\epsilon > 0$, then there is a step function ρ such that $\|f - \rho\| = \operatorname*{ess\,sup}_{a \le x \le b} |f(x) - \rho(x)| < \epsilon$. This follows directly from the uniform continuity of f. Choose δ to be the δ of uniform continuity that corresponds to $\epsilon/2$. Then divide the interval $[a,b]$ into ν equal parts, where ν is chosen so that $(b - a)/\nu < \delta$. The function ρ is defined as follows:

Let $a = a_0, a_1, \ldots, a_\nu = b$, be the points of subdivision; each $(a_j - a_{j-1}) = (b - a)/\nu$. Let ρ be the step function which vanishes outside $[a,b]$, and on the interval is defined by

$$\rho(x) = \max_{a_{j-1} \le x \le a_j} f(x) \qquad \text{when} \qquad a_{j-1} < x \le a_j, \quad j = 1, \ldots, \nu.$$

It is easily seen that $\|f - \rho\| < \epsilon$. The function ρ may of course be chosen so that it takes on the minimum values of f on the subintervals (Exercise 1).

The choice of the approximating functions would of course depend upon the problem at hand. For integrating functions of a real variable, step functions or piecewise linear functions are desirable because of the simplicity of the areas that lie beneath their graphs. Approximating functions are also useful to obtain numerical values of certain quantities. For example, the reader will recall that the sequence of polynomials

$$P_n(x) = 1 - \frac{x^2}{2} + \frac{x^4}{4!} - \frac{x^6}{6!} + \cdots + (-1)^n \frac{x^{2n}}{(2n)!}$$

converges uniformly to $\cos x$ on any bounded interval. Therefore, if we wanted to compute $\cos x$ to, say six decimal places, for $|x| \le 1$, we would know from Taylor's Theorem that

$$\cos x = P_n(x) + R_n(x),$$

where the remainder or error term $R_n(x)$ is bounded by $\frac{1}{(2n)!}$ in this interval. Thus by choosing n so that the remainder does not exceed 10^{-7}, we obtain the desired approximation of $\cos x$. The computation of the approximate value of $\cos x$ using $P_n(x)$ offers no difficulties to either man or machine.

In Sec. 2, the solution of certain differential equations will be represented by power series, which is equivalent to saying that they may be approximated by polynomials.

2 Definition. A subset S of a metric space X is said to be *dense* in X if $\bar{S} = X$. ($\bar{S}$ is the closure of S.)

3 Example. (a) $\bar{Q} = R$.

(b) We shall show that $\Pi[a,b]$ is dense in $C[a,b]$. (Theorem 13 and Example 14(a)).

4 Definition. A metric space that contains a countable dense subset is said to be *separable.*

5 Examples. (a) The countability of Q implies that R is separable.

(b) Any totally bounded metric space is separable. (Exercise 3.4.3).

(c) R^n equipped with any of the standard metrics is separable. The subset $Q^n = \{x = (x_1, \ldots, x_n) : x_i \in Q,\ i = 1, \ldots, n\}$ is dense in R^n, and is countable. Similarly for $l_p, p \geq 1$.

(d) The reader is asked to show (Exercise 11) that the space $P^*[a,b]$ of piecewise linear functions with rational points of subdivision (except possibly a and/or b), is countable and dense in $C[a,b]$.

The remainder of this section is devoted to a proof of the Stone-Weierstrass theorem, which gives sufficient conditions for a subfamily of $C(X)$ to be a family of approximating functions.

Up to now, $C(X)$, (where X is compact), has been treated as a normed vector space. However, $C(X)$ is closed also under multiplication, for the product of real-valued continuous functions is continuous. Moreover, since $|f| \in C(X)$ whenever $f \in C(X)$, it follows that if $f, g \in C(X)$, then

$$\max(f,g) = \left(\frac{f+g}{2}\right) + \frac{|f-g|}{2} \quad \text{and}$$

$$\min(f,g) = \left(\frac{f+g}{2}\right) - \frac{|f-g|}{2}$$

are in $C(X)$.

These additional properties of $C(X)$ are generalized in Definitions 6 and 7.

6 Definition. A vector space A defined over a field F is called an *algebra* if the product of any pair α,β in A is defined so that

(i) $\alpha,\beta,\gamma \in A$ imply $\alpha(\beta\gamma) = (\alpha\beta)\gamma$ and $\alpha(\beta + \gamma) = (\alpha\beta) + (\alpha\gamma)$. (This says that A is a ring.)

(ii) $\alpha,\beta \in A$ and $a \in F$ imply $a(\alpha\beta) = (a\alpha)\beta = \alpha(a\beta)$.

7 Definition. A partially ordered set $(S, \prec)$ is called a *lattice*, if to every pair $s,t \in S$ there are elements $s \vee t$ and $s \wedge t$ in S that satisfy:

(i) $s \vee t \succ s,t$; and if $r \succ s,t$, then $r \succ s \vee t$ or $r = s \vee t$. $s \vee t$ is called the *least upper bound* of s and t.

(ii) $s \wedge t \prec s,t$; and if $r \prec s,t$, then $r \prec s \wedge t$ or $r = s \wedge t$. $s \wedge t$ is the *greatest lower bound of s and t.*

8 Examples. (a) R is clearly an algebra over either the field Q or R. Also, taking $\prec$ to be "less than or equal to," R is a lattice. Here $s \vee t = \max[s,t]$ and $s \wedge t = \min[s,t]$.

(b) The family of all subsets of a given set X may be made into a lattice by taking $\prec$ to be the inclusion relation. The reader may verify that if S and T are subsets of X, then $S \cup T$ and $S \cap T$ are respectively the least upper and greatest lower bounds.

(c) $C(X)$ is an algebra. It is also partially ordered by $\prec$, where

$$f \prec g \qquad \text{if and only if} \qquad f(x) \leq g(x), \text{ for all } x \in X.$$

$C(X)$ is a lattice with respect to this partial ordering if we define

$$f \vee g = \max[f,g] \qquad \text{and} \qquad f \wedge g = \min[f,g].$$

(d) $S[a,b]$ (or the space of equivalence classes) can be made into a lattice. (See Exercise 7.)

By a *subalgebra* (*sublattice*) of $C(X)$ we mean a subset A which is itself an algebra (lattice).

We seek conditions which when imposed upon a subset A of $C(X)$ will insure that the functions of $C(X)$ may be uniformly approximated by the functions of A. This is equivalent to saying that if $f \in C(X)$ and $\epsilon > 0$, there is a $\rho \in A$ such that $\|f - \rho\| < \epsilon$. That is, $\bar{A} = C(X)$. One of these conditions is that A (or its closure) be a sublattice, which in this case means that

$$\sigma,\tau, \in A \qquad \text{implies} \qquad \max[\sigma,\tau], \min[\sigma,\tau], |\sigma|, |\tau| \in A \text{ (or } \bar{A}).$$

The second condition, which in effect guarantees that the functions of A take on sufficiently many values is described in Theorem 9:

9 Theorem. Let X be a compact metric space which contains at least two points. If A is a sublattice of $C(X)$ which satisfies

[1] $\forall \alpha, \beta \in R, x \neq y \in X, \quad \exists\, \sigma \in A$ with $\sigma(x) = \alpha$ and $\sigma(y) = \beta$,

then $\bar{A} = C(X)$.

Proof: The reader may find it helpful while reading this proof to draw the corresponding pictures for $X = [a,b]$ and $A = P[a,b]$, the piecewise linear (or polygonal) functions. During the course of the proof some statements will be interpreted for this case parenthetically.

Let $f \in C(X)$ and $\epsilon > 0$ be given. We seek an element $\rho \in A$, satisfying $\|\rho - f\| = \max_{x \in X} |\rho(x) - f(x)| < \epsilon$.

We begin by approximating f at some fixed but arbitrary point ζ in X. From [1] it follows that to each $t \in X$, there is a function $f_t \in A$ such that,

$$f_t(\zeta) = f(\zeta) \qquad \text{and} \qquad f_t(t) = f(t).$$

(Take f_t in $P[a,b]$ to be the line that passes through $(\zeta, f(\zeta))$ and $(t, f(t))$.) The continuity of the functions f and f_t imply that for $t \neq \zeta$, the sets

[2] $$G_t = [(f_t - f) < \epsilon]$$

are open. Clearly $\zeta, t \in G_t$ for all $t \neq \zeta$. Therefore,

[3] $$\bigcup_{t \neq \zeta} G_t = X.$$

The compactness of X yields a finite subcover, $\{G_{t_1}, \ldots, G_{t_n}\}$. Since A is a lattice, the function

[4] $$\rho_\zeta = \min\,[f_{t_1}, \ldots, f_{t_n}]$$

is in A and $\rho_\zeta(\zeta) = f(\zeta)$. (For the special case, ρ_ζ is a piecewise linear function. It is simply the min of n linear functions.) From [2], [3] and [4], it follows also that,

[5] $$\forall x \in X, \qquad \rho_\zeta(x) < f(x) + \epsilon.$$

Thus by keeping the point ζ fixed, we obtain a function in A which does not exceed $f + \epsilon$ at any point of X. By varying this point ζ, and forming a second open covering, we shall obtain a function ρ in A that has the desired property: $\forall x \in X, f(x) - \epsilon < \rho(x) < f(x) + \epsilon$.

For each ζ in X, the set

[6] $$\Gamma_\zeta = [(f - \rho_\zeta) < \epsilon]$$

is open, and the union of the Γ_ζ covers X. Again, the compactness of X yields a finite subcover $\{\Gamma_{\zeta_1}, \Gamma_{\zeta_2}, \ldots, \Gamma_{\zeta_m}\}$. The function

[7] $$\rho = \max\,[\rho_{\zeta_1}, \ldots, \rho_{\zeta_m}]$$

is in the sublattice A, and [6] and [7] imply that $f(x) - \rho(x) < \epsilon$. The inequality $\rho(x) - f(x) < \epsilon$ follows from [5]. ∎

The proof of Theorem 9 is not applicable to the subset $\Pi[a,b]$ (of polynomial functions) of $C[a,b]$ since this set is not a lattice. (See Exercises 8 and 9.) This shows that it is not *necessary* for an approximating set to be a lattice. In fact the reader should convince himself that the proof works if it is assumed only that $\bar{A}$ is a sublattice. We state this as a theorem:

9′ Theorem. Let X be a compact metric space which contains at least two points. If A is a subset of $C(X)$ whose closure is a lattice and which satisfies [1], then $\bar{A} = C(X)$.

The proof is left to the reader.

In some cases it is rather difficult to verify that $\bar{A}$ is a lattice, for example if $A = \Pi[a,b]$, the set of polynomial functions on $[a,b]$. Propositions 10 and 11 imply that it suffices to verify instead that A is a subalgebra:

10 Proposition. The closure of a subalgebra A of $C(X)$ is a subalgebra.

Proof: $\bar{A}$ is a subalgebra if

(i) $f,g \in \bar{A}$ implies $f + g,\ fg \in \bar{A}$, and

(ii) $f \in \bar{A}$ and $c \in R$ imply $cf \in \bar{A}$.

If f and g are in $\bar{A}$, then there are sequences $\{\sigma_n\}$ and $\{\tau_n\}$ in A which converge to f and g respectively. Since A is a subalgebra, the sequences $\{\sigma_n + \tau_n\}$ and $\{c\sigma_n\}$ are in A and converge to $f + g$ and cf, which must therefore be in the closure of A. ∎

11 Proposition. If A is a subalgebra of $C(X)$, then $\bar{A}$ is a sublattice.

Proof: Proposition 10 implies that $\bar{A}$ is a subalgebra. The identities

[8]
$$\max[f,g] = \frac{(f + g) + |f - g|}{2}$$
$$\min[f,g] = \frac{(f + g) - |f - g|}{2},$$

imply that it suffices to verify that $|f| \in \bar{A}$ whenever $f \in \bar{A}$. But if $f \in \bar{A}$, then there are functions σ_n in A such that $\|f - \sigma_n\|$ does not exceed $1/n$. This implies that $\||f| - |\sigma_n|\| \le 1/n$, from which it follows that if the $|\sigma_n|$ are in $\bar{A}$, then so is f. Thus the proposition is proved if it can be shown that

[9]
$$\sigma \in A \quad \text{implies} \quad |\sigma| \in \bar{A},$$

or equivalently, if $\sigma \in A$, then

[10] $\forall \epsilon > 0$, there is a $\rho \in A$ with $\| \rho - |\sigma| \| < \epsilon$.

Set $\hat{\epsilon} = \dfrac{\epsilon}{2M}$, where $M = \max_{x \in X} |\sigma(x)| = \|\sigma\|$ (it is assumed that σ does not vanish everywhere, for in that case [9] is obviously true). The real function

$$F(t) = \sqrt{t + \hat{\epsilon}^2}$$

may be expanded in a Taylor series around any point $\zeta > -\hat{\epsilon}^2$. Setting $\zeta = 1/2$, it is easily seen that the Taylor series

$$\sum_{n=0}^{\infty} \frac{F^{(n)}(1/2)\,(t - 1/2)^n}{n!}$$

converges uniformly and absolutely in the interval [0,1]. (The radius of convergence is in fact equal to $1/2 + \hat{\epsilon}^2$.) Letting $P_n(t)$ denote the nth partial sum of this series, the uniform convergence implies that for sufficiently large values of ν,

$$\max_{0 \le t \le 1} |P_\nu(t) - \sqrt{t + \hat{\epsilon}^2}| < \hat{\epsilon}$$

A simple calculation, which is left to the reader, yields

$$|\sqrt{t + \hat{\epsilon}^2} - \sqrt{t}| \le \hat{\epsilon},$$

whenever $\hat{\epsilon} > 0$ and $0 \le t \le 1$. Using the triangle inequality, it follows that

[11] $$|P_\nu(t) - \sqrt{t}| \le |P_\nu(t) - \sqrt{t + \hat{\epsilon}^2}| + |\sqrt{t + \hat{\epsilon}^2} - \sqrt{t}| < 2\hat{\epsilon}.$$

Since A is an algebra that contains the given function σ, it must also contain all sums and products of powers of σ; in particular the polynomial

$$\rho(x) = MP_\nu\left(\frac{(\sigma(x))^2}{M^2}\right)$$

is in A. Moreover, since $0 \le \dfrac{\sigma^2}{M^2} \le 1$, the substitution of $t = \dfrac{\sigma^2}{M^2}$ in [11] yields

$$\frac{1}{M}\,|\rho(x) - |\sigma(x)|\,| < 2\hat{\epsilon},$$

for all $x \in X$. Finally, $\hat{\epsilon} = \dfrac{\epsilon}{2M}$ implies

$$\| \rho - |\sigma| \| < \epsilon,$$

which is [10]. ∎

We have just proved that Theorem 9 (or 9′) is true if it is assumed either that A is a sublattice or a subalgebra. We show next (the Stone-Weierstrass theorem) that [1] may be replaced by a condition that is frequently easier to verify.

12 Definition. A family $\mathfrak{F}$ of real valued functions defined on a set X which contains at least two points, is said to *separate points of* X if, for every pair x and y of distinct points of X, there is a function f in the family with $f(x) \neq f(y)$.

If moreover, there is a function f in $\mathfrak{F}$ such that $0 \neq f(x) \neq f(y) \neq 0$, then $\mathfrak{F}$ is said to *separate the points of* X *strongly*.

13 Theorem (Stone-Weierstrass). Let X be a compact metric space that contains at least two points. If A is a subalgebra of $C(X)$ whose closure separates points of X strongly, then $\bar{A} = C(X)$.

Proof: Proposition 11 implies that $\bar{A}$ is a sublattice. If it can be shown that $\bar{A}$ satisfies [1], then the conclusion follows from Theorem 9′.

Let α and β be real numbers, and $x \neq y$ be points of X. Since $\bar{A}$ separates points of X strongly, it contains a function ρ such that $0 \neq \rho(x) \neq \rho(y) \neq 0$. We seek a function f in $\bar{A}$ that satisfies $f(x) = \alpha$ and $f(y) = \beta$. It is shown next that the desired function $f(x)$ may be written as $c\rho + d\rho^2$, for real numbers c,d. Since f is required to take on the values α and β at the points x and y respectively, the numbers c and d must satisfy the pair of linear equations:

[12]
$$\begin{aligned} \alpha &= c\rho(x) + d\rho^2(x). \\ \beta &= c\rho(y) + d\rho^2(y). \end{aligned}$$

The assumption $0 \neq \rho(x) \neq \rho(y) \neq 0$ implies that the determinant of this system can not vanish, proving that [12] has a solution. ∎

The Stone-Weierstrass may be restated in a slightly different way:

13′ Theorem. Let X be a compact space that contains at least two points. If A is a subalgebra of $C(X)$ whose closure separates points, and if $\bar{A}$ contains a nonzero constant function, then $\bar{A} = C(X)$.

It is left to the reader to show that the conditions of Theorem 13′ imply that $\bar{A}$ separates points of X strongly.

14 Examples. (a) The subalgebras $C[a,b]$, $P[a,b]$, $\Sigma[a,b]$, and $\Pi[a,b]$, of $S[a,b]$ separate points of $[a,b]$ strongly.

(b) Let $\mathfrak{F}$ be the subset of $C[a,b]$ which consists of those functions which vanish at $\zeta = \dfrac{a+b}{2}$. Clearly $\mathfrak{F}$ separates the points of $[a,b]$, but not strongly. The reader will observe that $\mathfrak{F}$ does not contain a nonzero constant function, since every function in $\mathfrak{F}$ vanishes at $\dfrac{a+b}{2}$. It is easily seen that $\mathfrak{F}$ is equal to its own closure, and is a proper subset of $C[a,b]$. This means that functions which do *not* vanish at a point ζ, can not be approximated by functions which do vanish there.

(c) Let ξ and ζ be distinct points of $[a,b]$. The set of those functions in $C[a,b]$ which vanish at both ξ and ζ is a subalgebra which does not separate points.

Example 14(b) suggests an approximation theorem for functions in $C(X)$ which vanish at some fixed point of X. We begin by showing that any closed subalgebra of $C(X)$ which separates points of X, but not strongly, consists of functions, each of which vanishes at a single point ζ of X. Example 13(c) demonstrates that there can not be more than one such point if $\overline{A}$ is to separate points. To prove that there is at least one such point, let us assume that $\overline{A}$ separates points and that there is no single point of X at which every function in $\overline{A}$ vanishes. We shall show that this implies that $\overline{A}$ separates points strongly. If $\xi \neq \zeta$ are points of X, there are functions $f_\xi, f_\zeta \in \overline{A}$ with $0 \neq f_\xi(\xi)$, $0 \neq f_\zeta(\zeta)$. It is easily seen, by considering the possibilities listed below, that this implies the existence of a function h in $\overline{A}$ with $0 \neq h(\xi) \neq h(\zeta) \neq 0$:

Case 1. If $f_\xi(\zeta) \neq 0 \neq f_\xi(\xi) \neq f_\xi(\zeta)$, then we may set $h = f_\xi$. (A similar argument holds if the roles of ξ and ζ are reversed.)

Case 2. If $f_\xi(\zeta) = f_\zeta(\xi) = 0$, set $h = f_\xi + \alpha f_\zeta$, where $\alpha \neq 0$ is chosen so that $h(\xi) \neq h(\zeta)$. It is easily seen that neither $h(\xi)$ nor $h(\zeta)$ vanishes.

Case 3. If $f_\xi(\zeta) = 0$ and $f_\xi(\xi) = f_\zeta(\xi) \neq 0$, then it is possible to choose α so that $h = f_\xi + \alpha f_\zeta$ is the desired function (Exercise 14). A similar argument holds if ξ and ζ are interchanged.

Case 4. If $f_\xi(\xi) = f_\xi(\zeta) = f_\zeta(\xi) = f_\zeta(\zeta) \neq 0$, then we choose a function g in $\overline{A}$ satisfying $g(\xi) \neq g(\zeta)$. (Remember, $\overline{A}$ separates points.) If either $g(\xi)$ or $g(\zeta)$ vanishes, then set $h = f_\xi + \alpha g$ (or $f_\zeta + \alpha g$), where α is chosen so that neither $h(\xi)$ nor $h(\zeta)$ vanishes. The remaining case, $g(\xi) \neq 0 \neq g(\zeta)$ is left to the reader.

We restate these results as a proposition:

15 Proposition. A closed subalgebra A ($A = \overline{A}$) of $C(X)$ separates points of X, but not strongly, if and only if there is a point ζ of X at which every function in A vanishes; but if $x \neq \zeta$, then there is a function $f_x \in A$, with $f_x(x) \neq 0$.

We are now in a position to prove a variation of the Stone-Weierstrass theorem for functions which vanish at a fixed point of X.

16 Theorem. Let X be a compact metric space that has at least two points. If A is a subalgebra of $C(X)$ whose closure separates points of X, but not strongly, then there is a point $\zeta \in X$ at which all the functions of A vanish, and $\bar{A}$ is the subset of $C(X)$ which consists of those functions which vanish at ζ.

Proof: The idea of the proof is to include A in a larger subalgebra $\hat{A}$ which contains nonzero constant functions, and then to use Theorem 13′.

Let $\hat{A}$ be the subset of $C(X)$ whose members $\hat{\rho}$ can be written as a sum $\rho + c$, where $\rho \in A$ and c denotes the constant function whose value everywhere on X is the real number c. Clearly $\hat{A}$ is a subalgebra of $C(X)$ which contains A and nonzero constant functions. Theorem 13′ implies that the closure of $\hat{A}$ is $C(X)$. Let ζ be the point at which all the functions in A vanish (Proposition 15). If f is a function in $C(X)$ which vanishes at ζ, there is a function $\hat{\rho} = \rho + c \in A$, such that $\|\hat{\rho} - f\| < \epsilon/2$, where ϵ is any preassigned positive number. Since $f(\zeta) = \rho(\zeta) = 0$, it follows that

$$|c| = |f(\zeta) - \rho(\zeta) + c| < \|f - \rho + c\| < \epsilon/2,$$

and therefore

$$\|f - \rho\| < \|f - \rho + c\| + |c| < \epsilon.$$

Thus the continuous functions which vanish at ζ may be uniformly approximated by the functions in A.

On the other hand, if $f \in C(X)$ and $f(\zeta) \neq 0$, then it is easily seen, by taking $\epsilon = \dfrac{|f(\zeta)|}{2} > 0$, that there is no function ρ in A for which $\|f - \rho\| < \epsilon$, since

$$\epsilon < |f(\zeta)| = |f(\zeta) - \rho(\zeta)| \leq \|f - \rho\|. \quad \blacksquare$$

We shall return to the problem of approximating functions when we study Hilbert spaces. In particular, we shall study the possibility of approximating functions by trigonometric sums, (the Fourier sums) and give another proof, using the Fourier sums, of the classical Weierstrass theorem, which asserts that the functions in $C[a,b]$ may be uniformly approximated by polynomials.

Exercises

***1.** Show, in Example 1, that $\|f - \rho\| < \epsilon$. Show also that ρ can be chosen so that $\rho(x) \leq f(x)$.

***2.** Prove that $P[a,b]$, the piecewise linear functions, are dense in $C[a,b]$. (See Example 1.)

*3. Prove that $\|f - g\| = \operatorname*{ess\,sup}_{a \le x \le b} |f(x) - g(x)|$ is a norm on $S[a,b]$, or on the appropriate space of equivalence classes.

*4. Prove that l_p and R^n are separable spaces.

*5. Prove the assertion made in Example 5(a).

*6. Verify the statements made in Example 8.

*7. $S[a,b]$ may be partitioned into equivalence classes, $[f]$, by defining $f \sim g$ if $f(x) = g(x)$ except at finitely many points. Show that the following relation is a partial ordering on $S[a,b]$: $[f] \prec [g] \leftrightarrow f(x) \le g(x)$ except at finitely many points. Extend the partial ordering to the space of equivalence classes, define $[f] \vee [g]$ and $[f] \wedge [g]$, and show that the space of equivalence classes is a lattice.

8. Show, without using the Stone-Weierstrass theorem that $\Sigma[a,b]$, $P[a,b]$ and $C[a,b]$ are dense in $S[a,b]$,

9. Show that $\Pi[a,b]$ is not a lattice. In particular, show that $\left|x - \left(\frac{a+b}{2}\right)\right|$ is not equal to a polynomial function on $[a,b]$. (Remember, a polynomial is uniquely determined by its values at finitely many points.)

*10. Prove Theorem 9′.

*11. Prove that $P^*[a,b]$ and $\Sigma^*[a,b]$, the piecewise linear functions and step functions whose subdivision points (with the possible exception of a and/or b) are rational, and in the latter case taking on only rational values, are countable, thereby proving that both $C[a,b]$ and $S[a,b]$ are separable.

*12. Prove that if $\epsilon > 0$ and $0 \le t \le 1$, then

$$|\sqrt{t + \epsilon^2} - \sqrt{t}| \le \epsilon.$$

13. Use Dini's theorem (Theorem 3.4.4) to show that $\sqrt{t}$ may be approximated uniformly by polynomials on a compact interval. [*Hint:* Look at the sequence of polynomials $p_{n+1}(t) = p_n(t) + 1/2[t - (p_n(t))^2]$.]

14. Complete the argument in cases 3 and 4 of Proposition 15.

*15. Use Theorem 9′ to prove that the countable families $\Sigma^*[a,b]$ and $P^*[a,b]$ are dense in $S[a,b]$.

*Exercises which request the proof of a statement made in the text, as well as those referred to in the text, are starred.

2 Fixed-Point Theorems and Contracting Mappings

The reader was asked to prove that if f is a continuous mapping of $[0,1]$ into itself, then there is a point $x \in [0,1]$ such that $f(x) = x$ (Exercise 3.3.7). This simple consequence of the Intermediate Value Property (Corollary 3.3.6) has a simple geometric interpretation: The existence of a fixed point is equivalent to the existence of a point of intersection of the two curves $y = x$ and $y = f(x)$ (Figure 16). It is intuitively clear that since the graph of $f(x)$,

$$\{(x,y) : 0 \le x \le 1,\ y = f(x),\ 0 \le f(x) \le 1\}$$

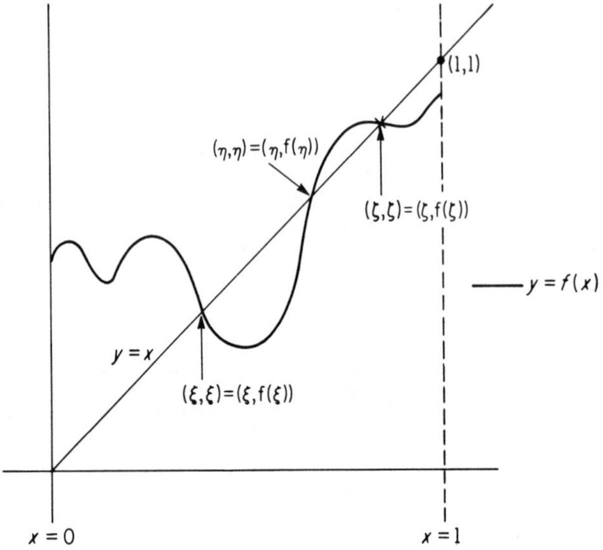

FIGURE 16

lies entirely in the unit square, it must intersect the line $x = y$ at least once. This theorem is easily seen to be true if $[0,1]$ is replaced by any closed bounded interval $[a,b]$. Moreover, a simple application of the theorem on the composition of continuous mappings yields a further generalization:

1 Theorem. Let f be a continuous injective mapping of $[0,1]$ into a metric space X (f is a path.) whose inverse, defined on the set $Y = f([0,1])$ is also continuous. Then, if

$$g : Y \to Y$$

is continuous, g has a fixed point.

The proof is left to the reader.

There are many variations of fixed-point theorems, most of which are beyond the scope of this book. Some simple applications to analysis appear in Bers [1], Mendelson [1], and Chinn and Steenrod [1]. A more algebraic fixed-point theorem, using homology theory, is the *Lefschetz fixed-point theorem* (Lefschetz [1]). A special case of this is the *Brouwer fixed-point theorem*, which asserts that a continuous mapping of the unit n-cube in R^n into itself has a fixed point. (See also Hocking and Young [1].)

Here we shall limit ourselves to the special case of *contracting* and *nonexpanding* mappings of complete metric spaces. The reader will recall that the mappings of the fixed-point theorems of Sec. 3 of Chapter 3 were defined on *connected* spaces; indeed each of these theorems was a consequence of the Intermediate Value Property. In this section it is not the connectedness of the space that is crucial, but its *completeness*.

2 Definition. A mapping f of a metric space X into itself is called a *contraction* (or a contracting map) if there is a nonnegative number $\lambda < 1$, such that

$$\forall x,y \in X, \qquad d(f(x),f(y)) \leq \lambda d(x,y).$$

f is said to be a *nonexpanding map* if

$$\forall x,y \in X \qquad d(f(x),f(y)) \leq d(x,y).$$

3 Proposition. A nonexpanding map (and therefore a contraction too) is uniformly continuous.

Proof: Set $\delta = \epsilon$. ∎

4 Example. If $0 < \rho < 1/2$, then $f(x) = x^2$ is a contraction of $[0,\rho]$ into itself. What is λ? On the other hand, $f(x) = x^2$ is a nonexpanding map of $[0,1/2]$ into itself.

We prove next that a contraction of a complete metric space has a fixed point.

5 Theorem. Let f be a contraction of a complete metric space. Then f has a unique fixed point.

Proof: We prove first that f cannot have more than one fixed point. For, suppose that both x and y were fixed points of f. Then

$$d(x,y) = d(f(x),f(y)) \leq \lambda d(x,y) < d(x,y),$$

which cannot be.

To prove that a fixed point exists, we shall construct a Cauchy sequence whose limit in the complete space X is the desired fixed point.

Let x_0 be any point in X. Form the sequence

$$x_1 = f(x_0),\ x_2 = f(x_1), \ldots, x_n = f(x_{n-1}), \ldots.$$

Since f is a contraction, it follows that for all values of n,

$$[1] \qquad d(x_{n+1},x_n) = d(f(x_n),f(x_{n-1})) \leq \lambda d(x_n,x_{n-1}) \\ \leq \lambda^2 d(x_{n-1},x_{n-2}) \leq \cdots \leq \lambda^n d(x_1,x_0),$$

where $0 < \lambda < 1$. (Of course if $\lambda = 0$ we are through.) If n and p are nonnegative integers, then the triangle inequality and [1] yield

$$\begin{aligned} d(x_{n+p},x_n) &\leq d(x_{n+p},x_{x+p-1}) + d(x_{n+p-1},x_{n+p-2}) + \cdots + d(x_{n+1},x_n) \\ &\leq \lambda^{n+p-1}d(x_1,x_0) + \lambda^{n+p-2}d(x_1,x_0) + \cdots \lambda^n d(x_1,x_0) \\ &= d(x_1,x_0)\lambda^n[1 + \lambda + \lambda^2 + \cdots + \lambda^{p-1}] < d(x_1,x_0)\lambda^n\left(\frac{1}{1-\lambda}\right). \end{aligned}$$

The final inequality is obtained by summing the geometric series. However, since $0 < \lambda < 1$, it follows that

$$\lim_{n\to\infty} d(x_{n+p},x_n) \leq \lim_{n\to\infty} \frac{\lambda^n d(x_1,x_0)}{1-\lambda} = 0,$$

proving that $\{x_n\}$ is a Cauchy sequence. The completeness of X implies that $\{x_n\}$ has a limit x in X. The continuity of f (Proposition 3) implies that f is sequence preserving (Proposition 3.2.9). Therefore $\lim f(x_n) = f(x)$. Recalling that $x_{n+1} = f(x_n)$, we obtain

$$x = \lim x_{n+1} = \lim f(x_n) = f(x). \quad \blacksquare$$

It is also true that if X is a Banach space, then every nonexpanding map has at *least* one fixed point. (See Exercise 10.)

The first application of Theorem 5 is a form of the Implicit Function Theorem:

6 Theorem (Implicit Function Theorem). Let $a < b$ be real numbers, and let

$$X = \{(x,y): a \leq x \leq b,\ y \in R\}.$$

Suppose that $F(x,y)$ is a continuous function defined on X, and that for every x in $[a,b]$ the partial derivative $F_y(x,y)$ exists. If there are numbers $0 < \mu < \lambda$ such that $\mu \leq F_y \leq \lambda$, for all $(x,y) \in X$, then there is a unique function $f(x)$ in $C[a,b]$ for which

$$F(x,f(x)) \equiv 0 \qquad \text{whenever } x \in [a,b].$$

That is, the equation $F(x,y) = 0$ may be solved uniquely for y as a function of x, whenever $a \leq x \leq b$.

Proof: We shall define a contraction ψ of $C[a,b]$ whose fixed point is the desired function. To motivate the definition of ψ, the reader will observe the rather obvious fact that $f(x)$ is a solution of the equation $F(x,y) = 0$ if and only if

$$f(x) \equiv f(x) - cF(x,f(x)), \tag{2}$$

where c is any nonzero real number. This suggests the following definition of ψ: If $g \in C[a,b]$, let

$$\psi(g) = g(x) - cF(x,g(x)). \tag{3}$$

The continuity of $F(x,y)$ implies that $\psi(g) \in C[a,b]$. Moreover, a fixed point of ψ satisfies [2], and is therefore a solution of the equation. We show next that if c is chosen appropriately, then ψ is a contraction.

Let g and h be in $C[a,b]$. Then

$$|\psi(g) - \psi(h)| = |g(x) - h(x) - c[F(x,g(x)) - F(x,h(x))]|.$$

The mean value theorem yields

$$|\psi(g) - \psi(h)| = |[g(x) - h(x)] - c[F_y(x,k(x))][g(x) - h(x)]|,$$

where $k(x) = g(x) + t[h(x) - g(x)]$ for some value of t in $(0,1)$. If c is chosen so that $0 < 1 - c\mu < 1$ (in this case $c = 1/\lambda$ will do), then factorization of the preceding expression gives

$$|\psi(g) - \psi(h)| \leq |g(x) - h(x)|(1 - c\mu) = \gamma \|g - h\|,$$

where $0 < \gamma < 1$. Since this inequality holds for all values of $x \in [a,b]$, we obtain

$$\|\psi(g) - \psi(h)\| < \gamma \|g - h\|,$$

thus proving that ψ is a contraction of $C[a,b]$. From Theorem 5 it follows that there is a function f in $C[a,b]$ which is a fixed point of ψ. [2] and [3] imply that $F(x,f(x)) \equiv 0$. Why is f uniquely determined? ∎

Theorem 6 may be stated for any number of variables. This is left to the reader as an exercise.

We turn our attention now to some existence theorems for differential and integral equations. Let $F(x,y)$ be a function which is defined on an open subset G of R^2. If $(x_0,y_0) \in G$, we seek a function $y = f(x)$ that satisfies the following conditions: (i) $y_0 = f(x_0)$, (ii) $f(x)$ is differentiable in some neighborhood of x_0, (iii) $f'(x) = F(x,f(x))$ in a neighborhood of x_0, and, (iv) f is the only function that satisfies (i), (ii) and (iii). That is, we wish to find a unique differentiable solution to the *initial-value problem*,

[4]
$$y' = F(x,y),$$
$$y(x_0) = y_0.$$

We begin by transforming [4] into an equivalent integral equation: If $F(x,y)$ is continuous in an open set which contains the point (x_0,y_0), then any function $y = f(x)$ which is a differentiable solution of [4] is also a solution of the integral equation

[5]
$$f(x) = y_0 + \int_{x_0}^{x} F(t,f(t))\,dt,$$

for values of x which are sufficiently close to x_0. Conversely, a solution of [5] is a solution of [4]. This follows from the Fundamental Theorem of Calculus. It can be shown that the continuity of $F(x,y)$ in an open subset of R^2 that contains (x_0,y_0) is sufficient to insure the existence of a solution to [4] or [5]. However, as the following example demonstrates, this solution may not be unique: The initial-value problem

$$\begin{aligned} y' &= y^{2/3} \\ y(0) &= 0 \end{aligned}$$

has at least two differentiable solutions, $y_1 \equiv 0$ and $y_2 = \dfrac{x^3}{27}$. (See Birkhoff-Rota [1].) To assure uniqueness, an additional condition is required. As we shall see, this condition permits us to use the theorem on contracting maps to obtain a *unique* solution. Further discussion of these topics may be found in Birkhoff-Rota [1], and White [1].

7 Definition. A function $F(x,y)$ defined on a subset S of R^2 is said to satisfy a *Lipschitz condition* if there is a nonnegative number L such that whenever (x,y) and (x,z) are in S,

$$|F(x,y) - F(x,z)| \leq L\,|y - z|.$$

8 Theorem. If $F(x,y)$ is continuous in an open subset G of R^2 that contains the point (x_0,y_0), and if $F(x,y)$ satisfies a Lipschitz condition in G, then there is one and only one solution $f(x)$ of the initial-value problem [4] in some neighborhood of x_0.

Proof: Let $S_r = \{(x,y): |x - x_0|, |y - y_0| < r\}$ be a square neighborhood whose closure is contained in G. For the remainder of the proof (x,y) will denote a point of S_r. We should like to show that a positive number $\gamma < r$ exists for which the expression

$$\psi(g) = y_0 + \int_{x_0}^{x} F(t,g(t))dt$$

is a contraction of a closed subspace of $C[x_0 - \gamma, x_0 + \gamma]$. Unlike the proof of Theorem 6, where $F(x,y)$ was defined for *all* y, and x in some interval, it is necessary to show here that γ may be chosen so that if g is in the closed subspace, then $(t,g(t)) \in G$ whenever $|t - x_0| < \gamma$. Otherwise $F(t,g(t))$ would not be defined, nor would the mapping $\psi(g)$.

The number γ may be chosen as follows: First choose a number M that exceeds $|F(x_0,y_0)|$. The continuity of $F(x,y)$ in G implies the existence of a neighborhood of (x_0,y_0) in which $|F(x,y)| < M$. Now let γ be a number satisfying the following three conditions:

(i) $0 < \gamma < 1/L$, where L is the Lipschitz constant (See Definition 7).

(ii) The set $S = \{(x,y): |x - x_0| < \gamma, \ |y - y_0| < \gamma M\}$ and its closure lie in G.

(iii) If $(x,y) \in S$, then $F(x,y) < M$.

Let $f_0(x) \equiv y_0$ denote the constant function defined on the interval $[x_0 - \gamma, x_0 + \gamma]$. We shall show that ψ is a contraction of the closed ball

$$B = \{g \in C[x_0 - \gamma, x_0 + \gamma]: \ d(g,f_0) \leq M\gamma\},$$

which is a subset of $C[x_0 - \gamma, x_0 + \gamma]$.

First, if $|t - x_0| \leq \gamma$ and $g \in B$, then $(t,g(t)) \in S$ (See (ii). Thus $F(t,g(t))$ is continuous for $|t - x_0| \leq \gamma$, and therefore ψ is a mapping of B into $C[x_0 - \gamma, x_0 + \gamma]$ whenever $|x - x_0| \leq \gamma$ and $g \in B$. It is easily seen that $\psi(B) \subset B$, for if $g \in B$, then

$$d(\psi(g),f_0) = \sup_{|x-x_0|\leq\gamma} [|\psi(g)(x) - y_0|]$$

$$\leq \sup_{|x-x_0|\leq\gamma} \left[\int_{x_0}^{x} |F(t,g(t))|\,dt\right] \leq \sup_{|x-x_0|\leq\gamma} [M|x - x_0|] \leq M\gamma.$$

Finally, it must be shown that ψ is a contraction of B. Indeed, if g_1, $g_2 \in B$ and if $|t - x_0| \leq \gamma$, then $(t,g_1(t)), (t,g_2(t)) \in S \subset G$, and

$$d(\psi(g_1),\psi(g_2)) = \sup_{|x-x_0|\leq\gamma} [|\psi(g_1)(x) - \psi(g_2)(x)|]$$

$$= \sup_{|x-x_0|\leq\gamma} \left| \int_{x_0}^{x} (F(t,g_1(t)) - F(t,g_2(t)))\,dt \right|$$

$$\leq \sup_{|x-x_0|\leq\gamma} L \int_{x_0}^{x} |g_1(t) - g_2(t)|\,dt \leq \gamma L d(g_1,g_2).$$

However, γ was chosen so that $\gamma L < 1$ [see (i)], thus proving that ψ is indeed a contraction of B. Again, as in Theorem 6, the unique fixed point of ψ is the desired solution of [5] and hence of [4]. ∎

The theorem on contracting maps not only assures the existence of a solution to [4] or [5], but provides an iterative procedure for obtaining that solution. This method is contained implicitly in the proof of Theorem 5 and is outlined in the exercises.

We conclude this section with an existence theorem for an integral equation.

9 Example. Let $K(s,t)$ be a continuous function in a square

$$S = \{(x,y): 0 \leq x,y \leq b\}.$$

Since S is compact, $K(s,t)$ is bounded in absolute value on this square by a positive number M. Let $F(x)$ be a function which is continuous on

$[0,b]$. We seek a continuous solution $f(x)$ of the integral equation

$$f(x) = F(x) + \int_0^\beta K(x,t) f(t)\, dt \tag{6}$$

for all values of x in a subinterval $[0,\beta]$ of $[0,b]$, where β is to be determined below. We begin by defining a mapping

$$\psi(g) = F(x) + \int_0^\beta K(x,t)\, g(t)\, dt \tag{7}$$

of $C[0,\beta]$ into itself. (The reader should verify this.) It will be shown that for sufficiently small β, ψ is a contraction. Its unique fixed point is the solution to [6].

If $g,h \in C[0,\beta]$, then

$$\begin{aligned} |\psi(g) - \psi(h)| &= \left| \int_0^\beta K(x,t)\,[g(t) - h(t)]\, dt \right| \\ &\le M\,\beta \max_{0 \le t \le \beta} |g(t) - h(t)| < \|g - h\|, \end{aligned}$$

provided that $M\,\beta < 1$. Thus, by restricting the size of β, ψ is a contraction, and the theorem is proved.

A highly readable and intuitive account of fixed point theorems may be found in an article by M. Shinbrot [1].

Exercises

1. Show that $y = \cos x$, $0 \le x \le 1$, has a fixed point. Find a sequence of approximations for this fixed point.
2. In what sense can it be said that the fixed point of a contraction is an "attracting point"? (See proof of Theorem 5.)
3. Show that if $f: X \to X$ is injective and has a fixed point, then the inverse mapping defined on $f(X)$ has also a fixed point.
4. Show that if f is differentiable on $[a,b]$, and if $0 \le f'(x) \le \lambda < 1$, then $y = f(x)$ has a fixed point.
5. Show that if $0 < \rho < 1/2$, then $f(x) = x^2$ is a contraction. Show also that if $\rho = 1/2$, then $f(x)$ is a nonexpanding map (Example 4).
6. Let $\rho > 0$, and let $f(x) = (1/2)(x + (\rho/x))$. Show that if f is defined on $[\sqrt{\rho}, \infty)$, then f has a fixed point.
7. Show directly (without using Theorem 5) that the *inhomogeneous* problem

$$\begin{aligned} y' &= y + h(x) \\ y(x_0) &= y_0 \end{aligned}$$

has a unique solution in some neighborhood of x_0. Assume that $h(x)$ is continuous wherever necessary.
8. Using the iterative procedure developed in Theorem 5, show that the unique solution of

$$y' = y + x$$
$$y(0) = 0$$

is $y = e^x - 1 - x$. (Write the corresponding integral equation and define the mapping ψ as in Theorem 5. The fixed point of ψ, which is the solution of the integral equation, is the limit of the sequence, $f_0 = 0, f_1 = \psi(f_0), \ldots, f_n = \psi(f_{n-1}), \ldots$. The f_n are called the *Picard iterants*.)

9. Using the Picard iterants, solve

(a) $\begin{cases} y' = xy \\ y(0) = 1 \end{cases}$ (b) $\begin{cases} y' = xy - e^x \\ y(0) = -1 \end{cases}$

***10.** Show that a nonexpanding map of a Banach space into itself has a fixed point. [*Hint:* Let ψ denote a nonexpanding map of X into itself, where X is a Banach space. The "approximating" contracting maps $\psi_n(x) = (1 - 1/n)\psi(x)$ will do it.]

11. Let S be a compact subset of a metric space X. Prove that a mapping f of S into itself has a unique fixed point if, for all $x,y \in S$,

$$d(f(x), f(y)) < d(x,y).$$

Using the circumference of the unit circle, show that this is false if it is assumed only that f is nonexpanding. Why doesn't this contradict Exercise 10? Show that if the circle is replaced by a line segment $[a,b]$, then there is a fixed point for every nonexpanding map.

12. Let $F(x_1, \ldots, x_n, z)$ be continuous for all values of z and for $a \le x_i \le b$. If F_z exists and is bounded from above and below by positive numbers $\mu < \lambda$, then there is a unique function $z = f(x_1, \ldots, x_n)$ such that

$$F(x_1, \ldots, x_n, f(x_1, \ldots, x_n)) \equiv 0,$$

whenever $a \le x_i \le b$. Prove this using a contraction.

13. Let $F(x,y)$ be a continuous function in R^2 which satisfies the following conditions:

(a) $F(x,y)$ satisfies a Lipschitz condition (in y).

(b) There is a number ρ for which $F(x + \rho, y) = F(x,y)$ for all values of x and y. ($F(x,y)$ is said to be *periodic* in the variable x and has period equal to ρ.)

(c) There are numbers ξ and ζ such that $F(x,\xi)F(x,\zeta) < 0$ for all values of x. Show that there is a periodic solution $f(x)$ of the equation $F(x,y) = y'$. (This means that $f(x + \rho) = f(x)$.) *Remark:* The preceding two problems may be found in Petrovski [1], Sec. 16 of Chapter 1. In Chapter 7, Brouwer's Fixed Point Theorem is proved, and applications are given to n-dimensional autonomous systems of differential equations.

*Exercises which request the proof of a statement made in the text, as well as those referred to in the text, are starred.

3 Complete Spaces and the Baire Category Theorem

None of the material covered in this section will be used in Part II, and with the exception of the Baire theorem itself, no subsequent discussion depends upon the results of this section. It suffices to know that the Baire theorem may be used to prove the existence of certain pathological functions. (See Theorems 20 and 25 and Corollaries 27 and 28.)

1 Definition. A subset S of a metric space X is said to be *nowhere dense* in X if $X - \overline{S}$ is dense in X; that is if $\overline{X - \overline{S}} = X$.

2 Examples. (a) The integers are nowhere dense in R.

(b) The space of constant functions is nowhere dense in any of the spaces $C[a,b]$, $S[a,b]$, $P[a,b]$ or $\Pi[a,b]$. This implies that a constant function may be uniformly approximated by nonconstant functions which are either continuous, sectionally continuous, piecewise linear, or polynomials.

3 Proposition. S is nowhere dense in X if and only if $\overline{S}$ contains no ϵ-neighborhoods of X; that is, if $x \in X$ and $\epsilon > 0$, then $S_\epsilon(x)$ contains some points of $X - \overline{S}$.

Proof: Suppose that $\overline{S}$ is nowhere dense in X. If $\overline{S}$ contains an ϵ-neighborhood of a point x, then $X - \overline{S} \subset X - S_\epsilon(x)$, and since the latter set is closed, it follows that

$$\overline{X - \overline{S}} \subset \overline{X - S_\epsilon(x)} = X - S_\epsilon(x) \subsetneq X,$$

proving that $X - S$ is *not* dense in X, or that it is false that S is nowhere dense in X.

Conversely, if $\overline{S}$ contains no neighborhoods, then every point x in $\overline{S}$ is an accumulation point of $X - \overline{S}$: For, if $x \in \overline{S}$, then every ϵ-neighborhood of x must contain points of $X - \overline{S}$, for otherwise $\overline{S}$ would contain a neighborhood. Thus x is in the closure of $X - \overline{S}$, proving that $\overline{X - \overline{S}} \supset S$. We therefore obtain

$$X = \overline{S} \cup X - \overline{S} = \overline{X - \overline{S}},$$

or that S is nowhere dense in X. ∎

We are now in a position to prove one version of the Baire Category theorem. The word "category," however, does not appear until the theorem is stated in a slightly different way. (See Definition 5 and Theorem 6.) The proof rests upon the Cantor Property of Nested Sets, which, it will be recalled, is equivalent to the completeness of a metric space (Theorem 3.1.2).

4 Theorem (Baire I). If $\{G_n\}$ is a sequence of open, dense, subsets of a complete metric space X, then $G = \bigcap G_n$ is dense in X.

Proof: We must prove that if $x \in X$, and $\delta > 0$, then there is a point $y_\delta \in G$ with $d(x,y_\delta) < \delta$. The point y_δ is obtained as the intersection of a sequence of closed, nested, spheres, whose diameters tend to zero.

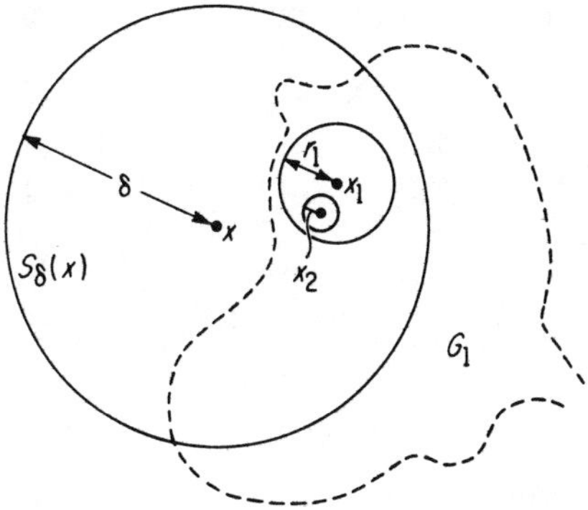

FIGURE 17

Let $x \in X$ and $\delta > 0$ be given. Since G_1 is dense in X, there is a point x_1 in the open set $G_1 \cap S_\delta(x)$. Let a number r_1 be chosen so that $0 < r_1 < \delta/2$ and

$$[1] \qquad S_1 = \overline{S_{r_1}(x_1)} \subset G_1 \cap S_\delta(x).$$

(See Figure 17.) Since G_2 is also dense in X, there is a point x_2 in the open set $G_2 \cap S_{r_1}(x_1)$ and a number r_2, $0 < r_2 < \delta/2^2$, with

$$[2] \qquad S_2 = \overline{S_{r_2}(x_2)} \subset G_2 \cap S_1.$$

Continuing in this way, a sequence $\{S_n\}$ of closed, nested spheres is constructed, whose diameters tend to zero. From [1] and [2], we obtain,

$$[3] \qquad S_n \subset G_1 \cap G_2 \cap \cdots \cap G_n \cap S_\delta(x).$$

Letting y_δ denote the (unique) point in the intersection of the S_n, it follows from [3] that $d(x,y_\delta) < \delta$. ∎

5 Definition. A metric space X is said to be *thin* or of *Category I* if it can be written as the countable union of subsets, each of which is nowhere dense in X.

A subspace S of X is called *thin* if S can be written as the countable union of sets which are nowhere dense in X.

If a metric space is not thin, it is said to be *thick* or of *Category II*.

Caveat: Any subset of the integers, when considered as a subspace of R (which is assumed to be equipped with the usual metric), is thin (why?). If however, we take $X = Z$, then no nonempty subset of the integers is thin, since each such subset possesses complete neighborhoods of all its points. Indeed, if n is a point of a subset S of Z, then $S_{1/2}(n) = \{n\}$, the set which consists of the single point n.

The Baire theorem may be restated as follows:

6 Theorem (Baire II). A (nonempty) complete metric space is thick (Category II).

Proof: If a metric space X is thin, there is a sequence of nowhere dense (in X) sets $\{S_n\}$ and $X = \bigcup S_n$. If each S_n is replaced by its closure, we obtain $X = \bigcup \overline{S}_n$; the $\overline{S}_n$ are closed nowhere dense subsets of X. It follows that the sets $G_n = X - \overline{S}_n$ are open and dense in X. The completeness of X and Theorem 4 imply that $G = \bigcap G_n$ is dense in X, and therefore not empty if X itself is not empty. However, $X = \bigcup \overline{S}_n$ implies that

$$\bigcap G_n = \bigcap (X - \overline{S}_n) = X - \bigcup (\overline{S}_n) = \emptyset,$$

which is a contradiction. ∎

Before giving some applications, we shall list some properties of thick (Category II) and thin (Category I) sets. Their verification is left to the reader.

7 Proposition. The empty set is of Category I.

8 Proposition. Every countable subspace of R is of Category I.

9 Proposition. A countable union of thin sets is thin.

10 Proposition. If X is thick and if $X = S \cup T$, then either S or T must be thick.

11 Proposition. If X contains a thick subspace, then X is thick.

12 Proposition. If S is a subset of R which contains an interval, then S is thick.

13 Proposition. If S is a dense subset of a complete metric space X, and if $S = \bigcap G_n$, where the G_n are open in X, then $X - S$ is thin.

Category arguments are frequently used to prove the existence of certain mathematical objects. For example, suppose that a subset S of a complete metric space X has been defined, and we wish to show that S is not empty. If it can be shown that $X - S$ is thin, then Proposition 10 implies that S is thick. It follows from Proposition 7 that $S \neq \emptyset$. This argument may be used to prove that there exist real numbers that are not rational. We use the fact that R has been defined as a *complete* metric space which contains Q, the rational numbers. Theorem 6 implies that R is thick. The countability of Q, and Propositions 8 and 10 imply that $R - Q$ is thick, and therefore not empty.

The proof just given for the existence of irrational numbers is certainly much shorter than the one given earlier which used decimals. However, the latter proof is constructive; that is, we can easily produce a decimal which "corresponds" to an irrational number. (See Sec. 4 of Chapter 0.) A Category argument, on the other hand, is nonconstructive, although it has the advantage that it leads to the conclusion that the subset of irrational numbers is *uncountable*; the constructive proof yielded but one irrational element. This will reoccur in Corollary 28, which asserts that the subset of $C[a,b]$ consisting of functions which are *nowhere* differentiable is dense in $C[a,b]$. This is very difficult to imagine, and the Category argument, being nonconstructive, does not immediately suggest a procedure for obtaining a concrete example. Examples of such pathological functions were, however, known in the nineteenth century (Exercises 17 and 18).

Several other types of Category arguments appear below, including one that uses Theorem 4 directly.

We state next a simple lemma about Riemann integrals which is needed in the proof of Theorem 16.

14 Lemma. If $f \in C[0,1]$, and if for all x in $[0,1]$,

$$F(x) = \int_0^x f(t)\,dt = 0,$$

then $f(x) \equiv 0$.

Proof: Let us assume that $x > 0$. A similar proof holds for $x = 0$. Choose a strictly increasing sequence of nonnegative numbers $\{x_n\}$ which converges to x. The mean value theorem for integrals implies

$$0 = F(x) - F(x_n) = \int_{x_n}^x f(t)\,dt = f(c_n)(x - x_n), \tag{4}$$

for some number c_n in the interval $[x_n,x]$. Since $(x - x_n) \neq 0$, it follows that $f(c_n) = 0$ for all n. Furthermore, $x_n \leq c_n \leq x$ implies that $\lim c_n = x$. Thus it follows from the continuity of f that

$$f(x) = \lim f(c_n) = 0. \quad \blacksquare$$

The principle of mathematical induction yields a generalization of Lemma 14:

15 Theorem. Suppose that $f \in C[0,1]$. Form the sequence of continuous functions,

$$f_1(x) = \int_0^x f(t)\,dt,\ f_2(x) = \int_0^x f_1(t)\,dt, \ldots, f_n(x) = \int_0^x f_{n-1}(t)\,dt, \ldots$$

If there is an integer ν for which $f_\nu(x) \equiv 0$, then $f(x) \equiv 0$.

The proof is left to the reader.

The more general theorem, which follows, requires a Category argument in its proof. The reader should convince himself that it is not possible to prove Theorem 16 by varying the induction proof of Theorem 15.

16 Theorem. Let f and f_n have the same meanings as in Theorem 15. If to each $x \in [0,1]$, there is an integer $k = k(x)$ such that $f_k(x) = 0$, then $f(x) \equiv 0$.

Proof: For each k, let

$$F_k = \{x : 0 \leq x \leq 1 \quad \text{and} \quad f_k(x) = 0\}.$$

We have assumed that if $x \in [0,1]$, then $f_k(x) = 0$, for some integer k which depends upon x. Thus

$$[0,1] = \bigcup F_k.$$

The continuity of f, and hence of the f_k, implies that the sets F_k are closed. Since $[0,1]$ is of Category II, it can not be that all the F_k are nowhere dense in R. Thus one of them, call it F_k, must contain an interval I_k, which may be assumed to be closed. Since $f_k(x) \equiv 0$ on I_k, it follows from Theorem 15 that $f(x) \equiv 0$ on I_k. If $I_k = [0,1]$ we are finished. If not, write $[0,1] = S \cup S_0$, where $S = [f \neq 0]$ and $S_0 = [f = 0] \supset I_k$. We wish to show that $S = \phi$. Since S_0 is a closed set, it follows that S is open in $[0,1]$. (This means that $S = G \cap [0,1]$, where G is an open subset of R.) If S is not empty, it must contain an interval $[a,b]$. By repeating the Category argument for $[a,b]$, we obtain $[a,b] = \bigcup F'_n$, where F'_n is the closed subset of $[a,b]$ on which $f_n(x) \equiv 0$. Again, one of the F'_n must contain an interval on which $f(x) \equiv 0$. But then S contains a nonempty subset on which f vanishes, contrary to assumption. Thus $S = \phi$ and $f(x) \equiv 0$ on $[0,1]$. ▌

We prove next the corresponding theorem about derivatives. Again, we begin with a theorem that may be proved by elementary calculus:

17 Theorem. Let $f \in C_\infty[0,1]$, the space of real-valued functions defined on $[0,1]$ which have continuous derivatives of all orders. If there is a positive number k such that

$$\frac{d^k f(x)}{dx^k} = f^{(k)}(x) = 0,$$

for all $x \in [0,1]$, then f is equal to a polynomial on $[0,1]$ whose degree does not exceed $k - 1$.

The proof is left to the reader.

18 Theorem. Let $f \in C_\infty[0,1]$. If to each $x \in [0,1]$, there is an integer $k = k(x)$, for which $f^{(k)}(x) = 0$, then $f(x)$ is equal to a polynomial function on $[0,1]$.

Proof: Again the sets $F_k = [f^{(k)} = 0]$ are closed and $[0,1] = \bigcup F_k$. Since $[0,1]$ is thick, at least one F_k must contain an interval I. We may assume that I possesses the following maximality property: If J is also an interval and if $I \subset J \subset F_k$, then $I = J$. The continuity of the derivatives insures that I is closed. From Theorem 17, it follows that there is a polynomial $P_{k-1}(x)$ whose degree does not exceed $k - 1$, and $f(x) \equiv P_{k-1}(x)$ on I.

Although I is not contained in a larger subinterval of F_k, it may be contained in a maximal interval J of some other F_q. In this case, since a polynomial is determined by its values at finitely many points, it follows that f is equal to a polynomial, on the larger of these two intervals, whose degree is not greater than min (k,q). More generally, if $I_{k_1} \subset I_{k_2} \subset \cdots$ is a chain of maximal intervals each contained in some F_{k_j}, then a similar argument implies that on the closed interval,

$$I = \overline{\left(\bigcup_{j=1}^{\infty} I_{k_j}\right)},$$

f is equal to a polynomial whose degree does not exceed $\nu = \min [k_1, k_2, \ldots]$, from which it follows that $I \subset F_\nu$. If I itself is not maximal, it is contained in a maximal interval of F_ν. We see therefore that it is possible to discard from the collection of all maximal intervals, those which are contained in larger ones. Let $\{I_n\}$ denote the sequence of closed, disjoint, maximal, intervals obtained by discarding all those which are contained in larger ones. Each I_n is contained in some $F_{i(n)}$, and on I_n, $f = P_{i(n)-1}$, a polynomial whose degree does not exceed $i(n) - 1$. It is easily seen that $\bigcup I_n$ is dense in $[0,1]$. If not, the relatively open set $[0,1] - \overline{\bigcup I_n}$ would contain an interval to which we could apply the same argument, thus obtaining additional intervals on which f may be expressed as a polynomial. Denoting the interior of each I_n by $\mathring{I}_n$, it follows that the set $\bigcup \mathring{I}_n$ is also dense in $[0,1]$ and therefore the set

$$H = [0,1] - \bigcup \mathring{I}_n$$

is both closed and nowhere dense in $[0,1]$. Furthermore each point of H is a point of accumulation of H. For if x is a point of H that has a neighborhood $(x - \epsilon, x + \epsilon)$ containing no other point of H, then x must be the common endpoint of two intervals I_j and I_k. These intervals are however closed and disjoint. Otherwise $I_j \cup I_k$ would be a larger interval on which f could be written as a polynomial function (Exercise 22). This would contradict the assumption that the I_j were not contained in larger intervals on which f is a polynomial.

We show next that the assumption $H \neq \phi$ leads to a contradiction:

Since H is a closed subset of a complete metric space, it is itself a complete metric space, and therefore the Baire theorem implies that it is of Category II. Thus, from $H = \bigcup(F_n \cap H)$, it follows that for some integer ν the set $H_\nu = F_\nu \cap H$ contains a neighborhood of H. This means that there is an interval J for which

$$\phi \neq J \cap H \subset H_\nu \subset F_\nu, \qquad \text{and} \qquad f^{(\nu)}(x) \equiv 0 \quad \text{on } J \cap H.$$

Since J is an interval and H is nowhere dense in $[0,1]$, it follows that $J - H$ consists of open intervals each of which is contained in or equal to one of the original $\mathring{I}_j$. However each $\mathring{I}_j$ is contained in some $F_{n(j)}$ so that $f^{(n(j))}(x) \equiv 0$ on $\mathring{I}_j$. There are two possibilities:

(i) If $n(j) \leq \nu$, then $f^{(\nu)}(x) \equiv 0$ on $\mathring{I}_j$ since

$$0 \equiv f^{(n(j))}(x) \equiv f^{(n(j)+1)}(x) \equiv \cdots \equiv f^{(\nu)}(x).$$

(ii) If $n(j) > \nu$, then $f^{(\nu)}(x) = f^{(\nu+1)}(x) = \cdots = f^{(n(j))}(x) = 0$ at the endpoints of $\mathring{I}_j$, since these endpoints are contained in the closed set $F_{n(j)}$. By integrating $f^{(n(j))}(x)$ sufficiently many times, we obtain $f^{(\nu)}(x) = 0$ in $\mathring{I}_j$. Thus $f^{(\nu)}(x) = 0$ in every open interval of $J - H$ as well as at the endpoints of $H \cap J$, and therefore in the entire interval J. Hence $J \subset F_\nu$, from which it follows that $J \cap H = \phi$, contrary to assumption.

This proves that H is empty and that $f(x)$ may be written as a single polynomial on $[0,1]$. ∎

19 Definition. A real valued function $f(x)$ defined on a subset S of R is said to be *nondecreasing* if

$$x_1 < x_2 \qquad \text{implies} \qquad f(x_1) \leq f(x_2).$$

f is an *increasing* function if

$$x_1 < x_2 \qquad \text{implies} \qquad f(x_1) < f_2(x).$$

A similar definition may be given for *nonincreasing* and *decreasing* functions.

All four types of functions just described are called *monotone* functions.

The reader will have no difficulty in giving examples of both monotone and nonmonotone functions. In particular, the function $y = \sin x$ is an example of a nonmonotone function if $S = R$. However on certain subsets of R, $\sin x$ is a monotone function. It is very likely that the reader will not be able to imagine a continuous function which is *not* monotonic on *any* interval. Such a function is said to be *nowhere monotonic* or *everywhere oscillating*. Discontinuous nowhere monotonic functions are readily given. For example,

$$f(x) = \begin{cases} 1 & \text{whenever } x \text{ is rational} \\ 0 & \text{whenever } x \text{ is irrational} \end{cases}$$

is not monotonic on any interval. It is however discontinuous at all points. We prove next that not only does $C[0,1]$ contain everywhere oscillating functions, but that the subset of these functions is dense in $C[0,1]$.

20 Theorem. The set of continuous everywhere oscillating functions is dense in $C[0,1]$.

Proof: Let $\{I_n\}$ be the sequence of all closed intervals of $[0,1]$ which have rational endpoints, and let G_n denote the subset of $C[0,1]$ of functions which are not monotone on I_n. The idea of the proof is to show that each G_n is both open and dense in $C[0,1]$, and then to use the Baire category theorem to obtain the desired result.

If $f \in G_n$, then there is a subinterval $[c,d]$ of I_n on which both of the following statements are false:

[i] $f(c) = \min\limits_{c \le x \le d} f(x)$ and $f(d) = \max\limits_{c \le x \le d} f(x)$; and

[ii] $f(c) = \max\limits_{c \le x \le d} f(x)$ and $f(d) = \min\limits_{c \le x \le d} f(x)$.

This means that $f(x)$ attains a relative maximum or minimum at an interior point ζ of (c,d). Let us assume that $(\zeta, f(\zeta))$ is a maximum point on the curve. (See Figure 18.) Set

$$3\epsilon = \min\,[(f(\zeta) - f(c)),\ (f(\zeta) - f(d))].$$

It is easily seen that if

$$\|f - g\| = \max_{0 \le x \le 1} |f(x) - g(x)| < \epsilon,$$

then $g(x)$ is not monotonic on $[c,d]$. Thus the neighborhood $S_\epsilon(f)$ is contained entirely in G_n, proving that G_n is an open set.

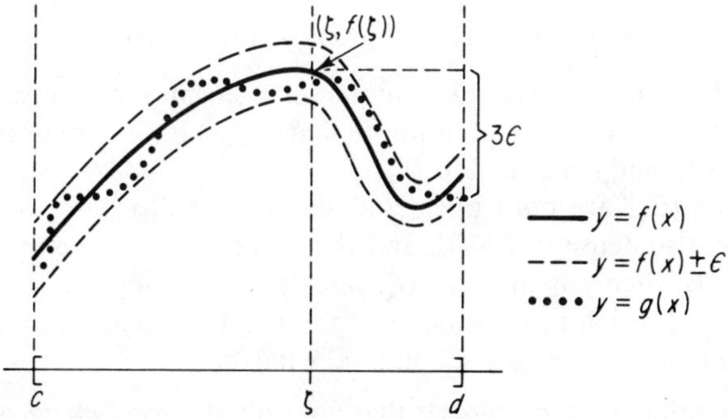

FIGURE 18

To prove that G_n is dense in $C[0,1]$, it suffices to prove that G_n is dense in $P[0,1]$, the subspace of piecewise linear functions, since $P[0,1]$ is dense in $C[0,1]$ (Exercise 4.1.8). Let $f(x)$ be in $P[0,1] - G_n$. Then $f(x)$ is piecewise linear and is monotone on I_n. We shall assume that $f(x)$ is nondecreasing on I_n (Figure 19). Let $[a,b]$ be a subinterval of I_n

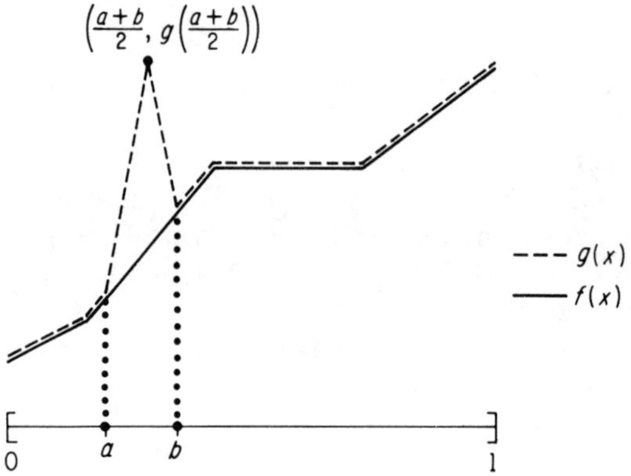

FIGURE 19

on which $f(x)$ is linear. If m is the slope of this linear segment, we define the function

$$\mu(x) = \begin{cases} 0 & \text{if} \quad 0 \leq x \leq a \\ 2m(x - a) & \text{if} \quad a < x \leq (a + b)/2 \\ -2m(x - b) & \text{if} \quad (a + b)/2 < x \leq b \\ 0 & \text{if} \quad b < x \leq 1 \end{cases}$$

which is also piecewise linear. The function $g(x) = f(x) + \mu(x)$ (the slashed line of Figure 19) is not monotone on I_n, since $g([a + b]/2)$ exceeds both $g(a)$ and $g(b)$. However, for all $x \in [0,1]$,

$$|g(x) - f(x)| = |\mu(x)| < m(b - a).$$

Therefore if $\epsilon > 0$ is given, the subinterval $[a,b]$ may be chosen so that $m(b - a) < \epsilon$, from which it follows that $\|f - g\| < \epsilon$ and that G_n is dense in $P[0,1]$, and hence in $C[0,1]$.

Since the $\{G_n\}$ are both open and dense in $C[0,1]$, the intersection $G = \bigcap G_n$ is also dense in $C[0,1]$, and therefore can not be empty. However, if $f \in G$, then f is in every G_n, and is not monotonic on any subinterval of $[0,1]$ which has rational endpoints. This clearly implies that f can not be monotonic on any subinterval whatsoever. ∎

The preceding theorem asserts that not only do everywhere oscillating functions exist, but that they are dense in $C[0,1]$. However, as noted earlier, no such function is constructed in the proof. In the Exercises, several examples of continuous nowhere differentiable functions are given. Each of these is also an everywhere oscillating function.

The following remarks and definitions are preparatory for the proof of a very general theorem, which has as a corollary, the existence of nowhere differentiable functions.

21 Definition. A continuous function $\nu_c(\delta)$ defined for nonnegative values of δ, is said to be a *modulus* of *continuity* of a *function* f at some point c of the interval I over which f is defined, if

(i) $\nu_c(0) = 0$,
(ii) $\delta < \delta'$ implies $\nu_c(\delta) \le \nu_c(\delta')$, and
(iii) $|f(x) - f(c)| \le \nu(|x - c|)$, for all x in I.

22 Proposition. If $f(x)$ is continuous on an interval I, then $f(x)$ has a modulus of continuity at each point c in I.

Proof: If $c \in I$, let $\nu_c(\delta) = \sup_{|x-c| \le \delta} |f(x) - f(c)|$; it is assumed that $x \in I$. Properties (i), (ii), and (iii) of Definition 21 follow immediately. ∎

23 Definition. If at some point c, $f(x)$ has a modulus of continuity of the form

$$\nu_c(\delta) = K\delta^\alpha,$$

where $K > 0$ and $0 < \alpha \le 1$, then f is said to be *Hölder continuous* at c with exponent α.

Verification of the following simple properties is left to the reader:

24 Proposition. (a) If $f(x)$ is Hölder continuous at c then $f(x)$ is continuous at c.

(b) If $f(x)$ is Hölder continuous at c with exponent $\alpha = 1$, then the difference quotients, $|[f(x) - f(c)]/[x - c]|$, are bounded.

Remark. If a continuous function of two variables, $f(x,y)$ satisfies a Lipschitz condition in y (Definition 4.2.7), then it is Hölder continuous for all y, and $\alpha = 1$. Furthermore the same K works at all points.

Query. Why have we not defined Hölder continuity with exponent $\alpha > 1$?

The principal result contained in Theorem 25 and Corollaries 26 and 27 is that the subset of functions in $C[0,1]$ which do not satisfy a Hölder condition at any point is dense in $C[0,1]$. In particular, by taking $\alpha = 1$, it follows that the continuous functions on $[0,1]$ whose difference quotients are nowhere bounded, are dense in $C[0,1]$. Clearly, a function cannot be differentiable at a point if its difference quotients are unbounded at that point. Thus the continuous nowhere differentiable functions are dense in $C[0,1]$.

The first step is to prove that if $\nu(\delta)$ is a given modulus of continuity, then the set of continuous functions on $[0,1]$ which do not have $\nu(\delta)$ as a modulus of continuity at any point of $[0,1]$, is dense in $C[0,1]$. This part of the argument does not require a Category argument. The second step, however, requires the Baire theorem: We show that the continuous, nowhere Hölder continuous, functions can be written as a countable intersection $\{G_n\}$ of sets which are open and dense in $C[0,1]$. The completeness of $C[0,1]$ implies that $\bigcap G_n \neq \phi$.

25 Theorem. Let $\nu(\delta)$ be any function. Then the subset G of $C[0,1]$ consisting of those functions for which $\nu(\delta)$ is not a modulus of continuity at any point of $[0,1]$, is both open and dense in $C[0,1]$.

Proof: We may assume that $\nu(\delta)$ satisfies the conditions of Definition 21, for otherwise it could not be the modulus of continuity of any function, and in that case we would have $G = C[0,1]$.

Set $F = C[0,1] - G$. This is the set of continuous functions for which $\nu(\delta)$ is a modulus of continuity at some point in $[0,1]$. We shall prove that F is closed by showing that if f is a point of accumulation of F, then it is in F. Let $\{f_n]$ be a sequence of functions in F which converges to f in norm; that is,

$$\lim \|f - f_n\| = \lim \max_{0 \leq x \leq 1} |f(x) - f_n(x)| = 0.$$

For each n, let x_n be a point at which f_n has $\nu(\delta)$ as a modulus of continuity. This means that for all $x \in [0,1]$,

$$|f_n(x_n) - f_n(x)| \leq \nu(|x_n - x|).$$

Since $\{x_n\}$ is a bounded sequence of real numbers, it contains a subsequence $\{x_{n_j}\}$ which converges to a point ξ in $[0,1]$. It is easily seen that $\nu(\delta)$ is a modulus of continuity for f at ξ:

$$[5] \qquad |f(\xi) - f(x)| \leq |f(\xi) - f(x_{n_j})| + |f(x_{n_j}) - f_{n_j}(x_{n_j})| \\ + |f_{n_j}(x_{n_j}) - f_{n_j}(x)| + |f_{n_j}(x) - f(x)|.$$

The second and fourth terms are bounded by $\|f_{n_j} - f\|$, which tends to zero as n_j tends to infinity. Since each f_{n_j} has $\nu(\delta)$ as a modulus of continuity at the point x_{n_j}, [5] becomes

$$|f(\xi) - f(x)| \leq |f(\xi) - f(x_{n_j})| + 2\|f_{n_j} - f\| + \nu(|x_{n_j} - x|).$$

This inequality holds for all n_j. Therefore, letting $n_j \rightarrow \infty$, it is easily seen that each of the three terms tends to zero: The continuity of f and $\lim x_{n_j} = \xi$ imply $\lim f(x_{n_j}) = f(\xi)$; the convergence of $\{f_n\}$ to f implies that the subsequence $\{f_{n_j}\}$ converges also to f; finally, the continuity of $\nu(\delta)$ yields $\lim \nu(|x_{n_j} - x|) = \nu(|\xi - x|)$, from which it follows that

$$|f(\xi) - f(x)| \leq \nu(|\xi - x|).$$

Thus $\nu(\delta)$ is a modulus of continuity for f at some point of $[0,1]$, proving that f belongs to F. Thus F is closed, and therefore its complement G is an open subset of $C[0,1]$.

We show next that G is dense in $P[0,1]$, the subspace of piecewise linear functions. As in the proof of Theorem 20, this implies that G is dense in $C[0,1]$. Let $p(x) \in P[0,1]$. Then $p(x)$ is linear on a set of subintervals $[x_i,x_{i+1}]$, $0 < x_0 < x_1 < \cdots < x_n = 1$. For a given positive number ϵ, we must find a function f in G such that

$$\|f - p\| = \max_{0 \leq x \leq 1} |f(x) - p(x)| < \epsilon.$$

This is done by adding to $p(x)$ a sufficiently steep "sawtooth" function, $\mu(x)$. (See Figure 20.)

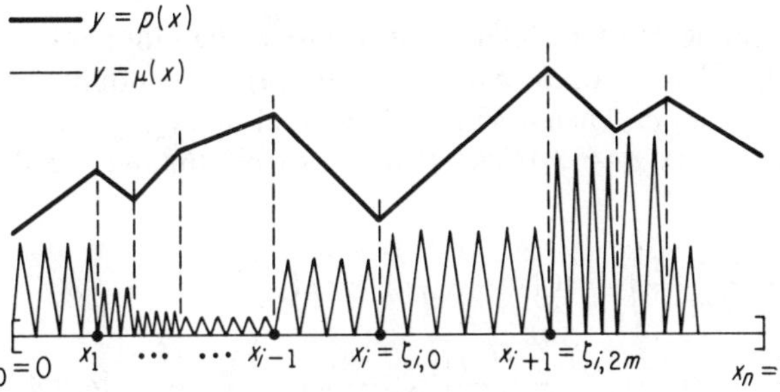

FIGURE 20

We begin by subdividing the intervals $[x_i,x_{i+1}]$, $i = 1,2,\ldots,n-1$, into an even number, say $2m$, of equal subintervals,

$$x_i = \zeta_{i,0} < \zeta_{i,1} < \cdots < \zeta_{i,2m} = x_{i+1}.$$

The integer m will be determined in the course of the proof. Let

$$\mu(x) = \begin{cases} \epsilon \left| \dfrac{x - \zeta_{i,j}}{\zeta_{i,j+1} - \zeta_{i,j}} \right| & \text{when } j \text{ is even and } \zeta_i \leq x \leq \zeta_{i,j+1}, \\ \epsilon \left(- \left| \dfrac{x - \zeta_{i,j}}{\zeta_{i,j+1} - \zeta_{i,j}} \right| \right) & \text{when } j \text{ is odd and } \zeta_i \leq x \leq \zeta_{i,j+1} \end{cases}$$

We estimate the slope of the "sawtooth" function $\mu(x)$, which is shown together with $p(x)$ in Figure 20:

$$|\text{slope of } \mu(x)| = \left| \frac{\epsilon}{\zeta_{i,j+1} - \zeta_{i,j}} \right| = \left| \frac{2\epsilon m}{(x_{i+1} - x_i)}, \right|$$

whenever $\zeta_{i,j} \leq x \leq \zeta_{i,j+1}$. Furthermore, $|\mu(x)| < \epsilon$.

Let $f(x) = p(x) + \mu(x)$. Certainly $f \in P[0,1]$, and

$$\max_{0 \le x \le 1} |f(x) - p(x)| = \max_{0 \le x \le 1} |\mu(x)| \le \epsilon.$$

It only remains to be shown that for sufficiently large values of m, f is in G; that is, f will not have $\nu(\delta)$ as a modulus of continuity at any point.

Let $M = \max |\text{slope of } p(x)|$. If x is in the interior of an interval of linearity of $f(x) = p(x) + \mu(x)$, then

$$[6] \qquad |\text{slope of } f(x)| > |\text{slope of } \mu(x)| - |\text{slope of } p(x)|$$
$$> \frac{2\epsilon m}{x_{i+1} - x_i} - M > \epsilon \cdot m \cdot A - M,$$

where A is chosen so that

$$\frac{2}{x_{i+1} - x_i} > A > 0, \qquad \text{for } i = 0, 1, \ldots, n-1.$$

If x is the endpoint of an interval of linearity, then [6] holds for both the left- and right-hand slopes. Suppose that $\nu(\delta)$ is a modulus of continuity of $f(x)$ at a point c. Then c is in some interval $[\zeta_{i,j}, \zeta_{i,j+1}]$, and $(c - \zeta_{i,j}) \le (\zeta_{i,j+1} - \zeta_{i,j}) = (x_{i+1} - x_i)/2m$. It follows from [6] and the definition of $\nu(\delta)$ that

$$[7] \qquad \nu\left(\frac{x_{i+1} - x_i}{2m}\right) \ge \nu(|c - \zeta_{i,j}|) \ge |f(\zeta_{i,j}) - f(c)|$$
$$= \left|\frac{f(\zeta_{i,j}) - f(c)}{(\zeta_{i,j} - c)}\right| |\zeta_{i,j} - c| > |\zeta_{i,j} - c| [\epsilon m A - M].$$

We see therefore that $\nu(\delta)$ is a modulus of continuity of $f(x)$ at a point c only if [7] is satisfied. This leads to a contradiction:

$$\lim_{m \to \infty} \frac{x_{i+1} - x_i}{2m} = 0;$$

but from [7] it follows that

$$\lim_{m \to \infty} \nu\left(\frac{x_{i+1} - x_i}{2m}\right) = \infty,$$

which violates the continuity of $\nu(\delta)$ at $\delta = 0$. ∎

26 Corollary. If $\{\nu^{(n)}(\delta)\}$ is a sequence of functions, then the subset G of $C[0,1]$ which consists of those functions which do not have any of the $\nu^{(n)}(\delta)$ as a modulus of continuity at any point of [0,1], is dense in $C[0,1]$.

Proof: For $n = 1, 2, \ldots$, let $\{G_n\}$ be the open, dense subsets of $C[0,1]$ of functions that do not have $\nu^{(n)}(\delta)$ as a modulus of continuity at any point of [0,1]. The Baire Theorem (4) implies that $\bigcap G_n$ is dense in $C[0,1]$. ∎

27 Corollary. The subset of $C[0,1]$ which consists of those functions that satisfy no Hölder condition at any point of $[0,1]$ is dense in $C[0,1]$.

Proof: The set of functions $\nu^{(n,m)}(\delta) = n\delta^{1/m}$ is countable, and if $f(x)$ is Hölder continuous at a point c in $[0,1]$, then one of these functions is a modulus of continuity for $f(x)$ at some c. The desired result follows from Corollary 26. ∎

If m is taken to be 1 in Corollary 27, we see that the set of continuous functions whose difference quotients are not bounded at any point, is dense in $C[0,1]$. These functions are however, a subset of the continuous nowhere differentiable functions, which proves the final corollary:

28 Corollary. The subset of continuous, nowhere differentiable functions in $C[0,1]$ is dense in $C[0,1]$.

Exercises

***1–7.** Prove Propositions 7–13.

***8.** Using Mathematical Induction, prove Theorem 15.

***9.** Prove Theorem 17.

10. Show that the constant functions are nowhere dense in $C[a,b]$.

11. Show that if S is thin and if $T \subset S$, then T is thin.

12. Suppose that X is a metric space and $X \supset T \supset S$. Show that:
(a) T is nowhere dense in X implies S is nowhere dense in X;
(b) S is nowhere dense in T implies S is nowhere dense in X;
(c) $\bar{S} - S$ is nowhere dense in X.

***13.** Prove that if X is of Category II, then if $\{G_n\}$ is a sequence of open, dense subsets of X, then $\bigcap G_n$ is dense in X. What can you say about $X - \bigcap G_n$?

14. Prove that a thick space is not necessarily complete.

15. Show that a nonempty closed subset of a complete metric space is thick.

16. Let the rational numbers be written as a sequence $\{q_n\}$. The subsets

$$\lambda_{n,m} = \{(x,y) : y = q_n x + q_m\}$$

of R^2 are the lines with rational slopes and rational y intercepts. Show that the union

$$\lambda = \bigcup_{n,m=-\infty}^{\infty} \lambda_{n,m}$$

is of Category I in R^2.

***17.** The continuous function $y = |x|$, which has no derivative at $x = 0$, suggests a method for defining a function which is continuous at all points of a bounded interval, but fails to have a derivative at points of a dense countable set. For convenience, let $[0,1]$ be the interval, and let $\{a_n\}$ be some countable subset of $[0,1]$, for example, the rational numbers in this interval. Then the series

$$\left| \sum_{1}^{\infty} \frac{x - a_n}{3^n} \right|$$

converges uniformly and absolutely in [0,1]. Prove this by verifying that the partial sums are bounded by a convergent geometric series. Letting $f(x)$ denote this function, show that if $x = a_n$, then the difference quotients

$$\frac{f(a_n + \Delta x) - f(a_n)}{\Delta x}$$

tend to different limits when Δx is positive or negative. Compare this with the behavior of $y = |x|$. (*Hint:* Write the difference quotient as a sum of three terms, the first a sum from 1 to $n - 1$, the second a single term corresponding to the nth term, and the third the remainder of the infinite sum.)

The preceding exercise, as well as the following two, appear in Titchmarsh [1].

18. *Weierstrass's Continuous Nondifferentiable Function.* Let a denote a positive odd integer and let $0 < b < 1$. Since the series

$$f(x) = \sum_{n=0}^{\infty} b^n \cos(a^n \pi x)$$

converges uniformly in any interval, and since its partial sums are continuous, it follows that f itself is continuous. However, if a and b are chosen so that $ab > 1$, the differentiated series diverges. This is proved in Titchmarsh [1] for $ab > 1 + (3/2)\pi$.

19. *Van der Waerden's Continuous Nondifferentiable Function.* For each real value of x, let $f_n(x)$ denote the value of the distance between x and the nearest number which can be written in the form $m/10n$, where m is an integer. The function

$$f(x) = \sum_{1}^{\infty} f_n(x)$$

is continuous and nondifferentiable at all points. The continuity may again be proved by taking into account the continuity of the f_n (why are they continuous?) and the uniform convergence of the series. The decimal representation of a real number may be used to prove that the difference quotients do not converge: If $x = .a_1 a_2 \ldots a_m \ldots$ is a real decimal in the interval (0,1), we define a sequence of real numbers $\{x_m\}$ which converges to x, but whose difference quotients diverge. Set

$$x_m = \begin{cases} x - 10^{-m} & \text{if } a_m = 4 \text{ or } 9 \\ x + 10^{-m} & \text{if } a_m \neq 4 \text{ or } 9. \end{cases}$$

Show that

$$f_n(x_m) - f_n(x) = \begin{cases} \pm(x_m - x) & \text{if } n < m \\ 0 & \text{if } n \geq m, \end{cases}$$

and that this implies that the difference quotients diverge.

Remark. It can be shown directly that the functions described in the preceding exercises are not monotone in any interval. The reader is asked to do this in Exercises 20 and 21. Later, in Chapter 9, we prove that if a function is monotone over an interval, then it must have derivatives at some points (in fact at "almost all" points) of the interval. Therefore the nondifferentiability of the preceding functions would imply that they are everywhere oscillating.

20. Show that the function described in Exercise 18 is everywhere oscillating.

21. Show that the function of Exercise 19 is everywhere oscillating.

22. Let I and J be intervals that have in common a single endpoint. If f is function which has continuous derivatives of all orders on an interval which contains both I and J, and if f is equal to a polynomial function P_1 on I and a polynomial function P_2 on J, then $P_1 \equiv P_2$. Prove it.

*Exercises which request the proof of a statement made in the text, as well as those referred to in the text, are starred.

PART II

Introduction

There are many ways to extend the concept of the Riemann integral of a real function of a single real variable. Leaving aside (temporarily) those generalizations that lead to integrals which are defined on more abstract spaces, there remain several standard extensions.

One familiar extension is the *Cauchy-Riemann integral*, or the *improper integral*. It is easily seen that the Riemann integral does not exist for unbounded functions (Definition 6.3.2 and Exercise 6.3.1). However, in some cases it is possible to assign a value to an integral of an unbounded function by taking limits. For example, the improper integral $\int_0^1 \frac{dx}{\sqrt{x}}$ is defined as $\lim_{\epsilon \to +0} \int_\epsilon^1 \frac{dx}{\sqrt{x}}$, which is easily seen to converge. Similar extensions can be made to integrals defined over unbounded intervals. Thus the Cauchy-Riemann integral is an extension which enlarges the class of integrable functions.

A second extension is the *Stieltjes integral*. Instead of enlarging the class of integrable functions, this integral introduces new "measures" on the real line. Again, the Riemann integral may be considered a special case of the Stieltjes integral. (See Sec. 4 of Chapter 11.)

The Lebesgue integral, however, may be thought of as an extension of the Riemann integral which, by enlarging the class of integrable functions, removes certain "defects" of the latter which arise from the incompleteness of the normed linear space of Riemann integrable functions. (The norm defined on this space is the integral norm; that is, if f is a Riemann integrable function that vanishes outside an interval $[a,b]$, then $\|f\| = \int_a^b |f|$.) A typical defect of the Riemann integral is the frequent failure of a limit of integrable functions to be itself integrable. For example, suppose that $\{f_n\}$ is a nondecreasing sequence of (Riemann) integrable functions whose integrals are uniformly bounded. Theorem 1.3.18 implies that the sequence of integrals $\left\{\int_a^b f_n\right\}$ converges to a real number L. However, it is not generally possible to draw any conclusions about the convergence of the sequence of functions, in particular, whether or not it converges to an integrable function. Only if it *assumed* that an integrable limit function exists, may we conclude that

$$\lim \int_a^b f_n = \int_a^b \lim f_n.$$

By completing the space in the norm defined above, the Lebesgue integrable functions are obtained, and in this larger complete space the limit of a monotone sequence of integrable functions whose integrals are uniformly bounded is integrable. The Lebesgue integral is simply the extension of the Riemann integral to the completion.

To avoid using any theorems about Riemann integrable functions, we shall begin by defining an integral on a simpler class of functions, the real step functions (See Sec. 2 of Chapter 2 and Example 2.3.19). The integral of a step function is taken to be the (finite) sum of the signed areas under the "steps." So far, this coincides with the construction of the Riemann integral, and in Sec. 3 of Chapter 6 we shall see that the Riemann integral may be defined as the "limit" of approximating sums; each sum is the integral of an approximating step function. We shall, however, begin by defining the Lebesgue integral, and later consider the Riemann integral as the restriction of the Lebesgue integral to a smaller class of functions.

We have already seen that the integral of the absolute value of a step function defines a norm on the space Σ of step functions, or, strictly speaking, on the space of equivalence classes of step functions. In Sec. 2 of Chapter 5, we shall show that the normed space of real step functions is incomplete. Its unique completion (Theorem 2.4.7) is described formally as the normed linear space of equivalence classes of Cauchy sequences of step functions! This sounds as formidable as our initial definition of a real number as an equivalence class of Cauchy sequences of rational numbers. However, in the latter case we were able to characterize the real numbers as a complete, Archimedean ordered field, and to describe a real irrational number as the *limit* of a sequence, in fact a monotone sequence, of rational numbers. Fortunately, it is possible to do something quite similar for the Lebesgue integrable functions. They will turn out to be "limits" of Cauchy sequences of step functions, and although it is not in general possible to find a monotone sequence which converges to a given integrable function, we shall see that every integrable function may be written as the difference of two integrable functions, each of which may be approximated by monotone sequences of step functions.

After proving the basic convergence theorems, (Sec. 1 and 2 of Chapter 6) as well as showing that the Lebesgue integral is truly an extension of the Riemann integral, in the sense that every Riemann integrable function is Lebesgue integrable, and that the two integrals coincide on the space of Riemann integrable functions, we shall introduce a *measure* on certain subsets of R, which agrees with the Riemann measure on a subclass of these sets. In particular, the measure of an interval is its length.

In Chapter 8 several alternative ways to define the Lebesgue integral and measure are discussed. Most important historically is Lebesgue's method. Here we begin with the concept of the measurability of a set, taking as our starting point the length of an interval to be its measure. From there we proceed in a manner reminiscent of the construction of the Riemann integral, but with one crucial change, to define the Lebesgue integral as a "limit" of approximating sums. Perhaps the most lucid and straightforward commentary on Lebesgue's method, as well as some general remarks on integration may be found in Lebesgue's expository writings. In particular, the reader is advised to look at Lebesgue's *Development of the Integral Concept*, which appears in an English translation in *Measure and the Integral.* (See Lebesgue [1].)

The concluding chapter of this section treats the subject of differentiation, and the appropriate "Fundamental" theorem, relating integration and differentiation, is proved (Theorems 9.4.7 and 9.4.12).

CHAPTER **5**

Lebesgue Integrable Functions

1 Step Functions and Their Integrals

The reader will recall that a step function $\sigma(x)$ is constant over finitely many open intervals, takes on arbitrary values at the endpoints of these intervals, and vanishes outside a bounded interval. That is, if $[a,b]$ is an interval outside of which $\sigma(x)$ vanishes, then there are finitely many points,

$$a = a_0 < a_1 < \cdots < a_n = b,$$

and

$$\sigma(x) = \begin{cases} \alpha_i & \text{whenever } a_{i-1} < x < a_i, \quad i = 1, 2, \ldots, n \\ \lambda_i & \text{whenever } x = a_i, \quad i = 0, 1, \ldots, n \\ 0 & \text{otherwise} \end{cases} \tag{1}$$

By defining two step functions to be equivalent if they are equal to each other except at finitely many points, the class Σ is partitioned into disjoint equivalence classes $[\sigma]$. We have already shown that the space of equivalence classes, $\bar{\Sigma}$, may be made into a normed vector space by defining the norm of the class which contains σ to be

$$\sum_{i=1}^{n} |\alpha_i| (a_i - a_{i-1}), \tag{2}$$

which is the sum of the areas under the "steps" of σ. Since [2] does not depend upon the values of $\sigma(x)$ at the end points a_i, it follows that the norm is well defined; that is, it is independent of the step function chosen to represent its class. To avoid the formality of equivalence classes, a "special" step function may be taken from each class as the representative. This step function takes on the average value at the end points a_i of the intervals over which $\sigma(x)$ is constant. That is, $\sigma(a_i) = (\alpha_i + \alpha_{i+1})/2$, $i = 1, \ldots, n-1$, $\sigma(a_0) = (\alpha_1/2)$, and $\sigma(a_n) = (\alpha_n/2)$. The reader was asked to show that $\bar{\Sigma}$, the space of equivalence classes, is isonormal to

Σ_0, the space of special step functions, both norms being given by [2]. Generally we shall fail to distinguish between Σ and Σ_0, or between Σ_0 and $\overline{\Sigma}$, and simply speak of the normed linear space of step functions. It is of course understood that equivalent step functions are considered to be identical.

In addition to being a normed linear space, Σ is a lattice (Definition 4.1.7). That is, if $\sigma, \tau \in \Sigma$, then $\max[\sigma,\tau]$ and $\min[\sigma,\tau] \in \Sigma$. From this it follows easily that σ^+, σ^-, and $|\sigma|$ are also step functions.

1 Definition. The *integral of a step function* σ, written $\int \sigma$, is given by the sum,

[3] $$\Sigma\, \alpha_i(a_i - a_{i-1});$$

The step function σ is defined by [1].

2 Proposition. $\sigma, \tau \in \Sigma$, and c is a real number.

(i) $\int(\sigma + \tau) = \int\sigma + \int\tau$ (additivity).
(ii) $\int c\sigma = c\int\sigma$ (homogeneity).
(iii) If $\sigma \geq 0$ except at finitely many points, then $\int\sigma \geq 0$ (positivity).
(iv) If $\sigma \geq \tau$ except at finitely many points, then $\int\sigma \geq \int\tau$ (monotonicity).
(v) $\|\sigma\| = \int|\sigma|$, and $\int|\sigma| = 0$ implies that $\sigma(x) = 0$ except at finitely many points.
(vi) $|\int\sigma| \leq \int|\sigma|$.
(vii) If $\{\sigma_n\}$ is a nonincreasing sequence of step functions that converges pointwise to zero, then

$$\lim_{n\to\infty} \int\sigma_n = 0 \qquad \text{(continuity)}.$$

Statements (i)–(vi) follow directly from the definitions and their verification is left to the reader.

Proof of (vii) (continuity of the integral): Dini's theorem (3.4.5) for step functions implies that the σ_n converge uniformly to zero. If $\sigma_1(x)$ vanishes outside an interval $[a,b]$, then it follows from the monotonicity of the sequence that every $\sigma_n(x)$ vanishes outside $[a,b]$. The uniform convergence implies that the sequence of real numbers, $s_n = \max_{a\leq x\leq b} \sigma_n(x)$, converges to zero. Hence,

$$0 \leq \int\sigma_n < \int\sigma_n\chi_{[a,b]} = s_n(b - a) \to 0,$$

as n tends to infinity. ($\chi_{[a,b]}$ denotes the characteristic function of the interval $[a,b]$. The reader will recall that if S is any set, then $\chi_S(x) = 1$ if $x \in S$ and is zero otherwise.) ∎

Remark. It is frequently more convenient to refer to properties (i) and (ii) of Proposition 2 as the *linearity* of the integral. This means that if σ and τ are step functions, and if a and b are real numbers, then

$$\int(a\sigma + b\tau) = a\int\sigma + b\int\tau.$$

Exercises

***1.** If the reader has not already done so, (Sec. 2 of Chapter 2), he should verify that Σ (more precisely, Σ_0 or $\bar{\Sigma}$), is a normed vector space and a lattice.

***2.** Show that every step function may be written as a finite sum of characteristic functions of intervals. Show also that if the intervals are all to be closed, then the sum may differ from the step function at finitely many points.

***3.** Let σ,τ,ρ be step functions. Let mid (σ,τ,ρ) be the function which takes on the middle value of $\sigma(x)$, $\tau(x)$ and $\rho(x)$ at each point x. Prove that mid(σ,τ,ρ) is a step function by verifying that

$$\text{mid}\,(\sigma,\tau,\rho) = \max\,[\min\,(\sigma,\tau), \min\,(\tau,\rho), \min\,(\sigma,\rho)].$$

Show that this identity remains true if max and min are interchanged.

***4.** Let S and T be subsets of a set X. Show that

(a) $\chi_{S\cup T} = \max\,(\chi_S, \chi_T)$, and (ii) $\chi_{S\cap T} = \min\,(\chi_S, \chi_T)$.

Generalize (i) and (ii) for finitely many sets and then for infinitely many sets.

5. It is not difficult to show that the continuity property, Proposition 2(vii), cannot be improved by requiring only that the σ_n converge everywhere to zero. Verify this statement by finding a sequence of step functions $\{\sigma_n\}$ which vanish outside [0,1], converge to zero at all points of [0,1] and yet the integral of each σ_n is equal to one.

6. Show now that a *bounded* sequence of step functions may converge to zero everywhere, while its integrals are all equal to one.

7. Prove that if $\{\sigma_n\}$ is a sequence of nonnegative step functions all of which vanish outside a fixed interval $[a,b]$, then, (i) $\sigma_n(x) < M$, for some real M, and (ii) lim $\sigma_n(x) = 0$, for all x in $[a,b]$, imply that $\lim_{n\to\infty} \int\sigma_n = 0$.

***8.** Let P denote the collection of all piecewise linear functions. Each of these functions vanishes outside a bounded interval, but there is no fixed bounded

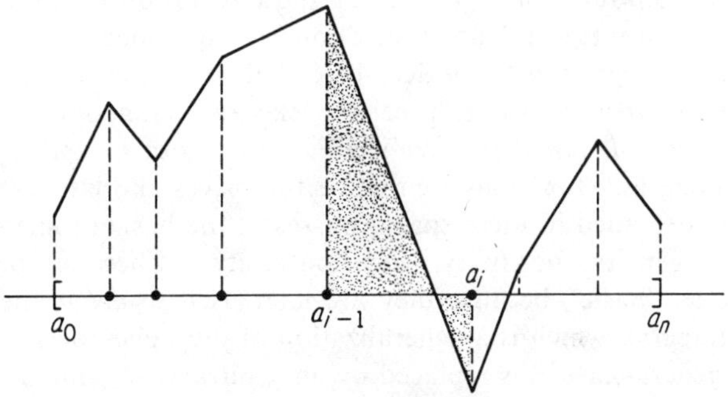

FIGURE 21

interval outside of which all of them vanish. Virtually everything that was said about step functions in this section is true also for the functions of P. By defining the integral of a piecewise linear function (see Figure 21) to be the sum of the signed areas of the trapezoids (or triangles) that are bounded by the vertical lines $x = a_i$ and $x = a_{i+1}$, the horizontal segment of $y = 0$ for which $a_i \le x \le a_{i+1}$, and the linear segment that begins at $(a_i, f(a_i))$ and terminates at $(a_{i+1}, f(a_{i+1}))$, we are able to prove that the integral satisfies properties (i)–(vii) of Proposition 2; the expression "except at finitely many points" may be omitted, since the functions of P, unlike those of Σ, are continuous. The reader is also asked to prove that

$$\|f\| = \int |f|$$

is a norm on P. Also prove Exercise 3 for functions in P. Restate and prove Exercises 5, 6, and 7.

***9.** Replace P of Exercise 8 by C_0, the collection of continuous functions, each of which vanishes outside a bounded interval. Let the integral be the Riemann integral. Prove the appropriate theorems for this class of functions.

It is also possible to do the same for the entire class of Riemann integrable functions (which is completely described in Sec. 3 of Chapter 6) and for the sectionally continuous functions. Which properties do not hold for the linear space of polynomials?

*Exercises which request the proof of a statement made in the text, as well as those referred to in the text, are starred.

2 Completion of the Space of Step Functions

In Sec. 1 we saw that the (Lebesgue) integral defined on Σ is an *additive*, *homogeneous*, *positive* and "*continuous*" mapping of Σ into the real numbers, in the sense described in Proposition 5.1.2. The sum $\Sigma \mid \alpha_i \mid (a_i - a_{i-1})$ ([2] of Sec. 1), which is used to define the norm of a step function, is simply the integral of $\mid \sigma \mid$. In this section we shall show that Σ is an incomplete normed vector space. Its abstract completion $\hat{L}_1$ is defined as the normed linear space of equivalence classes of Cauchy sequences of step functions. The norm and integral on $\hat{L}_1$ are extensions of the norm and integral defined on Σ, and are obtained by taking limits of norms and integrals in Σ. In Sec. 4 we shall show that $\hat{L}_1$ is isonormal to a space of functions (more precisely, classes of functions) called the *Lebesgue integrable functions.* We shall see that both $\hat{L}_1$ and the class of Lebesgue integrable functions are real vector spaces and lattices, and that the integral extended to these spaces possesses the basic properties: additivity, homogeneity, positivity, and continuity. These properties are referred to as "basic" because they are later used to *define* the abstract (Daniell) integral, which is a generalization of the Lebesgue integral. In this more general case R is replaced by an arbitrary set, and Σ by a class of real valued functions, which like Σ is a vector space and a lattice.

Before demonstrating the incompleteness of Σ, let us recall what is meant by convergence in the normed linear space Σ. "$\lim \sigma_n = \sigma$" means that the sequence $\{\sigma_n\}$ *converges* in *norm* to the step function σ; that is, $\lim_{n\to\infty} \|\sigma_n - \sigma\| = \lim_{n\to\infty} \int |\sigma_n - \sigma| = 0$. Similarly, $\{\sigma_n\}$ is a Cauchy sequence in Σ means that $\lim_{n,m\to\infty} \|\sigma_n - \sigma_m\| = 0$.

We wish to show that there is a Cauchy sequence of step functions which does *not* converge to a step function. Geometrically speaking, we seek a sequence of step functions $\{\sigma_n\}$ with the following properties: The sum of the rectangular areas that lie between the graphs of two functions σ_n and σ_m in the sequence will tend to zero as n and m tend to infinity. However, if σ is *any* step function whatsoever, the corresponding areas between σ and σ_n will *not* tend to zero as n tends to infinity. In the following example, this is done by constructing a Cauchy sequence $\{\sigma_n\}$ which converges *pointwise* to the function $y = x$.

1 Example. The sequence $\{\sigma_n\}$ defined below is nondecreasing and converges uniformly on $[0,1]$ to $y = x$. The functions σ_1, σ_2 and σ_3 are shown in Figure 22. Let

$$\sigma_1(x) = \begin{cases} 0, & 0 \le x \le 1/2 \\ 1/2, & 1/2 < x \le 1, \\ 0, & \text{elsewhere} \end{cases} \qquad \sigma_2(x) = \begin{cases} 0, & 0 \le x \le 2^{-2} \\ 2^{-2}, & 2^{-2} < x \le 2^{-1} \\ 2^{-1}, & 2^{-1} < x \le 3(2^{-2}), \\ 3(2^{-2}), & 3(2^{-2}) < x \le 1 \\ 0, & \text{elsewhere} \end{cases}$$

The nth term is given by

$$\sigma_n(x) = \begin{cases} 0, & 0 \le x \le 2^{-n} \\ 2^{-n}, & 2^{-n} < x \le (2)2^{-n} = 2^{n-1} \\ \vdots & \vdots \\ k(2^{-n}), & k(2^{-n}) < x \le (k+1)2^{-n} \\ \vdots & \vdots \\ 1 - 2^{-n}, & 1 - 2^{-n} < x \le 1 \\ 0, & \text{elsewhere} \end{cases}$$

where $k = 0, 1, \ldots, 2^n - 1$. Clearly $\sigma_n(x) \uparrow$ for all values of x, and $|x - \sigma_n(x)| < 2^{-n}$, proving that the convergence is monotone and uniform. We show next that $\{\sigma_n\}$ is a Cauchy sequence: Let $n = m + p$, where $n,m \ge 1, p > 0$. Then

$$\|\sigma_n - \sigma_m\| = \int |\sigma_{m+p} - \sigma_m| < 2^m(2^{-2m})(1/2) = 2^{-m-1}.$$

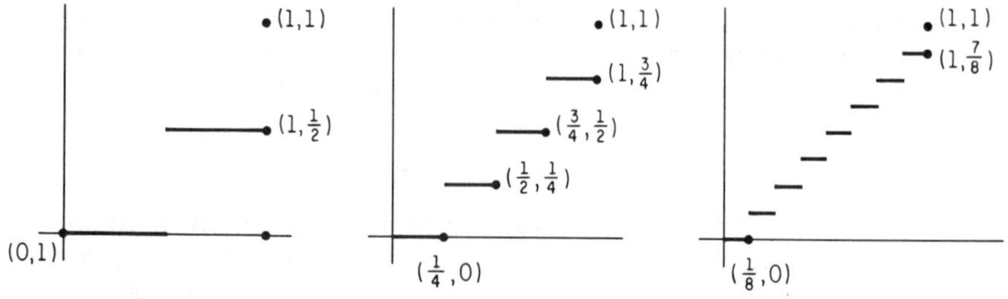

FIGURE 22

This is obtained by noting that there are 2^m triangles formed by the line $y = x$ and the step function $y = \sigma_m$. Each of these triangles has area equal to

$$(1/2)\,(\text{base})\,(\text{altitude}) = (1/2)\,(2^{-m})\,(2^{-m}).$$

We have used the fact that the sum of these areas is larger than the sum of the areas that lie between the graphs of $y = \sigma_m(x)$ and $y = \sigma_{m+p}(x)$, for all $p > 0$. Thus $\{\sigma_n\}$ is a Cauchy sequence.

Geometrically, it is easy to see that if $y = \sigma(x)$ is any step function whatsoever, then $\lim \|\sigma - \sigma_n\| \neq 0$. An analytic proof follows: Let us assume that there is a step function σ to which the given sequence converges. Choose any subinterval (a,b) of $[0,1]$ on which $\sigma(x)$ is equal to a constant, say γ. Exactly one of the following must be true: $a < \gamma$, $a = \gamma$, or $a > \gamma$. We shall assume first that $a < \gamma \leq 1$. (Why may we safely assume that $\gamma \leq 1$?) Then on the interval $[a,\gamma]$, $\sigma(x) - x \geq 0$, and since $\sigma_n(x) \uparrow$, it follows that

$$\int |\sigma_n - \sigma| > \int_a^\gamma |\sigma - \sigma_n| > \int_a^\gamma (\sigma - x) = \frac{(\gamma - a)^2}{2}.$$

The second and third integrals may be interpreted as Riemann integrals, although it is not necessary to do so. $[(\gamma - a)^2]/2$ is the area of the triangle bounded by $y = x$, $x = a$ and $y = \gamma$. Therefore if $a < \gamma$, the sequence $\{\sigma_n\}$ cannot converge to σ. Thus $\gamma \leq a$ follows. However, in this case the uniform convergence of $\{\sigma_n\}$ to x yields

$$\int |\sigma - \sigma_n| \geq \int_a^b (x - \sigma) - \int_a^b (x - \sigma_n) >$$

$$\frac{(b - a)^2}{2} - (b - a) \max_{a \leq x \leq b} (x - \sigma_n(x)) > \frac{(b - a)^2}{4}$$

for large enough values of n, thus proving that $\{\sigma_n\}$ cannot converge to a step function.

Theorem 2.4.7 asserts the existence of a unique completion $\hat{L}_1$ of Σ. An element of $\hat{L}_1$ is an equivalence class $[\{\sigma_n\}]$ of Cauchy sequences of step functions. Two Cauchy sequences are *equivalent*, written $\{\sigma_n\} \sim \{\sigma_n'\}$, if $\lim \|\sigma_n - \sigma_n'\| = 0$; that is, if $\{(\sigma_n - \sigma_n')\}$ is a *null* sequence. The vector space operations and the norm may be extended to $\hat{L}_1$: If $\hat{f} = [\{\sigma_n\}]$ and $\hat{g} = [\{\tau_n\}]$ are elements of $\hat{L}_1$, then their sum is defined to be the *class* $[\{\sigma_n + \tau_n\}]$, which is denoted by $\hat{f} + \hat{g}$. The product of $\hat{f}$ by a real number c is the class $[\{c\sigma_n\}]$. We have already shown that the addition and multiplication by a real number are *well defined* (Sec. 4 of Chapter 2). This amounts to proving that if $\{\sigma_n\}$ and $\{\tau_n\}$ are Cauchy sequences, and if c is a real number, then $\{\sigma_n + \tau_n\}$ and $\{c\sigma_n\}$ are Cauchy sequences; and if $\{\sigma_n\} \sim \{\sigma_n'\}$, $\{\tau_n\} \sim \{\tau_n'\}$, then $\{\sigma_n + \tau_n\} \sim \{\sigma_n' + \tau_n'\}$ and $\{c\sigma_n\} \sim \{c\sigma_n'\}$. This means that addition and multiplication of classes may be accomplished by choosing representatives from the respective classes. The result is independent of the representative that is chosen. Under these operations, $\hat{L}_1$ is a *vector space*. The reader will recall that the norm of $\hat{f} = [\{\sigma_n\}]$ is defined to be

$$\|\hat{f}\| = \lim \|\sigma_n\|.$$

Although this was done in the general case in Sec. 4 of Chapter 2, the reader should verify directly that the norm is well defined. This requires showing that $\lim \|\sigma_n\|$ exists, and remains unchanged if $\{\sigma_n\}$ is replaced by an equivalent sequence.

Along with the norm, the integral may also be extended by taking limits. The procedure is similar to the extension of the norm, and a brief outline follows.

2 Definition. The *integral* of an element $\hat{f} = [\{\sigma_n\}]$ in $\hat{L}_1$ is given by

[1] $$\hat{\int}\hat{f} = \lim \int \sigma_n.$$

Remark. The "cap" appearing over the integral sign may be removed after it is verified that this integral is indeed an extension of the integral over Σ. This means that it must coincide with the original integral when it is defined for an element in the subspace of $\hat{L}_1$ which is isonormal to Σ. Just as in the case of the real and rational numbers, the classes of $\hat{L}_1$ which correspond to the step functions are those which contain a constant sequence; that is, to each σ in Σ, there is an equivalence class in $\hat{L}_1$ which contains the constant (Cauchy) sequence $\sigma, \sigma, \ldots, \sigma, \ldots$. We denote this class by $\hat{\sigma} = [\{\sigma\}]$. From Definition 2 we obtain

$$\hat{\int}\hat{\sigma} = \lim \int \sigma = \int \sigma.$$

Thus the integral on $\hat{L}_1$ is indeed an extension of the integral on Σ, and we may remove the "cap" from the extended integral.

We show next that the integral [1] is well defined.

3 Proposition. (i) If $\{\sigma_n\}$ is a Cauchy sequence of step functions, then

$$\lim \int \sigma_n$$

exists.

(ii) If $\{\sigma_n\} \sim \{\sigma_n'\}$, then

$$\lim \int \sigma_n = \lim \int \sigma_n'.$$

Proof: (i) is verified by showing that the sequence $s_n = \int \sigma_n$ of real numbers is a Cauchy sequence:

$$|s_n - s_m| = \left|\int \sigma_n - \int \sigma_m\right| = \left|\int (\sigma_n - \sigma_m)\right|$$
$$\leq \int |\sigma_n - \sigma_m| = \|\sigma_n - \sigma_m\|.$$

Parts (i) and (vi) of Proposition 5.1.2 were used. Thus $\{s_n\}$ is a Cauchy sequence of real numbers and therefore has a limit.

If $\{\sigma_n\}$ and $\{\sigma_n'\}$ are equivalent Cauchy sequences of step functions, then

$$\left|\int \sigma_n - \int \sigma_n'\right| = \left|\int (\sigma_n - \sigma_n')\right| \leq \int |\sigma_n - \sigma_n'| = \|\sigma_n - \sigma_n'\|$$

tends to zero as n tends to infinity, which proves (ii). ∎

We show next that it is possible to define a partial ordering on $\hat{L}_1$ in such a way that $\hat{L}_1$ is a lattice, and that the monotonicity property of the integral is preserved. Let us start by defining, for each class $\hat{f}$ in $\hat{L}_1$, its absolute value, and its positive and negative parts.

4 Proposition. If $\{\sigma_n\}$ is a Cauchy sequence of step functions, then $\{|\sigma_n|\}$, $\{\sigma_n^+\}$, and $\{\sigma_n^-\}$ are Cauchy sequences of step functions; moreover, if $\{\sigma_n\} \sim \{\tau_n\}$, then

$$\{|\sigma_n|\} \sim \{|\tau_n|\}, \{\sigma_n^+\} \sim \{\tau_n^+\}, \quad \text{and} \quad \{\sigma_n^-\} \sim \{\tau_n^-\}.$$

Proof: The triangle inequality for the real numbers and the monotonicity of the integral (on Σ) yield

$$\||\sigma_n| - |\sigma_m|\| = \int \Big| |\sigma_n| - |\sigma_m| \Big| \leq \int |\sigma_n - \sigma_m| = \|\sigma_n - \sigma_m\|.$$

However, $\{\sigma_n\}$ is a Cauchy sequence, and therefore the last expression tends to zero as n and m tend to ∞. A similar argument may be used to prove that $\{\sigma_n^+\}$ and $\{\sigma_n^-\}$ are Cauchy sequences.

Suppose now that $\{\sigma_n\}$ is equivalent to $\{\tau_n\}$. Again,

$$\||\sigma_n| - |\tau_n|\| \leq \|\sigma_n - \tau_n\| \to 0, \text{ as } n \to \infty.$$

The argument holds also for the positive and negative parts of the sequence. ▎

5 Definition. Let $\hat{f} = [\{\sigma_n\}]$ be an element of $\hat{L}_1$. The *absolute value of* $\hat{f}$, the *positive part of* $\hat{f}$, and the *negative part of* $\hat{f}$, are respectively

$$|\hat{f}| = [\{\,|\sigma_n|\,\}], \quad \hat{f}^+ = [\{\sigma_n^+\}], \quad \text{and} \quad \hat{f}^- = [\{\sigma_n^-\}].$$

It follows from Proposition 4 that $|\hat{f}|$, $\hat{f}^+$, and $\hat{f}^-$ are well defined; that is, they do not depend upon the particular sequence that is chosen to represent $\hat{f}$.

6 Proposition. If $\hat{f} \in \hat{L}_1$, then $\hat{f} = \hat{f}^+ - \hat{f}^-$ and $|\hat{f}| = |-\hat{f}| = \hat{f}^+ + \hat{f}^-$.

The proof is left to the reader.

7 Definition. An element $\hat{f}$ in $\hat{L}_1$ is said to be *nonnegative*, written $\hat{f} \geq \hat{0}$, if $\hat{f} = |\hat{f}| = f^+$. ($\hat{0}$ is the additive identity of $\hat{L}_1$, that is, the class which contains the constant sequence of everywhere vanishing step functions.)

$\hat{f}$ is *nonpositive*, written $\hat{f} \leq \hat{0}$, if $\hat{f} = -|\hat{f}| = -\hat{f}^-$.

$\hat{f}$ is said to be *greater or equal* to $\hat{g}$ ($\hat{g}$ is *less or equal to* $\hat{f}$), written $\hat{f} \geq \hat{g}$, if $\hat{f} - \hat{g} \geq \hat{0}$.

8 Proposition. $\geq$ is a partial ordering on $\hat{L}_1$.

Proof: Reflexivity of $\geq$: Let $\hat{f} = [\{\sigma_n\}] \in \hat{L}_1$. Then

$[\{\sigma_n - \sigma_n\}] = [\{\bar{0}\}] = [\{\,|\sigma_n - \sigma_n|\,\}] \Rightarrow \hat{f} - \hat{f} = |\hat{f} - \hat{f}| \geq \hat{0} \Rightarrow \hat{f} \geq f.$

Antisymmetry of $\geq$: $\hat{f} \geq \hat{g}$ and $\hat{g} \geq \hat{f} \Rightarrow \hat{f} - \hat{g} = |\hat{f} - \hat{g}|$ and $\hat{g} - \hat{f} = |\hat{g} - \hat{f}| \Rightarrow \hat{f} - \hat{g} = \hat{g} - \hat{f} \Rightarrow \hat{f} = \hat{g}$.

Transitivity of $\geq$: Let $\hat{f} \geq \hat{g}$ and $\hat{g} \geq \hat{h}$, where $\hat{f} = [\{\sigma_n\}]$, $\hat{g} = [\{\tau_n\}]$ and $\hat{h} = [\{\rho_n\}]$.

$\hat{f} \geq \hat{g} \Rightarrow \hat{f} - \hat{g} \geq \hat{0} \Rightarrow \hat{f} - \hat{g} = |\hat{f} - \hat{g}| \Rightarrow \{(\sigma_n - \tau_n)\} \sim \{\,|\sigma_n - \tau_n|\,\}$.

Similarly,

$$\hat{g} \geq \hat{h} \Rightarrow \{(\tau_n - \rho_n)\} \simeq \{\,|\tau_n - \rho_n|\,\}.$$

The monotonicity of the integral and the inequality

$$0 \leq |\sigma_n - \rho_n| - (\sigma_n - \rho_n)$$

imply that

$$\begin{aligned} 0 &\leq \textstyle\int[\,|\sigma_n - \rho_n| - (\sigma_n - \rho_n)] \\ &\leq \textstyle\int[\,|\sigma_n - \tau_n| + |\tau_n - \rho_n| - (\sigma_n - \tau_n) - (\tau_n - \rho_n)] \\ &= \textstyle\int[\,|\sigma_n - \tau_n| - (\sigma_n - \tau_n)] + \int[\,|\tau_n - \rho_n| - (\tau_n - \rho_n)]. \end{aligned}$$

The integrals on the right are bounded from above by

$$\| |\sigma_n - \tau_n| - (\sigma_n - \tau_n) \| + \| |\tau_n - \rho_n| - (\tau_n - \rho_n) \| .$$

However, it follows from $\hat{f} - \hat{g} \geq \hat{0}$ and $\hat{g} - \hat{h} \geq \hat{0}$ that this sum tends to zero as n tends to infinity. ▮

9 Proposition. (Positivity and monotonicity of the integral on $\hat{L}_1$). If $\hat{f} \geq \hat{0}$ then $\int \hat{f} \geq 0$, and if $\hat{f} \geq \hat{g}$ then $\int \hat{f} \geq \int \hat{g}$.

Proof: $\hat{f} \geq \hat{0}$ implies that $\hat{f}$ contains a sequence $\{\sigma_n\}$ each of whose terms is nowhere negative. The monotonicity of the integral on Σ and Definition 2 yield

$$\int \hat{f} = \lim \int \sigma_n \geq 0.$$

If $\hat{f} \geq \hat{g}$, then $\hat{f} - \hat{g} \geq \hat{0}$. It follows therefore that

$$\int (\hat{f} - \hat{g}) \geq 0.$$

The additivity of the integral on $\hat{L}_1$ yields the desired result. ▮

To make $\hat{L}_1$ into a lattice, the maximum and minimum of two classes in $\hat{L}_1$ must be defined. This is done in the usual way, by using representative sequences. Again, it must be verified that the definition does not depend upon the particular sequence that is chosen from each class.

10 Proposition. If $\{\sigma_n\}$ and $\{\tau_n\}$ are Cauchy sequences of step functions, then both max $[\sigma_n, \tau_n]$ and min $[\sigma_n, \tau_n]$ are Cauchy sequences. Moreover, $\{\sigma_n\} \sim \{\sigma_n'\}$ and $\{\tau_n\} \sim \{\tau_n'\}$ imply

$$\{\max [\sigma_n, \tau_n]\} \sim \{\max [\sigma_n', \tau_n']\} \text{ and } \{\min [\sigma_n, \tau_n]\} \sim \{\min [\sigma_n', \tau_n']\}.$$

Proof: For each n, $\rho_n = \max [\sigma_n, \tau_n]$ is a step function. It is easily seen that for all x, and any pair of integers n and m,

$$|\rho_n(x) - \rho_m(x)| \leq \max [\, |\sigma_n(x) - \sigma_m(x)|, |\tau_n(x) - \tau_m(x)|$$
$$\leq |\sigma_n(x) - \sigma_m(x)| + |\tau_n(x) - \tau_m(x)|,$$

and therefore

$$\| \rho_n - \rho_m \| \leq \| \sigma_n - \sigma_m \| + \| \tau_n - \tau_m \|,$$

proving that $\{\rho_n\}$ is a Cauchy sequence.

Using a similar argument, it may be proved that min $[\sigma_n, \tau_n\}$ is a Cauchy sequence, and that replacing $\{\sigma_n\}$ and $\{\tau_n\}$ by equivalent sequences, does not change the class of either the max or min. The details are left to the reader. ▮

11 Definition. Let $\hat{f}, \hat{g} \in \hat{L}_1$. The *maximum* and *minimum* of $\hat{f}$ and $\hat{g}$ are given by

$$\max[\hat{f},\hat{g}] = [\{\max[\sigma_n, \tau_n]\}] \quad \text{and} \quad \min[\hat{f},\hat{g}] = [\{\min[\sigma_n, \tau_n]\}].$$

Proposition 10 asserts that max and min are well defined. The reader may easily verify that if $\hat{f}$ and $\hat{g}$ are any pair of classes in $\hat{L}_1$, then $\max[\hat{f},\hat{g}]$ and $\min[\hat{f},\hat{g}]$ are respectively their *least upper bound* and *greatest lower bound* (Definition 4.1.7 and Example 4.1.8). This proves that $\hat{L}_1$ is a lattice.

It is left to the reader to prove that the maximum and minimum of any finite number of classes in $\hat{L}_1$ may be defined in the same way.

Exercises

***1.** Replace the function $y = x$ of Example 1 by any continuous nonconstant function defined on $[0,1]$. Using the uniform continuity of this function (Theorem 3.2.16), show that there is a Cauchy sequence of step functions which converges pointwise to the continuous function, but does not converge in norm to a step function. (You may use the Riemann integral, but it isn't necessary.)

***2.** Prove that the integral [1] is additive and homogeneous.

***3.** Prove that if $\hat{f} \geq \hat{0}$ and $\int \hat{f} = 0$, then $\hat{f} = \hat{0}$.

***4.** Complete the proof of Proposition 4 by showing that if $\{\sigma_n\}$ is a Cauchy sequence of step functions, then the sequences of positive and negative parts of $\{\sigma_n\}$ are also Cauchy sequences. Show also that if $\{\sigma_n\} \sim \{\tau_n\}$, then their respective positive and negative parts are equivalent.

***5.** Show that $\leq$ (or $\geq$) could have been defined as follows:

$$\hat{f} \geq \hat{0} \text{ if } \hat{f} \text{ contains a sequence } \{\sigma_n\} \text{ satisfying } \sigma_n(x) \geq 0 \text{ for all } x.$$

Similarly for $\hat{f} \leq \hat{0}$.

***6.** Complete the proof of Proposition 10.

***7.** Let $\hat{f}_1, \hat{f}_2, \ldots, \hat{f}_k \in \hat{L}_1$. Define $\max[\hat{f}_1, \ldots, \hat{f}_k\}$ and $\min[\hat{f}_1, \ldots, \hat{f}_k]$. Verify that they are well defined and show that for $i = 1, \ldots, k$,

$$\min[\hat{f}_1, \ldots, \hat{f}_k] \leq \hat{f}_i \leq \max[\hat{f}_1, \ldots, \hat{f}_k].$$

The following exercises may be viewed as a continuation of Exercises 5.1.8 and 5.1.9.

***8.** Show that P, the normed linear space of piecewise linear functions, is incomplete. Complete the space and extend the integral. Show that all definitions and propositions stated in this section hold for P. In Sec. 4 you will be asked to prove that the completions of P and Σ are essentially the same, that is, isonormal. If you try to imagine the functions to which it will be shown that the classes in $\hat{L}_1$ correspond, you may be able to see why this is true.

***9.** Replace P in the preceding exercise first by C_0, the normed linear space of continuous functions each of which vanishes outside a bounded interval, and then by the Riemann integrable functions and polynomial functions which vanish outside a bounded interval. The latter space is not however a lattice, although its completion is.

*Exercises which request the proof of a statement made in the text, as well as those referred to in the text, are starred.

3 Null Sets

Neither the integral nor the norm of a step function ([3] and [2] of Sec. 1) depend upon the values of the step function at the endpoints of the intervals over which it is constant. As a consequence of this, both the integral and the norm of a step function σ remain unchanged if the values of σ are changed, in any way whatsoever, at finitely many points. This is of course also true of the Riemann integral. Thus function value changes on finite sets go unnoticed by the integral (both Lebesgue and Riemann), and it can be said that from the point of view of integration, finite sets are "negligible." The reader may know that even for the Riemann integral, there are infinite sets which are "negligible." An example of such a set is $\{1,1/2,1/3,\ldots,1/n,\ldots\}$. This means that if $f(x)$ is continuous whenever $0 \le x \le 1$, and if

$$g(x) = \begin{cases} f(x) & \text{if } x \neq 1/n, \quad n = 1,2,\ldots \\ a_n & \text{if } x = 1/n, \end{cases}$$

where the a_n are real numbers bounded from above and below, then both f and g are Riemann integrable, and

$$\int_0^1 f(x)\,dx = \int_0^1 g(x)\,dx.$$

(See Theorem 6.3.8 and Exercise 6.3.10.) There are even noncountable sets on which a Riemann integrable function may be changed without destroying its integrability or changing the value of the integral.

In this section *null* sets are defined and some of their elementary properties are derived. In Sec. 4 we shall see that the null sets are precisely those that are "negligible" in Lebesgue integration. This means that if two (Legesgue) integrable functions are equal to each other except on a null set, then their integrals are identical.

The Lebesgue null sets also play a role in the theory of Riemann integration. Theorem 6.3.8 asserts that a bounded function defined on a bounded interval $[a,b]$ is Riemann integrable, if and only if the set of discontinuities of the function is a null subset of $[a,b]$.

At first, a null set is defined as one that can be "covered" by countably many intervals of arbitrarily small length. Theorem 8 provides an alternate definition which suggests the role played by these sets in the theory of integration.

1 Definition. Let S be a subset of the real numbers, and let $\mathcal{I} = \{I_\alpha\}$ be a collection of intervals (not necessarily countable). S is said to be *covered by the system* $\mathcal{I} = \{I_\alpha\}$ if $\bigcup_\alpha I_\alpha \supset S$.

2 Definition. Let $\lambda(I)$ denote the length of an interval I. ($\lambda(I) = \infty$ is permitted.) If $\mathcal{I} = \{I_n\}$ is a countable system of intervals, then the sum, (finite or infinite),

$$\lambda(\mathcal{I}) = \Sigma\lambda(I_n)$$

is called the *length of the system* $\mathcal{I}$.

3 Example. It should be observed that if $\bigcup I_n$ is itself an interval, then it is not in general true that $\lambda(\bigcup I_n) = \lambda(\{I_n\})$.

The following examples illustrate this:

(a) If $\mathcal{I} = \{I_n\}$, where $I_n = (0,2^{-n})$, $n = 1,2,\ldots$, then $\lambda(\mathcal{I}) = \Sigma 2^{-n} = 1$. However, $\bigcup I_n = (0,1/2)$, which is an interval whose length is $1/2$.

(b) The length of the system $\mathcal{I} = \{I_n\}$, where $I_n = (0,1/n)$, is infinite, although $\bigcup I_n = (0,1)$.

Clearly, the length of the system, if finite, is equal to the length of the union (or the sum of the lengths of the disjoint intervals of the union) if and only if the intersection of any pair of intervals in the system is either empty or contains a single point.

4 Definition. A subset S of the real numbers is called a *null set* if it can be covered by a countable system of open intervals whose length is arbitrarily small. That is, for every $\epsilon > 0$, there is a system $\mathcal{I}^\epsilon = \{I_n^\epsilon\}$, and

$$\text{(i)}\quad \lambda(\mathcal{I}^\epsilon) < \epsilon, \qquad \text{(ii)}\quad \bigcup I_n^\epsilon \supset S.$$

Remark. Riemann "null" sets are those which can be covered by *finite* systems of open intervals whose length is arbitrarily small. Thus every set which is null in the Riemann sense is also a Lebesgue null set. This results in the existence of many functions which are Lebesgue integrable, but not Riemann integrable.

5 Examples of Null Sets. (a) Every finite collection $\{x_1, x_2, \ldots, x_m\}$ is null: If $\epsilon > 0$ is given, choose $I_n = [x_n - (\epsilon/2m),\, x_n + (\epsilon/2m)]$, $n = 1,2,\ldots,m$. $\mathcal{I} = \{I_n\}$ is an open covering whose length is equal to ϵ.

(b) Every countable set $\{x_1, x_2, \ldots\}$ is null: Choose $\{I_n\}$ to be a sequence of open intervals; each I_n contains x_n and $\lambda(I_n) = \epsilon(2^{-n})$. Summing a geometric series, we obtain, $\lambda(\mathcal{I}) = \epsilon$.

There also exist uncountable null sets.

6 Example. (The *Cantor set*; an uncountable null set) This set is constructed by removing from the interval [0,1] a sequence of open intervals whose total length is equal to one. The set of points that remain is the Cantor set. (See Figure 23.) Let $I_{11} = (1/3,2/3)$, the "middle

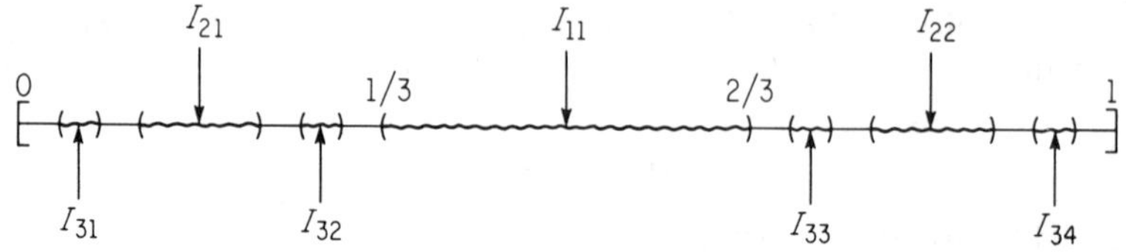

FIGURE 23

third" of the interval [0,1]; $I_{21} = (1/9,2/9)$, $I_{22} = (7/9,8/9)$ are the "middle thirds" of the remaining intervals; I_{31}, I_{32}, I_{33}, I_{34} are the "middle thirds" of the four intervals that remain after I_{21} and I_{22} are removed. In general, at the nth step, 2^{n-1} intervals $I_{n,m}$, $m = 1,2,\ldots,2^{n-1}$, are removed. Each $I_{n,m}$ has length equal to 3^{-n}. Since $\{I_{n,m}\}$ is a countable collection of open sets, the union

$$G = \bigcup_{n=1}^{\infty} \bigcup_{m=1}^{2^{n-1}} I_{n,m}$$

is an open set. The closed set $C = [0,1] - G$ is called the *Cantor set*. From $\lambda(I_{n,m}) = 3^{-n}$, $m = 1,2,\ldots,2^{n-1}$, we obtain

$$\sum_{n=1}^{\infty} \sum_{m=1}^{2^{n-1}} \lambda(I_{n,m}) = \sum_{n=1}^{\infty} 2^{n-1}3^{-n} = \frac{1}{3}\sum_{n=1}^{\infty} (2/3)^{n-1} = 1,$$

proving that the length of the system $\{I_{n,m}\}$ is 1. From the fact that the $I_{n,m}$ are disjoint, it is possible to draw the conclusion that C may be covered by countably many intervals whose total length does not exceed ϵ:

$n = 1$: Cover the end points of I_{11} and the two points 0 and 1 by four open intervals J_{1k}, $k = 1,2,3,4$, where $\lambda(J_{1k}) = \frac{1}{4}\left(\frac{\epsilon}{2}\right)$.

$n = 2$: Cover the four end points of I_{21} and I_{22} by four open intervals, J_{2k}, $k = 1,2,3,4$, where

$$\lambda(J_{2k}) = \frac{1}{4}\left(\frac{\epsilon}{2^2}\right).$$

$$\vdots \qquad\qquad \vdots$$

$n = \nu$: Cover the 2^ν end points of $I_{\nu 1}, I_{\nu 2}, \ldots, I_{\nu 2^{\nu-1}}$ by open intervals $J_{\nu k}$, $k = 1,2,\ldots,2^{\nu-1}$, where

$$\lambda(J_{\nu k}) = 2^{-\nu}\left(\frac{\epsilon}{2^\nu}\right) = 2^{-2\nu}\epsilon.$$

The following statements are easily verified:

(i) The only points which are *not* covered by the $J_{j,k}$ are in the interior of the $I_{n,m}$ and therefore these points are not in C. Thus

$$C \subset \bigcup_{j=1}^{\infty} \bigcup_{k=1}^{2^j} J_{jk}.$$

$$\text{(ii)}\quad \lambda(\{J_{j,k}\}) = \sum_{k=1}^{4} \lambda(J_{1k}) + \sum_{j=2}^{\infty} \sum_{k=1}^{2^j} \lambda(J_{j,k}) = \frac{\epsilon}{2} + \sum_{j=2}^{\infty} \frac{\epsilon}{2^j} = \epsilon.$$

The uncountability of C is left as an exercise.

7 Proposition. (a) A subset of a null set is null.

(b) The intersection (even uncountable) of null sets is a null set.

(c) A countable union of null sets is null.

Proof: Suppose that T is a subset of a null set S. Given $\epsilon > 0$, there is a system of open intervals $\{I_n\}$ whose union covers S, and therefore T, and whose total length is not greater than ϵ. Thus T is also null.

If $\{S_\alpha\}$ is a collection of null sets, then $\bigcap S_\alpha$ is contained in each S_α. The desired result follows from (a).

Example 5(b) suggests the proof for the union of countably many null sets. This is left to the reader as an exercise. ∎

The following theorem provides us with a second definition of a null set. This characterization of a null set suggests that it is indeed a set on which anything may happen to a function, without in any way affecting its integral. Later, when we discuss integration over abstract spaces (for which the idea of an open set may have no meaning), it will give a definition of a null set that does not require an open covering.

8 Theorem. A subset S of the real numbers is null if and only if there is a nondecreasing sequence $\{\sigma_n\}$ of step functions satisfying:

(i) $\sigma_n \uparrow \infty$ on S, and

(ii) there is a nonnegative number A and $\int \sigma_n < A$ for all values of n.

The proof requires the following lemma:

9 Lemma. If σ is a nonnegative step function, and if $\int \sigma < A$, then for all $\delta > 0$, $\sigma(x) < \delta$ except on a finite set of open intervals whose total length is less than A/δ.

Proof: Since $\sigma(x)$ takes on finitely many values, each on an open interval or at the endpoint of such an interval, there are finitely many intervals, $I_1, \ldots, I_k$ and points $a_1, \ldots, a_m$ on which $\sigma(x)$ may be greater or equal to δ. Since the finite set $\{a_1, \ldots, a_m\}$ may be covered by a system of intervals of arbitrarily small length, it suffices to prove that the system $\{I_j\}$ has total length that is less than A/δ. Suppose however that $\Sigma\lambda(I_j) \geq A/\delta$. Then

$$\int \sigma \geq \delta \Sigma \lambda(I_j) \geq A,$$

contradicting the hypothesis.

This proves the lemma.

Proof of Theorem 8: It may be assumed that the $\{\sigma_n\}$ are nonnegative. For if this were not the case, then we could replace the σ_n by $\tau_n = \sigma_n - \sigma_1$. The monotonicity of the σ_n implies that the τ_n are nonnegative. Once the theorem is proved for the nonnegative sequence τ_n, the linearity of the integral would give us the theorem for the σ_n.

Let n be a fixed but arbitrary integer. The divergence of the sequence $\{\sigma_j\}$ on S, implies that if $x \in S$, then for some integer j, $\sigma_j(x) \geq n$. Therefore every point x of S is in a set

$$S_{n,j} = \{x : \sigma_j(x) \geq n\} = [\sigma_j \geq n],$$

for sufficiently large values of j. From Lemma 9, $S_{n,j}$, which is the set on which $\sigma_j(x)$ is not less than n, must be contained in a finite system $\mathcal{I}_{n,j}$ of open intervals whose length is less than A/n. Therefore the set

$$S_n = \bigcup_{j=1}^{\infty} S_{n,j}$$

is contained in the countable system of intervals $\mathcal{I}_n = \bigcup_{j=1}^{\infty} \mathcal{I}_{n,j}$. The monotonicity of the sequence $\{\sigma_j\}$ implies that

$$S_{n,j} \subset S_{n,j+1},$$

and it may therefore be assumed that $\{\lambda(\mathcal{I}_{n,j})\}$ is a nondecreasing sequence of real numbers. Since each term of this sequence is bounded from above by A/n, it follows that

$$\lim_{j \to \infty} \lambda(\mathcal{I}_{n,j}) = \lambda(\mathcal{I}_n) \leq A/n.$$

Since S is contained in every S_n, it suffices to prove that for sufficiently large n, the system $\mathcal{I}_n$ of intervals which contains S_n, has length which is less than any preassigned $\epsilon > 0$. This is done by choosing n so that $A/n < \epsilon$ is satisfied, for in this case, $\lambda(\mathcal{I}_n) \leq A/n < \epsilon$. Thus the set S on which the sequence of step functions $\{\sigma_n\}$ diverges is a null set.

We show now that if S is a null set, there is a nondecreasing sequence of step functions which diverges on S but whose integrals remain uniformly bounded. For each positive integer n, let $\mathcal{I}_n$ be a system of open intervals $\{I_k^{(n)}\}$ whose length is less than 2^{-n} and whose union $\bigcup_{k=1}^{\infty} I_k^{(n)}$ contains S. Allowing n to range over the positive integers, we obtain a countable set $\{I_k^{(n)}\}$ of open intervals. Let $\{I_j\}$ denote some ordering of this countable set. If x is a point of S, then for each n, x is in some $I_k^{(n)}$. Therefore x is in infinitely many of the I_j. Set

$$\sigma_n = \sum_{j=1}^{n} \chi_{I_j}.$$

The nonnegativeness of the characteristic functions χ_{I_j} implies that $\{\sigma_n\}$ is nondecreasing, and if $x \in S$, then there are infinitely many values of j for which $\chi_{I_j}(x) = 1$. Thus $\sigma_n(x) \uparrow \infty$ on S. However,

$$\int \sigma_n = \int \sum_{j=1}^{n} \chi_{I_j} = \sum_{j=1}^{n} \int \chi_{I_j}$$
$$= \sum_{j=1}^{n} \lambda(I_j) \leq \sum_{n=1}^{\infty} \sum_{k=1}^{\infty} \lambda(I_k^{(n)}) < \sum_{n=1}^{\infty} 2^{-n} = 1,$$

proving that the integrals are uniformly bounded. ∎

Exercises

1. Show that every point in the interval [0,1] may be written as an infinite sum

$$\sum_{n=1}^{\infty} a_n 3^{-n}, \qquad a_n = 0,1, \text{ or } 2.$$

The *ternary* expansion of the number is denoted by

$$0\,.\,a_1 a_1 \ldots a_n \ldots .$$

(Compare with decimals. See Sec. 4 of Chapter 0.) Show that if a is a point in the Cantor set C, then its ternary expansion contains only 2's and 0's. Conversely, every ternary expansion whose terms $a_n = 0$ or 2, corresponds to a point in C. Use this to prove that C is uncountable. [*Hint:* Show that the ternary expansions of points in C may be put in a 1–1 correspondence with *all* the points of [0,1]. This may be done by representing each point of [0,1] by its binary expansion,

$$0\,.\,b_1 b_2 \ldots b_n \ldots, \qquad b_n = 0 \text{ or } 1.]$$

2. The preceding exercise suggests the following generalization: Let β be a positive integer. Then every point of [0,1] may be written as the sum

$$\sum_{n=1}^{\infty} a_n \beta^{-n}, \qquad a_n = 0, 1, \ldots, \beta - 1.$$

Let c denote a fixed but arbitrary integer satisfying $0 \leq c \leq \beta - 1$. Show that the subset of [0,1] consisting of those numbers which do *not* have the number c in their "β-ary" expansion,

$$0\,.\,a_1 a_2 \ldots a_n, \qquad 0 \leq a_n \leq \beta - 1 \qquad \text{and} \qquad a_n \neq c$$

is a null set.

***3.** Prove the following generalization of Exercise 2: Let $0\,.\,a_1 a_2 \ldots a_n \ldots$ be a β-ary expansion of a number in [0,1]. If c is a fixed but arbitrary integer satisfying $0 \leq c \leq \beta - 1$, let ν_c = number of occurrences of $c = a_n$ for $n \leq \nu$. If

$$\lim_{\nu \to \infty} \frac{\nu_c}{\nu} = \frac{1}{\beta}$$

for $c = 0, 1, \ldots, \beta - 1$, then the number whose β-ary expansion is given above is called *normal*.

Show that for any β, the set of nonnormal numbers is a null set. This implies that, except for a null set, all numbers are normal in all bases.

***4.** Prove that a countable union of null sets is countable. First do this using Definition 4 and patterning the proof after Example 5(b). Then show directly that Theorem 8 is satisfied; that is, show there is a nondecreasing sequence of step functions which diverges on $\bigcup S_n$, where each S_n is null, but whose integrals are uniformly bounded.

5. Show that S is a null set if and only if there exists a nonnegative sequence $\{\psi_n\}$ of step functions satisfying:

(i) $\sum_{i=1}^{n} \psi_i(x) \uparrow \infty$ for $x \in S$, and

(ii) there is a positive number A and

$$\int \sum_{i=1}^{n} \psi_i < A,$$

for all n.

6. Find a nondecreasing sequence, as in Theorem 8, or a series, as in Exericse 5, for the following null sets:

(a) the finite set $\{1,2,3\}$,

(b) the countable set $\{1, 1/2, \ldots, 1/n, \ldots\}$,

(c) the set $Q \cap [0,1]$,

(d) the Cantor set.

7. Show that if S is a null subset of a set T which is not itself null, then $T - S$ is not null.

8. If S and T are two sets of real numbers, then the set $S \oplus T$ is defined as the set of numbers which can be written as a sum $s + t$, where $s \in S$ and $t \in T$. Show that if S is a null set, then $S \oplus Q$ is null.

9. Let S^2 denote the set whose elements are squares of the numbers in a set S. Prove that if S is null then S^2 is null.

*Exercises which request the proof of a statement made in the text, as well as those referred to in the text, are starred.

4 Lebesgue Integrable Functions

Although the elements of the completion of a normed linear space X are defined to be classes of Cauchy sequences in X, it is often possible to find a concrete "representation" of the completion. For example, the completion of Λ_2, the space of finite sequences of real numbers with norm given by

$$\| x \| = (\Sigma x_i^2)^{1/2},$$

is the space l_2 of (infinite) sequences $x = (x_1, x_2, \ldots, x_n, \ldots)$ for which

$$\Sigma x_i^2 < \infty$$

(Example 2.4.11). In this section we shall define the normed linear space of *Lebesgue integrable functions*, and show, in what sense, it may be regarded as the completion of Σ.

It is convenient to introduce an abbreviation for the phrase "except on a null set," which appears frequently throughout the text: If two functions $f(x)$ and $g(x)$ are equal to each other except on a null set, we shall say that they are equal "almost everywhere," and write $f = g$ a.e. Similarly, $f \geq g$ a.e. stands for $f(x) \geq g(x)$ except on a null set. If the functions are defined only on a subset S of R, then we write

$$f = g \quad \text{a.e.} \quad \text{on } S$$

to stand for $f(x) = g(x)$ for $x \in S - E$, where E is a null set.

We begin by outlining the main results of the section: First it will be shown that every Cauchy sequence of step functions contains a subsequence which converges almost everywhere. Then we show that if $\{\sigma_n\}$ and $\{\tau_n\}$ are equivalent sequences, and if $\lim \sigma_{n_k} = f$ a.e. and $\lim \tau_{n_j} = g$ a.e., then $f = g$ a.e. Thus every equivalence class of Cauchy sequences of step functions determines a class $[f]$ of functions, each of which is the a.e. limit of a Cauchy sequence of step functions; and $g \in [f]$ if and only if g is also the limit a.e. of such a sequence, and $f = g$ a.e. These functions are called the *Lebesgue integrable functions*, and the integral of f is defined as

[1] $$\int f = \lim \int \sigma_n;$$

$\{\sigma_n\}$ is a Cauchy sequence which converges a.e. to f. The existence of this limit was proved in Sec. 2, and the reader will recall that the right-hand side of [1] is the integral of the class $\hat{f}$ in $\hat{L}_1$ which contains the sequence $\{\sigma_n\}$. We shall show that there is a bijection of the space $\hat{L}_1$ of classes $\hat{f} = [\{\sigma_n\}]$ onto the space of classes $[f]$ of integrable functions. This mapping will preserve the vector space operations and, as is suggested by [1], it will also preserve the norm and the integral.

The following lemma is needed to prove that every Cauchy sequence of step functions contains an almost everywhere convergent subsequence.

1 Lemma. Let $\{\sigma_n\}$ be a sequence of step functions which converges in norm to zero. Then there is a subsequence $\{\sigma_{n_k}\}$ which converges a.e. to zero.

Proof: $\lim \|\sigma_n\| = 0$ implies that there is a subsequence that satisfies $\|\sigma_{n_k}\| < 2^{-2k}$. Lemma 5.3.9 implies that $|\sigma_{n_k}(x)| < 2^{-k}$ except on a set $\mathcal{I}_k$ of intervals whose total length is less than 2^{-k}. (Simply take $A = 2^{-2k}$ and $\delta = 2^{-k}$.) For each value of k, set

$$S_k = \bigcup_{j \geq k} \mathcal{I}_j = \mathcal{I}_k \cup \mathcal{I}_{k+1} \cup \cdots.$$

If $x \in S_k$, then $j \geq k$ implies $| \sigma_{n_j}(x) | \leq 2^{-j}$, proving that $\lim \sigma_{n_j}(x) = 0$ outside S_k. Let E denote the set on which the subsequence $\{\sigma_{n_j}\}$ does not converge to 0. Clearly, $E \subset S_k$ for all values of k. However,

$$\lambda(S_k) = \sum_{j=k}^{\infty} \lambda(\mathcal{I}_j) < \sum_{j=k}^{\infty} 2^{-j} = 2^{-k+1},$$

from which it follows that E is a null set. ∎

2 Proposition. Every Cauchy sequence of step functions contains an almost everywhere convergent subsequence.

Proof: Let $\{\sigma_n\}$ be the given sequence. Choose a subsequence that satisfies

$$\| \sigma_{n_k} - \sigma_{n_{k+1}} \| < 2^{-2k}.$$

As in Lemma 1, $| \sigma_{n_k}(x) - \sigma_{n_{k+1}}(x) | < 2^{-k}$ except on a finite set of intervals $\mathcal{I}_k$ whose length does not exceed 2^{-k}. We shall show that for all values of k, this subsequence converges outside the set $S_k = \bigcup_{j \geq k} \mathcal{I}_j$, where $\lambda(S_k) < 2^{-k+1}$. This is done by proving that if $x \notin S_k$, then $\{\sigma_{n_k}(x)\}$ is a Cauchy sequence of real numbers. Let $p \geq 0$, $q \geq k$. Then if $x \notin S_k$,

$$\begin{aligned} | \sigma_{n_{p+q}}(x) - \sigma_{n_q}(x) | &= | (\sigma_{n_{p+q}} - \sigma_{n_{p+q-1}}) + (\sigma_{n_{p+q-1}} - \sigma_{n_{p+q-2}}) \\ &\quad + \cdots + (\sigma_{n_{q+1}} - \sigma_{n_q}) | < 2^{1-p-q} + 2^{2-p-q} + \cdots + 2^{-q} < 2^{1-q}. \end{aligned}$$

Thus the subsequence diverges on a set E which is contained in each S_k, $\lambda(S_k) < 2^{-k+1}$. This proves that E is a null set. ∎

The following example shows that Proposition 2 can not be "improved." It is an example of a Cauchy sequence of step functions which converges at no point of an interval.

3 Example. Let $\sigma_1 = \chi_{[0,1/2]}$, $\sigma_2 = \chi_{[1/2,1]}$, $\sigma_3 = \chi_{[0,1/3]}$, $\sigma_4 = \chi_{[1/3,2/3]}$, $\sigma_5 = \chi_{[2/3,1]}$, $\sigma_6 = \chi_{[0,1/4]}, \ldots$. (Draw a picture.) It is easily seen that for each integer n there are n step functions of the sequence equal to $\chi_{[0,1/n]}$, $\chi_{[1/n,2/n]}, \ldots, \chi_{[(n-1)/n,1]}$ respectively. The areas determined by each of these n functions is equal to $1/n$. From this it is not difficult to show that $\{\sigma_n\}$ is a Cauchy sequence. However, for each x in [0,1],

$$\underline{\lim}_{n\to\infty} \sigma_n(x) = 0 \qquad \text{and} \qquad \overline{\lim}_{n\to\infty} \sigma_n(x) = 1,$$

proving that the given sequence does not converge at any point of [0,1]. There are many ways to choose a subsequence that converges to zero at all points of the interval save one (Exercise 4).

Every Cauchy sequence is equivalent to each of its subsequences. It is therefore meaningful to talk about the subclass of $[\{\sigma_n\}]$ which consists of almost everywhere convergent sequences. We prove next that all the sequences of this subclass converge almost everywhere to the same function.

4 Proposition. Let $\{\sigma_n\}$ and $\{\tau_n\}$ be equivalent Cauchy sequences which converge almost everywhere to functions f and g respectively. Then $f = g$ a.e.

Proof: Lemma 1 and $\lim \|\sigma_n - \tau_n\| = 0$ assure the existence of a subsequence $\{\sigma_{n_k} - \tau_{n_k}\}$ which converges almost everywhere to zero. Let T be the null set on which this subsequence does not converge to zero, and let U and V be the null sets on which $\{\sigma_n\}$ and $\{\tau_n\}$ do not converge to f and g respectively. Then the subsequences $\{\sigma_{n_k}\}$ and $\{\tau_{n_k}\}$ converge also to f and g respectively except on the null sets U and V, and

$$|f - g| \leq |f - \sigma_{n_k}| + |\sigma_{n_k} - \tau_{n_k}| + |\tau_{n_k} - g|.$$

The right-hand side tends to zero except on the null set $T \cup U \cup V$. ∎

Let L denote the totality of functions each of which is the a.e. limit of a Cauchy sequence of step functions. We shall say that two functions in L are *equivalent* whenever $f = g$ a.e. The collection of equivalence classes $[f]$ of functions in L will be denoted by L_1. Propositions 2 and 4 imply that there is a mapping of $\hat{L}_1$, the set of equivalence classes of Cauchy sequences of step functions, into L_1. This mapping ψ is given by

$$\psi([\{\sigma_n\}]) = [f];$$

the function f is the a.e. limit of a sequence in the class $\hat{f} = [\{\sigma_n\}]$. Proposition 4 asserts that this mapping is well defined; that is, all a.e. convergent sequences in $\hat{f}$ converge to functions which are a.e. equal, and therefore in the same class $[f]$ in L_1. We prove next that ψ is injective. This requires verifying that if two Cauchy sequences $\{\sigma_n\}$ and $\{\tau_n\}$ converge a.e. to the same function, then they are equivalent. The sequence $\rho_n = |\sigma_n - \tau_n|$ is also a Cauchy sequence and $\lim \rho_n = 0$ a.e. The equivalence of $\{\sigma_n\}$ and $\{\tau_n\}$ is demonstrated by proving that $\lim \|\rho_n\| = 0$ (Proposition 7). First a lemma is needed.

5 Lemma. Let $\{\sigma_n\}$ be a nondecreasing sequence of nonnegative step functions, and let τ be another step function. If $\lim \sigma_n(x) \geq \tau(x)$ a.e., then $\lim \int \sigma_n \geq \int \tau$, provided that the limit exists.

Proof: Let $\rho_n = (\tau - \sigma_n)^+$, the positive part of the step function $\tau - \sigma_n$. From $\lim \sigma_n \geq \tau$ a.e., it follows that $\{\rho_n\}$ is a nonincreasing sequence of step functions that converges a.e. to zero. If the convergence of the sequence $\{\rho_n\}$ to zero took place everywhere, then the continuity of the integral would yield the desired result. For then we would have

$$\tau = (\tau - \sigma_n) + \sigma_n \leq (\tau - \sigma_n)^+ + \sigma_n = \rho_n + \sigma_n,$$

and it would follow that

$$\int \tau \leq \int \rho_n + \int \sigma_n \leq \int \rho_n + \lim \int \sigma_n,$$

and if $\int \rho_n \to 0$, we could conclude that

$$\int \tau \leq \lim \int \sigma_n.$$

It remains to be proved that the a.e. convergence of the sequence $\{\rho_n\}$ to zero implies that the sequence of integrals converges also to zero. This result is of sufficient importance to be stated formally:

6 Proposition. Let $\{\rho_n\}$ be a sequence of nonnegative step functions which tends monotonically to zero almost everywhere. Then

[2] $$\lim \int \rho_n = 0.$$

Proof: Let S be the null set on which $\{\rho_n\}$ does not converge to zero. From Theorem 5.3.8, it follows that there is a nondecreasing sequence of nonnegative step functions ϕ_n which diverges on S but whose integrals are uniformly bounded by a positive number A. Let $\epsilon > 0$. The sequence

$$\psi_n = (\rho_n - \epsilon\phi_n)^+$$

converges monotonically to zero for all x. The continuity of the integral [Proposition 5.1.2(vii)], yields

$$\lim \int \psi_n = 0.$$

The monotonicity of the integral and the inequality

$$\rho_n = (\rho_n - \epsilon\phi_n) + \epsilon\phi_n \leq \psi_n + \epsilon\phi_n$$

imply that

$$\lim \int \rho_n \leq \lim \int \psi_n + \lim \int \epsilon\phi_n \leq \epsilon A.$$

Since ϵ was chosen arbitrarily, [2] follows.

This completes the proof of both Proposition 6 and Lemma 5.

7 Proposition. If $\{\rho_n\}$ is a Cauchy sequence of step functions which converges a.e. to zero, then $\lim \| \rho_n \| = 0$.

Remark. This is not the same situation as the one described in Proposition 6. There the convergence was monotonic.

Proof: If it can be shown that a subsequence of $\{\rho_n\}$ converges in norm to zero, then the sequence itself must converge also to zero. For this reason, we shall assume that $\| \rho_{k+1} - \rho_k \| < 2^{-2k}$. Otherwise a subsequence could be selected for which this was true.

For each value of n, we write

$$\rho_n = \rho_1 + (\rho_2 - \rho_1) + \cdots + (\rho_n - \rho_{n-1}) = R_1 + R_2 + \cdots + R_n.$$

It has been assumed that

$$[3] \qquad 0 = \lim_{n\to\infty} \rho_n = \lim_{n\to\infty} \sum_{i=1}^{n} R_i = \sum_{i=1}^{\infty} R_i = \sum_{i=1}^{\infty} (R_i^+ - R_i^-), \text{ a.e.}$$

In order to rearrange the infinite series [3], we must show that except on a null set, $\Sigma R_i(x)$ converges absolutely. This is equivalent to proving that the series of positive (or negative) parts ΣR_i^+ (or ΣR_i^-) converges a.e. Let

$$\gamma_n^+ = \sum_{i=1}^{n} R_i^+.$$

The sequence $\{\gamma_n^+\}$ of step functions is nondecreasing, nonnegative, and

$$\int \gamma_n^+ = \int \sum_{i=1}^{n} R_i^+ = \sum_{i=1}^{n} \int R_i^+ \leq \sum_{i=1}^{\infty} \int |R_i| < \|\rho_1\| + 1.$$

To obtain the last inequality, we used $\|\rho_i - \rho_{i-1}\| = \|R_i\| < 2^{-i}$, and summed the geometric series. Theorem 5.3.8 implies that the series $\Sigma R_i^+(x)$ must converge a.e. A similar argument holds for the series of negative parts. Since both

$$\gamma_n^+ = \sum_{i=1}^{n} R_i^+ \qquad \text{and} \qquad \gamma_n^- = \sum_{i=1}^{n} R_i^-$$

are nonnegative, nondecreasing sequences of step functions, it follows from [3] that for any integer ν,

$$[4] \qquad \lim_{n\to\infty} \gamma_n^+ \geq \gamma_\nu^-.$$

Since $\{\gamma_n^+\}$ is a Cauchy sequence (why?), $\lim_{n\to\infty} \int \gamma_n^+$ exists. From [4] and Lemma 5 we obtain

$$\lim_{n\to\infty} \int \gamma_n^+ \geq \int \gamma_\nu^-,$$

for all integers ν. Taking the limit on the right yields

$$\lim_{n\to\infty} \int \gamma_n^+ \geq \lim_{n\to\infty} \int \gamma_n^-.$$

By reversing the argument, beginning at [4], the opposite inequality is derived, and it follows therefore that

$$0 = \lim_{n\to\infty} \left[\sum_{i=1}^{n} \int \gamma_i^+ - \sum_{i=1}^{n} \int \gamma_i^-\right] = \lim_{n\to\infty} \int \rho_n,$$

which is what we set out to prove.

We return to the original problem. $\{\sigma_n\}$ and $\{\tau_n\}$ are Cauchy sequences of step functions that converge a.e. to the same function. This implies that their difference, $\sigma_n - \tau_n$, converges a.e. to zero. The equivalence of the two sequences follows from Proposition 7. This completes the proof that the mapping $\psi([\{\sigma_n\}]) = [f]$, which takes the class of Cauchy sequences of step functions (in $\hat{L}_1$) into the class of functions $[f]$ (in L_1) which are the a.e. limits of sequences in $[\{\sigma_n\}]$, is a bijection.

The next step is to show that if the proper vector space operations are defined in L_1, then ψ will be a vector space isomorphism.

8 Definition. Let $[f]$ and $[g]$ be classes in L_1 and let $\{\sigma_n\}$ and $\{\tau_n\}$ be Cauchy sequences of step functions which converge a.e. to f and g respectively. The *sum* of $[f]$ and $[g]$ and the *product* of $[f]$ by a real number c are given by

(i) $[f] + [g] = [f + g]$, and

(ii) $c \cdot [f] = [cf]$.

The addition and multiplication are well defined. In this case it means that

[a] Both $\{\sigma_n + \tau_n\}$ and $\{c\sigma_n\}$ are Cauchy sequences of step functions, and $\lim (\sigma_n + \tau_n) = f + g$, $\lim c\sigma_n = cf$ a.e.

[b] If $f = f'$ and $g = g'$ a.e., then $[f + g] = [f' + g']$ and $[cf] = [cf']$.

The verification of [a] and [b] is left to the reader.

The vector space isomorphism follows easily: If $\hat{f} = [\{\sigma_n\}]$, $\hat{g} = [\{\tau_n\}]$ are elements of $\hat{L}_1$ and if $\lim \sigma_n = f$, $\lim \tau_n = g$ a.e., then $\psi(\hat{f}) = [f]$ and $\psi(\hat{g}) = [g]$. Then

$$\psi(\hat{f} + \hat{g}) = \psi([\{\sigma_n + \tau_n\}] = [f + g] = [f] + [g] = \psi(\hat{f}) + \psi(\hat{g}).$$

The first equality is a consequence of the definition of addition in $\hat{L}_1$, the second and fourth use the definition of the mapping ψ, and the third is simply the addition of two elements in L_1. A similar argument yields $\psi(c\hat{f}) = c\psi(\hat{f})$.

Finally, we define an integral and a norm on L_1, and show that under the mapping ψ, $\hat{L}_1$ and L_1 are isonormal.

9 Definition. Let f, $[f]$, $\hat{f} = [\{\sigma_n\}]$ have the same meaning as in the preceding paragraphs. The integral and norm of $[f]$ are given by

[5] $$\int_{L_1} [f] = \lim \int \sigma_n, \qquad \text{and} \qquad \| [f] \|_{L_1} = \lim \| \sigma_n \| .$$

Since the integral and norm of $[f]$ are defined by taking limits of the integrals and norms of the σ_n, it follows from Definition 5.2.2 that

$$\int_{L_1} [f] = \int \hat{f}, \qquad \text{and} \qquad \| [f] \|_{L_1} = \| \hat{f} \| .$$

This means that the integral and norm defined by [5] are well defined and that they inherit all the properties of the integral and the norm defined on $\hat{L}_1$. To avoid the unwieldy notation and language of equivalence classes, we shall hereafter write $\int f$ for $\int_{L_1} [f]$, remembering that if $f = f'$ a.e., then as integrable functions, they are to be considered identical. A similar statement holds for the norm. We shall therefore speak of L_1 as the Banach space of Lebesgue integrable *functions*, instead of as the space of equivalence classes of integrable functions. In this sense, L_1 is the completion of Σ.

For reference purposes, we shall list those properties of the integral that are inherited from the corresponding properties already proved for $\hat{L}_1$.

10 Proposition. Let f and g be Lebesgue integrable functions, and let c be a real number. Then

(i) $\int(f + g) = \int f + \int g$ (additivity).

(ii) $\int cf = c\int f$ (homogeneity).

(iii) $f \geq 0$ a.e. implies $\int f \geq 0$ (positivity).

(iv) $f \geq g$ a.e. implies $\int f \geq \int g$ (monotonicity).

(v) $\|f\| = \int |f|$; and $\int |f| = 0$ implies $f = 0$ a.e.

(vi) $|\int f| \leq \int |f|$.

The continuity of the integral [See Proposition 5.1.2 (vii)] will be proved as a special case of the more general theorem on the monotone convergence of integrals (Theorem 6.1.5).

It is also easily proved that L_1 is partially ordered by

[6] $$f \leq g \quad \text{if} \quad f(x) \leq g(x) \text{ a.e.},$$

and is a lattice. Furthermore, ψ preserves this ordering.

Exercises

***1.** Suppose that $\{\sigma_n\}$ is a sequence of step functions which are a.e. nonnegative, a.e. nonincreasing (for each n, $\sigma_n \geq \sigma_{n+1}$ a.e.) and converge a.e. to zero. Prove that
$$\int \sigma_n \downarrow 0.$$

2. Prove that if $\{\sigma_n\}$ is a nonincreasing sequence of step functions whose integrals converge to zero, then $\lim \sigma_n = 0$ a.e.

***3.** Suppose that $\{\sigma_n\}$ is a nondecreasing sequence of step functions satisfying $\lim \sigma_n(x) \geq 0$ for all x. ($\lim \sigma_n = \infty$ is permitted). Let $S = \{x : \lim \sigma_n(x) > 0\}$. Prove that if S is *not* a null set, then
$$\lim \int \sigma_n > 0.$$

***4.** Show that the sequence described in Example 3 is a Cauchy sequence. Find a subsequence which converges to zero on $(0,1]$.

***5.** Let $\mathfrak{F}$ be any collection of real valued functions defined on R. Prove that $f = g$ a.e. is an equivalence relation.

6. Prove that the class [0] of L_1 is a vector subspace and a sublattice of L, the totality of Lebesgue integrable functions. Describe L_1 as the quotient of L by [0].

7. Prove that if $\{f_n\}$ is a sequence of integrable functions and if $f_n = 0$ a.e., then

$$\sum_{i=1}^{\infty} f_i = 0 \text{ a.e.}$$

***8.** Prove that the product of an integrable function f and a step function σ is in L_1. (*Hint:* Find a Cauchy sequence of step functions which converges to σf a.e.)

***9.** Prove that [6] is a partial ordering of L_1 which is preserved under the mapping ψ. (See Definition 5.2.7.) Show also that L_1 is a lattice.

***10.** Suppose that $\{\sigma_n\}$ is a nondecreasing sequence of step functions converging everywhere to an integrable function f. If σ is any other step function, show that

$$\lim \int \sigma_n \sigma = \int f\sigma.$$

***11.** Prove that any continuous function defined on a closed bounded interval is Lebesgue integrable. Do the same thing for sectionally continuous functions that vanish outside a closed bounded interval.

This says that the functions in P, C_0, Π, and S_0 (sectionally continuous functions that vanish outside a closed and bounded interval) are Lebesgue integrable.

***12.** Discuss the concrete representation of the completions of the spaces mentioned in Exercise 11. (See Exercises 5.1.8, 5.2.8, 5.2.9.) Show that each one is equal to L_1. Using Exercise 11, it suffices to prove that if f is the a.e. limit of functions in either P, C_0, Π, or S_0 then f is the a.e. limit of step functions.

13. Prove that

$$f(x) = \begin{cases} x^{-\beta} & \text{if } 0 < x < 1 \\ 0 & \text{otherwise} \end{cases}$$

is integrable if and only if $\beta < 1$.

14. Prove that

$$f(x) = \begin{cases} x^{-\beta} & \text{if } x > b > 0 \\ 0 & \text{otherwise} \end{cases}$$

is integrable if and only if $\beta > 1$.

*Exercises which request the proof of a statement made in the text, as well as those referred to in the text, are starred.

CHAPTER **6**

Convergence Theorems and the Riemann Integral

Having defined a Lebesgue integrable function as the a.e. limit of a Cauchy sequence of step functions, it is reasonable to ask under what conditions the limit of a sequence $\{f_n\}$ of integrable functions will converge to an integrable function f, and if so, when can we conclude that

$$* \qquad \int f = \lim \int f_n.$$

In the theory of Riemann integration, some rather strong conditions must be imposed if the limit function is to be Riemann integrable, and in some cases the Riemann integrability of the limit function f must be *assumed* in order to conclude that the integral of the limit function is the limit of the integrals of the f_n. Earlier, it was stated that if the f_n are continuous, and if they converge uniformly on a bounded interval, then * follows. These are rather strong conditions, but they are necessary to insure the Riemann integrability of the limit function. For, as a consequence of the incompleteness of the space of Riemann integrable functions, some relatively "decent" sequences of step functions (Example 3) may fail to converge to Riemann integrable functions.

In this chapter the three principal convergence theorems of Lebesgue integration are proved. In each case we shall see that the space of Riemann integrable functions does not share these convergence properties with L_1.

The relation of the Riemann integral to the Lebesgue integral is discussed in the final section, and it is proved there that the Lebesgue integral is indeed an extension of the Riemann integral, in the sense that every Riemann integrable function is Lebesgue integrable, and that the two integrals coincide on the space of Riemann integrable functions.

The principal result of Sec. 3 is the proof that a bounded function is Riemann integrable if and only if it is a.e. continuous.

1 Monotone Sequences of Integrable Functions

In this section, we shall prove that a nondecreasing (nonincreasing) sequence $\{f_n\}$ of integrable functions whose integrals are uniformly bounded, converges a.e. to an integrable function f, and that * (see the introductory remarks) is satisfied. In the course of proving this theorem, we shall focus our attention upon a special subset Γ of L_1, the subset of monotone limits of sequences of step functions whose integrals are uniformly bounded. Although it will be shown that Γ is a proper subset of L_1, that is, there are integrable functions which cannot be obtained as the a.e. limits of monotone sequences of step functions, we shall prove that an integrable functions is "almost" the limit of such a monotone sequence. More precisely, it will be proved that every integrable function may be written as the difference $g - h$ of functions in Γ, and that it may be assumed that the norm of h is as small as we like. This characterization of L_1 as the collection of differences of functions in Γ, will be used in Chapter 12 to define the integrable functions in the abstract case.

By a *monotone sequence* of functions, is meant one that is either nondecreasing or nonincreasing. If $f_n(x) < f_{n+1}(x)$ (or $f_n(x) > f_{n+1}(x)$) for all n, then we shall say that the sequence $\{f_n\}$ is *strictly* increasing (or *strictly* decreasing). As before, we write $f_n \uparrow f$ (or $f_n \downarrow f$) to denote a nondecreasing (or nonincreasing) sequence of functions converging to a function f.

Theorem 5 asserts that L_1 is "closed" under monotone convergence of sequences of integrable functions whose integrals are uniformly bounded. We begin with the simplest case, assuming first that the elements of the monotone sequence are step functions whose integrals are uniformly bounded. We shall prove that the a.e. limit of this sequence exists, is integrable, and * is satisfied.

1 Theorem (Monotone Convergence Theorem for step functions). If $\{\sigma_n\}$ is a nondecreasing sequence of step functions, and if there is a positive number A such that $\int \sigma_n < A$, for all n, then

(i) $\lim \sigma_n = f$ exists a.e., and is integrable, and

(ii) $\int f = \lim \int \sigma_n \leq A$.

Proof: It may be assumed (as in the proof of Theorem 5.3.8) that the σ_n are nonnegative for all n. The monotonicity of the integral implies that $s_n = \int \sigma_n$ is a nondecreasing sequence of real numbers. The boundedness of this sequence (ii), and Theorem 1.3.18 imply that $\{s_n\}$ converges in R, and is therefore a Cauchy sequence of real numbers. As a consequence of this, we are able to show that $\{\sigma_n\}$ is a Cauchy sequence of step functions:

$$\| \sigma_m - \sigma_n \| = \int | \sigma_m - \sigma_n | = \left| \int (\sigma_m - \sigma_n) \right| = | s_m - s_n |,$$

and the last term tends to zero as n and m tend to infinity. Proposition 5.4.2 implies that a subsequence $\{\sigma_{n_k}\}$ converges a.e. to an integrable function f, and $\int f = \lim \int \sigma_{n_k}$.

It remains to be shown that the entire sequence $\{\sigma_n\}$ converges monotonically to f. This is a simple consequence of the monotonicity of the sequence: Take n to be any positive integer, and n_k to be an index of the subsequence which converges a.e. to f. The triangle inequality yields

$$|f - \sigma_n| \le |f - \sigma_{n_k}| + |\sigma_{n_k} - \sigma_n|.$$

The first term on the right side tends a.e. to zero, and the monotonicity implies that for $n \ge n_k$, $|\sigma_{n_k} - \sigma_n| \le |f - \sigma_{n_k}|$, which like the first term tends to zero. ▎

A similar theorem holds for nonincreasing sequences of step functions whose integrals are bounded from below.

Let Γ denote the set of integrable functions which are the a.e. limits of nondecreasing sequences of step functions whose integrals are bounded from above, and let $-\Gamma$ denote the set of integrable functions which are the negatives of functions in Γ. It is readily seen that $-\Gamma$ consists of those integrable functions that are the a.e. limits of nonincreasing sequences of step functions whose integrals are bounded from below. Their union, $\Gamma \cup -\Gamma$, is the set of integrable functions which are monotone limits of step functions with uniformly bounded integrals. Example 3 demonstrates that $\Gamma \cup -\Gamma$ is a proper subset of L_1. But first we show that Γ is closed under monotone convergence:

2 Theorem (Monotone Convergence Theorem for functions in Γ). Let $\{f_n\}$ be a nondecreasing sequence of functions in Γ satisfying $\int f_n < A$. Then

(i) $\{f_n\}$ converges a.e. to a function f in Γ, and
(ii) $\int f = \lim \int f_n$.

Proof: For each value of n, there is a nondecreasing sequence $\{\sigma_{n,i}\}$ of step functions which converges a.e. to f_n. We shall define a nondecreasing sequence $\{\sigma_n\}$ of step functions which converges, together with f_n, to a function in Γ. Let

$$[1] \qquad \sigma_n = \max_{i,j \le n} \{\sigma_{i,j}\}.$$

Each σ_n is the maximum of finitely many step functions, and is therefore itself a step function. Moreover,

$$\sigma_n = \max_{i,j \le n} \{\sigma_{i,j}\} \le \max_{i,j \le n+1} \{\sigma_{i,j}\} = \sigma_{n+1},$$

and since $\sigma_{i,j} \le f_i \le f_n$ whenever $i,j \le n$, it follows that

[2] $$\sigma_n = \max_{i,j \le n} \{\sigma_{i,j}\} \le f_n.$$

The monotonicity of the integral yields

$$\int \sigma_n \le \int f_n < A.$$

Theorem 1 implies that $\{\sigma_n\}$ converges a.e. to an integrable function f and (ii) is satisfied. It follows easily from [2] and the monotonicity of all the sequences that $\lim f_n = f$ a.e. and (ii) is satisfied for the sequence $\{f_n\}$ as well as for $\{\sigma_n\}$. ∎

A similar theorem is true for $-\Gamma$.

We show next that there are integrable functions that are not limits of monotone sequences of step functions with uniformly bounded integrals.

3 Example. Any function which is not bounded from above cannot be the limit of a nonincreasing sequence of step functions, nor can a function which is unbounded from below be the limit of a nondecreasing sequence of step functions. Hence our task is to find an integrable function which is neither bounded from below or above. Such a function must be the a.e. limit of a sequence of step functions whose steps are becoming unbounded, but whose corresponding areas are uniformly bounded. Let

$$f(x) = n \qquad \text{whenever} \qquad a_{n-1} < x < a_n,$$

where $a_0 = 0, a_1 = 1, a_2 = a_1 + \dfrac{1}{2\cdot 2}, a_3 = a_2 + \dfrac{1}{3\cdot 2^2}, \ldots, a_n = a_{n-1} + \dfrac{1}{n\cdot 2^{n-1}}, \ldots$. Also $f(a_n) = 0$ for all n. The function is defined for negative values of x by reflection through the origin. That is, $f(x) = -f(-x)$. (See Figure 24.) To show that $f(x)$ is integrable, we must show it is the a.e. limit of a Cauchy sequence of step functions. Set

$$\sigma_n(x) = \begin{cases} f(x) & \text{if } |x| < a_n \\ 0 & \text{otherwise} \end{cases}$$

If n and p are positive integers,

$$\|\sigma_{n+p} - \sigma_n\| = 2\left\{\sum_{j=1}^{n+p} (a_j - a_{j-1})j - \sum_{j=1}^{n} (a_j - a_{j-1})j\right\}$$

$$= 2 \sum_{j=n+1}^{n+p} \frac{j}{j(2^{j-1})} = \sum_{j=n+2}^{n+p} \frac{1}{2^{j-1}},$$

which tends to zero as n approaches infinity, proving that $\{\sigma_n\}$ is a Cauchy

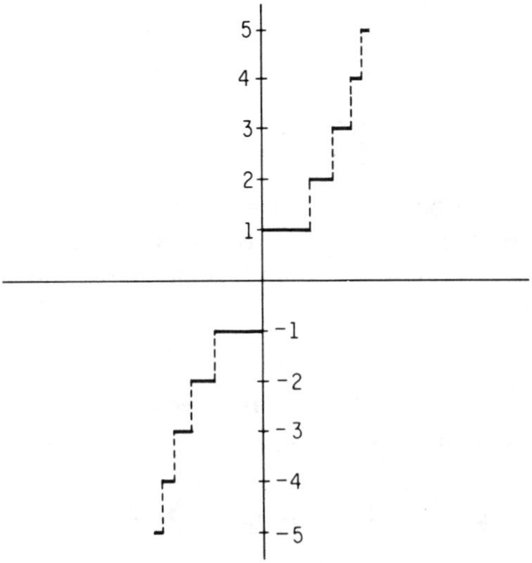

FIGURE 24

sequence of step functions. Thus f, which is the a.e. limit of the σ_n, is in L_1, but in neither Γ nor $-\Gamma$.

However, it should be observed that $f = f^+ - f^-$, and both f^+ and f^- are in Γ. Theorem 4 asserts that *any* integrable function may be written as the difference of functions in Γ.

4 Theorem. If $f \in L_1$ and $\epsilon > 0$, then there are functions g,h in Γ, satisfying

(i) $f = g - h$ and (ii) $\| h \| < \epsilon$.

Proof: Let $\{\sigma_n\}$ be a Cauchy sequence of step functions that converges a.e. to f. We may assume that whenever $n \geq m$, $\| \sigma_n - \sigma_m \| < 2^{-m}$. If this were not the case, a subsequence $\{\sigma_{n_k}\}$ could be chosen that satisfies

$$\begin{array}{ll} \| \sigma_n - \sigma_m \| < 2^{-1} & \text{whenever } n,m \geq n_1 \\ \| \sigma_n - \sigma_m \| < 2^{-2} & \text{whenever } n,m \geq n_2 \\ \vdots & \vdots \\ \| \sigma_n - \sigma_m \| < 2^{-k} & \text{whenever } n,m \geq n_k. \\ \vdots & \end{array}$$

To avoid using a double subscript, we shall assume that the sequence $\{\sigma_n\}$ itself has this property. The idea of the proof is to write f as an infinite series whose partial sums are the σ_n. (See proof of Proposition 5.4.7.)

$$f = \lim \sigma_n = \lim [\sigma_1 + (\sigma_2 - \sigma_1) + \cdots + (\sigma_n - \sigma_{n-1})] \quad \text{a.e.}$$

Setting $\sigma_1 = s_1$ and $(\sigma_n - \sigma_{n-1}) = s_n$ for $n \geq 2$, we obtain

$$f = \sum_{n=1}^{\infty} s_n = \sum_{n=1}^{\infty} [s_n^+ - s_n^-],$$

where s_n^+, s_n^- are the positive and negative parts of s_n. In order to rearrange the series on the right, and to write it as the difference of the series of positive parts and the series of negative parts, we must show that the latter two series,

$$\Sigma s_n^+ \qquad \text{and} \qquad \Sigma s_n^-,$$

converge a.e. The proof is given for the series of positive parts; it is the same for the series of negative parts. Let

$$\psi_\nu = \sum_{n=1}^{\nu} | \sigma_{n+1} - \sigma_n | = \sum_{n=1}^{\nu} | s_{n+1} |.$$

The sequence of step functions $\{\psi_\nu\}$ is nondecreasing and

$$\int \psi_\nu = \sum_{n=1}^{\nu} \int | \sigma_{n+1} - \sigma_n | < \sum_{n=1}^{\nu} 2^{-n} < 1.$$

Thus Theorem 1 yields

$$\lim \psi_n = \psi \quad \text{a.e.} \qquad \text{and} \qquad \psi \in \Gamma.$$

Theorem 1 may also be applied to the sequence,

$$\phi_\nu = \sum_{n=1}^{\nu} s_{n+1}^+ \leq \psi_\nu,$$

which is also nondecreasing and has uniformly bounded integrals. Thus

$$\lim_{\nu \to \infty} \phi_\nu = \lim_{\nu \to \infty} \sum_{n=1}^{\nu} s_{n+1}^+ = \sum_{n=1}^{\infty} s_{n+1}^+$$

exists a.e. and is in Γ. This permits the rearrangement of the series, and we have

$$f = \sum_{n=1}^{\infty} s_n^+ - \sum_{n=1}^{\infty} s_n^- = \left\{ \sum_{n=1}^{\infty} s_n^+ - \sum_{n=1}^{\nu} s_n^- \right\} - \left\{ \sum_{n=\nu+1}^{\infty} s_n^- \right\} = g - h;$$

g and h are the bracketed functions. We prove next that for all values of ν, g and h are in Γ. Let

$$\xi_m = \sum_{n=1}^{m} s_n^+ - \sum_{n=1}^{\nu} s_n^-, \qquad \zeta_m = \sum_{n=\nu+1}^{\nu+m} s_n^-.$$

Both $\{\xi_m\}$ and $\{\zeta_m\}$ are nondecreasing sequences of step functions and their integrals are uniformly bounded. Therefore the Theorem on Monotone Convergence for step functions (Theorem 1), implies that the monotone limits g and h of these sequences are in Γ. The details are left to the reader.

Finally, by choosing ν to be sufficiently large, we have

$$\|h\| = \lim_{m\to\infty} \int \sum_{n=\nu+1}^{\nu+m} s_n^- \le \lim_{m\to\infty} \int \sum_{n=\nu+1}^{\nu+m} |s_n|$$

$$\le \sum_{n=\nu+1}^{\infty} \|\sigma_n - \sigma_{n-1}\| < 2^{-\nu} < \epsilon. \quad \blacksquare$$

Remark. From the construction of g and h, it is clear that if $f > 0$ then g and h may be chosen so that they are nonnegative.

Although Theorem 4 was included mainly because it is needed in proving the general theorem on monotone convergence, it also provides an alternate description of the class of Lebesgue integrable functions. Instead of defining an integrable function as the a.e. limit of a Cauchy sequence of step functions, we could have instead first enlarged Σ by taking monotone limits, thus obtaining Γ, and then taken differences of functions in Γ to obtain L_1 (Exercise 9).

5 Theorem (Monotone Convergence Theorem for Functions in L_1). Let $\{f_n\}$ be a nondecreasing sequence of integrable functions with uniformly bounded integrals. Then the sequence $\{f_n\}$ converges a.e. to an integrable function f, and

$$\int f = \lim \int f_n.$$

Proof: As in the proof of Theorem 4, we write

$$f_n = f_1 + (f_2 - f_1) + \cdots + (f_n - f_{n-1}) = F_1 + F_2 + \cdots + F_n.$$

Since the F_n are integrable, there are functions G_n, H_n in Γ satisfying (i) $F_n = G_n - H_n$, and (ii) $\|H_n\| < 2^{-n}$ (Theorem 4). The monotonicity of the f_n implies that the F_n, (with the possible exception of F_1), are nonnegative, and it therefore follows that the G_n and H_n may be chosen also to be nonnegative. Therefore

$$\phi_n = \sum_{i=1}^{n} G_i \quad \text{and} \quad \psi_n = \sum_{i=1}^{n} H_i$$

are nondecreasing sequences in Γ. Furthermore,

$$\int |\psi_n| = \int \left|\sum_{i=1}^{n} H_i\right| = \sum_{i=1}^{n} \|H_i\| < \sum_{i=1}^{n} 2^{-i} < 1;$$

$$\int |\phi_n| = \int \left|\sum_{i=1}^{n} G_i\right| = \int \sum_{i=1}^{n} |H_i + F_i|$$

$$\leq \sum_{i=1}^{n} \|H_i\| + \int \sum_{i=1}^{n} F_i = \sum_{i=1}^{n} \|H_i\| + \int f_n < 1 + A,$$

where A is a bound for the integrals of the f_n. Thus Theorem 2 applies to the sequences $\{\phi_n\}$ and $\{\psi_n\}$, and so they converge a.e. to functions g and h in Γ. We have therefore

$$g - h = \lim \phi_n - \lim \psi_n = \sum_{i=1}^{\infty} G_i - \sum_{i=1}^{\infty} H_i$$

$$= \sum_{i=1}^{\infty} (G_i - H_i) = \sum_{i=1}^{\infty} F_i = \lim f_n \text{ a.e.}$$

The third equality follows from the absolute convergence of the two series at almost all points. And finally, since $g,h \in \Gamma$, we have

$$\int g = \lim \int \phi_n = \lim \int \sum_{i=1}^{n} G_i = \sum_{i=1}^{\infty} \int G_i$$

and

$$\int h = \lim \int \psi_n = \lim \int \sum_{i=1}^{n} H_i = \sum_{i=1}^{\infty} \int H_i.$$

These equalities, together with the additivity of the integral, and the absolute convergence of the series $\sum_{i=1}^{\infty} \int G_i$ and $\sum_{i=1}^{\infty} \int H_i$, yield

$$\int f = \int (g - h) = \int g - \int h = \sum_{i=1}^{\infty} \int G_i - \sum_{i=1}^{\infty} \int H_i$$

$$= \sum_{i=1}^{\infty} \int (G_i - H_i) = \sum_{i=1}^{\infty} \int F_i = \lim \int f_n. \quad \blacksquare$$

Exercises

*1. State and prove Theorems 1 and 2 for the case that the sequences are a.e. nondecreasing.
*2. State and prove Theorems 1 and 2 for nonincreasing functions.
*3. Prove that the space Γ of monotone limits of step functions is not a vector

space, but that if $f,g \in \Gamma$, and if a,b are nonnegative numbers, then $af + bg \in \Gamma$. Prove also that Γ is a lattice. What can be said about $-\Gamma$?

4. Show that the function

$$f(x) = \begin{cases} 0 & \text{if } |x| < 1 \\ 1/x^3 & \text{if } |x| \geq 1 \end{cases}$$

is integrable, but is not in either Γ or $-\Gamma$.

5. Show by example that the uniform boundedness of the integrals in Theorems 1,2 and 5 is an essential hypothesis.

6. Use the theorem on monotone convergence to prove that the Lebesgue integral is continuous; that is, if $\{f_n\}$ is a nonincreasing sequence of Lebesgue integrable functions which converges a.e. to zero, then the sequence of integrals converges to zero.

7. Prove that if $\{f_n\}$ is a nonincreasing sequence of integrable functions, and if the sequence of integrals tends to zero, then $\lim f_n = 0$ a.e. (See Exercise 5.4.2.)

8. Do Exercises 5.4.3, 5.4.10, 5.4.11, 5.4.13, and 5.4.14 by using the theorem on monotone convergence.

***9.** Describe carefully how you would define L_1 by taking differences of monotone limits. (See the remark following Theorem 4.)

*Exercises which request the proof of a statement made in the text, as well as those referred to in the text, are starred.

2 Fatou's Lemma and the Theorem on Dominated Convergence

In many applications and problems, a function is given which is the a.e. limit of a sequence of integrable functions. If the sequence is monotone, and if the integrals are uniformly bounded, then Theorem 6.1.5 insures the integrability of the limit function. However, in many cases the sequence is not monotone, nor is it always possible to substitute an equivalant monotone sequence. In this section, we shall prove a theorem which may be applied to nonmonotone sequences. The monotonicity of the sequence $\{f_n\}$ is replaced by the assumption that $\lim f_n$ exists a.e., and the uniform boundedness of the integrals is replaced by the condition that the f_n are bounded in absolute value by some fixed integrable function g. The conclusion is the same.

It was remarked in Sec. 1 that the monotone convergence theorem is not applicable to sequences of Riemann integrable functions, unless it is *assumed* that the limit function is integrable. A similar statement can be made here. In Sec. 3, this "defect" of the Riemann integral is discussed more fully.

We begin by restating a pair of equivalent definitions of the *limit inferior* (written $\underline{\lim}$) of a sequence of numbers.

1 Definition. The *limit inferior* of a sequence $\{a_n\}$ of real numbers is given by

$$\underline{a} = \underline{\lim}\, a_n = \lim_{n\to\infty} \inf\,[a_n, a_{n+1}, \ldots],$$

where $\inf\,[a_n, a_{n+1}, \ldots] = \gamma_n$ is the greatest lower bound of the set of numbers $\{a_n, a_{n+1}, \ldots\}$.

It is easily seen that either

(i) *all* $\gamma_n = -\infty$, in which case $\underline{a} = -\infty$, or

(ii) $\gamma_n \uparrow \underline{a}$. (It is of course possible that $\underline{a} = \infty$.)

2 Definition.

(i) A finite number $\underline{a}$ is said to be the *limit inferior* of a sequence of real numbers $\{a_n\}$ if: (a) given $\epsilon > 0$, almost all (all but finitely many) $a_n > \underline{a} - \epsilon$, and (b) for arbitrarily large values of N, there is an integer $n > N$ and $a_n < \underline{a} + \epsilon$.

(ii) We say that $\underline{\lim}\, a_n = \infty$ if, given any number M, almost all $a_n > M$. (Note that in this case $\underline{\lim}\, a_n = \overline{\lim}\, a_n$).

(iii) If for any value of M, there is an $a_n < M$, then we say that $\underline{\lim}\, a_n = -\infty$.

The equivalence of Definitions 1 and 2 is proved in Proposition 1.3.31.

3 Theorem (Fatou's Lemma). If $\{f_n\}$ is a sequence of nonnegative integrable functions, and if $\underline{\lim} \int f_n$ exists, then $\underline{\lim}\, f_n(x)$ exists a.e., and $f(x) = \underline{\lim}\, f_n(x)$ is integrable, and

$$\int \underline{\lim}\, f_n \le \underline{\lim} \int f_n.$$

Proof: If x is a point at which $\underline{\lim}\, f_n(x)$ is finite, then $\underline{\lim}\, f_n(x) = \lim_{n\to\infty} g_n(x)$, where

$$g_n(x) = \inf\,[f_n(x), f_{n+1}(x), \ldots].$$

The nonnegativeness of the f_n implies that the g_n are also nonnegative. Furthermore, $\{g_n\}$ is nondecreasing. In order to be able to apply the theorem on monotone convergence (Theorem 6.1.5), it must be shown that each g_n is integrable. To prove this, we shall show that each g_n is itself the monotone limit of a sequence of integrable functions. For this purpose, we define for each integer n, a sequence,

$$g_{n,k} = \inf\,[f_n, f_{n+1}, \ldots, f_{n+k}] = \min\,[f_n, f_{n+1}, \ldots, f_{n+k}].$$

Since L_1 is a lattice, each $g_{n,k}$ is integrable, and as $k \to \infty$, $g_{n,k}(x) \downarrow g_n(x)$.

Moreover, $g_{n,k} \geq 0$ implies that the sequence of integrals is bounded from below by 0. Therefore, the limit function g_n is integrable, and

$$\int g_n = \lim_{k\to\infty} \int g_{n,k} \geq 0.$$

Thus $\{g_n\}$ is a nondecreasing sequence of integrable functions. We show now that its integrals are uniformly bounded: If $k \geq n$, $g_n = \inf [f_n, f_{n+1}, \ldots] \leq f_k$, and therefore $\int g_n \leq \int f_k$. This yields

$$\int g_n \leq \lim_{n\to\infty} \inf \left[\int f_n, \int f_{n+1}, \ldots \right] = \underline{\lim} \int f_n,$$

proving that the nondecreasing sequence $\{g_n\}$ has uniformly bounded integrals. A second application of the monotone convergence theorem gives

$$\lim g_n = \underline{\lim}\, f_n \quad \text{is an integrable function}$$

and

$$\int \underline{\lim}\, f_n = \int \underline{\lim}\, g_n \leq \underline{\lim} \int f_n. \quad \blacksquare$$

Variations of Fatou's Lemma may be found in the exercises.

4 Theorem (Lebesgue's Theorem on Dominated Convergence). Let $\{f_n\}$ be a sequence of integrable functions that converge a.e. to a function f. If there is an integrable function g, and $|f_n| \leq g$, then f is integrable, and

$$\int f = \lim \int f_n.$$

Proof: Since the f_n and g are integrable, and $0 \leq g - f_n \leq 2g$ for all n, the monotonicity of the integral implies that

$$0 \leq \int (g - f_n) \leq 2 \int g.$$

This in turn yields the existence of both $\underline{\lim} \int (g - f_n)$ and $\overline{\lim} \int (g - f_n)$ However, it is assumed that $\lim f_n = f$ a.e., implying that $\underline{\lim}\, (g - f_n) = \lim\, (g - f_n)$ a.e.. Applying Fatou's Lemma (Theorem 3), we obtain the integrability of

$$g - f = \lim\, (g - f_n) = \underline{\lim}\, (g - f_n),$$

and the inequality

$$\int (g - f) \leq \underline{\lim} \int (g - f_n).$$

The additivity of the integral and the identity $\underline{\lim}\, (-a_n) = -\overline{\lim}\, a_n$ yield

$$\int g - \int f \leq \int g + \underline{\lim} \int (-f_n) = \int g - \overline{\lim} \int f_n,$$

or equivalently,

$$\overline{\lim} \int f_n \leq \int f.$$

On the other hand, $0 \leq g + f_n \leq 2g$. Applying Fatou's Lemma a second time, we get

$$\int (g + f) \leq \int g + \underline{\lim} \int f_n, \quad \text{or} \quad \int f \leq \underline{\lim} \int f_n.$$

Combining these two inequalities, we obtain

$$\int f \leq \underline{\lim} \int f_n \leq \overline{\lim} \int f_n \leq \int f.$$

Therefore, $\lim \int f_n$ exists and is equal to $\int f$. ∎

Remark. It is easily seen that if the dominating function g is absent, the sequence of integrals will diverge; that is, the limit function will not be integrable. This may occur in essentially two ways: First, the functions f_n may become unbounded in such a way that the integrals will diverge. An example of this is given by the sequence of functions

$$f_n(x) = \begin{cases} \dfrac{1}{x} & \text{if } \dfrac{1}{n} \leq x \leq 1 \\ 0 & \text{elsewhere.} \end{cases}$$

Or, it may happen that the sets upon which the f_n are not equal to zero are becoming unbounded in such a way that the integrals diverge. For example, look at

$$g_n(x) = \begin{cases} \dfrac{1}{x} & \text{if } 1 \leq x \leq n \\ 0 & \text{elsewhere.} \end{cases}$$

Exercises

*1. Let $\{a_n\}$ be a sequence of real numbers. Show that

$$\underline{\lim}\,(-a_n) = -\overline{\lim}\, a_n.$$

2. Give an example of a sequence of integrable functions which satisfies Fatou's lemma, and for which the strict inequality holds.
3. Let $\{f_n\}$ be a sequence of integrable functions whose absolute values are bounded by an integrable function g. Show that $\underline{\lim}\, f_n$ and $\overline{\lim}\, f_n$ exist a.e. and are integrable functions, and that

$$\int \underline{\lim}\, f_n \leq \underline{\lim} \int f_n \leq \overline{\lim} \int f_n \leq \int \overline{\lim}\, f_n.$$

4. Show that if Fatou's lemma is assumed, it may be used to prove the theorem on monotone convergence.

5. Use the theorem on dominated convergence to prove that

$$f(x) = \begin{cases} 0 & \text{if } |x| \leq 1 \\ \dfrac{\sin x}{x^2} & \text{if } |x| > 1 \end{cases}$$

is integrable.

6. Use the theorem on dominated convergence to do Exercises 5.4.7, 5.4.8, and 5.4.11.

***7.** Prove the following variation of Theorem 4: Suppose that $\{f_n\}$ is a sequence of integrable functions which converge a.e. to a function f. If there is an integrable g, and $|f| \leq g$ a.e., then f is integrable and $\int f_n = \int f$. [*Hint:* Use the fact that L_1 is a lattice to obtain a sequence $\{g_n\}$ of integrable functions that also converges to f, and all of whose terms are bounded in absolute value by g.]

***8.** Show by example that the product of integrable functions may not be integrable. If however, one of them is a.e. bounded, then their product is integrable.

***9.** Show that $f,g \in L_1$ implies $\sqrt{fg} \in L_1$. (*Hint:* $\sqrt{fg} \leq |f| + |g|$. Find a sequence of integrable functions that converges a.e. to $\sqrt{fg}$ and use Exercise 7.)

10. Show that $e^{-|x|}$, $(\sin x)e^{-|x|}$ are integrable.

*Exercises which request the proof of a statement made in the text, as well as those referred to in the text, are starred.

3 The Riemann Integral

In this section we shall discuss the relationship between the Riemann integral and the Lebesgue integral. First it will be demonstrated that the Lebesgue integral is an *extension* of the Riemann integral in the sense that if $f(x)$ is Riemann integrable over the interval $[a,b]$, then the function $f\chi_{[a,b]}$ is Lebesgue integrable, and the two integrals are identical.

Following this is a brief discussion of improper Riemann integrals.

And finally, we shall prove that a bounded function which vanishes outside a bounded interval I is integrable over I if and only if f is a.e. continuous on I.

We begin by defining the Riemann integral, and proving a Theorem that characterizes Riemann integrable functions as those which can be approximated from above and below by step functions.

1 Definition. Let $\mathcal{S}$ denote the subdivision of the interval $[a,b]$ which is determined by the points

$$a = a_0 < a_1 < \cdots < a_n = b.$$

The *norm* of $\mathcal{S}$, written $|\mathcal{S}|$, is the length of the largest subinterval (a_{i-1},a_i), $i = 1,2,\ldots,n$.

2 Definition. Let $f(x)$ be a function defined on $[a,b]$ and let $\mathcal{S}$ be a subdivision of this interval. Choose points c_i in each interval (a_{i-1},a_i). The *Riemann approximating sum* corresponding to this subdivision and choice of the points c_i is defined as

$$S(f,\mathcal{S},c_i) = \sum_{i=1}^{n} f(c_i)\Delta a_i,$$

where $\Delta a_i = (a_i - a_{i-1})$.

The function f is said to be *integrable* over $[a,b]$ if a number S exists having the following properties: To each $\epsilon > 0$, there is a number $\delta > 0$ such that whenever $|\mathcal{S}| < \delta$, then

$$|S - S(f,\mathcal{S},c_i)| < \epsilon,$$

for all choices of c_i in the subintervals of $\mathcal{S}$. We shall write

$$\lim_{|\mathcal{S}|\to 0} |S - S(f,\mathcal{S},c_i)| = 0.$$

The number S is called the (*Riemann*) *integral of f over* $[a,b]$ and is denoted by $\int_a^b f(x)\,dx$.

It is easily seen that if f is integrable, it is bounded. For, if f were an unbounded function defined on the interval $[a,b]$, then if M is any preassigned number, and $\mathcal{S}$ any subdivision, then it is possible to choose the c_i so that the approximating sum exceeds M (Exercise 1).

The criteria for Riemann integrability which are presented in Theorem 4, rely upon the Riemann integrability of real step functions. This is proved next.

3 Proposition. The Riemann integral of a real step function exists and is equal to its Lebesgue integral.

Proof: Let σ be a step function defined by

$$\sigma(x) = \begin{cases} \sigma_i & \text{if } s_{i-1} < x < s_i, \quad i = 1,2,\ldots,n \\ \lambda_i & \text{if } x = s_i, \quad i = 0,1,\ldots,n; \end{cases}$$

$a = s_0 < s_1 < \cdots s_n = b$. If $\epsilon > 0$ is given, choose $\delta = \dfrac{\epsilon}{2(n+1)M}$, where $M > |\lambda_i|, |\sigma_i|$ for $i = 0,1,\ldots,n$. Then if $\mathcal{S}$ is a subdivision whose norm is less than δ,

$$|S(\sigma,\mathcal{S},c_i) - \sum_{i=1}^{n} \sigma_i \Delta s_i| \leq \delta M(n+1) = \epsilon/2 < \epsilon. \quad \blacksquare$$

4 Theorem. Let f be a function which vanishes outside $[a,b]$. Then f is integrable over $[a,b]$ if and only if, for every $\epsilon > 0$, there are step functions σ and τ satisfying,

$$\text{(i)} \quad \sigma \le f \le \tau \quad \text{on } [a,b] \qquad \text{and} \qquad \text{(ii)} \quad \int_a^b (\tau - \sigma) < \epsilon.$$

Proof: Let us suppose first that the integral of f exists and is equal to a number S. Then for sufficiently small $\delta > 0$,

$$S - \frac{\epsilon}{3} < S(f,\mathcal{S},c_i) < S + \frac{\epsilon}{3}$$

whenever $|\mathcal{S}| < \delta$. Since f is bounded, the numbers

$$\sigma_i = \inf_{a_{i-1} \le x \le a_i} f(x), \qquad \tau_i = \sup_{a_{i-1} \le x \le a_i} f(x)$$

exist, for all $i \le n$. We define two step functions whose intervals of constancy are (a_{i-1}, a_i). They are called the *upper* and *lower* functions belonging to $\mathcal{S}$.

$$\sigma(x) = \begin{cases} \sigma_i & \text{if } a_{i-1} < x < a_i, \quad 1 \le i \le n \\ f(x) & \text{if } x = a_i, \quad 0 \le i \le n, \quad \text{or if } x \notin [a,b], \end{cases}$$

$$\tau(x) = \begin{cases} \tau_i & \text{if } a_{i-1} < x < a_i, \quad 1 \le i \le n \\ f(x) & \text{if } x = a_i, 0 \le i \le n, \quad \text{or if } x \notin [a,b]. \end{cases}$$

It is readily seen that (i) is satisfied. As for (ii), we have,

$$S - \frac{\epsilon}{3} \le \int_a^b \sigma\,dx = \sum_{i=1}^n \sigma_i \Delta a_i \le \sum_{i=1}^n f(c_i)\Delta a_i \le \sum_{i=1}^n \tau_i \Delta a_i$$
$$= \int_a^b \tau\,dx \le S + \frac{\epsilon}{3},$$

from which $\int_a^b (\tau - \sigma)dx \le \frac{2}{3}\epsilon < \epsilon$ follows.

Assume now that f is a function for which (i) and (ii) are satisfied. Since $\sigma \le f \le \tau$ for any pair of upper and lower step functions as defined above, even if σ and τ are defined for different subdivisions, it follows that

$$\underline{S} = \sup\left[\int_a^b \sigma\,dx\right] \le \inf\left[\int_a^b \tau\,dx\right] = \overline{S};$$

the sup and inf are taken over all subdivisions. However, since the number ϵ is arbitrary, it follows that $\underline{S} = \overline{S}$, which we shall hereafter denote by S.

Let us suppose that $\epsilon > 0$ is given. We wish to show that there is a number $\delta > 0$ for which $|\mathcal{S}| < \delta$ implies

$$|S - S(f,\mathcal{S},c_i)| < \epsilon,$$

for all possible choices of c_i. Let σ and τ be chosen so that (i) and (ii) are satisfied, and

$$\int_a^b \sigma\,dx > S - \frac{\epsilon}{2}, \qquad \int_a^b \tau\,dx < S + \frac{\epsilon}{2}.$$

It may also be assumed that σ and τ are defined over the same subdivision (why?). Since σ and τ are Riemann integrable, there is a $\delta > 0$ such that whenever $|\mathcal{S}| < \delta$, we have

$$\sum_{i=1}^{n} \sigma(c_i)\Delta a_i > \int_a^b \sigma\,dx - \frac{\epsilon}{2}$$

and

$$\sum_{i=1}^{n} \tau(c_i)\Delta a_i < \int_a^b \tau\,dx + \frac{\epsilon}{2},$$

for all choices of the c_i. The inequality $\sigma(c_i) \le f(c_i) \le \tau(c_i)$ yields

$$S - \epsilon < \int_a^b \sigma\,dx - \frac{\epsilon}{2} < \sum_{i=1}^{n} \sigma(c_i)\Delta a_i \le \sum_{i=1}^{n} f(c_i)\Delta a_i$$

$$\le \sum_{i=1}^{n} \tau(c_i)\Delta a_i < \int_a^b \tau\,dx + \frac{\epsilon}{2} < S + \epsilon,$$

from which it follows that $\int_a^b f\,dx = S$. ∎

Besides serving as an integrability test for particular functions, Theorem 4 permits us to prove some very general theorems about Riemann integration: Theorem 6 asserts that Riemann integrable functions are Lebesgue integrable, and Theorem 8 states that Riemann integrable functions may be described as the class of bounded functions that vanish outside a bounded interval, and are a.e. continuous. Before proving these theorems, we shall give a very simple example of a function which is Lebesgue integrable but not Riemann integrable.

5 Example. Let

$$f(x) = \begin{cases} 0 & \text{if } x < 0, \quad x > 1, \text{ or } x = m/2^n, \quad n,m = 0, 1, \ldots \\ 1 & \text{otherwise.} \end{cases}$$

It is readily seen that if σ,τ are step functions which bound f from below and above respectively, then $\tau - \sigma \geq 1$ for all x, from which it follows that $\int(\tau - \sigma) \geq 1$. Thus f is not Riemann integrable. On the other hand, the nonincreasing sequence of step functions

$$\sigma_n(x) = \begin{cases} 0 & \text{if } x \notin [0,1] \text{ or if } x = k \cdot 2^{-n}, \quad k = 0,1,\ldots,2^n - 1 \\ 1 & \text{otherwise} \end{cases}$$

converges to f, and since $0 \leq \sigma_n(x) \leq 1$, and $\displaystyle\int_0^1 \sigma_n\,dx = 1$, it follows, using either the theorem on dominated convergence or the theorem on monotone convergence, that f is (Lebesgue) integrable and that

$$\int f = \lim \int \sigma_n.$$

This example also illustrates that the theorems on dominated and monotone convergence are not true for Riemann integrable functions. It may also be used to show that the normed linear space of Riemann integrable functions (the norm is defined by the integral) is incomplete. The details are left to the reader.

6 Theorem. If f is a function which vanishes outside $[a,b]$, and is Riemann integrable over $[a,b]$, then f is Lebesgue integrable, and the two integrals are equal.

Proof: The upper and lower step functions of Theorem 4 will provide us with a monotone sequence of step functions whose integrals are uniformly bounded, and which converge to f.

Let $\mathcal{S}_n$ denote the subdivision of $[a,b]$ obtained by n consecutive bisections; that is, $a = a_0 < a_1 < \cdots < a_{2n-1} = b$, and $|\mathcal{S}| = a_i - a_{i-1} = 2^{-n}$, for $i = 1,2,\ldots,2^{n-1}$. If τ_n and σ_n are the corresponding upper and lower step functions, then for all values of m and n,

$$[1] \qquad \sigma_n \leq \sigma_{n+1} \leq f \leq \tau_m \leq \tau_{m+1}$$

Hence the integrals $\int \sigma_n$ are all bounded from above by $\int \tau_1$, and the sequence $\{\sigma_n\}$ converges monotonically to a Lebesgue integrable function f'. However $\{\sigma_n\}$ and $\{\tau_n\}$ are equivalent Cauchy sequences since $\lim \int(\tau_n - \sigma_n) = 0$, and so $\tau_n \downarrow f'$ also. It follows from [1] that $f = f'$ a.e., proving that f is Lebesgue integrable. It follows from the theorem on monotone convergence that

$$\int f = \int f' = \lim \int \sigma_n;$$

these integrals are Lebesgue integrals. However, Proposition 3 asserts that σ_n is also Riemann integrable, and that its Riemann integral is equal to its Lebesgue integral. Thus Definition 2 yields

$$\int f = \lim \int \sigma_n = \lim \int_a^b \sigma_n \, dx = \int_a^b f \, dx,$$

the latter two integrals being Riemann integrals. ∎

We prove next that the class of Riemann integrable functions coincides with the class of a.e. continuous, bounded, functions that vanish outside a bounded interval. The following lemma uses the upper and lower step functions to describe continuity.

7 Lemma. Let f be a bounded function defined on the interval $[a,b]$; $\{\mathcal{S}_n\}$ is a sequence of subdivisions whose corresponding upper and lower functions are denoted by τ_n and σ_n, and for all values of n, $|\mathcal{S}_n| < 1/n$. Then if θ is a point that is interior to *all* the subintervals of every $\mathcal{S}_n$, $f(x)$ is continuous at $x = \theta$, if and only if $\lim_{n\to\infty} \sigma_n(\theta) = \lim_{n\to\infty} \tau_n(\theta)$.

Remark. This theorem states criteria for the continuity of a function only at points that are interior to all subintervals of all the subdivisions $\mathcal{S}_n$. However, the set of points which are endpoints of at least one subinterval is countable, and therefore has measure zero. Hence, the continuity or discontinuity of f at any of these points will not affect the main theorem.

Proof of Lemma 7: Let θ be a point of continuity of $f(x)$. Then given $\epsilon > 0$, there is a $\delta > 0$, such that

$$|x - \theta| < \delta \qquad \text{implies} \qquad |f(x) - f(\theta)| < \epsilon.$$

Choose an integer $N > 0$ which satisfies $1/N < \delta$. Then if $n \geq N$ $|\mathcal{S}_n| < \delta$, and if θ is an interior point of a subinterval (a_{i-1}, a_i) of $\mathcal{S}_n$, it follows that $(a_i - a_{i-1}) \leq |\mathcal{S}_n| < 1/n < \delta$, and so

$$a_{i-1} < x < a_i \qquad \text{implies} \qquad |f(x) - f(\theta)| < \epsilon.$$

Let σ_n and τ_n denote the lower and upper step functions of $\mathcal{S}_n$. This means that if $a_{i-1} < x < a_i$, $i = 1,2,\ldots,i(n)$, then

$$\sigma_n(x) = \sigma_i = \inf_{a_{i-1} \leq x \leq a_i} f(x), \quad \tau_n(x) = \tau_i = \sup_{a_{i-1} \leq x \leq a_i} f(x).$$

Since n was chosen so that $|\mathcal{S}_n| < \delta$, it follows that

$$|f(\theta) - \sigma_n(\theta)| < \epsilon \qquad \text{and} \qquad |f(\theta) - \tau_n(\theta)| < \epsilon,$$

and therefore $(\tau_n(\theta) - \sigma_n(\theta)) < 2\epsilon$, implying that $\lim \tau_n(\theta) = \lim \sigma_n(t)$.

Let us assume now that θ is an interior point of every interval of all subdivisions of the sequence $\{\mathcal{S}_n\}$, and that $\lim \sigma_n(\theta) = \lim \tau_n(\theta)$. We wish to prove that f is continuous at θ. Let $\epsilon > 0$ be given. Since $\sigma_n(\theta) \leq f(\theta) \leq \tau_n(\theta)$ for all n, it follows that $f(\theta) = \lim \sigma_n(\theta) = \lim \tau_n(\theta)$, and therefore an integer N exists for which $|\sigma_N(\theta) - \tau_N(\theta)| < \epsilon$. Let (a_{i-1},a_i) denote the interval of $\mathcal{S}_N$ which contains θ. Set

$$\delta = \min[(\theta - a_{i-1}),(a_i - \theta)].$$

That is, δ is the smaller of the two distances of θ to the endpoints of the interval of $\mathcal{S}_N$ in which it is contained. It is readily seen that

$$|x - \theta| < \delta \qquad \text{implies} \qquad |f(x) - f(\theta)| < \epsilon. \quad \blacksquare$$

8 Theorem. A bounded function f defined on a bounded interval $[a,b]$ is Riemann integrable if and only if it is a.e. continuous.

Proof: If f is a.e. continuous, then the monotone sequences $\{\sigma_n\}$ and $\{\tau_n\}$, which correspond to the sequence of subintervals obtained by successive bisection, converge a.e. to f, and since the integrals are uniformly bounded (why?), f is Lebesgue integrable and

$$\int f = \lim \int \sigma_n = \lim \int \tau_n.$$

However, the pairs σ_n, τ_n satisfy the conditions of Theorem 4, proving that f is Riemann integrable. It follows from Theorem 6 that $\int f = \int_a^b f\,dx$.

Conversely, if f is Riemann integrable, Theorem 4 asserts the existence on a subdivision $\mathcal{S}_n$ and a pair of step functions σ_n, τ_n that satisfy, (i) $\sigma_n \le f \le \tau_n$, and (ii) $\int(\tau_n - \sigma_n) < 1/n$. From Theorem 6, it follows that f, σ_n and τ_n are Lebesgue integrable, and since $\{\sigma_n(x)\}$ and $\{\tau_n(x)\}$ are monotone, bounded, sequences of real numbers for all x, it follows that the limit

$$g(x) = \lim(\tau_n(x) - \sigma_n(x))$$

is integrable, and $g(x) = 0$ a.e. The integrability of g comes from the Theorem on dominated convergence and $g(x) = 0$ a.e. is implied by the equivalence of the sequences $\{\sigma_n\}$ and $\{\tau_n\}$. From Lemma 7, we obtain the a.e. continuity of $f(x)$. $\blacksquare$

We conclude this section with a brief discussion of improper Riemann integrals.

In some cases, the Riemann integral may be extended, by taking limits, to certain unbounded functions, or to functions defined on unbounded intervals. This leads to two types of improper integral:

Type I: Let $f(x)$ be continuous for all $x > a$ and $\lim_{x \to a+} f(x)$ be finite, where a is a real number. Then for all $b > a$, $\int_a^b f(x)\,dx$ exists (Theorem 8). If $\lim_{b \to \infty} \int_a^b f(x)\,dx$ is finite, then we say that the *improper integral*,

$$\int_a^\infty f(x)\,dx = \lim_{b\to\infty} \int_a^b f(x)\,dx$$

converges.

Type II: If $f(x)$ is continuous on the half-closed interval $(a,b]$, but $\lim_{x\to a+} f(x)$ does not exist, then whenever $a < c < b$, $\int_c^b f(x)\,dx$ exists. If $\lim_{c\to a} \int_c^b f(x)\,dx$ is finite, then the *improper integral,*

$$\int_a^b f(x)\,dx = \lim_{c\to a} \int_c^b f(x)\,dx$$

converges. (A corresponding definition may be given if the discontinuity occurs in the interior of $[a,b]$.)

Having proved that all Riemann integrable functions are Lebesgue integrable, it may at first come as a surprise that there are functions whose improper Riemann integrals converge, but which are not Lebesgue integrable. We have already seen that if $f \in L_1$, then its absolute value $|f|$ is also in L_1. (The reader will recall that L_1 is a lattice.) However, there are functions whose improper Riemann integrals converge, while the improper integrals of their absolute values diverge. For example, consider the function,

$$f(x) = \begin{cases} \dfrac{\sin x}{x} & \text{if } x \geq \pi \\ 0 & \text{otherwise.} \end{cases}$$

For any real number b which exceeds π, $\int_\pi^b \frac{\sin x}{x}\,dx$ exists. Integration by parts yields

$$\int_\pi^b \frac{\sin x}{x}\,dx = \frac{-\cos x}{x}\bigg|_\pi^b + \int_\pi^b \frac{\cos x}{x^2}\,dx = -\frac{\cos b}{b} + \int_\pi^b \frac{\cos x}{x^2}\,dx.$$

The inequality

$$\left|\frac{\cos x}{x^2}\right| \leq \frac{1}{x^2}$$

implies that

$$\lim_{b\to\infty} \int_\pi^b \frac{\cos x}{x^2}\,dx$$

is finite, and therefore

$$\lim_{b \to \infty} \int_{\pi}^{b} \frac{\sin x}{x}\, dx$$

exists. On the other hand,

$$\int_{\pi}^{n\pi} \left| \frac{\sin x}{x} \right| dx = \sum_{j=1}^{n-1} \int_{j\pi}^{(j+1)\pi} \left| \frac{\sin x}{x} \right| dx = \sum_{j=1}^{n-1} \int_{0}^{\pi} \frac{\sin t}{j\pi + t}\, dt$$

$$\geq \sum_{j=1}^{n-1} \frac{1}{(j+1)\pi} \int_{0}^{\pi} \sin t\, dt = \sum_{j=1}^{n-1} \frac{2}{\pi}\left(\frac{1}{j+1}\right).$$

The last term is a partial sum of a divergent series, and it follows that the improper integral $\lim_{n \to \infty} \int_{\pi}^{n\pi} \left| \frac{\sin x}{x} \right| dx = \infty$. The reader should convince himself that $|f(x)|$ is not Lebesgue integrable. In Exercise 4 the reader is asked to prove that if f is continuous and if the improper integral of $|f|$ converges, then both f and $|f|$ are Lebesgue integrable. It is not, however, true that the integrability of $|f|$ implies that f is also integrable. For it is entirely possible that $|f|$ is integrable, but f^+ and f^- are not integrable.

Exercises

*1. Prove that if the Riemann integral of $f(x)$ over the interval $[a,b]$ exists, then $f(x)$ is bounded.

2. Without using Theorem 8, show that $f(x)$ is Riemann integrable over $[a,b]$ if it is continuous on $[a,b]$. (Use Theorem 4.)

*3. Using Example 5, show that the normed linear space of Riemann integrable functions is incomplete. ($\|f\| = \int_a^b |f|\, dx$; $[a,b]$ is an interval outside of which f vanishes.)

*4. Prove that if $f(x)$ is continuous and if $|f(x)|$ has a convergent improper Riemann integral, then $f \in L_1$. [It is not true that if $|f| \in L_1$, then $f \in L_1$. Although this cannot be proved now, it happens because f^+ and f^- are not integrable (See Exercise 7.4.1).]

*5. Let $f(x)$ be a bounded nondecreasing function defined on the interval $[a,b]$. Show that the set of discontinuities of f is countable. [*Hint:* If θ is a point of discontinuity, then $\lim_{x \to \theta+} f(x) - \lim_{x \to \theta-} f(x) = d > 0$. Show that for each positive integer n, the set S_n of discontinuities for which $d > 1/n$ is a finite set. $\bigcup S_n$ is the complete set of discontinuities.]

6. Show that a bounded monotone function defined on a bounded interval $[a,b]$ is Riemann integrable.

*7. Prove that L_1 is the completion of the normed linear space of Riemann integrable functions.

*8. What are the theorems about series of real numbers that correspond to the theorems on improper integrals? State one major difference.

9. For each of the following functions, discuss the corresponding improper

Riemann integral and its Lebesgue integral: (a) $f(x) = \dfrac{1}{e^x + 1}$, $x \geq 0$; (b) $f(x) = x^p \log x$, $x \geq 1$; (c) $f(x) = \dfrac{\cos x}{x^2}$, $x \geq 1$.

10. Prove directly, without using Theorem 8, that if $f(x)$ is a continuous function on the interval $[0,1]$, then

$$g(x) = \begin{cases} f(x) & \text{if } x \neq 1/n, \quad n = 1, 2, \ldots \\ a_n & \text{if } x = 1/n, \quad n = 1, 2, \ldots, \end{cases}$$

where $\{a_n\}$ is any bounded sequence of real numbers, is Riemann integrable.

*Exercises which request the proof of a statement made in the text, as well as those referred to in the text, are starred.

CHAPTER **7**

Measurable Functions and Measurable Sets

1 Measurable Functions

In Secs. 1 and 2 of Chapter 6 we were able to show that with the addition of certain boundedness conditions, an a.e. convergent sequence of integrable functions will converge to an integrable function. It was also shown that these conditions were necessary to insure the integrability of the limit function. In this section we shall study the class of functions which are the a.e. limits of sequences of integrable functions; no boundedness conditions are assumed. These functions, called *measurable* functions, will at first be defined as the a.e. limits of step functions. However, we shall see that the a.e. limit of a sequence of integrable functions is also measurable.

1 Definition. A function f is called *measurable* if it is the a.e. limit of a sequence of step functions. The class of measurable functions is denoted by M.

It is readily seen that M is a real vector space which contains L_1 (or L), the space of integrable functions (Exercise 1). That this inclusion is proper, is demonstrated in Examples 2 and 3.

2 Example. The sequence of step functions

$$\sigma_n(x) = \begin{cases} n/i & \text{if } (i-1)/n < x \le i/n, \quad i = 1, 2, \ldots, n \\ 0 & \text{elsewhere} \end{cases}$$

increases monotonically to the function

$$f(x) = \begin{cases} 1/x & \text{if } 0 < x \le 1 \\ 0 & \text{elsewhere.} \end{cases}$$

Hence f is measurable. However, the sequence of integrals,

$$s_n = \int \sigma_n = 1 + 1/2 + \cdots + 1/n,$$

diverges, proving that f is not integrable.

3 Example. The sequence $\sigma_n = \chi_{[0,n]}$ converges monotonically to the function $\chi_{[0,\infty)}$, which is therefore measurable. Of course, this limit function is not integrable.

The preceding examples illustrate the remarks made in Sec. 2 of Chapter 6 about the necessity of the boundedness conditions to insure integrability of the limit function. Intuitively, we see that the measurable functions which are not integrable, are of two types: Either, as in Example 2, they become "too" unbounded, or as in Example 3, they are nonzero on too large a set. (Of course, there *are* integrable functions which are either unbounded, or do not equal zero outside a bounded set.) In both Examples 2 and 3, integrable functions would be obtained if the functions were "truncated" in the following sense:

4 Definition. Let $f(x)$ be defined for all values of x. By the *truncation of f at $A > 0$*, written $f_A(x)$, we mean the function

$$f_A(x) = \begin{cases} 0 & \text{if } |x| \geq A \\ A & \text{if } |x| < A \text{ and } f(x) \geq A \\ -A & \text{if } |x| < A \text{ and } f(x) \leq -A \\ f(x) & \text{otherwise.} \end{cases}$$

See Figure 25.

It is readily seen that the truncations of the measurable functions of the preceding examples are integrable. This is a special case of Theorem 6 (below) which provides a useful alternate definition of a measurable function. First a lemma is needed.

5 Lemma. The a.e. limit of a sequence of integrable functions $\{f_n\}$ is measurable.

Proof: Let $\{\sigma_n\}$ be a sequence of step functions satisfying, $\| \sigma_n - f_n \| < 2^{-n}$. The integrability of the f_n implies that such a sequence exists. The functions $g_n = \sum_{k=1}^{n} |f_k - \sigma_k|$ are nondecreasing, integrable, and

$$\int g_n = \sum_{k=1}^{n} \int |f_k - \sigma_k| < \sum_{k=1}^{n} 2^{-k} < 1.$$

It follows from the theorem on monotone convergence that $\{g_n\}$

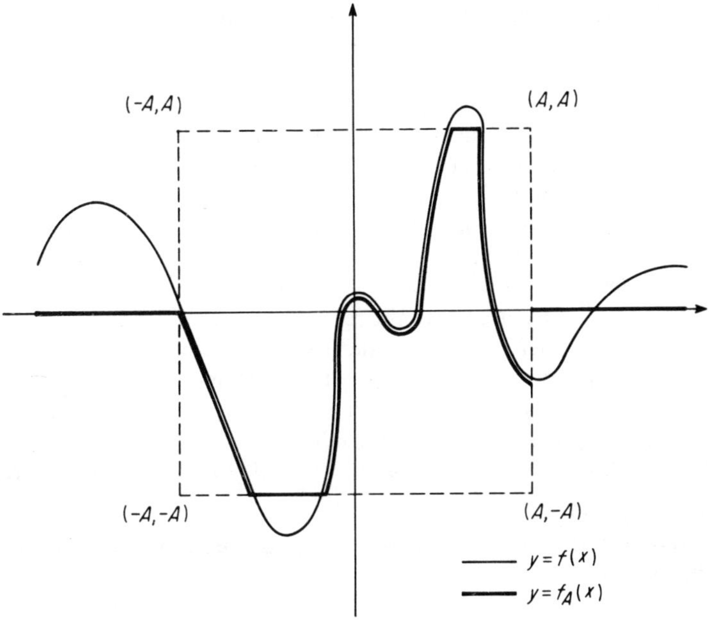

FIGURE 25

converges a.e. to an integrable function g, which is equal to the sum $\sum_{k=1}^{\infty} |f_k - \sigma_k|$. The a.e. convergence of this sum implies that $\lim_{k\to\infty} |f_k - \sigma_k| = 0$ a.e. This yields,

$$\lim_{n\to\infty} |f - \sigma_n| \leq \lim_{n\to\infty} |f - f_n| + \lim_{n\to\infty} |f_n - \sigma_n| = 0 \text{ a.e.},$$

proving that f is the a.e. limit of step functions, and therefore measurable. ∎

6 Theorem. $f(x)$ is measurable if and only if all its truncations are integrable.

Proof: If all the truncations of f are integrable, then in particular the truncations f_N, where N is an integer, are integrable. Since the sequence $\{f_N\}$ converges monotonically to f, it follows, by Lemma 5, that f is measurable.

Conversely, if $f(x)$ is measurable, there is a sequence $\{\sigma_n\}$ of step functions which converges a.e. to f. The inequality, $|f_A - (\sigma_n)_A| \leq |f - \sigma_n|$, which is true for all x, n and A, implies that $f_A = \lim_{n\to\infty} (\sigma_n)_A$ a.e. However, each $(\sigma_n)_A$ is integrable, and $|(\sigma_n)_A| \leq A\chi_{[-A,A]}$, from which it follows that the limit function f_A is integrable (Theorem 6.2.4). ∎

7 Theorem. (Properties of measurable functions)

(i) If $f \in M$, then f^+, f^- and $|f| \in M$.

(ii) If $f,g \in M$, then $\max(f,g)$ and $\min(f,g) \in M$.

(iii) If $\{f_n\}$ is a sequence of measurable functions that converges a.e., then the limit function is measurable. (Thus, unlike L_1, M is closed under monotone convergence, without adding any boundedness conditions.)

(iv) If f_n is a sequence of measurable functions, then $\sup [f_n(x)]$, $\inf[f_n(x)]$, $\underline{\lim}[f_n(x)]$ and $\overline{\lim}[f_n(x)]$ are measurable functions if they are finite a.e.

Proof: (i) and (ii) are left to the reader. They imply that M is a lattice.

To prove (iii), let $\{f_n\}$ be a sequence of measurable functions which converges a.e. to a function f. We shall use Theorem 6 to prove that f is measurable. The measurability of the f_n implies that their truncations $(f_n)_A$ are integrable. The inequality $|(f_n)_A| \leq A$, implies that $\lim_{n\to\infty} (f_n)_A = f_A$ is integrable (Theorem 6.2.4). It follows, using Theorem 6 again, that f is measurable.

Proof of (iv): Since $\max[f_1, f_2, \ldots, f_n]$ is measurable (Property (ii)), it follows that

$$\sup[f_n] = \lim_{n\to\infty} \max[f_1, f_2, \ldots, f_n\}$$

is measurable, (Property (iii)), provided that this limit exists a.e. Similarly for $\inf[f_n]$. This implies, however, that

$$\overline{\lim_{n\to\infty}}\ [f_n] = \lim_{n\to\infty} \sup[f_n, f_{n+1}, \ldots]$$

is measurable if the limit exists a.e. Similarly for $\underline{\lim}\,[f_n]$. ∎

8 Theorem. A bounded measurable function that vanishes outside a bounded interval is integrable.

Proof: Let A be a positive real number satisfying, $|f| \leq A$ and $f = 0$ outside $[-A,A]$. Then $f = f_A$. From Theorem 6, it follows that f is integrable. ∎

9 Theorem. If f is measurable and g is integrable, and if $|f| \leq g$ a.e., then f is integrable.

The proof is left as an exercise.

Exercises

***1.** Prove that every integrable function is measurable.

***2.** Prove that every continuous function is measurable. (Why isn't Exercise 1 applicable directly?)

***3.** Prove Theorem 9.

***4.** Let f be an integrable function and let g be a measurable function which is bounded a.e. (This means that there is a positive number A and $|g| < A$ a.e.) Prove that fg is integrable.

*5. Prove that f is measurable if and only if mid $(-\sigma, f, \sigma)$ is integrable for all nonnegative step functions σ. (mid (a,b,c), where a, b and c are real numbers, is equal to either a, b or c, depending upon which of these numbers neither exceeds nor is exceeded by the other two, that is, the "middle" value.)

*6. Replace σ of Exercise 5, by a nonnegative integrable function.

*7. Show that if f is measurable, then for all positive numbers p, $|f|^p$ is measurable.

*8. Show that if $f(x)$ is measurable, then if a is any number, $f(x + a)$ is also measurable.

*9. A function f, which can be written as a finite sum $\sum_{i=1}^{n} \alpha_i \chi_{S_i}$, where the α_i are real numbers and the S_i are sets whose characteristic functions are integrable, is called a *generalized step function*. Prove that the generalized step functions form a vector space and a lattice.

Define the integral of f to be

$$\int f = \sum_{1}^{n} \alpha_i m(S_i).$$

Show that the generalized step functions are Lebesgue integrable and that the Lebesgue integral agrees with the integral just defined.

Define the norm of a generalized step function f to be $\int |f|$. Show that the normed linear space of generalized step functions is not complete, and its completion is L_1.

10. Let $\{f_n\}$ be a sequence of measurable functions and let S be a null set. Prove that if $\{f_n\}$ converges to a function f on $R - S$, and if f is constant on S, then f is measurable.

*Exercises which request the proof of a statement in the text, as well as those referred to in the text, are starred.

2 Measurable Sets I: Basic Properties

The Lebesgue integral was defined so that for particularly simple nonnegative functions $f(x)$, such as step functions or piecewise linear functions, the integral gave the area of the plane region bounded by the graphs of $y = f(x)$, the x axis, and the vertical lines $x = a$ and $x = b$. (It is assumed that $f(x) = 0$ outside the interval $[a,b]$.) The "basic" properties of the integral: additivity, homogeneity, monotonicity, and continuity, correspond to properties that we would require of an area function, for each of them has a very simple geometric interpretation. Thus, starting with the areas of "simple" regions, such as rectangles in the case of step functions, or trapezoids for the piecewise linear functions, we proceed, by taking limits, to obtain the Legesgue integral, which may be thought of as a generalization of an area function.

In much the same way, a "measure" can be assigned to certain subsets of the real numbers. Starting with the simplest "measurable" set, an interval, the most natural thing to do is to assign to it its length as its measure. We should like to extend this measure function to a larger class of sets, in such a way that the basic geometric properties of length are preserved. For example, the measure of a set should be a real valued, nonnegative function which coincides with the simple notion of length when the set is an interval; it should be *additive*, in the sense that if two measurable sets are disjoint, then the measure of their union is the sum of their measures; it should be a *monotone set function*, which means that if $S \subset T$, and if both sets are measurable, then the measure of S should not exceed that of T. Finally, we should require the measure of a null set to be zero.

In this section the measure of a set will be defined by a Lebesgue integral. The appropriate definition is suggested by the following identity:

$$b - a = \int \chi_{[a,b]};$$

A finite sum of such integrals gives the sum of the lengths of a finite union of bounded intervals. It is therefore natural to define the measure of a set to be the integral of its characteristic function, if that integral exists (Definition 1). The basic properties of the measure function: nonnegativeness, additivity, and monotonicity, will follow directly from the corresponding properties of the integral (Theorem 2).

Later, (Sec. 1 of Chapter 8), we shall show that this measure function can be defined without using the integral. In fact, the more standard approach to Lebesgue integration and measure theory is to begin with the idea of measurable sets, and to proceed to measurable functions and the integral. This approach is discussed in Chapter 8, where it is demonstrated that the two methods yield the same classes of measurable and integrable functions and measurable sets. The approach taken in this text lends itself to a generalization of the integral to abstract spaces. This method is due primarily to Daniell, [1] and [2]. (See also Chapter 12.)

1 Definition. A subset S of the real numbers is called *measurable* if its characteristic function χ_S is a measurable function.

S is said to be *finitely measurable* or *integrable* if χ_S is integrable, and the *measure of* S is given by $m(S) = \int \chi_S$.

If χ_S is measurable but not integrable, then letting $(\chi_S)_A$ denote the truncation of χ_S at A, we have

$$\int (\chi_S)_A \uparrow \infty \qquad \text{as} \qquad A \uparrow \infty.$$

In this case, we write $m(S) = \infty$.

Thus the measure function m is a mapping of the class $\mathfrak{M}$ of measurable subsets of R into the *extended* real numbers, $R \cup \{\infty\}$.

It follows immediately that all intervals, bounded or unbounded, open, closed, or neither, are measurable, and that if $I = [a,b]$, then

$$m(I) = \int \chi_I = b - a.$$

2 Theorem. (Properties of the measure function).

(i) If S is a measurable set, then $m(S) \geq 0$.

(ii) If S and T are measurable, and if $S \subset T$, then $m(S) \leq m(T)$.

(iii) If $\{S_i\}$ is a sequence of measurable sets, then $\bigcup S_i$ and $\bigcap S_i$ are measurable.

(iv) If S and T are measurable sets, then $S - T$ is measurable.

(v) If $\{S_i\}$ is a sequence of measurable sets, and if the sum $\Sigma m(S_i)$ converges, then $\bigcup S_i$ is finitely measurable, and

$$m(\bigcup S_i) \leq \Sigma m(S_i).$$

(vi) S is a null set if and only if $m(S) = 0$.

(vii) If S is a bounded measurable set, then $m(S) < \infty$.

(In (i)–(iv), the value $m(S) = \infty$ is permitted. It does not seem necessary to set forth formally the rules of arithmetic for the set of extended real numbers. It suffices to say that if $m(S) = \infty$, $m(T) < \infty$, then $m(S - T) = \infty$; that is, $\infty - r = \infty$, $r \in R$. Also, $m(S \cup T) = \infty$, if the measure of either S or T is infinite. No definite value may be given to $\infty - \infty$. Why?)

Proof: The proofs of (i) and (ii) are left to the reader.

Proof of (iii): Let $S = \bigcup S_i$; the sets S_i are measurable. Definition 1 implies that for each value of i, χ_{S_i} is a measurable set. Thus

$$\chi_S = \sup[\chi_{S_i}]$$

is measurable (Theorem 7.1.7), and therefore the set $S = \bigcup S_i$ is measurable (Definition 1). A similar statement can be made for the intersection.

The measurability of the set $S - T$, where both S and T are measurable and $S \supset T$, follows from the identity, $\chi_{S-T} = \chi_S - \chi_T$. If S does not contain T, then $\chi_{S-T} = \chi_S - \chi_{S\cap T}$.

Proof of (v): Set $T_n = \bigcup_{i=1}^{n} S_i$ and $S = \bigcup_{i=1}^{\infty} S_i$. Then the sequence $\{\chi_{T_n}\}$ is nondecreasing, and

$$\chi_{T_n} = \max_{i \leq n} [\chi_{S_i}] \leq \sum_{i=1}^{n} \chi_{S_i}.$$

This implies

$$\int \chi_{T_n} \leq \int \sum_{i=1}^{n} \chi_{S_i} = \sum_{i=1}^{n} m(S_i) \leq \sum_{i=1}^{\infty} m(S_i).$$

The theorem on monotone convergence implies that $\chi_S = \lim \chi_{T_n}$ is an integrable function, and

$$m(S) = \int \chi_S = \lim_{n\to\infty} \int \chi_{T_n} \le \sum_{i=1}^{\infty} m(S_i).$$

Proof of (*vi*): If $m(S) = 0$, then

$$0 = \int \chi_S = n \int \chi_S = \int n\chi_S.$$

Thus $\sigma_n = n\chi_S$ is a nondecreasing sequence of step functions whose integrals are bounded (they all vanish) and which diverges on S. It follows from Theorem 5.3.8, that S is a null set.

Conversely, if S is a null set, then $\chi_S = 0$ a.e., and therefore

$$m(S) = \int \chi_S = \int \chi_\emptyset = 0.$$

Proof of (*vii*): If S is bounded, then there is a positive number A, and $S \subset [-A,A]$. The measurability of S implies that χ_S is measurable, and the inequality $\chi_S \le \chi_{[-A,A]}$ and Theorem 7.1.9 yield $m(S) \le 2A < \infty$. ▎

Properties (iii) and (iv) assert that the class $\mathfrak{M}$ of measurable subsets of R is closed under complementation and the operations of countable union and intersection.

3 Definition. Let $\mathfrak{F}$ be a family of subsets of a fixed set X. $\mathfrak{F}$ is said to be a *σ-ring* if it is closed under complementation and under countable union and intersection. If $X \in \mathfrak{F}$, then it is called a *σ-algebra* or an *additive class*.

We have just proved that the class $\mathfrak{M}$ of measurable sets is a σ-algebra. Since $\mathfrak{M}$ contains all intervals, including R itself, it follows therefore that $\mathfrak{M}$ contains all sets which are obtainable from intervals by taking complements and countable unions and intersections. In Sec. 3 we shall show that $\mathfrak{M}$ also contains all open and closed sets, as well as those sets which are "generated" from open and closed sets by taking complements and countable unions and intersections. That there are measurable sets which are not obtained in this way is also discussed in Sec. 3.

We conclude this section by showing that the Lebesgue measure is *countably* additive. This property sets it apart from the Riemann measure, which is only finitely additive (Exercise 10).

4 Theorem. (Countable additivity of measure) If $S = \bigcup_{i=1}^{\infty} S_i$, where the S_i are measurable and disjoint sets, then

$$m(S) = \sum_{i=1}^{\infty} m(S_i). \tag{1}$$

Proof: If for any value of i, $m(S_i) = \infty$, then it is easily seen that both sides of [1] are equal to ∞.

Let us assume now that each S_i is integrable. Since the sets are disjoint, it follows that

$$\chi_S = \sum_{i=1}^{\infty} \chi_{S_i}.$$

Setting $T_n = \bigcup_{i=1}^{n} S_i$, it follows from the (finite) additivity of the integral that for all values of n,

$$m(T_n) = \int \chi_{T_n} = \int \sum_{i=1}^{n} \chi_{S_i} = \sum_{i=1}^{n} \int \chi_{S_i} = \sum_{i=1}^{n} m(S_i).$$

The sequence of partial sums $F_n = \chi_{T_n} = \sum_{i=1}^{n} \chi_{S_i}$ is nondecreasing, and each F_n is integrable. If

$$\int F_n < A,$$

then the theorem on monotone convergence yields the integrability of the limit function, $\lim F_n = \sum_{i=1}^{\infty} \chi_{S_i}$, and

$$m(S) = \int \chi_S = \lim_{n\to\infty} \int \sum_{i=1}^{n} \chi_{S_i} = \lim_{n\to\infty} \sum_{i=1}^{n} \int \chi_{S_i} = \sum_{i=1}^{\infty} m(S_i).$$

On the other hand, if $\int F_n \uparrow \infty$, then it is easily seen that $m(S) = \infty$. For, if $m(S) = d < \infty$, then it would follow that

$$m(T_n) = \int F_n \leq m(S) = d,$$

for all values of n (Theorem 2(ii)), which cannot be if $\int F_n \uparrow \infty$. ∎

If S is a measurable set, and if f is integrable, then $|f\chi_S| \leq |f|$. Theorem 7.1.9 implies that $|f\chi_S|$ is integrable.

5 Definition. If f is an integrable function and S is a measurable set, then the *integral of f over S* is defined to be

$$\int_S f = \int f\chi_S.$$

It is easily seen that if S is a finite interval, and f is Riemann integrable, then this coincides with the definition of the Riemann integral of f over I.

The proof of the following theorem is left to the reader.

6 Theorem. (Countable Additivity of the Integral) If $\{S_n\}$ is a sequence of disjoint integrable sets, and if $m(\bigcup S_n) < \infty$, then for every integrable function f,

$$\int_S f = \sum_{n=1}^{\infty} \int_{S_n} f,$$

where $S = \bigcup_{n=1}^{\infty} S_n$.

Exercises

*1. Verify (i) and (ii) of Theorem 2.

*2. Prove that all open and closed sets are measurable.

*3. Let f be an a.e. positive integrable function. Prove that if S is measurable and if $\int_S f = 0$, then $m(S) = 0$.

*4. Prove that if f and g are integrable and if $\int_S f = \int_S g$, for all measurable S, then $f = g$ a.e.

*5. If f is integrable and if d is any positive number, then the set $[f \geq d]$ has finite measure.

*6. Let $\{S_n\}$ be a sequence of measurable sets satisfying $S_n \supset S_{n+1}$, and let $S = \bigcap_{i=1}^{\infty} S_n$. Prove that if for some n, $m(S_n) < \infty$, it follows that

$$m(S) = \lim m(S_n).$$

Why is this false if it is not assumed that at least one of the sets has finite measure? (*Hint:* Use a sequence of unbounded intervals.)

7. Let S be a closed, bounded, measurable set whose measure is positive. Prove that there is a positive number ϵ, such that,

$$|x| < \epsilon \quad \text{implies} \quad \exists\, s,t \in S, \text{ and } x = s - t.$$

[*Hint:* Let $G_n = \{x \in R : \exists\, s \in S \text{ and } |x - s| < 1/n\}$. Show that $S = \bigcap_{i=1}^{\infty} G_i$, and that for sufficiently large n, we may take $\epsilon = 1/n$.]

8. $\mathfrak{F}$ is said to be a *ring* of subsets of a fixed set X if $\mathfrak{F}$ is closed under complementation and finite union and intersection. $\mathfrak{F}$ is an *algebra* if $X \in \mathfrak{F}$. (See Definition 3.) Prove that it suffices to define an algebra (or σ-algebra) as a family of sets closed under complementation and finite (or countable) union.

9. In each case, decide whether is a ring, algebra, σ-ring, or a σ-algebra.
   (a) $X = Z$, the set of integers, and $\mathfrak{F}$ is the class of subsets with the following property: $S \in \mathfrak{F}$ if either S or $Z - S$ is finite.
   (b) X is any metric space, and $\mathfrak{F}$ is the family of bounded subsets of X.
   (c) X is a metric space and $\mathfrak{F}$ is the family of sets, each of which is either closed or open in X.
   (d) $X = R$ and $\mathfrak{F}$ is the collection of bounded measurable sets.
   (e) $X = R$ and $\mathfrak{F} = \mathfrak{M}$, the collection of all measurable sets.

***10.** A bounded set of real numbers is *Riemann measurable* if its characteristic function is Riemann integrable, and in this case, its Riemann measure is equal to the Riemann integral of its characteristic function. Show that every Riemann measurable set is Lebesgue measurable, and that the two measures coincide on this class of sets. Prove that there are bounded Lebesgue measurable sets which are not Riemann measurable. The example you give to illustrate this should also demonstrate that the Riemann integral is not countably additive.

11. Let $a = a_0 < a_1 < \cdots < a_n = b$ be a subdivision of an interval $[a,b]$. Prove, using Theorem 4 and Definition 5, that

$$\int_a^b f(x)\,dx = \sum_{i=1}^{n} \int_{a_{i-1}}^{a_i} f(x)\,dx.$$

12. Let f be an integrable function and let S and T be measurable sets whose intersection is a null set. Show that

$$\int_{S \cup T} f = \int_S f + \int_T f.$$

State and prove a generalization for finitely many measurable sets.

***13.** Prove Theorem 6.

***14.** Let f be a nonnegative function which is integrable over a measurable set S. Prove that if $\epsilon > 0$ is given, then there is a $\delta > 0$, such that

$$T \subset S \quad \text{and} \quad m(T) < \delta \qquad \text{imply} \qquad \int_T f < \epsilon.$$

*Exercises which request the proof of a statement in the text, as well as those referred to in the text, are starred.

3 Measurable Sets II: More Properties

Let us return to the remarks that follow Definition 7.2.3.

Let $\mathcal{C}$ be any collection of subsets of a fixed nonempty set X, and let $\mathfrak{F}$ denote the family of all subsets of X, including X itself. Clearly, $\mathfrak{F}$ is a σ-algebra (Definition 7.2.3) which contains $\mathcal{C}$. Denoting by $\mathcal{B}$ the intersection of all σ-algebras which contain $\mathcal{C}$, we note the following: The existence of the intersection $\mathcal{B}$ is assured by the existence of at least one such σ-algebra, namely $\mathfrak{F}$; $\mathcal{B}$ is itself a σ-algebra; in fact it is the *smallest* σ-algebra that contains the given collection $\mathcal{C}$, in the sense that it is contained in every such σ-algebra. $\mathcal{B}$ is called the *σ-algebra generated by* $\mathcal{C}$. The σ-algebra generated by the open intervals of R is called the family of *Borel sets*. As a consequence of Proposition 1 (proved below) and the properties of a σ-algebra, it can easily be demonstrated that the Borel sets may be described as the σ-algebra generated by any of the following families of subsets of R: (i) the set of all intervals; (ii) the set of all closed intervals; (iii) the family of all compact subsets of R; (iv) the family of all open subsets of R; (v) the family of all closed subsets of R (Exercise 1).

Although it was assigned earlier as an exercise, we prove next that every open subset of R may be written as the disjoint union of countably many open intervals. This proposition is useful in proving the equivalence of the six statements listed above, as well as in the proofs of other theorems.

1 Proposition. Every nonempty open subset G of R can be written as the disjoint countable union of open intervals.

Proof: Each point x of the open set G is contained in an open interval (a,b) which is itself contained in G. To each such x, there is a largest interval $I_x = (a_x,b_x)$ contained in G. It is left to the reader to verify that the end points of this interval are given by

$$[1] \qquad a_x = \inf_{(a,x)\subset G} \{a\}, \qquad b_x = \sup_{(x,b)\subset G} \{b\}.$$

Furthermore, it is easily seen that if x and y are points of G, then either $I_x = I_y$ or $I_x \cap I_y = \phi$. For, if $I_x \cap I_y \neq \phi$, then one of the following statements must be true:

(i) If $I_x \subset I_y$, then [1] implies that they must be equal, for I_x is the *largest* open interval that contains x and is contained in G.

(ii) If I_x and I_y intersect and neither one is included in the other, then one of the following inequalities holds: Either $a_x \leq a_y < b_x < b_y$, $a_x < a_y < b_x \leq b_y$, $a_y \leq a_x < b_y < b_x$, or $a_y < a_x < b_y \leq b_x$. If the first inequality holds, (Figure 26), then (a_x,b_y) is an interval containing x, and

$$(a_x,b_y) = (a_x,b_x) \cup (a_y,b_y) \subset G.$$

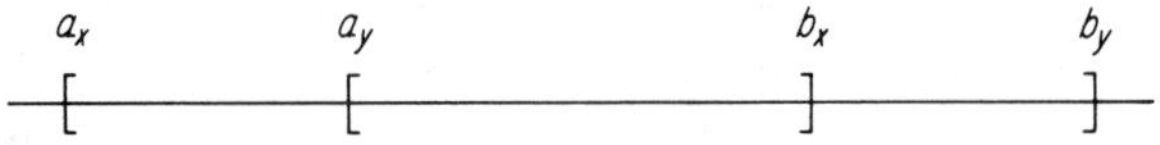

FIGURE 26

Thus (a_x,b_x) is a *proper* subset of (a_x,b_y) and is therefore not the largest interval containing x which is contained in G. A similar argument may be given for the remaining cases.

Thus G may be written as the disjoint union $\bigcup_{x\in G} I_x$. Since each of these intervals contains a rational point, and since no rational point may be in two distinct intervals, it follows that the set of distinct intervals may be put in a 1–1 correspondence with a subset of the countable set of rational numbers, thus proving that the set of intervals is itself countable. ∎

We introduce next a convenient abbreviation for sets which are obtained by taking countable unions and intersections of open and closed sets: F-set = closed set, G-set = open set, F_σ-set = countable union of F-sets, G_δ-set = countable intersection of open sets, $F_{\sigma\delta}$-set = countable intersection of F_σ-sets, etc. (The letters G and F come from the German words for open and closed respectively.)

2 Proposition. G, F, G_δ, F_σ, $G_{\delta\sigma}$, $F_{\sigma\delta}$, . . . , sets are measurable.

Proof: Every G-set is the countable union of open intervals, each of which is a measurable. An F-set is the complement of a G-set, and is therefore also measurable. A similar argument may be given for the remaining cases. ∎

Remark. The sets described in Proposition 2 are the Borel sets.

It would be convenient if every measurable set was a Borel set. Although this is not the case it will be shown (Theorem 3) that every bounded measurable set differs from a G_δ (or a F_σ)-set by a null set. (See Gelbaum and Olmstead, Burrill and Knudsen, or Exercises 7.4.3, 7.4.4 and 7.4.5 for examples of such sets. See also Hewitt and Stromberg for a cardinality argument that such sets exist.) This result is not only intrinsically interesting but suggests a way to define a measure function without using the Lebesgue integral. (See Sec. 1 of Chapter 8.)

3 Theorem. If S is a bounded measurable set, then there is a G_δ-set $\hat{S}$ satisfying

(i) $\hat{S} \supset S$, and

(ii) $m(\hat{S}) = m(S)$, (or equivalently, $m(\hat{S} - S) = 0$).

Proof—Case 1. If $m(S) = 0$, then S is a null set, (Theorem 7.2.2), and for each value of n, there is a countable system $\mathcal{G}_n$ of open intervals, whose union $G_n \supset S$, and whose length, or measure, does not exceed $1/n$. Clearly, $S \subset \bigcap_{i=1}^{\infty} G_n = \hat{S}$, and it follows from the inequality $m(G_n) < 1/n$, which holds for all n, and the monotonicity of measure, that $m(\hat{S}) = 0$.

Case 2. $m(S) = \int \chi_S > 0$. Since S is bounded, $m(S) < \infty$, and χ_S is an integrable function. Let $\{\sigma_n\}$ be a Cauchy sequence of step functions which converges a.e. to χ_S. The idea of the proof is to replace $\{\sigma_n\}$ by an equivalent sequence whose terms are the characteristic functions of open sets whose unions "almost" contain S (except for a null set). The intersection of these unions is "almost" the desired G_δ-set. Let

$$\tau_n = \begin{cases} 1 & \text{if } \sigma_n > 1/2 \\ 0 & \text{otherwise.} \end{cases}$$

Except on a null subset of S, $\sigma_n > 1/2$, for sufficiently large values of n. And, except on a null subset of $R - S$, $\sigma_n \leq 1/2$ for large enough values of n. Therefore $\lim \tau_n = \chi_S$ a.e. The theorem on dominated convergence implies further that $m(S) = \int \chi_S = \lim \int \tau_n$.

Since each σ_n is a step function, the set S_n of points on which $\sigma_n > 1/2$, consists of a finite union of intervals and points. Let T_n be the interior of the set S_n; that is, T_n is the set that remains after the isolated points and endpoints of intervals of S_n have been removed. Clearly, $\chi_{T_n} = \tau_n$ except at finitely many points, and it follows that $\{\chi_{T_n}\}$ is also a Cauchy sequence that converges a.e. to χ_S. Let

$$G_n = \bigcup_{j=n}^{\infty} T_j.$$

Since the T_j are open, each G_n is open and therefore measurable (Proposition 3), and $G_n \subset I$ implies that $m(G_n) \leq m(I) < \infty$. Moreover, it follows from $\lim \chi_{T_n} = \chi_S$ a.e., that for each n, there is a null set E_n, and

$$G_n \supset S - E_n.$$

Therefore

$$S' = \bigcap_{i=1}^{\infty} G_i \supset S - E,$$

where $E = \bigcup E_i$ is also a null set, and S' is a G_δ-set. From case 1, the null set E is contained in a null G_δ-set $\hat{E} = \bigcap H_i$, where the H_i are open. Clearly, the set $\hat{S} = S' \cup E$ contains S, and

$$\hat{S} = S' \cup \hat{E} = (\bigcap G_n) \cup (\bigcap H_n) = \bigcap (G_n \cup H_n),$$

proving that $\hat{S}$ is a G_δ-set.

It remains to be proved that $m(\hat{S}) = m(S)$. The inequality

$$\chi_{G_1} \geq \chi_{G_2} \geq \cdots \geq \chi_{G_n} \geq \chi_{G_{n+1}} > \cdots,$$

and $\bigcap G_n = S - E$, where E is a null set, imply that

$$\begin{aligned} m(\hat{S}) &= m(S') = \int \chi_{S'} = \int \min\,[\chi_{G_1}, \chi_{G_2}, \ldots] \\ &= \lim \int \chi_{G_n} = \int \chi_S = m(S). \quad \blacksquare \end{aligned}$$

4 Corollary. If S is a bounded measurable set, and if $\epsilon > 0$, then there is an open set G which contains S, and $m(G - S) < \epsilon$.

Proof: The sets G_n of the preceding proof are nested, and $\lim m(G_n) = m(S)$. By choosing n to be sufficiently large, $G_n = \bigcap_{i=1}^{n} G_i$ is the desired open set. ▌

By taking complements and using De Morgan's laws, the following additional corollaries may be proved.

5 Corollary. If S is a bounded measurable set, then there is a F_σ-set T, contained in S satisfying, $m(T) = m(S)$.

6 Corollary. If S is a bounded measurable set and if $\epsilon > 0$, then there is a closed set F contained in S, and $m(S - F) < \epsilon$.

The proofs of Corollaries 5 and 6 are left to the reader.

We prove next a theorem which characterizes measurable functions in terms of measurable sets. Later, this theorem will serve as a basis for defining measurable functions in terms of measurable sets. Let us start with a lemma.

7 Lemma. Let f be a function defined on a measurable set S. If, for each real value of A the set $[f < A]$ is measurable, then the following sets are also measurable: $[f \leq A]$, $[f > A]$, $[f \geq A]$, and $[A \leq f < B]$ (B is a real number too).

Proof: $[f \leq A] = \bigcap_{n=1}^{\infty} [f < A + 1/n]$. The result follows from Theorem 7.2.2.

$[f > A] = S - [f \leq A]$. Again, we use Theorem 7.2.2. The remaining cases are left to the reader. ∎

8 Theorem. A function $f(x)$ defined on a measurable set S is measurable if and only if the sets $[f < A]$ are measurable for all real values of A.

Proof: Let us assume first that the sets $[f < A]$ are measurable for all real values of A. From Lemma 7, it follows that the sets

$$S_{n,k} = [k/n < f \leq (k + 1)/n], \qquad n = 1, 2, \ldots, \qquad k = 0, \pm 1, \pm 2, \ldots,$$

are measurable. Let n be a fixed but arbitrary positive integer. Then, every point of S is in exactly one $S_{n,k}$, since the $S_{n,k}$ are obviously disjoint. That is, if $x \in S$, then to each n there is exactly one integer k satisfying, $k/n \leq f(x) < (k + 1)/n$. Thus, $S = \bigcup_{k=-\infty}^{\infty} S_{n,k}$, where $S_{n,k} \cap S_{n,j} = \phi$ if $k \neq j$. The measurability of f will be proved by showing that f is the a.e. limit of a sequence of measurable functions. Let

$$g_n(x) = k/n, \qquad \text{if} \quad x \in S_{n,k}.$$

The definition of the $S_{n,k}$ implies that $|f - g_n| < 1/n$ for all n and $x \in S$. Thus it only remains to be shown that the g_n are measurable functions. This follows from the identity

$$g_n = \frac{k}{n} = \sum_{k=-\infty}^{\infty} \frac{k}{n} \chi_{S_{n,k}} \qquad \text{if} \quad x \in S_{n,k}.$$

Suppose now that f is a measurable function. The measurability of the sets $[f < A]$ will follow from the measurability of their characteristic functions. We shall show that $\chi_{[f<A]}$ is measurable by proving that it is the limit of a sequence of measurable functions. If $\epsilon > 0$, then the function

$$F_\epsilon(x) = \begin{cases} 0 & \text{if } f \leq A - \epsilon < A \\ 1 & \text{if } f \geq A > A - \epsilon \\ 1 + (f - A)/\epsilon & \text{if } A - \epsilon < f < A, \end{cases}$$

is measurable, since it can also be written as

$$F_\epsilon(x) = \frac{\min\,[f, A - \epsilon] - \min\,[f, A]}{-\epsilon}$$

(Theorem 7.1.7). It is easily seen that $0 \leq F_\epsilon \leq 1$, and that

$$\lim_{\epsilon \to 0} F_\epsilon = \chi_{[f \geq A]}.$$

Taking $\epsilon = 1/n$, we obtain $\chi_{[f \geq A]}$ as the limit of the sequence $\{F_{1/n}\}$ of measurable functions, from which it follows that the set $[f \geq A]$ is measurable. Taking complements, the measurability of $[f < A]$ is obtained. ∎

Exercises

*1. Prove that the six descriptions of the Borel sets given in the opening paragraph of this section are equivalent.

2. Complete the proof of Proposition 2.

3. A function is said to be *Borel measurable* if, for real values of A, the sets $[f < A]$ are Borel sets.
   (i) Prove Lemma 7 for Borel sets and Borel measurable functions.
   (ii) Prove that if f and g are Borel measurable and if c denotes the constant function, then $f \pm g$, $f + c$, and fg are Borel measurable.
   (iii) Prove that every Borel measurable function is (Lebesgue) measurable.
   (iv) Prove that if f and g are Borel measurable, then $|f|$, f^+, f^-, max $[f,g]$ and min $[f,g]$ are Borel measurable.
   (v) Prove that if S is a Borel set and if f is Borel measurable, then the set $f^{-1}(S)$ is also a Borel set.

*4. Complete the proof of Lemma 7. Then show that any of the inequalities of that lemma imply to all the others.

*5. Prove Corollaries 5 and 6.

*6. Prove that the interval (a_x, b_x) defined in Proposition 1 is the largest open interval which contains x and which is contained in G. Can a closed set be written as a disjoint countable union of closed sets? Prove that a closed set may be written as a countable intersection, $\bigcap F_n$, where each F_n is either an unbounded closed interval or the union of two such intervals.

   Illustrate this for the Cantor set (Example 5.3.6).

7. If $0 < \lambda < 1$, a Cantor set may be constructed whose measure is $1 - \lambda > 0$. The intervals I_{jk} which are to be removed from $[0,1]$ are as follows: I_{11} has

center at $1/2$ and length equal to $\lambda/2$. I_{12} and I_{22} are removed from the remaining closed intervals. Their centers coincide with the centers of the closed intervals of $[0,1] - I_{11}$, and their lengths are $\lambda/8$. Continuing in this way, a finite set of open intervals of total length $\lambda(1/2 + 1/4 + \cdots 2^{-n})$ will have been removed by the nth stage. Show that the Cantor set which remains after the union of these open intervals has been removed, has measure $1 - \lambda$.

8. Give an example of an open subset G of R, which is dense in R, but $R - G$ has nonzero measure.

9. Give an example of a subset of $[0,1]$ whose measure is equal to one but which is of Category I. [*Hint:* For each positive integer n, let C_n be the Cantor set of Exercise 7 whose measure is $(n - 1)/n$. Show that $\bigcup_{n=1}^{\infty} C_n$ is the desired set.]

10. Give an example of a subset of $[0,1]$ whose measure is zero and which is of Category II. [*Hint:* Use Exercise 9.]

11. Show that the set of measure zero obtained in Exercise 10 is not an F_σ set. (Exercises 7-11 and other similar examples are discussed in Gelbaum and Olmstead [1].)

*Exercises which request the proof of a statement in the text, as well as those referred to in the text, are starred.

4 Nonmeasurable Sets and Functions

Having proved that measurability is retained under complementation and countable union and intersection, the reader may have begun to wonder how nonmeasurable sets are obtained, if indeed there are any such sets. Whether or not one is willing to believe that nonmeasurable sets exist, depends upon one's willingness to accept "constructions" that are based on the Axiom of Choice. In 1965, it was proved that nonmeasurable sets (and functions) can not be constructed without using the Axiom of Choice or some other "inadmissible" procedure. (See Solovay [1].) It is interesting to note that Lebesgue in a paper written in 1926, (Legesgue [1]) did not accept these nonconstructive procedures. He went so far as to say that up to that moment, no one had succeeded in exhibiting a nonmeasurable function or a nonmeasurable set.

Nevertheless, we shall include a standard example of a nonmeasurable set. Its credibility rests with the reader.

1 Definition. Let S be a set of real numbers, and let a be any real number. Then the set obtained from S by a *translation of a units* is denoted by

$$S + [a] = \{x \in R : \exists y \in S \text{ and } x = y + a\}.$$

From Exercise 7.1.8 it follows that if S is measurable, then $S + [a]$ is measurable for all real values of a, and $m(S) = m(S + [a])$. Writing this as an integral, we have

[1] $$\int \chi_S = \int \chi_{S+[a]}.$$

It is left to the reader to generalize [1], first to step functions and then to arbitrary integrable functions. For reference purposes, we shall state this formally:

2 Proposition. If f is an integrable function, then

$$\int f(x) = \int f(x + a),$$

for all real values of a.

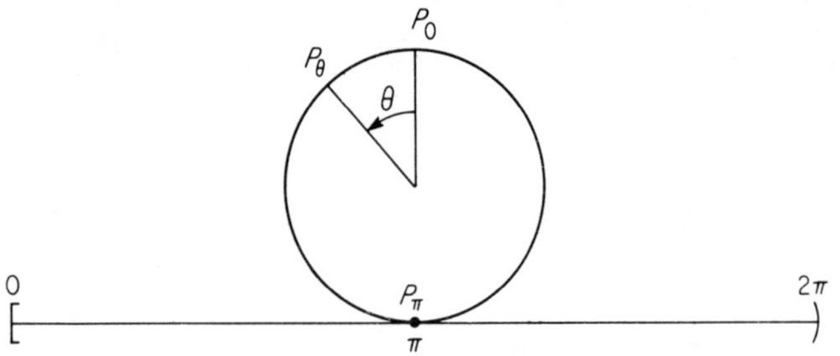

FIGURE 27

3 Example of a Nonmeasurable Set. The points on the unit circle $x^2 + y^2 = 1$ may be put in a 1–1 correspondence with the points of the interval $[0,2\pi)$, by setting $x = \cos\theta$ and $y = \sin\theta$, $0 \le \theta < 2\pi$. Let P_θ denote the point of the unit circle that corresponds to a counterclockwise rotation of θ radians, $\theta > 0$. We begin by defining an equivalence relation between points on the circle: The points P_θ and $P_{\theta'}$ are said to be equivalent, written

$$P_\theta \equiv P_{\theta'}$$

if there are integers j and k such that $\theta - \theta' = j + 2\pi k$. In order to prove that there are infinitely many distinct equivalence classes, it suffices to show that if $0 \ne n \ne m \ne 0$, then $P_{1/n} \equiv P_{1/m}$. Let us suppose that $P_{1/n} \equiv P_{1/m}$ for some pair of distinct nonzero integers. Then it follows that there are integers j and k satisfying

$$1/m - 1/n = j + 2\pi k.$$

Assuming that $m < n$, one of the following two cases must hold:

(i) If $k = 0$, then $0 < 1/m - 1/n = j$. This is a contradiction since $j \ge 1$ and $1/m - 1/n < 1$.

(ii) If $k \ne 0$, then $\pi = (1/m - 1/n - j)/2k$, which contradicts the irrationality of π.

Thus the collection of disjoint equivalence classes is infinite. Let S_0 be a set obtained by choosing a single element from each equivalence class (Axiom of Choice). For each integer n, let $S_n = \{P_{\theta+n}: P_\theta \in S_0\}$. ($P_{\theta+n}$ is the point obtained from P_θ by a rotation of n radians, counterclockwise or clockwise, depending upon the sign of n.) It is readily seen that every point on the circle is in exactly one of the S_n. This means that the points of the unit circle may be written as a disjoint countable union:

$$\{P_\theta : 0 \leq \theta < 2\pi\} = \bigcup_{n=-\infty}^{\infty} S_n.$$

Let $\{T_n\}$ be the corresponding sets on the interval $[0,2\pi)$. Then

$$[0,2\pi) = \bigcup_{n=-\infty}^{\infty} T_n;$$

the T_n are disjoint. If the set T_0 is measurable, then so are the remaining T_n, and their measures are identical (Proposition 2). The countable additivity of measure (Theorem 7.2.4) yields

$$2\pi = m([0,2\pi)) = \sum_{n=-\infty}^{\infty} m(T_n) = \begin{cases} 0 & \text{if } m(T_0) = 0 \\ \infty & \text{if } m(T_0) > 0, \end{cases}$$

from which it follows that the set T_0 can not be measurable.

In view of Definition 7.2.1, the preceding example provides also a nonmeasurable function, namely, χ_{T_0}, the characteristic function of the nonmeasurable set.

In closing, we point out that measurability is not a property that is preserved under topological mappings. Indeed, there are topological mappings (bicontinuous), that carry measurable sets into nonmeasurable sets, and sets having measure zero into sets with positive measure. Examples of such sets and mappings may be found in Gelbaum and Olmstead [1] and Exercise 5.

Exercises

***1.** We have already shown that if f is a measurable function, then f^2 and $|f|$ are measurable. Why is the converse false? Show that, f is measurable if and only if f^2 is measurable and $[f > 0]$ is a measurable set.

***2.** Prove Proposition 2.

***3.** Show that if F is continuous and strictly increasing, and maps R onto R, then F carries Borel sets into Borel sets.

***4.** Let $f(x)$ be the Cantor function, which is defined in Example 9.4.2. $f(x)$ is a continuous, nondecreasing function defined on $[0,1]$, which is constant on the interiors of the intervals I_{jk} which appear in the definition of the Cantor set (Example 5.3.6). Extend f to R by setting $f(x) = 1$ whenever $x > 1$, and $f(x) = 0$ when $x < 0$. Show that $F(x) = x + f(x)$ is a strictly increasing, continuous function that maps R onto R.

*5. Using Exercises 3 and 4, show that there exist measurable sets which are not Borel sets. [*Hint:* Let (a,b) be an interval that contains no points of the Cantor set C. Prove first that $F((a,b))$, where F is defined in Exercise 4, is an interval whose length is $b - a$. It follows that the measure of the image of $[0,1] - C$ under F is one, from which we may conclude that $m(F(C)) = 1$. Let S be a nonmeasurable subset of $F(C)$ (Show they exist). Since $F^{-1}(S)$ is a subset of C, it is measurable (Why?). The rest follows from Exercise 3.]

*Exercises which request the proof of a statement made in the text, as well as those referred to in the text, are starred.

5 Egoroff's Theorem and Convergence in Measure

In Sec. 3 we showed that a bounded measurable set S is "almost" an open or closed set, in the sense that if $\epsilon > 0$ is given, there is an open set G and a closed set F satisfying

(i) $G \supset S \supset F$, and (ii) $m(G - F) < \epsilon$.

A similar theorem for measurable functions is provided in this section. It asserts that a measurable function is continuous except on a set of arbitrarily small measure. Before proving this theorem about the nature of measurable functions, we shall obtain an even more useful and interesting result: If $\{f_n\}$ is a sequence of measurable functions converging a.e. on a finitely measurable set S, then the convergence is uniform except on a set whose measure is arbitrarily small.

This section concludes with a very brief discussion of *convergence in measure*. It may be omitted if the reader so desires, since no subsequent discussion depends upon the results there obtained.

1 Theorem (Egoroff I). If $\{f_n\}$ is a sequence of measurable functions converging a.e. to f on a finitely measurable set S, then for each $\epsilon > 0$, there is a subset $S(\epsilon)$ of S satisfying

(i) $m(S(\epsilon)) < \epsilon$, and

(ii) $\{f_n\}$ converges uniformly to f on the set $S' = S - S(\epsilon)$.

Proof: It may be assumed that the sequence converges everywhere on S. For if this were not the case, we could remove the null set on which $\{f_n\}$ does not converge to f and include it in $S(\epsilon)$.

Let p be a positive integer. Since $\lim f_n = f$ on S, there corresponds to each x in S, an integer $n(x)$ for which

$$j > n(x) \qquad \text{implies} \qquad |f_j(x) - f(x)| < 1/p.$$

Thus, keeping p fixed, there is an integer n for each x such that

$$x \in S_{n,p} = \{|f_j - f| < 1/p \quad \text{if} \quad j \geq n\}. \qquad [1]$$

It is readily seen that

[2] $$S_{1,p} \subset S_{2,p} \subset \cdots \subset S_{n,p} \subset S_{n+1,p} \subset \cdots \subset S,$$

and

[3] $$S = \bigcup_{n=1}^{\infty} S_{n,p}.$$

It follows from [2] and [3] that for some integer $N(p)$, the set

$$S_{N(p),p} = \bigcup_{n=1}^{N(p)} S_{n,p}$$

satisfies

$$m(S_{N(p),p}) > m(S) - \epsilon \cdot 2^{-p}.$$

We show now that the set

$$S' = \bigcap_{p=1}^{\infty} S_{N(p),p}$$

satisfies the conditions of the theorem. Indeed, setting

$$S(\epsilon) = S - S' = S - \bigcap_{p=1}^{\infty} S_{N(p),p} = \bigcup_{p=1}^{\infty} (S - S_{N(p),p}),$$

it follows that

$$m(S(\epsilon)) \leq \sum_{p=1}^{\infty} m(S - S_{N(p),p}) < \epsilon \sum_{p=1}^{\infty} 2^{-p} = \epsilon.$$

It remains to be proved that $\{f_n\}$ converges uniformly on S'. Let $\Delta > 0$ be given. We seek an integer N such that

$$j > N \qquad \text{implies} \qquad |f_j - f| < \Delta \qquad \text{on } S'.$$

Choose an integer $p > 1/\Delta$. Since S' is contained in $S_{N(p),p}$, it follows that if $x \in S'$, then

$$j > N(p) \qquad \text{implies} \qquad |f_j(x) - f(x)| < 1/p < \Delta,$$

thus proving that the convergence is uniform on S'. ▮

Egoroff's theorem can not be improved by removing the condition that S be finitely measurable. (See Exercise 1.) There is however a variation of this theorem in which a boundedness condition on the sequence $\{f_n\}$ is substituted for $m(S) < \infty$:

2 Theorem (Egoroff II). Let $\{f_n\}$ be a sequence of measurable functions converging a.e. to a function f on a measurable set S. If there is function g which is integrable over S, and if $|f_n| \le g$ for all n and $x \in S$, then given $\epsilon > 0$, there is a subset $S(\epsilon)$ of S satisfying,

(i) $m(S(\epsilon)) < \epsilon$, and

(ii) $\{f_n\}$ converges uniformly to f on the set $S' = S - S(\epsilon)$.

Proof: Again it may be assumed that the sequence converges everywhere on S to f. (See the proof of Theorem 1.) Since g is integrable, the sets $S_k = [g > 1/k]$ are measurable and $m(S_k) < \infty$. For, if there were an integer k for which $m(S_k) = \infty$, then it would follow that

$$\int_S g \ge \int_{S_k} g > \int_{S_k} \frac{1}{k}\, \chi_{S_k} = \frac{1}{k}\, m(S_k) = \infty,$$

thus contradicting the integrability of g. Theorem 1 asserts that a set $S_k(\epsilon)$ may be removed from each S_k, satisfying

(i) $m(S_k(\epsilon)) < \epsilon \cdot 2^{-k}$,

(ii) $\{f_n\}$ converges uniformly on $S_k - S_k(\epsilon)$ to f.

We show next that $\bigcup_{k=1}^{\infty} S_k(\epsilon)$ is the desired set $S(\epsilon)$:

(i) $m\left(\bigcup_{k=1}^{\infty} S_k(\epsilon)\right) \le \sum_{k=1}^{\infty} m(S_k(\epsilon)) < \epsilon \sum_{k=1}^{\infty} 2^{-k} = \epsilon.$

To verify (ii), let $\Delta > 0$ be given. Setting

$$S' = S - \bigcup_{k=1}^{\infty} S_k(\epsilon) = \bigcap_{k=1}^{\infty} (S - S_k(\epsilon)),$$

it follows that for all values of k, in particular for $k > 2/\Delta$, we have $S' \subset (S - S_k(\epsilon))$. Since

$$S - S_k(\epsilon) = (S_k - S_k(\epsilon)) \cup (S - S_k),$$

there are two possible cases:

1. If $x \in (S_k - S_k(\epsilon))$, then the uniform convergence on this set yields an integer $N(\Delta)$ for which

$$j \ge N(\Delta) \qquad \text{implies} \qquad |f_j(x) - f(x)| < \Delta.$$

2. If $x \in (S - S_k)$, then $g(x) \le 1/k < \Delta/2$, which yields

$$|f_k - f| < |f_k| + |f| < 2g < \Delta.$$

Combining the two cases, we see that for any integer $j > \max\,[2/\Delta, N(\Delta)]$,

$$x \in (S_j - S_j(\epsilon)) \qquad \text{implies} \qquad |f_j - f| < \Delta.$$

However, S' is contained in every $S_j - S_j(\epsilon)$, proving that the convergence is uniform on S'. ▎

To prove Lusin's theorem, which characterizes measurable functions defined on finitely measurable sets as being "almost" continuous, two lemmas are needed. Since both appeared earlier as exercises, we shall merely outline their proofs, and leave the details to the reader.

3 Lemma. Let f be a continuous function defined on a closed, bounded subset S of a bounded interval $I = [a,b]$. Then there is a function g, which is continuous on I, and $g = f$ on S. Moreover, if $|f| \leq A$, then g may be chosen so that $|g| \leq A$ also.

Proof: The open set $(a,b) - S$ may be written as a disjoint union of open intervals $\bigcup_{k=1}^{\infty} I_k$. If $I_k = (a_k,b_k)$, we define g on the closure of I_k to be the linear function

$$g(x) = \left(\frac{f(b_k) - f(a_k)}{b_k - a_k}\right) x + c_k,$$

where c_k is chosen so that $g(a_k) = f(a_k)$ and $g(b_k) = f(b_k)$. It is easily verified that this can be done and that $|g| \leq A$ if $|f| \leq A$ (Exercise 2). ▌

4 Lemma. A measurable function is the a.e. limit of a sequence of continuous functions.

Proof: Let $\{\sigma_n\}$ be a sequence of step functions which converges to a measurable function f. For each value of n, remove finitely many intervals which cover the discontinuities of σ_n. It may be assumed that the total length of the system $\mathcal{I}_n$ of intervals removed, is in each case less than $1/n$. Let σ_n' denote the restriction of σ_n to $I_n - \mathcal{I}_n$; I_n is an interval outside of which $\sigma_n \equiv 0$. Use Lemma 3 to extend σ_n' to a function f_n which is continuous. Show that $\lim f_n = f$ a.e. ▌

5 Theorem (Lusin I). If f is a measurable function defined on a finitely measurable set S, then for any $\epsilon > 0$, there is a subset $S(\epsilon)$ of S satisfying

(i) $m(S(\epsilon)) < \epsilon$, and

(ii) f is continuous on $S' = S - S(\epsilon)$.

Proof: Let $\{f_n\}$ be a sequence of continuous functions which converges to f a.e. on S (Lemma 4). Egoroff's theorem (I) asserts that $\{f_n\}$ converges uniformly to f on a set $S' = S - S(\epsilon)$, where $m(S(\epsilon)) < \epsilon$. Thus on S', f is the uniform limit of a sequence of continuous functions and is therefore itself continuous. ▌

6 Corollary. If f is a measurable function defined on a finite interval I, then for any $\epsilon > 0$ there is a continuous function g defined on I, and $m[f \neq g] < \epsilon$.

The proof is left to the reader as an exercise.

We have so far encountered sequences of functions that converge pointwise, almost everywhere, in the (integral) norm, and uniformly. There is yet another type of convergence which is of considerable interest in probability theory. We shall not attempt here to discuss any of these applications, and will limit the presentation to the basic definitions and properties. For further discussion, the reader is urged to see Halmos [1] and Munroe [1].

7 Definition. Let $\{f_n\}$ be a sequence of measurable functions defined on a measurable set S. The sequence is said to *converge in measure to a function* f, denoted by $f_n \xrightarrow{m} f$, if for any $\epsilon > 0$,

$$\lim_{n\to\infty} m[\,|f_n - f| > \epsilon] = 0.$$

8 Example. (a) The sequence $\chi_{[n,n+1]}$ converges pointwise to zero, but does not converge in measure to the function which is always equal to zero. Indeed, if $\epsilon = 1/2$, then $m[\,|\chi_{[n,n+1]}| > 1/2] = 1$ for all values of n.

(b) The sequence $\chi_{[n,n+1/n]}$ converges to zero both pointwise and in measure.

(c) See Exercise 5 for an example of a sequence which converges in measure but not pointwise or a.e.

9 Proposition. Suppose that $f_n \xrightarrow{m} f$ and $g_n \xrightarrow{m} g$. Then,

(i) $(f_n + g_n) \xrightarrow{m} (f + g)$;

(ii) If the measure of the set on which the sequences converge is finite, then $f_n g_n \xrightarrow{m} fg$.

Proof of (*i*): If at some point y,

$$|(f_n(y) + g_n(y)) - (f(y) + g(y))| > \epsilon,$$

then the triangle inequality yields

$$|f_n(y) - f(y)| + |g_n(y) - g(y)| > \epsilon,$$

so that either

$$|f_n(y) - f(y)| > \epsilon/2 \quad \text{or} \quad |g_n(y) - g(y)| > \epsilon/2.$$

Thus

$$[\,|(f_n + g_n) - (f + g)| > \epsilon] \subset [\,|f_n - f| > \epsilon/2] \cup [\,|g_n - g| > \epsilon/2],$$

whose measure tends to zero as n tends to infinity.

The proof of (ii) is left to the reader. ▌

10 Proposition. If $\{f_n\}$ is a sequence of measurable functions that converges in measure to both f and g, then $f = g$ a.e.

Proof: As in the preceding proof, the set

$$[\,|f - g| > \epsilon] \subset [\,|f - f_n| > \epsilon/2] \cup [\,|g - f_n| > \epsilon/2],$$

for all values of n. The measure of the quantity on the right, however, tends to zero as n tends to infinity, proving that the set on which f and g are unequal is a null set. ▌

11 Proposition. If $f_n \xrightarrow{m} f$, and if $f = g$ a.e., then $f_n \xrightarrow{m} g$.

The proof is left to the reader.

12 Theorem. Let $\{f_n\}$ be a sequence of measurable functions defined on a measurable set S which converges a.e. on S to a function f. Then if either $m(S) < \infty$, or if there is an integrable function g such that $|f_n| \leq g$, for all n, then $f_n \xrightarrow{m} f$.

Proof: If $m(S) < \infty$ or $|f_n| \leq g$, we may use Egoroff's theorem in one of the two forms (Theorems 1 and 2). Therefore if $\Delta > 0$ is given, we may remove from S a set $S(\Delta)$ whose measure is less than Δ, and on the complement of this set, the sequence converges uniformly. Thus, if $\epsilon > 0$ is given,

$$[\,|f_n - f| > \epsilon] \subset ([\,|f_n - f| > \epsilon] \cap (S - S(\Delta)) \cup S(\Delta),$$

so that the monotonicity of measure implies that

$$m[\,|f_n - f| > \epsilon] < m([f_n - f| > \epsilon] \cap (S - S(\Delta)) + \Delta.$$

The uniform convergence of the sequence on the set $S - S(\Delta)$ implies that for sufficiently large values of n, the first term on the right side is equal to zero; the second term is Δ, which may be assumed to be as small as we like. ▌

A partial converse of Theorem 12 may be proved:

13 Theorem. If $f_n \xrightarrow{m} f$ on a measurable set S, then there is a subsequence $\{f_{n_k}\}$ which converges a.e. to f.

Proof: For convenience, we shall assume that $f \equiv 0$. For if this were not the case, then the f_n could be replaced by the sequence $g_n = f_n - f$, which converges in measure to zero. The existence of a subsequence $\{g_{n_k}\}$ which converges a.e. to zero implies that the corresponding $\{f_{n_k}\}$ converge a.e. to f.

The hypothesis assures the existence of an increasing sequence $\{n_k\}$ of positive integers for which

$$m[\,|f_{n_k}| > 2^{-k}] < 2^{-k}.$$

Setting $S_k = [\,|f_{n_k}| > 2^{-k}]$, we form the union, $T_p = \bigcup_{k=p}^{\infty} S_k$. The sets T_p are measurable, and

$$m(T_p) \leq \sum_{k=p}^{\infty} m(S_k) < \sum_{k=p}^{\infty} 2^{-k} = 2^{1-p}.$$

For all $k > p$ and $x \in S - T_p$,

$$|f_{n_k}(x)| < 2^{-k}.$$

Thus $\{f_{n_k}\}$ converges uniformly on all T_p. This implies that the set T on which $\{f_{n_k}\}$ does *not* converge to zero, is contained in every T_p. The monotonicity of the measure function implies

$$m(T) \leq m(T_p) < 2^{1-p}$$

for all positive values of p, from which it follows that $m(T) = 0$. ▎

The reader may have noticed that Propositions 9, 10, and 11, and Theorems 12 and 13 remain true if "$f_n \xrightarrow{m} f$" is replaced by "f_n converges in norm to f." It is even true that Theorem 13 may not be improved, in the sense that there exist sequences which converge in measure but do not converge at any points. (See Example 5.4.3 and Exercise 5.) It is also left to the reader to verify that the theorems on dominated and monotone convergence, and Fatou's lemma remain true if "convergence a.e." is replaced by "convergence in measure" (Exercises 10, 11, and 12). And finally, it is possible to define a Cauchy sequence in measure and to prove a "completeness" theorem. This also appears in the exercises.

Exercises

1. Show that the boundedness conditions of Egoroff's theorem are necessary, by giving an example of a sequence of measurable functions which converges a.e., but for which it is not possible to remove a set of arbitrarily small measure on whose complement the sequence converges uniformly.
*2. Complete the proof of Lemma 3.
*3. Prove Lemma 4.
*4. Prove Corollary 6.
5. Give an example of a sequence of measurable functions which converges in measure but does not converge at any point. [*Hint:* Use a sequence which converges in norm to zero, but which does not converge at any point.]

6. Show that if $\{f_n\}$ is a sequence of measurable functions which converges in measure to a function f, then every subsequence of $\{f_n\}$ converges in measure to f.

7. Let $f_n \xrightarrow{m} f$. Prove that if there is an integrable function g and $|f_n| \leq g$, for all n, then $\lim \|f - f_n\| = 0$; that is, the sequence converges also in norm to f.

***8.** Prove (ii) of Proposition 9 and show that the assumption $m(S) < \infty$ is necessary.

***9.** Prove Proposition 11.

10. Show that the theorem on monotone convergence remains true if "converges a.e." is replaced by "converges in measure."

11. Do the same for Fatou's lemma.

12. Do the same for the Lebesgue theorem on dominated convergence.

13. A sequence $\{f_n\}$ of measurable functions is said to *Cauchy in measure* if, given $\epsilon > 0$,

$$\lim_{j,k\to\infty} m[\,|f_j - f_k| > \epsilon] = 0.$$

Show that if $\{f_n\}$ converges in measure then it is Cauchy in measure.

14. Prove the following "completeness" theorem: If $\{f_n\}$ is a Cauchy sequence in measure, then there is a measurable function f, and $f_n \xrightarrow{m} f$.

*Exercises which request the proof of a statement made in the text, as well as those referred to in the text, are starred.

CHAPTER **8**

An Alternate Approach to Lebesgue Integration and Measure

1 Exterior Measure, Measurable Sets and Functions

In this section we shall show that Lebesgue measure and the class $\mathfrak{M}$ of measurable sets may be defined without using the Lebesgue integral. It is of course quite possible to give the definition of measure without any preliminary remarks, and to prove later that it coincides with the measure defined in Sec. 2 of Chapter 7. However, we shall try to motivate this definition by looking at certain known properties of measurable sets.

In Sec. 3 of Chapter 7 it was shown that if S is a bounded measurable set, then it is contained in a G_δ-set $\hat{S}$, and $m(S) = m(\hat{S})$; that is, there exist open sets $G_n \supset G_{n+1} \supset \cdots \supset S$, and $\hat{S} = \bigcap_{n=1}^{\infty} G_n$. The monotonicity of Lebesgue measure implies that for all n, $m(G_n) \geq m(\hat{S}) = m(S)$. This suggests that we first try to define a measure for open sets, (which, in view of Proposition 7.3.1 is quite easy to do), and then to extend it by taking limits. Eventually it will be proved that the measure defined in this section coincides with the measure defined earlier. However, until the equivalence is demonstrated, we shall refer to the "old" measure as the *m-measure* and the one defined in this section as the *μ-measure*.

We begin by defining the μ-measure of an interval to be equal to its length. If G is an open set of real numbers, then by writing $G = \bigcup_{n=1}^{\infty} I_n$, where $\{I_n\}$ is a disjoint sequence of open intervals (Proposition 7.3.1), we take the μ-measure of G, written $\mu(G)$, to be the sum of the lengths of the I_n, $\sum_{i=1}^{\infty} \lambda(I_n)$, which coincides with the sum of the m-measures. Thus, for open sets, $m(G) = \mu(G)$. The extension of μ to a larger class of sets is suggested by the remarks of the preceding paragraph: If S is an m-measurable set, then the proper extension is given by

$$\mu(S) = \inf \mu(G),$$

where G ranges over the open sets which contain S; the value $\mu(S) = \infty$ is permitted. It is easily seen that *if S is m-measurable, then $m(S) = \mu(S) = \inf \mu(G)$*. For, surely $\inf \mu(G) \geq m(S)$, since $S \subset G$. Conversely, Corollary 7.3.4 asserts there is a sequence $\{G_n\}$ of open sets which contain S, and

$$0 \leq \mu(G_n) - m(S) = m(G_n) - m(S) < 1/n,$$

from which it follows that $m(S) = \inf \mu(G_n)$. However, since every set S is contained in at least one open set, say R, it follows that $\inf \mu(G)$, taken over all open sets G which contain S, is defined for *all* subsets of R (the value ∞ is permitted), including the nonmeasurable sets, so that it can not be taken as a definition of the measure of a set. Thus, our problem is to decide which subsets of R will be called μ-measurable.

1 Definition. The *exterior measure* of a subset S of the real numbers, written $\mu_e(S)$, is taken to be $\inf \mu(G)$, where G ranges over all open sets which contain S. ($\mu_e(S) = \infty$ is permitted.)

It was observed above that the exterior measure is defined for any set of real numbers, and that if the set is m-measurable, then $m(S) = \mu_e(S)$. As an immediate consequence of Definition 1, it follows that the exterior measure has all but one of the basic properties of measure. This missing property is used to define μ-measurable sets.

2 Proposition. (Properties of exterior measure)

(i) For all $S \subset R$, $\mu_e(s) \geq 0$ (nonnegativity).

(ii) $S \subset T$ implies $\mu_e(S) \leq \mu_e(T)$ (monotonicity).

(iii) For all S and T, $\mu_e(S \cup T) \leq \mu_e(S) + \mu_e(T)$ (*subadditivity*).

The proof is left to the reader.

The nonadditivity of the exterior measure is easily verified by using nonmeasurable sets (See Example 7.4.3). It turns out that if a μ-measurable set is defined as one for which μ_e is additive, then this class of sets coincides with the m-measurable sets, which were defined as those sets whose characteristic functions are measurable. Let us carry this out formally, starting with bounded sets.

3 Definition. A set S contained in a bounded interval $I = [a,b]$ is said to be *μ-measurable* if

$$\mu_e(S) + \mu_e(I - S) = \mu_e(I) = b - a.$$

The *μ-measure of S*, written $\mu(S)$, is equal to its exterior measure.

The following theorem asserts that the bounded μ-measurable sets coincide with the bounded m-measurable sets:

4 Theorem. A subset S of a bounded interval $I = [a,b]$ is m-measurable if and only if $\mu_e(S) + \mu_e(I - S) = b - a$.

Proof: If S is an m-measurable subset of I, then its complement in I, $I - S$, is also m-measurable. The additivity of m-measure and the identity $m(S) = \mu_e(S)$, which holds for m-measurable sets, yields

$$b - a = m(S \cup (I - S)) = m(S) + m(I - S) = \mu_e(S) + \mu_e(I - S).$$

Conversely, if $\mu_e(S) + \mu_e(I - S) = b - a$, then there are sequences $\{A_n\}$ and $\{B_n\}$ of open sets that satisfy:

$$[1] \qquad \begin{aligned} &A_n \supset A_{n+1} \supset \cdots \supset S \text{ and } m(A_n) \downarrow \mu_e(S), \\ &B_n \supset B_{n+1} \supset \cdots I - S \text{ and } m(B_n) \downarrow \mu_e(I - S). \end{aligned}$$

For all n, $S \supset I - B_n$, and it follows that

$$[2] \qquad \chi_{A_n} \geq \chi_S \geq \chi_{I-B_n} = 1 - \chi_{B_n}, \qquad \text{if } x \in I.$$

The proof rests upon the equivalence of the m-measurability of the set S and the measurability of its characteristic function χ_S (Definition 7.2.1). We shall show that the sequence of integrable functions $\chi_{A_n} \downarrow \chi_s$ a.e. From [2] we obtain

$$[3] \qquad 0 \leq \chi_{A_n} - \chi_S \leq \chi_{A_n} - 1 + \chi_{B_n} = \sigma_n \downarrow .$$

The monotonicity of the integral, and [1] and [3], imply that as n tends to infinity,

$$[4] \qquad 0 \leq \int_I \sigma_n = m(A_n) + m(B_n) - (b - a) \downarrow 0.$$

The theorem on monotone convergence yields $\sigma_n \downarrow g$ a.e., where g is an integrable, a.e. nonnegative function, and

$$\int g = \lim \int \sigma_n = 0.$$

Thus $g = 0$ a.e., and it follows from [3] that $\lim \chi_{A_n} = \chi_S$ a.e., proving that χ_S is a measurable function, and therefore S is measurable and

$$\mu_e(S) = \lim m(A_n) = \lim \int \chi_{A_n} = m(S). \quad \blacksquare$$

The proper extension of the μ-measure to unbounded sets is suggested by the following properties of m-measurable sets: A set S is m-measurable if and only if its characteristic function χ_S is measurable. Moreover, χ_S is measurable if and only if all its truncations are integrable (Theorem 7.1.6). However, in this case, each truncation $(\chi_S)_A$ is the characteristic function of the *truncated* set $S_A = S \cap [-A,A]$, which is the intersection of two m-measurable sets and is therefore itself m-measurable. We proved earlier that every bounded m-measurable set is μ-measurable, and that their measures coincide. This suggests the following definition for unbounded sets:

5 Definition. A set of real numbers is said to be *μ-measurable* if all its truncations $S_A = S \cap [-A,A]$ are bounded (m or μ) measurable sets, and the μ-measure of S is defined by taking the limit

[5] $$\mu(S) = \lim_{A\to\infty} \mu(S_A).$$

The value $\mu(S) = \infty$ is permitted.

6 Theorem. A subset of the real numbers is m-measurable if and only if it is μ-measurable, and the two measures coincide.

Proof: This has already been verified for bounded sets. It S is unbounded, and if it is μ-measurable, then

[6] $$\mu(S) = \lim_{N\to\infty} \mu(S_N) = \lim_{N\to\infty} m(S_N) = \lim_{N\to\infty} \int \chi_{S_N} = \int \chi_S = m(S),$$

proving that S is m-measurable and that $m(S) = \mu(S)$. Reading [6] from right to left, gives the converse. ∎

Since μ and m are identical, we shall henceforth write $m(S)$ to denote the measure of a set of real numbers.

The next step is to define μ-measurable functions so that they coincide with the m-measurable functions that were discussed in Sec. 1 of Chapter 7. Theorem 7.3.8 characterizes m-measurable functions in terms of m-measurable sets, and in view of the fact that the μ and m-measurable sets are identical, we are assured of obtaining the same class of measurable functions if this theorem about m-measure is taken as the definition of a μ-measurable function:

7 Definition. A function f defined on a measurable set S, is said to be *μ-measurable*, if for all real values of A, the sets $[f < A]$ are measurable.

Since Theorem 7.3.8 asserts that this class of functions coincides with M, we shall simply refer to them as "measurable functions" hereafter. All the properties of measurable functions proved in Sec. 1 of Chapter 7 hold. However, the reader is asked, in the exercises, to reprove these properties using only Definition 7.

Exercises

Whenever the reader is asked to verify properties of measurable functions and sets, in the exercises of this section, it is understood that only the definitions and theorems that are proved in this section are to be used. An exception is made in Exercise 15, where it is assumed that Lusin's theorem has been proved.

*1. Prove that if $\mu_e(S) = 0$, then S is μ-measurable and $\mu(S) = 0$.
2. Prove that if S is countable, then $\mu(S) = 0$.
3. Prove that the set $S = \{x: 0 \leq x < 1\}$ is uncountable.
4. Prove that if S is μ-measurable and if $\mu(T) = 0$, then $\mu(S \cup T) = \mu(S)$.
5. Prove that if S and T are μ-measurable, then $S - T$ is μ-measurable.
*6. Prove that T is μ-measurable if and only if, for all sets S, $\mu_e(S) = \mu_e(S \cap T) + \mu_e(S - T)$.
7. Show that all open and closed sets are μ-measurable.
8. Riemann measurable sets are often defined in the following way: Let S be a subset of a bounded interval $I = [a,b]$, and let $\mathcal{S}$ denote a (finite) subdivision of I. The sum of the lengths of the intervals of $\mathcal{S}$ which are contained *entirely* in S is called the *inner sum* that corresponds to $\mathcal{S}$, and is denoted by $\lambda_I(S,\mathcal{S})$. The sum of the lengths of the intervals that intersect S, is called the *outer sum* and is denoted by $\lambda_0(S,\mathcal{S})$. It is readily seen that if $\mathcal{S}$ and $\mathcal{S}'$ are a pair of subdivisions, then $\lambda_I(S,\mathcal{S}) \leq \lambda_0(S,\mathcal{S}')$. We define the *inner and outer (Riemann)* measure of S to be $\sup_{\mathcal{S}} \lambda_I(S,\mathcal{S})$ and $\inf_{\mathcal{S}} \lambda_0(S,\mathcal{S})$ respectively. A set S is said to be *Riemann measurable* if the two quantities are equal. Show that every Riemann measurable set if μ-measurable. Why is the converse false, even for bounded sets? State carefully the differences between Riemann measure and Lebesgue measure.
9. Show that if f,g are μ-measurable functions and if c is a real number, then $f + g$, cf, $|f|$, f^+, f^-, $\max[f,g]$, $\min[f,g]$, are μ-measurable. (Remember, use only Definition 7.)
10. If $\{f_n\}$ is a sequence of μ-measurable functions, show that $\sup[f_1, f_2, \ldots]$, $\inf[f_1, f_2, \ldots]$, $\underline{\lim}[f_1, f_2, \ldots]$, and $\overline{\lim}[f_1, f_2, \ldots]$ are μ-measurable functions.
11. Prove that a set S is μ-measurable if and only if its characteristic function χ_S is μ-measurable.
12. Prove that if f is μ-measurable, and if $f = g$ except on a set S for which $\mu(S) = 0$, then g is μ-measurable.
13. Let f be μ-measurable on the interval $I = [a,b]$. Prove that for every $\epsilon > 0$, there is a positive number A, and a subset S of I such that, $\mu(S) < \epsilon$, and $|f| \leq A$ except on S.
14. Let f be a generalized step function which vanishes outside $I = [a,b]$. $\left(f = \sum_{n=1}^{N} c_n \chi_{S_n}\right.$, where the S_n are μ-measurable subsets of I, and $\left. c_n \in R.\right)$ Prove that if $\epsilon > 0$ is given, then there is a subset S of I, such that $\mu(S) < \epsilon$, and a step function σ satisfying $\sigma = f$ except on S.
15. Let f be a μ-measurable function defined on $I = [a,b]$. Prove that if $\epsilon > 0$ is given, there is a subset S of I, and a step function σ such that, $\mu(S) < \epsilon$ and $|f - \sigma| < \epsilon$ except on S. (*Hint:* First use Lusin's Theorem to get a continuous approximation.)

*Exercises which request the proof of a statement made in the text, as well as those referred to in the text, are starred.

2 The Lebesgue Integral (A Second Definition)

The Lebesgue integral of a bounded measurable function is defined in this section as the "limit" of approximating sums. Although it has already been shown that the Lebesgue integral is an "extension" of the Riemann integral (See Sec. 3 of Chapter 6.) it must be made clear at the outset, that the Lebesgue approximating sums will differ radically from the Riemann sums. Indeed, the latter are not adequate to define the Lebesgue integral for those functions which are in L but whose Riemann integrals do not exist. This is illustrated by the following example: Let S denote the subset of irrational points of $[0, 1]$. Then $\chi_S = 1$ a.e. on $[0, 1]$ and therefore

$$\int \chi_S = \int \chi_{[0,1]} = 1.$$

However, if we were to form *all* the Riemann approximating sums for a given subdivision, we would obtain sums that were equal to 1, and sums that were equal to 0. For, no matter how small the norm of the subdivision is taken to be, each interval of the subdivision would contain both rational and irrational points. (See Example 6.3.5.) It is therefore useless to attempt to approximate the Lebesgue integral by subdividing the x axis into smaller and smaller intervals, since there are functions in L that are everywhere discontinuous. (See Theorem 6.3.8) and for which it is not possible to find suitably fine subdivisions whose approximating sums will be "close" to the value of the integral. The idea of Lebesgue is to *collect values of $f(x)$ which are very nearly equal, even if the corresponding values of x may be distributed over the entire interval.* Thus, in place of a Riemann approximating sum, $\sum_{i=1}^{n} f(c_i)(a_i - a_{i-1})$, for a given subdivision of (a,b) and function f, we shall subdivide an interval $(-A,A)$ which contains the *function values* of a bounded function $f(x)$; that is, let $-A = y_0 < y_1 < < \cdots < y_n = A$. Then points λ_i are chosen in each subinterval $[y_{i-1}, y_i]$, and an approximating sum $\Sigma\, \lambda_i\, m[y_{i-1} \le f < y_i]$ is formed which will converge for a much larger class of functions, (which includes χ_S), than the class of Riemann integrable functions.

The reader may wonder what advantage there is in being able to integrate functions which are as badly discontinuous as χ_S, since this integral does not represent an area in the usual sense of that word. Again, the reader is urged to reread Sec. 3 of Chapter 6. For although χ_S and other similar functions are not Riemann integrable, they may be obtained as the a.e. limits of sequences of Riemann integrable functions. The absence of their limit functions from the normed space of Riemann integrable functions (incompleteness) accounts for the "defects" of the Riemann integral under convergence.

Let us begin by defining the Lebesgue integral of a nonnegative, bounded, measurable function defined on a bounded interval. Following this, the integral is extended to unbounded functions, and it is verified that the integral so obtained is the same as the integral defined in Chapter 5.

1 Definition. Let $f(x)$ be a bounded, nonnegative, measurable function defined on a bounded, measurable set S. If $0 \leq f < A$, we subdivide the interval $[0,A]$: $0 = y_0 < y_1 < \cdots y_n = A$, choose a point λ_i in each $[y_{i-1}, y_i]$, $i = 1,2,\ldots,n$ and form the *Lebesgue approximating sum*

$$S(f,\mathcal{S},\lambda_i) = \sum_{i=1}^{n} \lambda_i\, m[y_{i-1} \leq f < y_i].$$

(The measurability of the function $f(x)$ assures the measurability of the sets $[y_{i-1} \leq f < y_i]$.) The *upper* and *lower Lebesgue sums* for a particular subdivision $\mathcal{S}$ are given by

$$\overline{S}(f,\mathcal{S}) = \sum_{i=1}^{n} y_i\, m[y_{i-1} \leq f < y_i]$$

and

$$\underline{S}(f,\mathcal{S}) = \sum_{i=1}^{n} y_{i-1}\, m[y_{i-1} \leq f < y_i]$$

respectively.

Since $\overline{S}(f,\mathcal{S}) \geq \underline{S}(f,\mathcal{S}) \geq 0$, for all subdivisions $\mathcal{S}$, it follows that the greatest lower bound of the set of values $\{\overline{S}(f,\mathcal{S})\}$ and the least upper bound of the set $\{\underline{S}(f,\mathcal{S})\}$, taken over all subdivisions, exist, and

$$\underline{S}(f) = \sup_{\mathcal{S}}\, [\underline{S}(f,\mathcal{S})] \leq \inf_{\mathcal{S}}\, [\overline{S}(f,\mathcal{S})] = \overline{S}(f)$$

2 Theorem. If f is a bounded, nonnegative measurable function defined on a bounded measurable set S, then $\underline{S}(f) = \overline{S}(f)$.

Proof: If $\mathcal{S}$ is any subdivision, then

$$\overline{S}(f,\mathcal{S}) - \underline{S}(f,\mathcal{S}) = \sum_{i=1}^{n} (y_i - y_{i-1})\, m[y_{i-1} \leq f < y_i]$$

$$\leq \max_{i \leq n}\{(y_i - y_{i-1})\} \sum_{i=1}^{n} m[y_{i-1} \leq f < y_i] = \max_{i \leq n}\{(y_i - y_{i-1})\}\, m(S).$$

If $\epsilon > 0$ is any preassigned number, then by requiring that $\max_{i \leq n}$

$(y_i - y_{i-1}) = |\mathcal{S}| < \dfrac{\epsilon}{m(S)}$, we obtain $\overline{S}(f,\mathcal{S}) - \underline{S}(f,\mathcal{S}) < \epsilon$, thus proving that $\overline{S}(f) = \underline{S}(f)$. ∎

3 Definition. Let f be a bounded nonnegative, measurable function defined on a bounded measurable set S. The number J is said to be the *integral of f over S* if, given $\epsilon > 0$, there is a $\delta > 0$ such that

$$[1] \qquad |\mathcal{S}| < \delta \qquad \text{implies} \qquad |J - S(f,\mathcal{S},\lambda_i)| < \epsilon,$$

for all choices of the λ_i. The integral is denoted by

$$\hat{\int_S} f.$$

(The circumflex will be removed after it is proved that this integral coincides with the one defined in Chapter 5.)

Instead of [1], we shall write

$$[2] \qquad \hat{\int_S} f = \lim_{|\mathcal{S}|\to 0} S(f,\mathcal{S},\lambda_i).$$

If $\mathcal{S}$ is any subdivision of $[0,A]$, then for any choice of λ_i in $[y_{i-1} \le f < y_i]$, we have

$$\underline{S}(f,\mathcal{S}) \le S(f,\mathcal{S},\lambda_i) \le \overline{S}(f,\mathcal{S}),$$

proving that [1] and [2] are equivalent to $\underline{S}(f) = \overline{S}(f)$. Thus Theorem 2 asserts that if f is a bounded, nonnegative, measurable function, and if S is a bounded measurable set, then $\hat{\int_S} f$ exists. Before extending the integral to a larger class of functions and sets, we shall prove that [2] coincides with $\int_S f$, as given by Definition 7.2.5.

4 Theorem. If f is a bounded, nonnegative, measurable function defined on a bounded measurable set S, then

$$\hat{\int_S} f = \int_S f.$$

Proof: We shall show that if $\epsilon > 0$ is any preassigned number, then by choosing $|\mathcal{S}|$ to be sufficiently small, the Lebesgue sum will differ from $\int_S f$ by less than ϵ.

Let $\mathcal{S}$ be a subdivision of the interval $[0,A]$ as described in Definition 1, and let $S_i = [y_{i-1} \le f < y_i]$, $i = 1,2,\ldots,n$. Then the set S, on which f is defined, may be written as the disjoint union $\bigcup_{i=1}^{n} S_i = S$. The additivity of the integral implies that,

$$\left| \int f - S(f, \mathcal{S}, \lambda_i) \right| = \left| \sum_{i=1}^{n} \int_{S_i} f - \sum_{i=1}^{n} \lambda_i m(S_i) \right|$$

[3]

$$= \left| \sum_{i=1}^{n} \int_{S_i} (f - \lambda_i \chi_{S_i}) \right| \leq m(S) \max_{\substack{x \in S_i \\ i \leq n}} |f(x) - \lambda_i| = m(S) \, |\mathcal{S}| ,$$

thus proving that [3] can be made to be less than ϵ if $|\mathcal{S}| < \dfrac{\epsilon}{m(S)}$. ∎

The extension of the integral to unbounded functions and sets is made by taking truncations (Definition 7.1.4): If f is a nonnegative, measurable function defined on a measurable set S, then its truncations f_A, for real positive values of A, are integrable over the truncated sets $S_A = S \cap [-A, A]$, since they satisfy the conditions of Theorem 2.

5 Definition. If f is a nonnegative, measurable function defined on a measurable set S, then

$$\hat{\int_S} f = \lim_{A \to \infty} \int_{S_A} f_A$$

provided that this limit exists.

A measurable function f (not necessarily nonnegative) is integrable over a measurable set if both f^+ and f^- are integrable, and the integral of f is given by

$$\hat{\int_S} f = \hat{\int_S} f^+ - \hat{\int_S} f^-.$$

The reader should have no difficulty verifying that the extension of the integral given by Definition 5, agrees with the integral as defined earlier. It is also left to the reader to show that if S is a set whose measure is finite, then $m(S) = \hat{\int} \chi_S$.

Exercises

1. Prove that $\hat{\int_S} f$ exists only if $\hat{\int_S} |f|$ exists.
2. Show that $m(S) = \int \chi_S$ for all finitely measurable sets S.
3. Let f be integrable over S. Using only the definitions and theorems proved in this section, prove that if $\epsilon > 0$, there is a *bounded* measurable function g

such that

$$\hat{\int_S} |f - g| < \epsilon.$$

4. Prove that if f is a bounded measurable function, then there is a generalized step function σ such that $\max_{x \in R} |f(x) - \sigma(x)| < \epsilon$, where ϵ is any preassigned positive number.
5. Why isn't the function $\dfrac{\sin x}{x} \chi_{[1,\infty)}$ integrable?
6. Prove, directly from the definitions given in this section, that the integral is additive, positive, and homogeneous.
7. Using only the definitions given in this section, prove that if f and g are integrable over S, then so are max $[f,g]$ and min $[f,g]$.
8. Prove that $\left| \hat{\int_S} f \right| \leq \hat{\int_S} |f|$.

An Epilogue to Chapters 5–8

What follows is a brief sketch of three methods of obtaining the Lebesgue integral.

Method I. In Chapter 5, we began by defining an integral on the linear space Σ of real step functions (Definition 5.1.1). The integral expression $\int |\sigma|$ was shown to be a norm on Σ (or, more precisely on the space of equivalence classes). The incomplete normed space Σ was extended to its abstract completion $\hat{L}_1$, along with the integral and norm (Sec. 2 of Chapter 5). In Sec. 4 of Chapter 5, we saw that the space $\hat{L}_1$ of equivalence classes of Cauchy sequences of step functions could be put in a 1–1 correspondence with a space L_1 whose elements are the classes of *integrable functions* (Sec. 4 of Chapter 5). An *integrable function* was shown to be the a.e. limit of a Cauchy sequence of step functions.

After defining *measurable functions* as the a.e. limits of sequences of step functions (not necessarily Cauchy), we define a *measurable set* S as one whose characteristic function χ_S is a measurable function. S is called *finitely measurable* if χ_S is integrable, and the measure of S, denoted by $m(S)$, is given by $\int \chi_S$. Otherwise, $m(S) = \infty$.

Method II. Again we start with the normed linear space Σ of step functions. However, instead of completing the space as in Method I, we first extend Σ to a space Γ which contains the monotone limits of sequences of step functions whose integrals are uniformly bounded (Sec. 1 of Chapter 6). The space L_1 is seen to be the set of differences of functions in Γ (Theorem 6.1.4). Measurable functions and sets are defined as in method I.

Remark: Variations of Methods I and II are obtained if Σ is replaced by any dense subspace of L_1, in particular, the piecewise linear functions, the continuous functions that vanish outside a bounded inverval, the Riemann integrable functions, etc. (See Exercises 5.1.9 and 5.2.9.)

Method III. Here the order of definition is reversed. We began by defining the *measure of a bounded interval* to be its length, and then define the *exterior measure of a bounded set* S to be the greatest lower bound of the lengths of systems of open intervals whose union covers S. A *bounded measurable set* S was one that satisfied: $b - a = m(I) = \mu_e(S) + \mu_e(I - S)$; I is some interval which contains the bounded set S, and μ_e denotes the exterior measure. The measure of S is given by $m(S) = \mu_e(S)$. A function f defined on a measurable set is said to be *measurable* if the sets $[f < A]$ are measurable for all real A. The integral is defined first for nonnegative, bounded, measurable functions by

$$\hat{\int_S} f = \lim_{|\mathcal{S}| \to 0} \sum_{i=1}^{n} \lambda_i \, m[y_{i-1} \leq f < y_i].$$

(See Definition 8.2.3.) Finally these definitions are extended to unbounded functions and sets, and it is shown that the classes of measurable and integrable functions, and measurable sets obtained in this manner are the same as those of Methods I and II.

CHAPTER **9**

Differentiation

In the elementary calculus, integration and differentiation are generally introduced at the outset as independent operations. The integral appears first in the context of solving an area problem, and the derivative of a function $y = f(x)$ is presented first as the slope of the tangent line to the curve $y = f(x)$ at the point $(x,f(x))$. Eventually, it is shown that these two operations (up to additive constants) are "inverses" of each other. The importance of this result is reflected in the rather imposing title that is given to it: *the Fundamental Theorem of Calculus.*

In this chapter we shall characterize those functions which have derivatives almost everywhere (a.e.) and prove the corresponding "Fundamental Theorem" which relates differentiation and Lebesgue integration. We shall have no difficulty proving that if $f(x)$ is integrable on $[a,b]$, then the function,

[1] $$F(x) = \int_a^x f(t)\,dt, \qquad a \leq x \leq b,$$

is a.e. differentiable, and $F'(x) = f(x)$ a.e.. (The integrals that appear in this chapter, unless otherwise stated, are Lebesgue integrals. We have adopted the Riemann integral notation in [1] for notational convenience.) It is, however, *false* that if $F(x)$ is a.e. differentiable, then $F(x) = \int_a^b F'(t)\,dt + c$, where c is a constant (Example 9.4.2). For this to be the case, additional conditions must be imposed on $F(x)$ (Theorem 9.4.12).

In Sec. 1, the four *derivates* of a function at a point are defined. These are the "right" and left" limits superior and inferior of the difference quotients. The differentiability of a function at a point is equivalent to the existence and equality of these limits.

In order to prove the principal theorem of Sec. 3, which asserts that monotone functions are a.e. differentiable, *Vitali coverings* are introduced in Sec. 2. Although it is possible to obtain the results of Sec. 3 without the Vitali Covering Theorem, the alternative methods seem equally lengthy. Moreover, the method used here is readily adaptable to the more general theory of the differentiation of additive set functions (measures). Although we shall touch upon this again in the final Chapters of the text, the reader is urged to consult other texts for a more complete discussion of the relation between differentiation and integration in R^n as well as in abstract spaces. (See Munroe [1], chap. vii; Saks [1], pp 114–116; Natanson [1]; Shilov and Gurevich [1].)

1 The Derivates of a Function

Every first year calculus student is familiar with examples of continuous functions that do not possess derivatives at all points of continuity. Two such functions are

[1] $$y = |x|,$$

and

[2] $$y = \begin{cases} 0 & \text{if } x = 0 \\ x \sin \dfrac{1}{x} & \text{if } x \neq 0. \end{cases}$$

The curve described by [1] has a "spike" at the origin, where the right and left difference quotients have unequal limits. The second has the property that in every neighborhood of the origin, there are points at which the slope of the tangent is $+1$, and points at which it is -1, proving that the derivative cannot be continuous at the origin. To show that it does not even exist there, we form the difference quotient

$$\frac{x \sin \dfrac{1}{x} - 0}{x - 0} = \sin \frac{1}{x},$$

which has no limit at the origin.

One can easily imagine continuous functions whose derivatives fail to exist at finitely many points. Both [1] and [2] may be used to construct such functions. The picture clouds when we attempt to imagine continuous functions whose derivatives do not exist at points of a dense subset of an interval, or do not exist anywhere. Earlier, we proved that such functions do indeed exist, and in fact, they are dense in the space of continuous functions on an interval $[a,b]$ (Corollary 4.3.28). Examples of functions which are continuous on an interval but fail to have derivatives at points of a dense subset of that interval, or at any point of the interval appear in Exercises 4.3.17, 4.3.18, and 4.3.19. It is strongly recommended that the reader become acquainted with these functions if he has not already done so. He will observe that these functions are everywhere oscillating (nowhere monotone). This will perhaps make the assertion of the a.e. existence of derivatives for monotone functions seem more natural (Theorem 9.3.6).

The derivative of a function $f(x)$ at a point θ is defined by the limit

$$f'(\theta) = \lim_{x \to \theta} \frac{f(x) - f(\theta)}{x - \theta}.$$

It is assumed of course that $f(x)$ is defined in a neighborhood $(\theta - \Delta, \theta + \Delta)$ of the point θ. There are several ways that the difference quotient may fail to have a limit: The left and right hand limits could exist, but be different; one or both of these limits could be $\pm \infty$; one or both could oscillate as x tends to θ, etc. To deal with these cases, we introduce the limits superior and inferior of a function at a point. These definitions are very similar to the ones given for sequences (Definition 1.3.28).

Let $g(x)$ be defined in the open interval $(\theta, \theta + \Delta)$. Allowing $+\infty$ as a value, set

$$\gamma^{\delta} = \sup_{0 < x - \theta < \delta} g(x),$$

where $\delta < \Delta$. As δ tends to zero, γ^{δ} is nonincreasing. If it is bounded from below, then we write $\lim_{\delta \downarrow 0} \gamma^{\delta} = \gamma^{+}$. Otherwise, $\lim_{\delta \downarrow 0} \gamma^{\delta} = -\infty$.

1 Definition. Let $g(x)$ and γ^{δ} be defined as in the preceding paragraph. The number γ^{+}, which may be $\pm\infty$, is called the *right limit superior* of $g(x)$ at θ, and is denoted by

$$\overline{\lim_{x \to +\theta}}\, g(x) = \gamma^{+}.$$

Similarly, the *left limit superior* is given by

$$\overline{\lim_{x \to -\theta}}\, g(x) = \gamma^{-} = \lim_{\delta \downarrow 0} \sup_{0 < \theta - x < \delta} g(x).$$

The corresponding *right* and *left limits inferior* are defined to be

$$\underline{\lim_{x \to +\theta}}\, g(x) = \gamma_{+} = \lim_{\delta \downarrow 0} \inf_{0 < x - \theta < \delta} g(x)$$

and

$$\underline{\lim_{x \to -\theta}}\, g(x) = \gamma_{-} = \lim_{\delta \downarrow 0} \inf_{0 < \theta - x < \delta} g(x)$$

respectively.

2 Example. (a) If

$$g(x) = \begin{cases} 18 & \text{if } x = 0 \\ \sin \dfrac{1}{x} & \text{if } x \neq 0, \end{cases}$$

then

$$\gamma^+ = \gamma^- = 1, \text{ and } \gamma_+ = \gamma_- = -1.$$

(b) If

$$g(x) = \begin{cases} 1 & \text{if} \quad x > 0 \\ -3 & \text{if} \quad x = 0 \\ 5 & \text{if} \quad x < 0, \end{cases}$$

then

$$\gamma^+ = \gamma_+ = 1 \qquad \text{and} \qquad \gamma^- = \gamma_- = 5.$$

Later we shall give an example of a function for which γ^+, γ_+, γ^-, and γ_- are all unequal.

We state next some properties of the limits superior and inferior. Since their verification varies only slightly from the corresponding properties for sequences, the proofs are left to the reader in the exercises.

3 Proposition. Let $g(x)$ be defined in a "deleted" neighborhood of θ: $(\theta - \Delta, \theta) \cup (\theta, \theta + \Delta)$. Then

(i) $\gamma_- \leq \gamma^-$ and $\gamma_+ \leq \gamma^+$;

(ii) $\lim_{x \to \theta} g(x)$ exists if and only if $\gamma_+ = \gamma^+ = \gamma_- = \gamma^-$ and all these quantities are finite. If they are equal to $+\infty$, then $g(x) \uparrow +\infty$ as x tends to θ, and if they are equal to $-\infty$, then $g(x) \downarrow -\infty$ as x tends to θ.

(iii) If γ^+ is finite, then $\overline{\lim}_{x \to +\theta} g(x) = \gamma^+$ if and only if

(a) to each $\epsilon > 0$, there is a $\delta > 0$, such that

$$0 < x - \theta < \delta \qquad \text{implies} \qquad g(x) < \gamma^+ + \epsilon, \text{ and}$$

(b) to each pair of positive numbers, ϵ and δ, there is a point x in the interval $(\theta, \theta + \delta)$ satisfying, $g(x) > \gamma^+ - \epsilon$. (Similar statements may be made about the remaining limits.)

(iv) If $\gamma^+ = \infty$, then for all positive M and δ, there are points x in $(\theta, \theta + \delta)$ for which $g(x) > M$.

(v) If $\gamma^+ = -\infty$, then for each positive M, there is number $\delta > 0$ such that

$$0 < x - \theta < \delta \qquad \text{implies} \qquad g(x) < -M.$$

(Again, similar statements may be made for γ_+, γ^-, and γ_-.)

Let $f(x)$ be defined in a neighborhood $(\theta - \delta, \theta + \delta)$ of the point θ, and let h be a real number satisfying $0 < |h| < \delta$. Then the difference quotient

$$\Delta_\theta(h) = \frac{f(\theta + h) - f(\theta)}{h}$$

is a function of h in the deleted neighborhood, $(-\delta,0) \cup (0,\delta)$.

4 Definition. The set of numbers (including possibly $+\infty$ and/or $-\infty$)

$$D^+(f(\theta)) = \overline{\lim_{h\to+0}}\, \Delta_\theta(h), \qquad D_+(f(\theta)) = \underline{\lim_{h\to+0}}\, \Delta_\theta(h)$$

$$D^-(f(\theta)) = \overline{\lim_{h\to-0}}\, \Delta_\theta(h), \qquad D_-(f(\theta)) = \underline{\lim_{h\to-0}}\, \Delta_\theta(h)$$

are called the *derivates* or *Darboux derivatives* of $f(x)$ at θ.

For convenience, these numbers will often be referred to as the right upper and lower derivates, and the left upper and lower derivates. If the function $f(x)$ and the point θ are fixed throughout a discussion, we shall write simply D^+, D_+, D^- and D_-.

5 Definition. If θ is a point at which the four derivates of a function $f(x)$ are equal and finite, then their common value

$$\lim_{h\to 0} \Delta_\theta(h)$$

is called the *derivative* of the function $f(x)$ at θ, and is denoted by $f'(\theta)$.

6 Example. (a) At $\theta = 0$, the derivates of $f(x) = |x|$ are,

$$D^+ = D_+ = 1 \qquad \text{and} \qquad D^- = D_- = -1.$$

(b) At $\theta = 0$, the four derivates of the function

$$f(x) = \begin{cases} 0 & \text{if } x = 0 \\ x \sin \dfrac{1}{x} & \text{if } x \neq 0 \end{cases}$$

are $D_+ = D_- = -1$ and $D^+ = D^- = 1$.

(c) Let a, b, c and d be real numbers, $a \le b$, $c \le d$. Then the function defined below has derivates equal to

$$D_+ = a \le b = D^+ \qquad \text{and} \qquad D_- = c \le d = D^-:$$

$$f(x) = \begin{cases} ax\left(\sin \dfrac{1}{x}\right)^2 + bx\left(\cos \dfrac{1}{x}\right)^2 & \text{if } x > 0 \\ 0 & \text{if } x = 0 \\ cx\left(\sin \dfrac{1}{x}\right)^2 + dx\left(\cos \dfrac{1}{x}\right)^2 & \text{if } x < 0. \end{cases}$$

To prove this, take $\theta = 0, h > 0$. Then

$$\Delta_0(h) = \frac{h[a(\sin 1/h)^2 + b(\cos 1/h)^2]}{h} = a + (b - a)\left(\cos \frac{1}{x}\right)^2.$$

It is readily seen that $a \le \Delta_0(h) \le b$, and that

$$h = 1/\pi, 1/2\pi, \ldots, 1/n\pi, \ldots, \text{implies } \Delta_0(h) = b,$$

and

$$h = 2/\pi; 2/3\pi, \ldots, 2/(2n - 1)\pi, \ldots \text{implies } \Delta_0(h) = a,$$

from which it follows that $D_+ = a$ and $D^+ = b$. A similar analysis may be used for the remaining two cases.

See the Exercises for additional examples.

Exercises

1. Do Exercises 4.3.17, 18 and 19 if you have not already done so.
2. Show that if $f(x)$ is continuous on $[a,b]$ and if any one of its derivates, say D^+, is nonnegative on $[a,b]$, then $f(a) \le f(b)$. [*Hint:* Suppose that $f(b) - f(a) = -2\delta < 0$. Form $H(x) = f(x) - f(a) + \delta$. Use the Intermediate Value Theorem to obtain $H(y) = 0$ for some point y in $[a,b]$. Let c be the largest value of y for which $H(y) = 0$. Show that $D^+(f(c)) < 0$.]
3. Prove that if $f(x)$ has a relative maximum at θ, then $D^+(f(\theta)) \le 0$ and $D_-(f(\theta)) \ge 0$. What can be said about the remaining derivates?
4. Using [1], give an example of a function which is continuous on an interval $[a,b]$ but fails to have derivatives at interior points, $\theta_1 < \theta_2 < \cdots < \theta_n$.
5. Let θ_i, $i = 1,2,\ldots,n$ be points in an interval $[a,b]$. Find a function $f(x)$ which is continuous on $[a,b]$ but whose derivatives do not exist at the points θ_i, and at these points the right upper and lower derivates and the left upper and lower derivates are equal to $a_i \le b_i$ and $c_i \le d_i$, $i = 1,2,\ldots,n$.
6. Prove that the following function has a derivative at $\theta = 0$, but that the derivative at that point is not continuous.

$$f(x) = \begin{cases} x^2 \sin \dfrac{1}{x} & \text{if } x \neq 0 \\ 0 & \text{if } x = 0. \end{cases}$$

*7. Prove Proposition 3.

8. Find the derivates of the following function at $\theta = 0$:

$$f(x) = \begin{cases} x^\alpha \sin x^{-\beta} & \text{if } x \neq 0 \\ 0 & \text{if } x = 0 \end{cases}$$

α and β are positive numbers.

*Exercises which request the proof of a statement made in the text, as well as those referred to in the text, are starred.

2 Vitali Coverings

The Vitali covering theorem and its corollary, which are proved in this section, are used in Sec. 3 to prove that monotone functions are a.e. differentiable.

1 Definition. Let S be a subset of the real numbers. A family $\mathfrak{F}$ of closed intervals is said to be a *Vitali Covering of S* if to every $x \in S$ and $\epsilon > 0$, there is an interval I in $\mathfrak{F}$, such that

(i) $x \in I$, and (ii) $m(I) < \epsilon$.

2 Example. Let $S = [a,b]$ and let $\{x_n\}$ be some enumeration of the rational numbers in S. Then the collection of closed intervals,

$$I_{n,j} = \left[x_n - \frac{1}{j}, x_n + \frac{1}{j}\right], \qquad n,j = 1,2,\ldots,$$

is a Vitali covering of S.

3 Theorem (Vitali). If $\mathfrak{F}$ is a Vitali covering of a bounded set S of real numbers, then there is a sequence of disjoint intervals $\{I_n\} \subset \mathfrak{F}$ such that $S - \bigcup I_n$ is a null set.

Proof: Since S is bounded, it is contained in an interval $[a,b]$. Replacing each interval I of $\mathfrak{F}$ by $I' = I \cap [a,b]$, we obtain a family of intervals $\mathfrak{F}' = \{I'\} = \{I \cap [a,b], I \in \mathfrak{F}\}$. It is easily seen that $\mathfrak{F}'$ is also a Vitali Covering of S, and it follows readily that if the theorem can be proved for $\mathfrak{F}'$, then it is true also for $\mathfrak{F}$. Therefore we shall simply assume that the intervals of $\mathfrak{F}$ are contained in $[a,b]$.

We begin by choosing an interval of $\mathfrak{F}$ that contains a point of S, calling it I_1. If I_1 contains S, then we are through. If not, choose a second interval I_2 as follows.

Consider all I in $\mathfrak{F}$ that satisfy

[1] $$I \cap S \neq \emptyset \qquad \text{and} \qquad I \cap I_1 = \emptyset.$$

Let us prove first that there are intervals I satisfying [1]. Suppose that $x \in S - I_1$. Then since I_1 is closed, $d(x,I_1) = \delta > 0$. Moreover, $\mathfrak{F}$ is a Vitali covering, and therefore there is an interval I in $\mathfrak{F}$ that contains x and whose length does not exceed $\delta/2$. Thus [1] is satisfied for this interval. Let

$$d_1 = \sup\,[m(I)],$$

where the supremum is taken over all intervals I for which [1] holds. The interval I_2 is chosen from this class of intervals by requiring that $m(I_2) > d_1/2$. If $I_1 \cup I_2$ fails to cover S, we repeat the process. In this way, a sequence (perhaps finite) of disjoint, closed intervals $\{I_n\}$ is chosen according to the following prescription: If $I_1 \cup I_2 \cup \cdots \cup I_n$ does not cover S, let $d_n = \sup [m(I)]$; the supremum is taken over the intervals I in $\mathfrak{F}$ that satisfy

[2] $$I \cap S \neq \emptyset \quad \text{and} \quad I \cap \left(\bigcup_{j=1}^{n} I_j \right) = \emptyset.$$

I_{n+1} is chosen from the collection of intervals in $\mathfrak{F}$ that satisfy [2] and the additional requirement that $m(I_{n+1}) > d_n/2$. Naturally, if for some value of n, $\bigcup_{j=1}^{n} I_j \supset S$, then we are finished. Let us assume that this never happens. Then we must show that

[3] $$E = S - \bigcup_{n=1}^{\infty} I_n$$

is a null set. This is proved by deriving a contradiction from the assumption that the measure of E is positive.

To each $I_n = [a_n, b_n]$, we associate the interval $J_n = [\alpha_n, \beta_n]$ whose midpoint coincides with that of I_n, and whose length,

[4] $$m(J_n) = 5\, m(I_n).$$

Since the I_n are disjoint and contained in the interval $[a,b]$, it follows that

[5] $$\sum_{n=1}^{\infty} m(J_n) = 5 \sum_{n=1}^{\infty} m(I_n) \leq 5(b-a).$$

Let an integer N be chosen so that,

[6] $$\sum_{n=N+1}^{\infty} m(J_n) < m(E).$$

The monotonicity of measure and [6] imply that it is false that $E \subset \bigcup_{n=N+1}^{\infty} J_n$, from which it follows that $E - \bigcup_{n=N+1}^{\infty} J_n$ contains at least one point, say x. [3] yields $x \notin \bigcup_{n=1}^{\infty} I_n$. However, since $\mathfrak{F}$ is a Vitali cover, it contains an interval I for which,

$$x \in I \quad \text{and} \quad I \cap \left(\bigcup_{n=1}^{N} I_n \right) = \emptyset.$$

The definition of the d_n yields

$$m(I) \leq d_N.$$

On the other hand, it is easily seen that $d_n \downarrow 0$. For if this were not the case, it would follow that

$$\sum_{n=2}^{\infty} m(I_n) > \frac{1}{2}\sum_{n=1}^{\infty} d_{n-1} = \infty,$$

which contradicts [5]. Thus for sufficiently large values of n, $d_n < m(I)$, from which it follows that $I \cap I_n \neq \emptyset$. Let p be the smallest integer for which $I \cap I_p \neq \emptyset$. Recalling that I was chosen so that $I \cap \left(\bigcup_{n=1}^{N} I_n\right) = \emptyset$, it follows that $p - 1 \geq N$. Moreover, $I \cap \left(\bigcup_{n=1}^{p-1} I_n\right) = \emptyset$ implies that

$$m(I) \leq d_{p-1} < 2m(I_p). \tag{7}$$

The contradiction is obtained by showing that

$$m(I) > 2m(I_p).$$

Since $x \notin \bigcup_{n=N+1}^{\infty} J_n$, and $p \geq N + 1$, it follows that $x \notin J_p$ (see Figure 28). Thus $|x - a_p| > 2m(I_p) = 2(b_p - a_p)$. (Remember, $m(J_p) = \beta_p - \alpha_p = 5(b_p - a_p)$.) But $x \in I$, and p was chosen so that $I \cap I_p \neq \emptyset$.

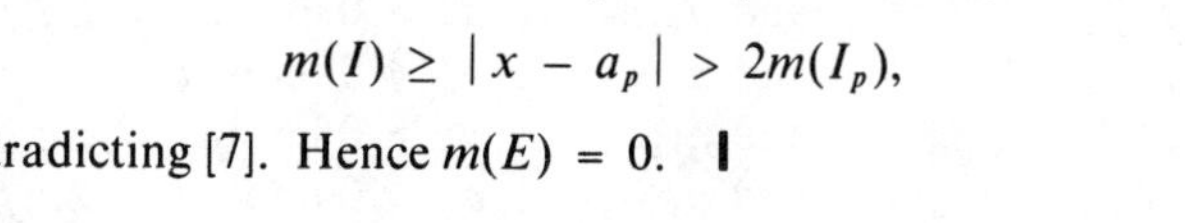

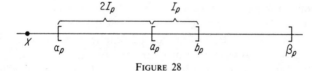

FIGURE 28

Therefore I contains an endpoint of I_p, say a_p, which implies that

$$m(I) \geq |x - a_p| > 2m(I_p),$$

thus contradicting [7]. Hence $m(E) = 0$. ∎

The following corollary of the Vitali Covering Theorem is used directly in the proof of the main theorem of Sec. 3:

4 Corollary (Vitali). If $\mathfrak{F}$ is a Vitali covering of a bounded subset S, then to every $\epsilon > 0$, there is a finite collection of disjoint intervals, $I_1, I_2, \ldots, I_N$ in $\mathfrak{F}$ satisfying

$$m\left(S - \bigcup_{n=1}^{N} I_n\right) < \epsilon.$$

(We shall say that "almost all" of S is covered by the I_n.)

Proof: Let $\{I_n\}$ be a disjoint sequence of intervals of $\mathfrak{F}$ that satisfy

$$m\left(S - \bigcup_{n=1}^{\infty} I_n\right) = 0$$

(Theorem 3). Choose an integer N so that $\sum_{n=N+1}^{\infty} m(I_n) < \epsilon$. Then

$$S - \bigcup_{n=1}^{N} I_n = \left[S \cap \left(\bigcup_{n=N+1}^{\infty} I_n\right)\right] \cup \left[S - \bigcup_{n=1}^{\infty} I_n\right],$$

from which we obtain

$$m\left(S - \bigcup_{n=1}^{N} I_n\right) = m\left(S \cap \left(\bigcup_{n=N+1}^{\infty} I_n\right)\right) + 0$$

$$\leq m\left(\bigcup_{n=N+1}^{\infty} I_n\right) = \sum_{n=N+1}^{\infty} m(I_n) < \epsilon. \quad \blacksquare$$

3 Monotone Functions and Functions of Bounded Variation

The reader will recall that a function defined on a set S is said to be monotone on S, if it is either nondecreasing or nonincreasing on S. These are denoted by $f(x) \uparrow$ on S and $f(x) \downarrow$ on S respectively.

The main theorem of this section asserts that monotone functions are a.e. differentiable. Since differentiation is a linear operation, it will follow that any linear combination of monotone functions is a.e. differentiable. Of course, a linear combination of monotone functions need not itself be monotone. Indeed, only combinations of the following type will always be monotone:

(i) $f_1, f_2, \ldots, f_n \uparrow$ and $a_1, a_2, \ldots, a_n \geq 0$ imply $f = \Sigma a_i f_i \uparrow$.
(ii) $f_1, f_2, \ldots, f_n \uparrow$ and $a_1, a_2, \ldots, a_n \leq 0$ imply $f = \Sigma a_i f_i \downarrow$.
(iii) $f_1, f_2, \ldots, f_n \downarrow$ and $a_1, a_2, \ldots, a_n \geq 0$ imply $f = \Sigma a_i f_i \downarrow$.
(iv) $f_1, f_2, \ldots, f_n \downarrow$ and $a_1, a_2, \ldots, a_n \leq 0$ imply $f = \Sigma a_i f_i \uparrow$.

The verification of (i)–(iv) is left to the reader in the exercises.

More generally, if $\Sigma a_i f_i$ is any finite linear combination of monotone functions, then it is easily seen that it can be expressed as a difference $F - G$ of two nondecreasing functions. This is done by letting F be the sum of those $a_i f_i$ which are nondecreasing, that is, which satisfy one of the following sets of conditions:

(a) $a_i \geq 0$ and $f_i \uparrow$, or (b) $a_i \leq 0$ and $f_i \downarrow$. (See (i) and (iv) above.) The remaining terms of the sum are nonincreasing, and therefore their negative, G is also nondecreasing.

This leads us quite naturally to the class of functions that can be expressed as the difference of two nondecreasing functions. These functions play a major role in the theory of differentiation. An alternate description of this class, which appears in the proof of the main theorem as well as in subsequent discussions (Stieltjes integration, Sec. 4 of Chapter 11) is given below:

1 Definition. Let $f(x)$ be defined on an interval $I = [a,b]$. For each subdivision $\mathcal{S}$ of I, $a = a_0 < a_1 < \cdots < a_n = b$, the sum

[1] $$V(f,\mathcal{S}) = \sum_{j=1}^{n} |f(a_j) - f(a_{j-1})|,$$

is called the *variation of f over the set of intervals* $\{(a_{j-1},a_j)\}$, $j = 1,2,\ldots,n$.

If the set of numbers $\{V(f,\mathcal{S})\}$ taken over all possible subdivisions of I is bounded, then f is said to be of *bounded variation over I*, and the number

$$V(f) = \sup_{\mathcal{S}} [V(f,\mathcal{S})]$$

is called the *total variation of f over I*.

2 Theorem. f is a function of bounded variation over an interval $I = [a,b]$ if and only if there are nondecreasing functions g and h such that $f = g - h$.

Proof: Let us assume first that $g,h \uparrow$ on I and that $f = g - h$. We must show that the set of numbers $V(f,\mathcal{S})$ is bounded from above. A typical term of [1] satisfies the inequality

$$\begin{aligned} |f(a_j) - f(a_{j-1})| &= |[g(a_j) - g(a_{j-1})] - [h(a_j) - h(a_{j-1})]| \\ &\le [g(a_j) - g(a_{j-1})] + [h(a_j) - h(a_{j-1})]. \end{aligned}$$

This follows from the monotonicity of the functions g and h and the triangle inequality. Summing, we obtain

$$\begin{aligned} V(f,\mathcal{S}) &= \sum_{j=1}^{n} |f(a_j) - f(a_{j-1})| \\ &\le \sum_{j=1}^{n} [g(a_j) - g(a_{j-1})] + \sum_{j=1}^{n} [h(a_j) - h(a_{j-1})] \\ &= [g(b) - g(a)] + [h(b) - h(a)]; \end{aligned}$$

the last term is a bound for $V(f,\mathcal{S})$ which is independent of the subdivision.

To prove the converse, let us assume that the set of nonnegative variations $V(f,\mathcal{S})$ is bounded from above. Associated with each subdivision are the two sums

$$P(f,\mathcal{S}) = \sum_{j=1}^{n} [f(a_j) - f(a_{j-1})]^+$$

and

[2] $$N(f,\mathcal{S}) = \sum_{j=1}^{n} [f(a_j) - f(a_{j-1})]^-$$

of the positive and negative parts of the differences $[f(a_j) - f(a_{j-1})]$. It follows easily that

[3] $$P(f,\mathcal{S}) + N(f,\mathcal{S}) = V(f,\mathcal{S})$$

and

[4] $$P(f,\mathcal{S}) - N(f,\mathcal{S}) = \sum_{j=1}^{n} [f(a_j) - f(a_{j-1})] = f(b) - f(a).$$

First adding, and then subtracting [3] and [4], we obtain

[5] $$V(f,\mathcal{S}) = 2P(f,\mathcal{S}) + f(a) - f(b)$$

[6] $$V(f,\mathcal{S}) = 2N(f,\mathcal{S}) + f(b) - f(a).$$

The boundedness of the set of variations implies that the "positive" and "negative" parts of the variation, $P(f,\mathcal{S})$ and $N(f,\mathcal{S})$, are also bounded. Letting $V(f)$, $P(f)$ and $N(f)$ denote, respectively, the supremum of the sets $\{V(f,\mathcal{S})\}$, $\{P(f,\mathcal{S})\}$ and $\{N(f,\mathcal{S})\}$, taken over all subdivisions, it is easily verified that

[7] $$V(f) = P(f) + N(f).$$

The proof of this is left as an exercise. The quantities $V(f)$, $P(f)$ and $N(f)$ are called the *total*, the *positive*, and the *negative variations* of f over $I = [a,b]$.

Everything in the preceding paragraph remains true if the interval $[a,b]$ is replaced by $I_x = [a,x] \subset [a,b]$. Letting the subscript x denote the dependency of the quantities on the variable x, [5], [6], and [7] become, upon substitution of x for b, and taking the supremum in [5] and [6],

[5′] $$V_x(f) = 2P_x(f) + f(a) - f(x),$$

[6′] $$V_x(f) = 2N_x(f) + f(x) - f(a),$$

[7′] $$V_x(f) = P_x(f) + N_x(f).$$

Eliminating $V_x(f)$ from [5′] and [6′], we obtain

[8] $$f(x) = (f(a) + P_x(f)) - N_x(f) = g(x) - h(x).$$

It is readily seen that this is the desired decomposition of f, for both $P_x(f)$ and $N_x(f)$ are nondecreasing functions of x. ▎

Remark: If $f(x)$ is continuous on the interval $[a,b]$, then the *length of the curve* $y = f(x)$, $a \le x \le b$, is defined to be

$$L(f) = \sup [L(f,\mathcal{S})]$$

where

$$L(f,\mathcal{S}) = \sum_{j=1}^{n} \sqrt{(\Delta x_j)^2 + (\Delta y_j)^2}$$

is the length of the polygonal curve obtained by joining (x_{j-1}, y_{j-1}) to $(x_j, y_j), j = 1, 2, \ldots, n$. It is left to the reader to show that $L(F)$ is finite if and only if f is of bounded variation.

If $f(x)$ is defined in a neighborhood of a point c, then the four limits,

$$\gamma^+ = \overline{\lim_{x \to +c}} f(x), \; \gamma_+ = \underline{\lim_{x \to +c}} f(x), \; \gamma^- = \overline{\lim_{x \to -c}} f(x), \; \gamma_- = \underline{\lim_{x \to -c}} f(x)$$

may be distinct, and some or all may take on values $+\infty$ or $-\infty$. However, if f is a monotone function in a neighborhood $(c - \delta, c + \delta)$, then it is easily seen that

(i) $\gamma_+, \gamma^+, \gamma_-$, and γ^- are finite, and

(ii) $\gamma_+ = \gamma^+$ and $\gamma_- = \gamma^-$.

Indeed, the monotonicity of f implies, if $f \uparrow$, that

$$f(c - \delta/2) \le \gamma_- \le \gamma^- \le f(c) \le \gamma_+ \le \gamma^+ \le f(c + \delta/2),$$

from which (i) follows. This inequality also implies (ii). The details are left to the reader, as well as the case $f \downarrow$. Thus, for monotone functions, right and left limits exist, and we write

$$f(c + 0) = \lim_{x \to +c} f(x) \qquad \text{and} \qquad f(c - 0) = \lim_{x \to -c} f(x).$$

Also, the monotonicity implies that $f(c - 0) \le f(c) \le f(c + 0)$. If f is continuous at c, then $f(c + 0) = f(c - 0)$. Otherwise, the difference $f(c + 0) - f(c - 0)$ measures the magnitude of the "jump discontinuity" at $x = c$. In this manner, there is associated with each nondecreasing function $f(x)$ defined on $[a,b]$, the function

$$d(x) = \begin{cases} f(a + 0) - f(a) & \text{if } x = a \\ f(x + 0) - f(x - 0) & \text{if } a < x < b \\ f(b) - f(b - 0) & \text{if } x = b. \end{cases}$$

From the preceding remarks, $f(x)$ is continuous at $x = c$ if and only if $d(c) = 0$.

3 Theorem. If $f(x)$ is monotone on an interval $[a,b]$, then the set of discontinuities is countable.

Proof: Assume that $f(x) \uparrow$. The set of discontinuities of $f(x)$ is the subset of $[a,b]$ on which $d(x) > 0$. Let

$$S_1 = \{x : d(x) > 1\}.$$

The set S_1 contains only finitely many points, for otherwise, there would be a sequence x_n of points in $[a,b]$ at which $d(x_n) > 1$. This implies that

$$n < d(x_1) + d(x_2) + \cdots + d(x_n) \leq f(b) - f(a),$$

which is a contradiction for sufficiently large n. Similarly, if m is any positive integer,

$$S_m = \{x : d(x) > 1/m\}$$

is finite, from which it follows that the countable union, $\bigcup_{m=1}^{\infty} S_m = S$ is a countable subset of $[a,b]$. However, S is precisely the set of discontinuities of $f(x)$, since it contains the points at which $d(x) > 0$. ▎

Let $\{x_n\}$ be the countable set of discontinuities of the monotone function $f(x)$, defined on the interval $[a,b]$. Then

$$\sum_{n=1}^{\infty} d(x_n) \leq f(b) - f(a).$$

This sum measures that part of the increase of $f(x)$ over $[a,b]$ which occurs at the jumps. If $f(x)$ is everywhere continuous, then the sum vanishes. If b is replaced by a point x in $[a,b]$, then the sum of the jumps of $f(x)$ over $[a,x]$ is given by

[9] $$J(x) = \sum_{x_n \leq x} d(x_n).$$

The function $J(x)$ is called the *jump function of* $f(x)$.

4 Theorem. If $f(x) \uparrow$ on $[a,b]$ and $f(x) = f(x + 0)$ at each point of discontinuity, then the function $f(x) - J(x)$, where $J(x)$ is the jump function of $f(x)$, is a monotone continuous function.

Proof: We prove first the continuity of the function $g(x) = f(x) - J(x)$. Let c be a point of continuity of $f(x)$. Then $d(c) = 0$ and $\lim_{\delta \to 0} J(c + \delta) = J(c)$. (See [9].) This yields,

$$[f(c + \delta) - J(c + \delta)] - [f(c) - J(c)] = [f(c + \delta) - f(c)] + [J(c) - J(c + \delta)] \to 0 \quad \text{as} \quad \delta \to 0,$$

proving that the function $g(x)$ is continuous wherever $f(x)$ is continuous. Now let c be a point of discontinuity of $f(x)$. Then

$$d(c) = f(c + 0) - f(c - 0) = f(c) - f(c - 0) > 0.$$

We must show that both the right- and left-hand limits of $g(x)$ tend to the same value as x tends to c. First, if $\delta > 0$,

$$\lim_{\delta \to +0} [f(c + \delta) - f(c)] = f(c + 0) - f(c) = 0,$$

and

[10] $$\lim_{\delta \to +0} [J(c + \delta) - J(c)] = \lim_{\delta \to +0} \left[\sum_{c < x_n \le c + \delta} J(x_n) \right] = 0.$$

For negative values of δ,

[11] $$\lim_{\delta \to -0} [f(c + \delta) - f(c)] = -d(c)$$

and

$$\lim_{\delta \to -0} [J(c + \delta) - J(c)] = -\lim_{\delta \to -0} \left[\sum_{c + \delta \le x_n < c} d(x_n) \right] = -J(c).$$

It follows from [10] and [11] that for all points,

$$\lim_{\delta \to 0} [g(c + \delta) - g(c)] = 0,$$

proving that $g(x)$ is continuous at the points of discontinuity of $f(x)$ as well as at the points of continuity.

We prove next that $g(x)$ is monotone. Suppose that $a \le x < y \le b$. Then $g(x) \le g(y)$ if and only if $f(x) - J(x) \le f(y) - J(y)$. The latter inequality is equivalent to $J(y) - J(x) \le f(y) - f(x)$. However this inequality follows immediately from the fact that the increase in the function f over the interval $[x,y]$ is not exceeded by that part of the increase that is due to the jumps. Thus $g(x) = f(x) - J(x)$ is a nondecreasing function on $[a,b]$. ▌

5 Theorem. The derivates of a monotone function defined on an interval I (bounded or unbounded) are measurable functions.

Proof: Since the function $f(x)$ will remain constant throughout the proof, we shall denote the derivates of f at x by, $D^+(x)$, etc. The proof is given for $D^+(x)$ only. The remaining cases are proved in a similar manner.

Let $x \in I$; x is not an end point of I. For all $\delta > 0$ for which $[x, x + \delta] \subset I$, the function

[12] $$\lambda(x,\delta) = \sup_{0 < h < \delta} \frac{f(x + h) - f(x)}{h}$$

is well defined, if we allow $+\infty$ as a value, and

[13] $$D^+(x) = \lim_{\delta \to 0} \lambda(x,\delta).$$

The idea of the proof is to show that for all values of δ, $\lambda(x,\delta)$ is a measurable function. Then the sequence of measurable functions $\{\lambda(x, 1/n)\}$ converges to the measurable function $D^+(x)$. (The space of measurable functions is closed under a.e. convergence.) We begin by comparing $\lambda(x,\delta)$ with the function

$$\hat{\lambda}(x,\delta) = \sup_{0<h_i<\delta} \left[\frac{f(x+h_i) - f(x)}{h_i}\right],$$

where the h_i are the rational points of $(0,\delta)$. Theorem 7.1.7 and Exercise 7.1.2 yield the measurability of $\hat{\lambda}(x,\delta)$. The inclusion $\{h_i\} \subset (0,\delta)$ implies that $\hat{\lambda}(x,\delta) \leq \lambda(x,\delta)$. To prove the equality of these two functions, we shall show that for any $\epsilon > 0$, there is a rational number $r \in (0,\delta)$ such that

$$\frac{f(x+r) - f(x)}{r} > \lambda(x,\delta) - \epsilon. \tag{14}$$

From [12], it follows that there is a point $h \in (0,\delta)$ for which

$$\frac{f(x+h) - f(x)}{h} > \lambda(x,\delta) - \epsilon/2.$$

If h is a rational number, then we are through. If not, let r be a rational number in the interval (h,δ). Since $f(x)$ is nondecreasing, it follows that

$$\frac{f(x+r) - f(x)}{h} \geq \frac{f(x+h) - f(x)}{h} > \lambda(x,\delta) - \epsilon/2. \tag{15}$$

Multiplying [15] by h/r, we obtain

$$\frac{f(x+r) - f(x)}{r} > \frac{h}{r}[\lambda(x,\delta) - \epsilon/2] = \lambda(x,\delta) + \left[\frac{h}{r} - 1\right]\lambda(x,\delta) - \frac{\epsilon h}{2r}.$$

By choosing r to be sufficiently close to h, we can make

$$[1 - h/r]\lambda(x,\delta) < \epsilon/2.$$

Taken together with $\epsilon h/2r < \epsilon/2$, this yields [14]. Thus $\lambda(x,\delta)$ is measurable, and it follows that

$$D^+(x) = \lim_{n\to\infty} \lambda(x, 1/n)$$

is also measurable. ∎

We are now in a position to prove the main theorem of this section, which asserts that a monotone function is a.e. differentiable. The proof rests upon several theorems that have been proved earlier, including the Vitali Covering Theorem (9.2.3) and Theorem 7.3.8 which states that a function g is measurable if and only if the sets $[g < A]$ are measurable, for all real values of A.

6 Theorem. If $f(x)\uparrow$ on an interval I, then $f'(x)$ exists a.e. on I, and is a measurable function.

Proof: The Vitali covering theorem requires the assumption that I is a bounded interval. The extension to unbounded intervals is easily obtained from this special case, and is therefore left to the reader (Exercise 3).

It has already been observed that if x is an interior point of I, then $D^+(x) \geq D_+(x)$ and $D^-(x) \geq D_-(x)$, and that $f'(x)$ exists if and only if the four derivates are equal and finite. We shall show that the set $T = [D^+ > D_+]$ is a null set. A similar argument may be given for the left hand derivates.

If $x \in T$, there are rational numbers α and β such that

$$x \in [D^+ > \beta > \alpha > D_+] = T_{\alpha\beta}.$$

Hence

$$T = [D^+ > D_+] = \bigcup_{\substack{\alpha,\beta \in Q \\ \alpha < \beta}} [D^+ > \beta > \alpha > D_+] = \bigcup_{\substack{\alpha,\beta \in Q \\ \alpha < \beta}} T_{\alpha\beta}$$

If T is not a null set, then for some pair $\alpha < \beta$, $m(T_{\alpha\beta}) = \mu_{\alpha\beta} > 0$. We shall show that this leads to a contradiction.

Let G be an open set which satisfies the following properties:

[16] $$G \supset T_{\alpha\beta}$$

and

[17] $$m(G) < \mu_{\alpha\beta} + \epsilon,$$

where ϵ is a preassigned positive number (Corollary 7.3.4). Proposition 7.3.1 asserts that the open set G may be written as a disjoint union of open intervals,

[18] $$G = \bigcup_{n=1}^{\infty} I_n.$$

From the definition of $D_+(x)$, it follows that if $x \in T_{\alpha\beta}$, there are arbitrarily small values of $h > 0$ such that,

[19] $$\frac{f(x+h) - f(x)}{h} < \alpha.$$

Let $\mathfrak{F}$ be the collection of closed intervals $\{[x,x+h]\}$ which satisfy the following properties:

(i) $x \in T_{\alpha\beta}$,

(ii) the inequality [19] holds for x and h, and

(iii) $[x,x+h]$ is contained in one of the I_n of [18].

It is readily seen that $\mathfrak{F}$ is a Vitali covering of the set $T_{\alpha\beta}$, since each point x in $T_{\alpha\beta}$ is the left-hand end point of arbitrarily small intervals of the type $[x,x+h]$. Corollary 9.2.4 asserts the existence of a finite disjoint subset $\{J_1,\ldots,J_n\}$ of $\mathfrak{F}$ which covers "almost all" of the $T_{\alpha\beta}$; that is,

[20] $$m\left(T_{\alpha\beta} - \bigcup_{i=1}^{N} J_i\right) < \epsilon,$$

where $\epsilon > 0$ is the number that appears in [17]. This inequality may also be written as

[20′] $$m\left(T_{\alpha\beta} \cap \left(\bigcup_{i=1}^{N} J_i\right)\right) > \mu_{\alpha\beta} - \epsilon.$$

Writing $J_i = [x_i, x_i + h_i]$, we have

[21] $$S(h) = \sum_{i=1}^{N} [f(x_i + h_i) - f(x_i)] = \sum_{i=1}^{N} \left[\frac{f(x_i + h_i) - f(h_i)}{h_i}\right] h_i$$

$$< \alpha \sum_{i=1}^{N} h_i < \alpha(\mu_{\alpha\beta} + \epsilon).$$

The last inequality follows from [20] and the fact that each J_i is contained in some I_n, where $m(G) = m(\bigcup I_n) < \mu_{\alpha\beta} + \epsilon$ (See [17]).

The contradiction is arrived at by using a similar argument for β: If $y \in T_{\alpha\beta} \cap \left(\bigcup_{i=1}^{N} J_i\right)$, then $D^+(y) > \beta$. Let $\mathfrak{F}'$ be the family of intervals $\{[y, y + k]\}$ that satisfy the following conditions:

(i′) $y \in T_{\alpha\beta} \cap \left(\bigcup_{i=1}^{N} J_i\right) = T'_{\alpha\beta}$,

(ii′) $\dfrac{f(y + k) - f(y)}{k} > \beta$,

and

(iii′) $[y, y + k]$ is contained in one of the J_i.

Again, $\mathfrak{F}'$ is a Vitali covering of $T'_{\alpha\beta}$, and therefore contains a finite disjoint subset $J'_1, \ldots, J'_p$ such that

$$m\left(T'_{\alpha\beta} - \bigcup_{i=1}^{p} J'_i\right) < \epsilon.$$

It follows therefore that

$$m\left(T_{\alpha\beta} - \bigcup_{i=1}^{p} J'_i\right) \le m(T_{\alpha\beta} - T'_{\alpha\beta}) + m\left(T'_{\alpha\beta} - \bigcup_{i=1}^{p} J'_i\right) < 2\epsilon,$$

proving that the J'_i cover a subset of $T_{\alpha\beta}$ whose measure exceeds $\mu_{\alpha\beta} - 2\epsilon$. Summing, as in [21], and writing $J'_j = [y_j, y_j + k_j]$, we obtain,

[22] $$S(k) = \sum_{j=1}^{p} [f(y_j + k_j) - f(y_j)] > \beta \sum_{j=1}^{p} k_j > \beta(\mu_{\alpha\beta} - 2\epsilon).$$

But since each J'_j is contained in some J_i (condition (iii′)), and since

$f(x) \uparrow$, it follows from [21] and [22] that

$$\beta(\mu_{\alpha\beta} - 2\epsilon) < S(k) \leq S(h) < \alpha(\mu_{\alpha\beta} + \epsilon),$$

for all $\epsilon > 0$. This, however, yields $\beta \leq \alpha$, which is contrary to the original assumption.

Thus except for a null set T, $D^+(x) = D_+(x)$. A similar argument yields the corresponding equalities for the left derivates, from which it follows that the derivative exists a.e. The measurability of f' follows from Theorem 5. ▎

7 **Corollary.** If $f(x)$ is a function of bounded variation on an interval I, then $f'(x)$ exists a.e. and is a measurable function.

The proof is left to the reader.

Exercises

***1.** Prove that a monotone function is measurable.

***2.** Prove that a monotone function defined on a closed bounded interval is Lebesgue integrable.

***3.** Extend Theorem 6 to unbounded intervals by using truncations.

***4.** Prove that a function of bounded variation is a.e. differentiable (Corollary 7).

5. Prove that

$$f(x) = \begin{cases} 0 & \text{if } x = 0 \\ x^2 \sin 1/x & \text{if } 0 < x \leq 1 \end{cases}$$

is of bounded variation on [0,1], but that

$$g(x) = \begin{cases} 0 & \text{if } x = 0 \\ x \sin 1/x & \text{if } 0 < x \leq 1 \end{cases}$$

is not.

6. Express the function $f(x)$ of Exercise 5 as a difference of two monotone functions.

7. Find the derivates of $g(x)$ of Exercise 5, at $x = 0$.

8. What can be said of the derivates of $h(x)$ at $x = 0$?

$$h(x) = \begin{cases} 0 & \text{if } x = 0 \\ x^a \sin x^{-b} & \text{if } 0 < x < 1. \end{cases}$$

9. Let

$$f(x) = \begin{cases} 0 & \text{if } x \text{ is rational} \\ 1 & \text{if } x \text{ is irrational.} \end{cases}$$

Compute the four derivates of $f(x)$ at all points.

***10.** Prove (i)–(iv) at the beginning of this section.

***11.** Show that if $f \uparrow$, then $\gamma_+ = \gamma^+$ and $\gamma_- = \gamma^-$.

12. Let f and g be nondecreasing functions defined on an interval I. Give examples for which $f - g$ is (i) nondecreasing, (ii) nonincreasing, (iii) not monotone.

***13.** Verify equation [7] of Theorem 2.

***14.** Show that if $f(x)$ is continuous and of bounded variation on $[a,b]$, then the total variation of f over a subinterval $[a,x]$ of $[a,b]$, $V_x(f)$, is a continuous function on $[a,b]$. Prove also that there exist continuous, nondecreasing functions g and h such that $f = g - h$.

15. Let $f(x)$ be continuous on $[a,b]$. Show that the curve $y = f(x)$, $a \leq x \leq b$, has finite length if and only if $f(x)$ is of bounded variation on the interval $[a,b]$. [*Hint:* Let $\mathcal{S}_n$ denote the subdivision of $[a,b]$ obtained by n consecutive bisections of that interval; that is, $a_i = a + [i(b - a)]/2^n$, $i = 0, 1, \ldots, 2^n$. Show that the corresponding sequences $V_n = V(f,\mathcal{S}_n)$ and $L_n = L(f,\mathcal{S}_n)$ are nondecreasing, and that the convergence of either implies the convergence of the other.]

*Exercises which request the proof of a statement made in the text, as well as those referred to in the text, are starred.

4 Differentiation and Integration

We turn our attention to the relationship between differentiation and integration. As stated earlier, in the introductory paragraphs of this chapter, we shall see in what sense differentiation and integration are inverse processes.

1 Theorem. If $f(x)\uparrow$ on $I = [a,b]$, then $f'(x)$ is integrable on I, and

[1]
$$\int_a^b f' \leq f(b) - f(a).$$

Proof: For convenience, we shall extend the function $f(x)$ to R by setting $f(x) = f(b)$ whenever $x \geq b$, and $f(x) = f(a)$ if $x \leq a$. The extended function, which we continue to denote by f, remains monotonic, and $f'(x) \equiv 0$ for all $x \in R - I$.

The monotonicity of f on R implies that the sequence

$$g_n(x) = n[f(x + 1/n) - f(x)],$$

is nonnegative for all x and n, and $\lim_{n\to\infty} g_n(x) = f'(x)$ a.e. Moreover, $f(x)$ is integrable (Exercise 9.3.2), and

$$\underline{\lim} \int_a^b g_n(x) = \underline{\lim}\left[n \int_a^b (f(x + 1/n) - f(x))\right]$$
$$= \underline{\lim}\, n \left[\int_{a+1/n}^{b+1/n} f(x) - \int_a^b f(x)\right]$$
$$= \underline{\lim}\, n \left[\int_{a+1/n}^{b+1/n} f(x) + \int_b^{a+1/n} f(x)\right.$$

$$- \int_b^{a+1/n} f(x) - \int_a^b f(x)\Bigg]$$

$$= \underline{\lim}\, n \left[\int_b^{b+1/n} f(x) - \int_a^{a+1/n} f(x) \right]$$

$$\leq \underline{\lim}\, n \left[\frac{1}{n} f(b) - \frac{1}{n} f(a) \right]$$

$$= f(b) - f(a).$$

The third integral is obtained by a simple change of variables in the second integral; the inequality results from the monotonicity of f. It follows from Fatou's lemma (Theorem 6.2.3) that $\underline{\lim}\ g_n(x) = f'(x)$ (a.e.) is integrable, and

$$\int_a^b f'(x) = \int_a^b \underline{\lim}\, g_n(x) \leq \underline{\lim} \int_a^b g_n(x) \leq f(b) - f(a). \quad \blacksquare$$

The following example demonstrates that the conclusion of Theorem 1 can not be strengthened, for there do indeed exist monotone functions for which the strict inequality holds. It is therefore necessary to make additional assumptions about $f(x)$ in order to conclude that it can be written as the "indefinite" integral of $f'(x)$.

2 Example. Let C be the Cantor set (Example 5.3.6), and let $I_{n,m}$, $n = 1, 2, \ldots$, $m = 1, 2, \ldots, 2^{n-1}$, denote the removed middle thirds. We shall define a sequence of step functions converging monotonically to a function which is constant on each of the intervals $I_{n,m}$, and which is monotone and continuous on [0,1]. Let

$$f_1(x) = \begin{cases} 1/2 & \text{if } x \in I_{11} = [1/3, 2/3] \\ 0 & \text{otherwise;} \end{cases}$$

$$f_2(x) = \begin{cases} 1/4 & \text{if } x \in I_{21} = [1/9, 2/9] \\ 3/4 & \text{if } x \in I_{22} = [7/9, 8/9] \\ f_1(x) & \text{otherwise,} \end{cases}$$

$$\vdots \qquad \vdots \qquad \vdots$$

$$f_n(x) = \begin{cases} 2^{-n} & \text{if } x \in I_{n1} = [3^{-n}, 2 \cdot 3^{-n}] \\ 3 \cdot 2^{-n} & \text{if } x \in I_{n2} = \cdots \\ \vdots & \vdots \\ (2^n - 1) \cdot 2^{-n} & \text{if } x \in I_{n, 2^{n-1}} = \cdots \\ f_{n-1}(x) & \text{otherwise.} \end{cases}$$

If $x \in [0,1] - C$, then x is in one of the removed intervals. Hence, there is a smallest integer N, satisfying

$$n \geq N \qquad \text{implies} \qquad f_n(x) = f_N(x).$$

Furthermore, whenever $n < N$, $f_n(x) = 0$. Thus $\{f_n\}$ is a nondecreasing sequence of step functions which converges a.e. to

$$f(x) = \begin{cases} f_N(x) & \text{if } x \in [0,1] - C \text{ and } x \in I_{N,m} \text{ for some } m \\ \sup\limits_{\substack{t<x \\ t \in [0,1]-C}} f(t) & \text{if } x \in C. \end{cases}$$

We shall prove that $f(x)$ is nondecreasing and continuous on $[0,1]$: Let $x < y$. Then there are four cases to be considered:

(i) If x and y are in a single, or in two different removed intervals, then either $f(x) = f(y)$ or $f(x) < f(y)$.

(ii) If $x \in C$ and $y \in I_{n,m}$, then $f(x) = \sup\limits_{t<x} f(t)$; the points t are assumed to be in the complement of C, and it follows from $t < x < y$ that $f(x) = \sup\limits_{t<x} f(t) \leq f(y)$.

(iii) If $x \in I_{n,m}$ and $y \in C$, the argument is similar to (ii).

(iv) If both x and y are in C, then

$$f(x) = \sup_{t<x<y} f(t) \leq \sup_{t<y} f(t) = f(y).$$

Since $f(x)$ is constant on the interior of each $I_{n,m}$, it is clearly continuous at points of the interior of any $I_{n,m}$. The remaining points are in C. If $c \in C$, then every neighborhood $(c - \delta, c + \delta)$ contains infinitely many of the $I_{n,m}$, in particular intervals with arbitrarily large values of n. (Why?) Suppose that $f(x)$ is discontinuous at c. Then the "jump" of $f(x)$ at $x = c$ cannot exceed the jumps of any of the $f_n(x)$ in this interval. (Remember, the f_n are themselves monotone, and $f_n \uparrow f$ a.e.) But for each n, the jumps of $f_n(x)$ at all its points of discontinuity are equal to 2^{-n}. Thus, the jump of $f(x)$ at $x = c$ does not exceed 2^{-n} for all positive integers n, and therefore $f(x)$ must be continuous at $x = c \in C$.

Finally, we prove that this function f makes [1] an inequality: Since f is monotone, it is a.e. differentiable. In fact $f' = 0$ at all points which are interior to the $I_{n,m}$: that is $f'(x) = 0$ a.e. Therefore

$$0 = \int_0^1 0 = \int_0^1 f' < f(1) - f(0) = 1 - 0 = 1.$$

The goal of this section is the complete characterization of the class of functions which may be expressed as the indefinite integrals of their derivatives. We begin by proving that the indefinite integral,

$$F(x) = \int_a^x f(t)$$

is an a.e. differentiable function on $[a,b]$, and that $F'(x) = f(x)$ a.e.

3 Lemma. If f is integrable over $[a,b]$, then

$$F(x) = \int_a^x f(t)$$

is continuous on $[a,b]$.

Proof: It suffices to prove the Lemma for nonnegative functions. For, if f is any function, then $f = f^+ - f^-$, and both f^+ and f^- are nonnegative. The result is extended to f by the linearity of the integral.

If f is bounded from above by A, then

$$F(x + h) - F(x) = \int_x^{x+h} f(t) \leq Ah \to 0 \tag{2}$$

as h tends to zero, which proves that $F(x)$ is continuous.

If f is not bounded, we consider its truncations f_A. From the preceding paragraph, the indefinite integral $F_{(A)}$ of each truncated function f_A is continuous. Therefore,

$$F(x + h) - F(x) = \int_x^{x+h} f(t) = \int_x^{x+h} [f(t) - f_A(t)] + \int_x^{x+h} f_A(t). \tag{3}$$

Since, $\lim_{A\to\infty} \int_a^b f_A(t) = \int_a^b f(t)$ (The theorem on dominated convergence), it is possible to choose a value of A for which

$$\int_a^b [f(t) - f_A(t)] < \epsilon/2. \tag{4}$$

Since the first term on the right-hand side of [3] is not greater than the integral in [4] it can be made to be less than $\epsilon/2$ if A is properly chosen. The second term on the right side of [3] is bounded by Ah, and therefore tends to zero with h, thus proving that $F(x)$ is continuous. ∎

A slight generalization of Lemma 3 follows. The proof is very similar to that of Lemma 3, and is left to the reader.

4 Proposition. Let f be integrable over a measurable set S. Then if $\epsilon > 0$ is given, there is a $\delta > 0$ such that

$$m(S - T) < \delta \qquad \text{implies} \qquad \int_{S-T} f < \epsilon.$$

5 Lemma. If f is integrable over $[a,b]$, then

$$F(x) = \int_a^x f(t)$$

is a function of bounded variation over $[a,b]$, and is a.e. differentiable.

Proof: By writing $f = f^+ - f^-$, we obtain

$$F(x) = \int_a^x f^+(t) - \int_a^x f^-(t) = F_1(x) - F_2(x).$$

It follows from the nonnegativeness of f^+ and f^- that both F_1 and F_2 are nondecreasing functions on the interval $[a,b]$. Thus their difference $F_1 - F_2$ is a function of bounded variation (Theorem 9.3.2) and is a.e. differentiable (Corollary 9.3.7). ∎

6 Lemma. Let f be integrable over $[a,b]$. If, for all x in $[a,b]$,

[5] $$F(x) = \int_a^x f(t) \equiv 0,$$

then $f(x) = 0$ a.e. in $[a,b]$.

Proof: The measurability of f implies that $S^+ = [f > 0]$ is a measurable subset of $[a,b]$ (Theorem 7.3.8). If $m(S^+) = \delta > 0$, then there is a closed subset $F \subset S^+$ such that $m(F) > \delta/2$; or equivalently, $m(S^+ - F) < \delta/2$ (Corollary 7.3.6). The set $G = (a,b) - F$ is open, and may be written as the disjoint union of open intervals $\bigcup I_n$. From [5], we have

$$0 = \int_a^b f = \int_F f + \int_G f = \int_F f + \sum_{n=1}^{\infty} \int_{I_n} f.$$

Writing $I_n = (x_n, y_n)$,

$$\int_{x_n}^{y_n} f = \int_a^{y_n} f - \int_a^{x_n} f = F(y_n) - F(x_n) = 0,$$

from which it follows that $\int_F f = 0$. However, $F \subset [f > 0]$ and $m(F) > \delta/2 > 0$, which implies that $\int_F f > 0$. Thus $S^+ = [f > 0]$ does not contain a closed subset whose measure is positive. It follows that $m(S^+) = 0$. A similar argument may be given for $S^- = [f < 0]$. ∎

7 Theorem. If f is integrable over $[a,b]$, then

$$F(x) = \int_a^x f(t), \qquad a < x < b,$$

is a.e. differentiable, and $F' = f$ a.e.

Proof: Again, we assume that $f \geq 0$. The general case, $f = f^+ - f^-$ is left to the reader.

Case 1. Assume that $0 \leq f \leq A$. Then $F(x)\uparrow$ and therefore $F'(x)$ exists a.e. The sequence

$$g_n(x) = n[F(x + 1/n) - F(x)] \geq 0,$$

and

$$\lim g_n(x) = F'(x) \text{ a.e. on } [a,b]$$

Moreover,

$$g_n(x) = n\left[\int_x^{x+1/n} f\right] \leq A.$$

The theorem on dominated convergence yields the integrability of $F' = \lim g_n$ a.e., and

$$[6] \quad \int_a^x F'(t) = \int_a^x \lim g_n(t) = \lim \int_a^x g_n(t)$$

$$= \lim n \int_a^x [F(t + 1/n) - F(t)]$$

$$= \lim n\left[\int_x^{x+1/n} F(t) - \int_a^{a+1/n} F(t)\right]$$

$$= \lim n\left[\frac{1}{n} F\left(x + \frac{\theta}{n}\right) - \frac{1}{n} F\left(a + \frac{\theta'}{n}\right)\right]$$

$$= F(x) - F(a) = F(x) = \int_a^x f(t).$$

The second equality of [6] follows from the theorem on dominated convergence, and the fifth from the mean value theorem for integrals (Exercise 5). Subtracting the last term of [6] from the first, we obtain

$$\int_a^x [F'(t) - f(t)] = 0, \qquad a < x < b.$$

From Lemma 6, it follows that $F' = f$ a.e. on $[a,b]$.

Case 2. Suppose now that f is a nonnegative, integrable (not necessarily bounded) function defined on $[a,b]$. Let $\{f_n\}$ denote the sequence of truncations of f at n. The theorem on dominated convergence yields

$$\int_a^b f = \lim \int_a^b f_n.$$

As in the proof of Lemma 3, we write

$$F(x) = \int_a^x [f - f_n] + \int_a^x f_n,$$

and use case 1 for $0 \le f_n \le n$. This yields

$$[7] \qquad F'(x) = \frac{d}{dx}\int_a^x [f(t) - f_n(t)] + f_n(x).$$

(Remember, it is known from Lemma 5 that $F'(x)$ exists a.e.) The inequality $0 \le f - f_n$ implies that $\int_a^x [f - f_n]$ is a nondecreasing function, from which it follows that its a.e. derivative is nonnegative. Thus from [7] we obtain $f_n \le F'$ a.e., for all values of n, and therefore,

$$[8] \qquad f = \lim f_n \le F' \text{ a.e.}$$

Combining [8] with Theorem 1 yields

$$F(b) - F(a) = \int_a^b f \le \int_a^b F' \le F(b) - F(a),$$

from which it follows that

$$\int_a^b [F' - f] = 0.$$

However, [8] and $m([a,b]) = b - a > 0$ imply that $F' = f$ a.e. ∎

We have just proved that integration and differentiation, taken in that order, are inverse processes. However, Example 2 shows that when the order is reversed, this is no longer true if it is assumed only that the function to be differentiated is monotone or of bounded variation. The additional assumption required to conclude that a given function $F(x)$ is the indefinite integral of a second function $f(x)$, is that $F(x)$ be *absolutely continuous*. The definition and the main theorem follow.

8 Definition. A function $F(x)$ is said to be *absolutely continuous* on an interval $[a,b]$, if to each $\epsilon > 0$, there is a $\delta > 0$, such that whenever $\mathcal{I} = \{[x_1, y_1], [x_2, y_2], \ldots, [x_n, y_n]\}$ is a finite set of subintervals of $[a,b]$ whose total length is less than δ, then the variation of F over these intervals,

$$\sum_{i=1}^{n} | F(y_i) - F(x_i) | = V(F,\mathcal{I}) < \epsilon.$$

Taking $n = 1$, we see that an absolutely continuous function is uniformly continuous. Example 2 may be used to show that the converse is false. The function constructed in that example is continuous on a closed, bounded interval, and hence, uniformly continuous. However, if $\epsilon < 1$, and δ is any positive number, a set of intervals may be found whose total length is less than δ, but the variation of f over this set of intervals exceeds ϵ. The details are left to the reader in the exercises.

9 Lemma. If $F(x)$ is absolutely continuous on $[a,b]$, then it is of bounded variation.

Proof: Let $\delta > 0$ be chosen so that the conditions of Definition 8 are satisfied for $\epsilon = 1$. There is a unique integer N that satisfies

$$\frac{b-a}{\delta} \leq N < \frac{b-a}{\delta} + 1 \tag{9}$$

We shall show that for any subdivision $\mathcal{S}$, the variation of F over the intervals of $\mathcal{S}$, $V(F,\mathcal{S})$, is not greater than N.

First, we add to $\mathcal{S}$, if necessary, a finite number of additional points to obtain a new subdivision $\mathcal{S}'$, which has the following properties. The intervals of $\mathcal{S}'$ are so small that they may be partitioned into *exactly* N subsets of intervals, each subset having total length no greater than δ. The existence of this subdivision is guaranteed by [9]. Let the points of $\mathcal{S}'$ be denoted by, $a = b_0 < b_1 < \cdots < b_m = b$. Since each point a_i of the original subdivision $\mathcal{S}$, appears as a b_j, the triangle inequality yields

$$\begin{aligned} |F(a_i) - F(a_{i-1})| &= |F(b_j) - F(b_{j-p})| \\ &\leq |F(b_j) - F(b_{j-1})| + |F(b_{j-1}) - F(b_{j-2})| + \cdots \\ &\qquad + |F(b_{j-p+1}) - F(b_{j-p})|, \end{aligned}$$

from which it follows that

$$V(F,\mathcal{S}) \leq V(F,\mathcal{S}') < N \cdot 1 = N,$$

since ϵ was taken as 1. ∎

10 Proposition. If $F(x)$ is absolutely continuous on $[a,b]$, then $F'(x)$ exists a.e. and is an integrable function.

Proof: From Lemma 9, it follows that $F(x)$ is of bounded variation, and hence a.e. differentiable. Theorem 1 yields the integrability of $F'(x)$. ∎

11 Lemma. If $F(x)$ is absolutely continuous on $[a,b]$, and $F'(x) = 0$ a.e., then $F(x)$ is a constant.

Remark. It is precisely this property that a function of bounded variation may lack. Indeed, the function described in Example 2 has a derivative that vanishes a.e., but it is not a constant function.

Proof: We must prove that for all $c \in [a,b]$, $F(c) = F(a)$. Let us assume that $c > a$. Let $S_c = [a,c) \cap [F' = 0]$. Choose $\delta > 0$ so that the conditions of Definition 8 are satisfied for a given $\epsilon > 0$. Then if $x \in S_c$, and if γ is a positive number, there are arbitrarily small intervals $[x,x+h] \subset [a,c]$ such that

$$|F(x+h) - F(x)| < \gamma |h|. \tag{10}$$

This follows from $F'(x) = 0$ for $x \in S_c$. Keeping γ fixed, and allowing x

to range over S_c, we obtain a Vitali covering $\mathfrak{F}$ of S_c by intervals $[x,x+h] \subset [a,c]$ for which [10] holds. Corollary 9.2.4 implies that there is a disjoint subset $\{I_i = [x_i, x_i + h_i]\}$, $i = 1, 2, \ldots, n$, of intervals in $\mathfrak{F}$ such that $m(S_c - \bigcup I_i) < \delta$, or equivalently, since $[a,c] - S_c$ is a null set, $m([a,c] - \bigcup I_i) < \delta$. Let us assume that the intervals have been chosen so that

$$a = x_1 < x_1 + h_1 < x_2 < x_2 + h_2 < \cdots < x_n + h_n \leq c.$$

The triangle inequality yields

$$[11] \quad \begin{aligned} |F(c) - F(a)| \leq |F(x_n + h_n) - F(x_n)| \\ + |F(x_n) - F(x_{n-1} + h_{n-1})| + \cdots \\ + |F(x_2) - F(x_1 + h_1)| + |F(x_1 + h_1) - F(x_1)| = \theta_1 + \theta_2, \end{aligned}$$

where

$$\theta_1 = \sum_{i=1}^{n} |F(x_i + h_i) - F(x_i)|$$

and

$$\theta_2 = \sum_{i=2}^{n} |F(x_i) - F(x_{i-1} + h_{i-1})|.$$

In order to prove that $F(c) = F(a)$, it suffices to show that θ_1 and θ_2 can be made arbitrarily small: Since $m\left([a,c] - \bigcup_{i=1}^{n} I_i\right) < \delta$, where δ satisfies the conditions of Definition 8, it follows that the variation over the set $[a,c] - \bigcup_{i=1}^{n} I_i$, which is simply θ_2, is less than ϵ; ϵ was an arbitrarily chosen positive number. Furthermore, [10] yields

$$\theta_1 = \sum_{i=1}^{n} \frac{F(x_i + h_i) - F(x_i)}{h_i} |h_i| < \gamma \sum_{i=1}^{n} |h_i| \leq \gamma(b - a).$$

Again, γ was arbitrarily chosen, so that θ_1 can be made as small as desired. It follows from [11] that $F(c) = F(a)$. ∎

12 Theorem. (i) If $f(x)$ is integrable on $[a,b]$, then

$$F(x) = \int_a^x f(t), \qquad a \leq x \leq b,$$

is an absolutely continuous function.

(ii) If $F(x)$ is absolutely continuous on $[a,b]$, then

$$F(x) - F(a) = \int_a^x F'(t).$$

Proof: Let $I_j = [x_j, x_j + h_j]$, $j = 1, 2, \ldots, n$ be a disjoint set of subintervals of $[a,b]$. The variation of F over the union, $S = \bigcup_{j=1}^{n} I_j$, is given by

$$[12] \quad \sum_{j=1}^{n} | F(x_j + h_j) - F(x_j) |$$

$$= \sum_{j=1}^{n} \left| \int_{x_j}^{x_j+h_j} f(t) \right| \leq \sum_{j=1}^{n} \int_{x_j}^{x_j+h_j} | f(t) |$$

$$= \int_S | f(t) | .$$

Proposition 4 implies that

$$\lim_{m(S) \to 0} \int_S | f(t) | = 0.$$

This says that the variation of F over a set of subintervals of $[a,b]$ tends to zero as the total length of the set of intervals goes to zero, which is precisely the definition of absolute continuity. Thus (i) is proved.

Let us assume now that $F(x)$ is absolutely continuous. Then $F'(x)$ exists a.e. and is integrable (Proposition 10). The function,

$$[13] \qquad G(x) = \int_a^x F'(t) - F(x),$$

is a.e. differentiable on $[a,b]$, and

$$[14] \qquad G'(x) = F'(x) - F'(x) = 0 \text{ a.e.}$$

(Theorem 7). From part (i), it follows that $\int_a^x F'(t)$ is absolutely continuous. Taken together with the absolute continuity of $F(x)$, we obtain the absolute continuity of $G(x)$ (See [13]). Then [14] and Lemma 11 imply that $G(x)$ is a constant. By substituting $x = a$ in [13], we find that $G(a) = -F(a)$, from which the desired result follows. ▌

Exercises

***1.** Prove Lemma 3 for an arbitrary integrable function.

***2.** Prove Proposition 4.

***3.** Prove Theorem 7 for any integrable function.

4. Show directly from Definition 8 that the function of Example 2 is not absolutely continuous. [*Hint:* The measure of the Cantor set is zero.]

***5.** (Mean Value Theorem for Integrals) Let $f(x)$ be continuous for $a \leq x \leq b$. Then there is a (at least one) point $c \in [a,b]$, satisfying

$$f(c)(b - a) = \int_a^b f(x).$$

[*Hint:* A continuous function on a closed bounded interval attains its maximum and minimum values and all values that lie between them.]

***6.** Let f be of bounded variation over $[a,b]$. Show that the total variation of f over $[a,b]$, $V(f)$, (Definition 9.3.1) satisfies the inequality

$$\int_a^b |f'(t)| \leq V(f).$$

7. Show that $\int_a^b |f'(t)| = V(f)$ if and only if f is absolutely continuous.

8. Let f and g be absolutely continuous functions on $[a,b]$.
(i) Prove that $f \pm g$ are absolutely continuous.
(ii) Prove that $f \cdot g$ is absolutely continuous.

9. (Integration by Parts) Let f be integrable over $[a,b]$, and let G be absolutely continuous, Writing $F(x) = \int_a^x f(t)$, and $g(x) = G'(x)$ a.e., show that

$$\int_a^b (F \cdot G)' = \int_a^b Fg + \int_a^b fG.$$

Prove first that FG is absolutely continuous, so that you can conclude that it is a.e. differentiable.

10. Prove that if $F(x)$ is differentiable (everywhere) on $[a,b]$, and if $F'(x)$ is bounded, then

$$\int_a^b F'(t) = F(b) - F(a)$$

***11.** (Decomposition Theorem). Prove that if f is of bounded variation on $[a,b]$, then $f = g - h$, where g is absolutely continuous and h is of bounded variation, $h'(x) = 0$ a.e. Furthermore, g and h are determined uniquely up to an additive constant.

12. Let $f(x)$ be a function of bounded variation over $[a,b]$ that vanishes outside this interval. Show that there is a positive number γ such that,

$$* \qquad \int_a^b |f(x + h) - f(x)| < \gamma h,$$

for all real values of h. Prove that $*$ implies that f is a.e. equal to a function of bounded variation over $[a,b]$. (See Titchmarsh [1], p. 372.)

13. Let f be a nondecreasing function defined on an interval I (bounded or unbounded).
(i) Show that if c is a real number, then one and only one of the following statements is true:
(a) $f(x) \neq c$ for any $x \in I$.
(b) There is exactly one point $x \in I$ for which $f(x) = c$.
(c) The set of points at which $f(x) = c$ is an interval I_c.
(ii) Show that the set of intervals $\{I_c\}$ is disjoint and countable. Can it be concluded that they are open, closed?

*Exercises which request the proof of a statement made in the text, as well as those referred to in the text, are starred.

PART III

Banach and Hilbert Spaces

CHAPTER **10**

Banach Spaces, Hilbert Spaces and Fourier Series

We turn our attention now to the function spaces L_p, $1 \le p \le \infty$. A measurable function f is in L_p, $1 \le p < \infty$, if $|f|^p$ is an integrable function. L_∞ is the space of a.e. bounded measurable functions. It is possible to prove many theorems about the general L_p spaces which have already been proved for the special case L_1. In particular, we shall prove that each L_p space is a Banach space (complete normed linear space). In Sec. 2 and 3, the space L_2 is singled out for a separate investigation. Although it is not possible at this time to justify fully the special attention received by L_2, the analogy with the l_2 space and R^n suggests a partial answer. (See Examples 2.2.2(a), (b), (c) and Theorems 2.1.4, 2.1.6, and 2.1.8.) The reader will recall that in the Euclidean metric defined on R^n, $d(x,y) = \sqrt{(x_1 - y_1)^2 + \cdots + (x_n - y_n)^2}$, it is possible to define a *scalar* or *inner product* of two vectors in such a way that the idea of the "angle" determined by two vectors may be extended to spaces whose dimension exceeds 3. In a similar way, it is possible to measure angles in L_2, or in any *Hilbert space* (Definition 10.2.4) so that many geometrical properties of Euclidean spaces are retained. Because of these geometrical properties, we shall be able to derive some interesting and useful results about the approximation of elements in L_2 by (finite) linear combinations of mutually orthogonal "basis" vectors. This leads to a discussion of Fourier series in Sec. 4.

The reader who is primarily interested in measure and integration theory, may omit this chapter and the next, referring to them whenever necessary, while reading Chapters 12 and 13.

Only the results of Sec. 1 and 2 appear in Chapter 11.

1 L_p Spaces; The Riesz-Fischer Theorem

1 Definition. The space L_p, $1 \le p < \infty$, is the set of equivalence classes of real functions f for which the (Lebesgue) integral,

[1] $$\int |f|^p,$$

exists. (The reader will recall that two measurable functions are said to be equivalent if they differ only on a null set.)

L_∞ is the collection of equivalence classes of a.e. bounded measurable functions.

To avoid using the language of equivalence classes, we shall again adopt the convention of calling the elements of L_p *functions* which satisfy [1]. Two functions which are a.e. equal will be treated as identical functions, since the integral does not recognize the difference between such a pair of functions.

In this section we shall prove that each set L_p, $1 \leq p \leq \infty$, under the usual operations of addition of two functions and multiplication of a function by a real number, can be made into a real vector space (Proposition 2), and that it is also possible to define in each L_p space a norm, with respect to which it is complete. If $1 \leq p < \infty$,

[2] $$\|f\|_p = \left[\int |f|^p\right]^{1/p}$$

is the desired norm on L_p (Theorems 7 and 8). The norm on L_∞ is defined as follows: If $f \in L_\infty$, then there exist numbers M satisfying $|f(x)| \leq M$ a.e. The greatest lower bound of these numbers, denoted by ess sup $|f(x)|$ (read: essential supremum of f) is defined to be the L_∞ norm of the function f:

[3] $$\|f\|_\infty = \text{ess sup } |f(x)| = \inf\{M : |f(x)| \leq M \text{ a.e.}\}$$
$$= \inf\{M : m([f > M]) = 0\}.$$

(See Exercise 4). It is left to the reader to prove that $(L_\infty, \|\ \|_\infty)$ is a Banach space.

The case $p = 1$ has been discussed at great length in Part II, so that we may restrict ourselves to a discussion of the L_p spaces for $1 < p \leq \infty$.

2 Proposition. If $1 < p < \infty$, then L_p is a real vector space. (See Exercise 4 for $p = \infty$.)

Proof: If $f \in L_p$ and if $a \in R$, then $|af|^p = |a|^p |f|^p$, from which it follows that

$$\int |af|^p = |a|^p \int |f|^p < \infty,$$

and that $af \in L_p$.

For any pair of functions f and g,

$$|f + g|^p \leq \begin{cases} |2f|^p & \text{if } |f| \geq |g| \\ |2g|^p & \text{if } |g| \geq |f|. \end{cases}$$

Therefore, if $f,g \in L_p$, then $|f+g|^p$ is a measurable function (Why?) which is bounded by the integrable function

$$2^p\{|f|^p + |g|^p\}.$$

Theorem 7.1.9 yields the integrability of $|f+g|^p$, proving that L_p is closed under addition.

Verification of the remaining vector space properties is left to the reader. ∎

The next step is to prove that [2] defines a norm in L_p. This requires the Hölder and Minkowski inequalities for integrals. (See Sec. 1 of Chapter 2.)

3 Definition. A pair of real numbers p and q are said to be *conjugate* (or, p and q are *conjugate indices*) if $1 < p, q < \infty$ and $1/p + 1/q = 1$. (See Theorems 2.1.3, 2.1.5, and 2.1.7.)

4 Lemma. If a and b are positive numbers, and if p and q are conjugate indices, then

$$a^{1/p} b^{1/q} \le a/p + b/q.$$

Proof: This is simply a special case of Theorem 2.1.2, with $b_1 = a$, $b_2 = b$, $\lambda_1 = 1/p$, $\lambda_2 = 1/q$. ∎

5 Theorem (Hölder's inequality for integrals). Suppose that either p and q are conjugate indices, or $p = 1$ and $q = \infty$, and that $f \in L_p$, $g \in L_q$. Then fg is integrable, and

$$\|fg\|_1 = \int |fg| \le \|f\|_p \, \|g\|_q. \tag{4}$$

Proof: If $p = 1$ and $q = \infty$, then $|g(x)| \le \|g\|_\infty$ a.e.. Therefore, fg is a measurable function and $|fg| \le \|g\|_\infty |f|$ a.e.. The integrability of f and Theorem 7.1.9 yield the integrability of $|fg|$, and

$$\int |fg| \le \|g\|_\infty \int |f| = \|g\|_\infty \|f\|_1.$$

Suppose now that $1 < p, q < \infty$. For each x, one of the following is true:

(i) $|g| \le |f|^{p-1}$, and therefore $|fg| \le |f|^p$, or

(ii) $|g| > |f|^{p-1}$, and therefore $|fg| < |g|^{1+1/(p-1)} = |g|^q$.

(Remember, $1/p + 1/q = 1$.) It follows that for all values of x, $|fg| \le |f|^p + |g|^q$, proving that $|fg|$ is integrable (Theorem 7.1.9).

We may assume that neither f nor g is a.e. equal to zero. For, if this were the case, then $fg = 0$ a.e., and both sides of [4] would vanish. Let

$$a(x) = \left(\frac{|f(x)|}{\|f\|_p}\right)^p, \qquad b(x) = \left(\frac{|g(x)|}{\|g\|_q}\right)^q.$$

Both $a(x)$ and $b(x)$ are well defined, since neither $\|f\|_p$ nor $\|g\|_q$ vanishes, and both of these functions are nonnegative. Lemma 4 yields

$$\frac{|fg|}{\|f\|_p\|g\|_q} \le \frac{1}{p}\left[\frac{|f|}{\|f\|_p}\right]^p + \frac{1}{q}\left[\frac{|g|}{\|g\|_q}\right]^q.$$

Integrating both sides of this inequality, we obtain

$$\frac{\int |fg|}{\|f\|_p\|g\|_q} < \frac{1}{p} + \frac{1}{q} = 1.$$

Multiplication by $\|f\|_p\|g\|_q$ yields the Hölder inequality for $1 < p$, $q < \infty$. ▎

6 Theorem (Minkowski's inequality for integrals). If $f,g \in L_p$, $1 \le p \le \infty$, then

[5] $$\|f + g\|_p \le \|f\|_p + \|g\|_p.$$

Proof: The case $p = 1$ has already been proved. The reader is asked to verify [5] for $p = \infty$.

If $1 < p < \infty$, then the triangle inequality for real numbers yields

[6] $$|f + g|^p = |f + g|\,|f + g|^{p-1} \le |f|\,|f + g|^{p-1} + |g|\,|f + g|^{p-1}.$$

We show first that $|f + g|^{p-1}$ is in L_q, where q is conjugate to p. Then Hölder's inequality applied to the right-hand side of [6] will yield [5].

Since $|f + g|^p$ is integrable (Proposition 2), and $1/p + 1/q = 1$, it follows that

$$|f + g|^{(p-1)q} = |f + g|^{(p-1)(p/p-1)} = |f + g|^p,$$

proving that $|f + g|^{p-1} \in L_q$. Theorem 5 yields the integrability of the left-hand side of [6], and

[7] $$[\|f + g\|_p]^{p/q} = [\int |f + g|^p]^{1/q} = \||f + g|^{p-1}\|_q.$$

The identity $p/q = p - 1$ implies [5]. ▎

7 Theorem. The vector spaces L_p, $1 \le p \le \infty$, are normed vector spaces, with the norm of f, $\|f\|_p$ defined by [2] and [3].

Proof: For $p = 1$, the theorem has already been proved, and the case $p = \infty$ is left to the reader as an exercise. Proposition 2 asserts that each L_p is a real vector space. Therefore it only remains to be proved that

$$\|f\|_p = [\int |f|^p]^{1/p}$$

satisfies the properties of a norm.

(i) The positivity of the integral implies that $\|f\|_p \ge 0$, and the equality $0 = \|f\|_p^p = \| |f|^p \|_1$ implies that $f = 0$ a.e. (Remember, $\hat{L}_1$ is a normed linear space.)

(ii) For all c in R, $\|cf\|_p = \left[\int |cf|^p\right]^{1/p} = |c| \, \|f\|_p$.

(iii) The triangle inequality is simply Minkowski's inequality. ∎

8 Theorem (Riesz-Fischer). The normed vector spaces L_p, $1 \leq p \leq \infty$, are complete.

Proof: Again, we shall prove the theorem only for $1 < p < \infty$. (See Chapter 5 for $p = 1$ and Exercise 4 for $p = \infty$.)

The reader will recall that a normed linear space is complete if and only if every absolutely summable series is summable (Definition 2.4.9 and Theorem 2.4.10). Therefore it suffices to prove that if $\sum_1^\infty f_k$ is an absolutely summable series in L_p, then there is an element $f \in L_p$ such that $f = \sum_1^\infty f_k = \lim_{n \to \infty} \sum_1^n f_k$. Written in norm notation, this becomes

$$\lim_{n \to \infty} \|f - \sum_1^n f_k\|_p = 0. \tag{8}$$

The absolute summability of the series means that there is a number A, and $\sum_1^\infty \|f_k\|_p = A$. Minkowski's inequality (Theorem 6) applied to the function $\sigma_n = \sum_1^n |f_k|$ yields $\|\sigma_n\|_p \leq A$ for all values of n, or equivalently,

$$\int |\sigma_n|^p = \|\sigma_n\|_p^p \leq A^p.$$

Thus the nondecreasing sequence of nonnegative integrable functions $\{\sigma_n^p\}$ satisfies the conditions of Fatou's Lemma (Theorem 6.2.3), so that

$$\lim_{n \to \infty} \int \sigma_n^p = \underline{\lim}_{n \to \infty} \int \sigma_n^p \leq A^p.$$

Thus $\underline{\lim}\, \sigma_n^p = \lim \sigma_n^p = \sigma^p$ is an integrable function, and as a consequence it is a.e. finite. This implies that the pth root of the a.e. finite function σ^p,

$$\sigma = \sum_1^\infty |f_k|, \tag{9}$$

converges (or is finite) a.e., from which it follows that the series $\sum_1^\infty f_k$ also converges a.e. We define a function $f(x)$ in the following way:

$$f(x) = \begin{cases} \sum_1^\infty f_k(x) & \text{if } \sum_1^\infty |f_k(x)| < \infty \\ 0 & \text{otherwise.} \end{cases}$$

It is now quite easy to prove that $\sum_1^\infty f_k$ is summable to f in the L_p norm (See [8]).

Setting $s_n = \sum_1^n f_k$, we have $|s_n| \le \sigma_n \le \sigma$, since $\sigma_n \uparrow \sigma$. Furthermore, $\sigma \in L_p$, so that σ^p is integrable. Thus the theorem on dominated convergence yields the integrability of $|f|^p = \lim |s_n|^p \le \sigma^p$, and since $\lim |f - s_n| = 0$ everywhere, it follows also from the theorem on dominated convergence that

$$\lim_{n\to\infty} \|f - s_n\|_p = \lim_{n\to\infty} \left[\int |f - s_n|^p\right]^{1/p} = 0,$$

which is [8], proving that the absolutely summable series is summable in the L_p norm to f. ∎

Exercises

***1.** Let l_∞ be the space of bounded sequences of real numbers. That is, a sequence $x = \{x_n\}$ is in l_∞ provided there is a number M such that $|x_n| \le M$ for all n.

(i) Prove that l_∞ is a real vector space; the addition of two vectors is coordinatewise addition etc.

(ii) Prove that a norm on l_∞ is defined by

$$\|x\|_\infty = \sup\{|x_n|\}, \quad n = 1,2,\ldots.$$

(iii) Prove that $(l_\infty, \| \|_\infty)$ is complete.

(iv) Is this space separable? (Does it contain a dense sequence?)

***2.** Complete the proof of Proposition 2 by verifying the remaining vector space properties for L_p, $1 < p < \infty$.

***3.** Prove Theorem 6 (Minkowski's inequality) for $p = \infty$.

***4.** (i) Prove that L_∞ is a real vector space.

(ii) Prove that [3] defines a norm on L_∞.

(iii) Describe convergence in this norm.

(iv) Prove that if $\{f_n\}$ converges to f in L_∞, then it converges uniformly to f except on a null set.

(v) Describe Cauchy sequences in L_∞, and show that every Cauchy sequence in L_∞ converges to a function in L_∞.

5. Prove that if f is a bounded integrable function, then it is in every L_p space, $1 \le p \le \infty$.

6. Let $L_p[a,b]$ denote the subspace of L_p consisting of those functions which vanish outside the interval $[a,b]$.

(i) Prove that these spaces are also complete.

(ii) Prove that if $1 \le p < q < \infty$, then $L_p \subset L_q$.

***7.** Let χ be the set of real functions f satisfying,

(i) If $f \in \chi$, then there is a sequence $\{x_n\}$ in R, such that $f(x) = 0$ unless $x = x_n$, and

(ii) $\sum_n^\infty f^2(x_n) < \infty$.

(a) Prove that χ is a real vector space.

(b) Show that $\|f\| = \left[\sum_{1}^{\infty} f^2(x_n)\right]^{1/2}$ defines a norm on χ.

(c) Show that χ is complete.

(d) Prove that χ is not separable.

***8.** Let $1 \leq p < \infty$. Prove that if $\{f_n\}$ is a sequence of functions in L_p that converges a.e. to a function f which is also in L_p, then $\lim_{n\to\infty} \|f - f_n\|_p = 0$ if and only if $\lim_{n\to\infty} \|f_n\|_p = \|f\|_p$. Can you give an example of an a.e. convergent sequence in L_p which does *not* converge in the L_p norm?

***9.** Show that with the exception of $p = 2$, the product of functions in L_p is not always a function in L_p.

10. Let $\{f_n\}$ be a sequence of functions in L_p, $1 < p < \infty$, whose norms are uniformly bounded, and which converges a.e. to a function f in L_p. Show that if $g \in L_q$, where q is conjugate to p, then

$$\lim_{n\to\infty} \int f_n g = \int fg.$$

*Exercises which request the proof of a statement in the text, as well as those referred to in the text, are starred.

2 Hilbert Spaces (Geometry)

Among the L_p and l_p spaces, L_2 and l_2 enjoy a special status. In both cases, especially the latter, the norm closely resembles the Euclidean norm in R^n (Example 2.2.2). This suggests that these spaces may share additional properties with the Euclidean spaces; and in this section, the geometrical properties of L_2 and l_2 are explored. In particular, we shall define *orthogonality* in L_2 and l_2, and explore the possibility of expressing an element of L_2 (or l_2) as a (finite) linear combination of mutually orthogonal elements. We shall see that it is possible only to "approximate" (in the norm) elements of L_2 (or l_2) by linear combinations of orthogonal vectors. It should be stressed here that "approximation in the L_2 norm" is quite different from "approximation in the max norm of $C(X)$." In the latter case, "approximation" means *uniform* approximation of continuous functions on X by functions in a subalgebra or sublattice (See Sec. 1 of Chapter 4). Here the L_2 norm is defined by an integral, ([2] and [3] of Sec. 1), and the approximation is by no means uniform.

We begin by giving the definition of an *inner product* in a real vector space, and some examples of *inner product spaces*.

1 Definition. A mapping of pairs (x,y) of elements in a real vector space into the real numbers is called an *inner product* (or *scalar product*) if

(i) for all $x \in X$, $(x,x) \geq 0$; and $(x,x) = 0$ if and only if $x = \bar{0}$ ($\bar{0}$ is the additive identity of the vector space X);

(ii) for all $x,y \in X$, $(x,y) = (y,x)$;
(iii) for all $x, y \in X$ and $c \in R$, $(cx, y) = c(x, y)$;
(iv) for all $x,y,z \in X$, $(x + y,z) = (x,z) + (y,z)$.

It follows from (ii), that (iii) and (iv) are equivalent to:

(iii′) for all $x, y \in X$ and $c \in R$, $(x,cy) = c(x, y)$;
(iv′) for all $x,y,z \in X$, $(x,y + z) = (x,y) + (x,z)$.

A pair, consisting of a real vector space X and an inner product (x,y), is called an *inner product space.*

2 Examples. (a) Let $X = R^3$, the space of ordered triples of real numbers. If $x = (x_1,x_2,x_3)$ and $y = (y_1,y_2,y_3)$ are in R^3, then the expression

$$(x,y) = x_1y_1 + x_2y_2 + x_3y_3,$$

is easily seen to be an inner product on R^3.

(b) Again, let $X = R^3$. If γ_1,γ_2, and γ_3 are real numbers, then the expression

$$(x,y) = \gamma_1x_1y_1 + \gamma_2x_2y_2 + \gamma_3x_3y_3,$$

satisfies (ii), (iii), and (iv) of Definition 1. In order that (i) is satisfied, it is both necessary and sufficient that $\gamma_i > 0$, $i = 1,2,3$ (Exercise 2).

(c) It is left to the reader to generalize (a) and (b) for the real vector spaces R^n, $n = 1,2,\ldots$.

(d) Let $X = l_2$, and let

$$(x,y) = \sum_1^\infty x_iy_i.$$

Hölder's inequality yields the convergence of this sum (Theorem 2.1.5). The verification of (i)–(iv) of Definition 1 is left to the reader.

(e) Let $X = C[a,b]$, the real vector space of real continuous functions defined on $[a,b]$. It is left to the reader to verify that

$$(f,g) = \int_a^b fg$$

is an inner product.

(f) Similarly for $X = L_2$ and $(f,g) = \int fg$.

Remark: This is not the case for the other L_p spaces. In general, the product of a pair of functions in L_p, $p \neq 2$, is not integrable. For $p = 2$, the integrability of a product fg of square integrable functions is assured by Hölder's inequality for integrals (Theorem 10.1.5).

For each of the examples described above, the expression

$$\| x \| = \sqrt{(x,x)}$$

defines a norm on X. Indeed, if X is *any* real vector space upon which an inner product is defined, then X can be made into a normed vector space by setting $\|x\| = \sqrt{(x,x)}$. For definition 1 yields

(i) $\|x\| = \sqrt{(x,x)} \geq 0$; and $\|x\| = 0$ if and only if $x = \bar{0}$;

(ii) $\|cx\| = \sqrt{(cx,cx)} = \sqrt{c^2(x,x)} = |c| \sqrt{(x,x)} = |c| \|x\|$.

To verify the triangle inequality, we must prove first the *Cauchy-Schwarz inequality* [1] for inner products. The reader will observe that it is a special case of Hölder's inequality [see Example 2(a), (d) and (f)], for $p = q = 2$.

3 Theorem (Cauchy-Schwarz Inequality). If X is an inner product space, then for all $x, y \in X$,

[1] $$(x,y)^2 \leq (x,x)(y,y);$$

equality is attained in [1] if and only if one element is a multiple of the other.

Proof: If $x,y \in X$ and if $c \in R$, then Definition 1 yields $(x + cy, x + cy) \geq 0$; and $(x + cy, x + cy) = 0$ if and only if $x = -cy$. Expanding this expression, we obtain

[2] $$0 \leq (x + cy, x + cy) = (x,x) + 2c(x,y) + c^2(y,y).$$

If x and y are kept fixed, then [2] may be regarded as a quadratic expression in c which is nonnegative for all real values of c, and vanishes only when $x = -cy$. Therefore, the discriminant of [2] must be nonpositive:

[3] $$4(x,y)^2 - 4(x,x)(y,y) \leq 0.$$

This inequality, however, is equivalent to [1]. It is easily seen that equality is attained only if $x = -cy$; that is, the only real root of the quadratic expression [2] is a double root, which occurs when the discriminant vanishes. ∎

We return to the verification of the triangle inequality for $\|x\| = \sqrt{(x,x)}$: The Cauchy-Schwarz inequality yields

$$\begin{aligned}\|x + y\|^2 &= (x + y, x + y) = (x,x) + 2(x,y) + (y,y) \\ &\leq \|x\|^2 + 2\|x\| \cdot \|y\| + \|y\|^2 = [\|x\| + \|y\|]^2.\end{aligned}$$

The triangle inequality is obtained by taking the square root of both sides of this inequality. Therefore, we have proved that: If X is an inner product space, then

[4] $$\|x\| = \sqrt{(x,x)}$$

defines a norm on X.

4 Definition. A (real) inner-product space X which is complete with respect to the norm defined by the inner product [4], is called a (real) *Hilbert space.*

(The word "real" will be omitted hereafter, since complex Hilbert spaces will not appear in this text.)

A concrete interpretation of the inner product and the Cauchy-Schwarz inequality can be given for Example 2 (a). Here x and y may be thought of as three-dimensional vectors each having (0,0,0) as its initial point and (x_1,x_2,x_3), (y_1,y_2,y_3) as their respective terminal points. Using the law of cosines, it is easily verified that the expression

$$\frac{(x,y)}{\sqrt{(x,x)}\sqrt{(y,y)}} = \frac{(x,y)}{\|x\| \cdot \|y\|}$$

is the cosine of the angle θ determined by these two vectors. If the vectors are orthogonal ($\theta = 90°$), then $(x,y) = 0$. The ideas may be formally transferred to the other inner product spaces of Example 2 by *defining* the orthogonality of two vectors as the vanishing of their inner product. Many of the results of this section and the next will be preceded by heuristic arguments which are based on elementary Euclidean geometry. The reader may at first be surprised at how far his geometric intuition will carry him in Hilbert space theory.

Theorem 5, which is proved below, is a generalization of the so-called *parallelogram identity*: A pair of noncollinear vectors x and y in R^2 determine uniquely a parallelogram whose nonparallel sides are x and y respectively, and whose diagonals are $x + y$ and $x - y$. If we let $\|x\| = \sqrt{(x,x)} = \sqrt{x_1^2 + x_2^2}$, then it is not difficult to prove that

$$[5] \qquad \|x + y\|^2 + \|x - y\|^2 = 2\|x\|^2 + 2\|y\|^2.$$

We shall prove that not only does the Parallelogram Identity [5], hold in every Hilbert space, but that every Banach space for which [5] holds is also a Hilbert space:

5 Theorem. If H is a Hilbert space, then for any pair of elements $x, y \in H$,

$$[5] \qquad \|x + y\|^2 + \|x - y\|^2 = 2\|x\|^2 + 2\|y\|^2; \ (\|x\| = \sqrt{(x,x)}).$$

Conversely, if H is a Banach space in which [5] is true, then an inner product may be defined in H which is "compatible" with the given norm; that is, $(x,x) = \|x\|^2$.

Proof: Let us prove the easier part first. Suppose that H is a Hilbert space. If $x,y \in H$, then

$$\|x + y\|^2 + \|x - y\|^2 = (x + y, x + y) + (x - y, x - y)$$
$$= (x,x) + 2(x,y) + (y,y) + (x,x) - 2(x,y) + (y,y) = 2\|x\|^2 + \|y\|^2.$$

Thus the parallelogram identity holds in every Hilbert space.

The first step in proving the converse is to choose a proper definition of the scalar product (x,y). Again, a heuristic argument suggests the answer: If x and y are vectors in R^2, and if θ is the angle determined by these vectors (Figure 29), then it follows from the definition of the scalar

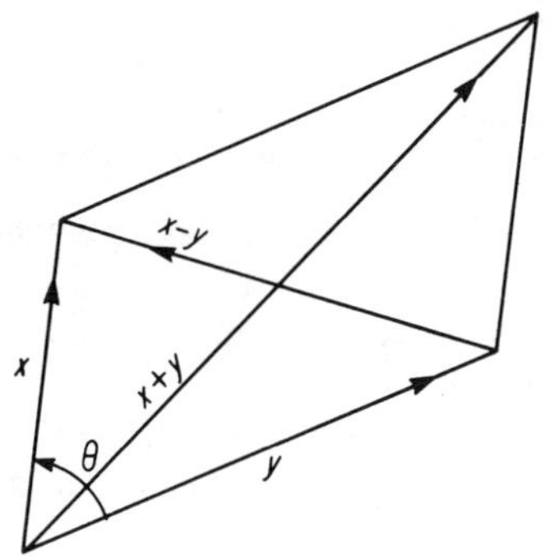

FIGURE 29

product in R^2 (Example 2(a)) and from the law of cosines that

$$\|x\|^2 + \|y\|^2 - 2\cos\theta\,\|x\|\cdot\|y\| = \|x - y\|^2,$$
$$(x,y) = \cos\theta\,\|x\|\cdot\|y\|.$$

Eliminating $\cos\theta$, we obtain

[6] $$(x,y) = \frac{\|x\|^2 + \|y\|^2 - \|x - y\|^2}{2}.$$

We show next that whenever the parallelogram identity holds in a Banach space H, then [6] may be used to *define* a scalar product in H which is compatible with the given norm.

There is no difficulty in verifying that the expression defined by [6] satisfies (i) and (ii) of Definition 1. To verify (iv), simply requires a straightforward but tedious calculation. Make the appropriate substitutions in [6] and use [5].

Proof of (iii): This is another example of a proof by "continuity" (See the proof of Theorem 2.1.2). First (iii) is verified for positive integers, then for all integers and rational numbers, and finally for irrational numbers:

Let $c = n = 1, 2, \ldots$. Repeated application of (iv) yields

$$(nx,y) = (\overbrace{x + x + \cdots + x}^{n \text{ times}},y) = \overbrace{(x,y) + \cdots + (x,y)}^{n \text{ times}} = n(x,y).$$

(A rigorous proof requires the Principle of Mathematical Induction.) Suppose now that $c = -n = -1, -2, \ldots$. Then

$$[7] \qquad (-nx,y) + (nx,y) = (\bar{0},y) = 0\cdot(\bar{0},y) = 0,$$

which implies

$$(-nx,y) = -(nx,y) = -n(x,y).$$

The proof for the case $n = 0$ is included in [7].

If $c = 1/n$, then $n\left(\frac{1}{n}x,y\right) = \left(\frac{n}{n}x,y\right) = (x,y)$. The desired result is obtained upon division by n. A similar argument holds for any rational number n/m.

For irrational values of c, a continuity argument is given. Let $\{a_n\}$ be a sequence of rational numbers that converges to c. If it can be shown that for any fixed but arbitrary pair of elements $x,y \in H$, that $F(c) = (cx,y)$ is a continuous function of c, then it would follow that

$$(cx,y) = \lim_{n\to\infty} (a_n x,y) = \lim_{n\to\infty} a_n(x,y) = c(x,y).$$

The continuity of $F(c)$ is proved by using [6]:

$$\begin{aligned} 2[F(c) - F(a)] &= 2[(cx,y) - (ax,y)] \\ &= \|x\|^2(c^2 - a^2) + (\|ax - y\|^2 - \|cx - y\|^2). \end{aligned}$$

Since the first term tends to zero as a tends to c, it suffices to show that the same is true of the second term. Factoring, and assuming that $|a| \le 2|c|$ (remember, we shall take a limit), we obtain an upper bound for the absolute value of this term:

$$(3c\|x\| + 2\|y\|)(\|ax - y\| - \|cx - y\|).$$

Calling the first term M, and using the triangle inequality yields the upper bound

$$M\|ax - y - cx + y\| = M|a - c|\|x\|,$$

which tends to zero as a tends to c. ∎

Exercises

1. Which of the following expressions defines an inner product on the space R^3?. For those that do not, indicate the properties that are not satisfied.
(a) $(x,y) = 2x_1y_2 - x_2y_3 + x_3y_1$.
(b) $(x,y) = x_1y_1 + x_2y_2 + x_3y_2 + x_2y_3 + x_3y_3$.

(c) $(x,y) = x_1y_1 + x_2y_2$.
(d) $(x,y) = (x_1y_1 + x_2y_2 + x_3y_3)^2$.

2. For what values of γ_{ij} is

$$(x,y) = \sum_{i=1}^{3}\sum_{j=1}^{3} \gamma_{ij}x_iy_j$$

an inner product on R^3?. The answer to this question can be found in most standard linear algebra texts. However, the reader is urged to work it out for himself. It might be useful to describe first all inner products on the real vector space R, and then to find the values of γ_{ij} which make

$$(x,y) = \gamma_{11}x_1y_1 + \gamma_{12}x_1y_2 + \gamma_{21}x_1y_2 + \gamma_{22}x_2y_2$$

an inner product on R^2.

3. Which of the spaces described in Exercises 1 and 2 and Example 2 are Hilbert spaces?

*4. Show that (x,y) as defined by [6] satisfies (iv) of Definition 1.

*5. Prove that χ of Exercise 10.1.7 is a Hilbert space.

*6. Let X be an inner product space. Show that if $(z,x) = 0$ for all $x \in X$, then $z = \overline{0}$.

7. Let

$$f(x) = \sum_{i=0}^{n} a_ix^i, \qquad g(x) = \sum_{i=0}^{m} b_ix^i$$

be elements of the real vector space $R[x]$ of polynomials in x with real coefficients. Show that

$$(f,g) = \sum_{i=0}^{n}\sum_{j=0}^{m} \frac{a_ib_j}{i+j+1}$$

is an inner product on $R[x]$.

8. Let X be an inner product space. Prove that (x,y) is a continuous function of two variables.

9. A sequence $\{x_n\}$ in a Hilbert space H is said to *converge weakly* to $x \in H$ if, for all $y \in H$, $\lim (x_n,y) = (x,y)$. This will be denoted by $x_n \dashrightarrow x$.
   (a) Show that convergence in norm ($\lim \|x_n - x\| = 0$) implies weak convergence ($x_n \dashrightarrow x$).
   (b) Show that if $x_n \dashrightarrow x$, and if $\lim \|x_n\| = \|x\|$, then $\lim \|x_n - x\| = 0$. (*Hint:* Use the parallelogram identity on $\|x_n - x\|^2$.)
   (c) Give an example of a sequence which converges weakly but not in norm. (See Exercises 10.1.8 and 10.)

*10. Prove that the Pythagorean theorem holds in a Hilbert space; that is, $(x,y) = 0$ if and only if $\|x + y\|^2 = \|x\|^2 + \|y\|^2$.

*11. Show that in a Hilbert space, $\|x + y\| = \|x\| + \|y\|$ if and only if $x = ay$ or $y = ax$, for some real number a.

*Exercises which request the proof of a statement in the text, as well as those referred to in the text, are starred.

3 Separable Hilbert Spaces and Orthonormal Sequences

Each vector of R^3 (equipped with the Euclidean norm and the conventional definition of angle) can be expressed uniquely as a linear combination of the mutually orthogonal unit vectors, $e_j = (\delta_{j1},\delta_{j2},\delta_{j3})$, $j = 1,2,3$; $\delta_{jk} = 0$ if $j \neq k$ and $\delta_{jj} = 1$ (δ_{jk} is called the *Kronecker delta*). Since orthogonality can be defined in any inner product space, it is reasonable to direct our inquiry toward determining the existence of a set of mutually orthogonal elements which will serve as a set of "basis" vectors. We shall prove first that if H is a separable Hilbert space which contains an orthogonal set, then that set is countable (Proposition 5), and as we shall see, finite combinations of such vectors will approximate the elements of H in the Hilbert space norm.

All of the Hilbert spaces of Example 10.2.2 are separable. The space χ (Exercise 10.1.7) is not separable.

One of the principal results of this section is the complete classification of separable Hilbert spaces: If a Hilbert space has (vector space) dimension n, then it is isonormal to R^n, which is equipped with the scalar product that gives rise to the Euclidean norm. If the separable space does not have finite dimension, then it is isonormal to l_2, the space of real sequences $x = \{x_n\}$ for which $\Sigma x_n^2 < \infty$, and $(x,y) = \Sigma x_n y_n$. This means that l_2, L_2 and all other *infinite dimensional separable Hilbert spaces are, as Hilbert spaces, indistinguishable.*

1 Definition. Two nonzero elements x,y of a Hilbert space H are said to be *orthogonal* if $(x,y) = 0$.

A subset $\{x_\alpha\}$ (not necessarily countable) of nonzero elements of H is called an *orthogonal set* if $(x_\alpha,x_\beta) = 0$ whenever $\alpha \neq \beta$. If moreover, $(x_\alpha,x_\alpha) = 1$ for all indices α, then the set of vectors is said to be *orthonormal.*

2 Example. (a) Let $H = R^n$, $(x,y) = \sum_{i=1}^{n} x_i y_i$. The subset $e_j = (\delta_{j1},\delta_{j2},\ldots,\delta_{jn})$, $j = 1,2,\ldots,n$, is an orthonormal set.

(b) Similarly, $e_j = (\delta_{j1},\delta_{j2},\ldots,\delta_{jn},\ldots)$, $j = 1,2,\ldots$ is an *orthonormal sequence* in l_2.

(c) If each δ_{jj} in the preceding examples is replaced by $r\delta_{jj} = r \neq 0,1$, then the set of vectors $\{re_j\}$ is an orthogonal subset of R^n which is not orthonormal.

(d) Let $H = L_2[0,2\pi]$, the space of square integrable functions that vanish outside the interval $[0,2\pi]$, and let

$$(f,g) = \int_0^{2\pi} fg.$$

It is left to the reader to show that the set of trigonometric functions:

$$\frac{1}{\sqrt{2\pi}}, \quad \frac{\cos x}{\sqrt{\pi}}, \quad \frac{\sin x}{\sqrt{\pi}}, \ldots, \frac{\cos nx}{\sqrt{\pi}}, \quad \frac{\sin nx}{\sqrt{\pi}}, \ldots,$$

is an orthonormal sequence in $L_2[0,2\pi]$.

3 Definition. A set of vectors $\{x_\alpha\}$ (not necessarily finite or even countable) is said to be *linearly independent* if, whenever $x_1, \ldots, x_n$ is a finite subset of $\{x_\alpha\}$, then

$$\sum_{i=1}^{n} a_i x_i = \bar{0}, \; a_i \in R, \qquad \text{implies} \qquad a_i = 0, \; i = 1,2,\ldots,n.$$

4 Proposition. Every orthogonal set of vectors is a linearly independent set.

Proof: Let $\{x_1, \ldots, x_n\}$ be a finite subset of an orthogonal set $\{x_\alpha\}$. If $a_1, \ldots, a_n$ are real numbers for which $\sum_{i=1}^{n} a_i x_i = \bar{0}$, then for $j = 1,2, \ldots, n$,

$$0 = (x_j,\bar{0}) = \left(x_j, \sum_{i=1}^{n} a_i x_i\right) = \sum_{i=1}^{n} a_i (x_j, x_i) = \sum_{i=1}^{n} a_i \delta_{ji} = a_j,$$

which asserts that the a_j are all equal to zero. ∎

The converse of Proposition 4 is obviously false: The vectors (1,0,0), (0,1,2) and (0,0,1) of the Euclidean space R^3 are linearly independent but not mutually orthogonal.

5 Proposition. Every orthonormal set in a separable Hilbert space is countable.

Proof: Let $\mathcal{O} = \{e_\alpha\}$ be an orthonormal subset of a separable Hilbert space H. If $e_\alpha \neq e_\beta$, then $(e_\alpha, e_\beta) = \delta_{\alpha\beta}$ implies

$$\|e_\alpha - e_\beta\|^2 = (e_\alpha - e_\beta, e_\alpha - e_\beta) = (e_\alpha, e_\alpha) - 2(e_\alpha, e_\beta) + (e_\beta, e_\beta) = 2,$$

proving that the distance between any distinct pair of elements of an orthonormal set is $\sqrt{2}$. Let $\{x_n\}$ be a dense sequence of the separable Hilbert space H. Then to each index α, there is an integer $n(\alpha)$, and $\|e_\alpha - x_{n(\alpha)}\| < 1/4$. (This is simply the definition of a dense set.) It is easily seen that the mapping $\alpha \to n(\alpha)$ is injective. For, if there were a pair of distinct indices α and β for which $n(\alpha) = n(\beta)$, and consequently $x_{n(\alpha)} = x_{n(\beta)}$, it would follow that

$$\|e_\alpha - e_\beta\| \leq \|e_\alpha - x_{n(\alpha)}\| + \|x_{n(\alpha)} - e_\beta\| < 1/2,$$

which cannot be unless $\alpha = \beta$. Thus the set $\{e_\alpha\}$ can be put into a 1–1 correspondence with a subset of $\{x_n\}$, and is therefore countable. ▎

Remark: Propositions 4 and 5 are true in any inner product space. Neither proof makes use of the completeness of H.

Hereafter we shall restrict the discussion to separable Hilbert spaces. The next theorem defines the role of orthonormal sets in approximating elements of H.

6 Theorem. Let $\{e_i\}$ be an orthonormal sequence (o.n.s.) in a Hilbert space H. Then for positive integral n, real c_i, and $x \in H$, the expression

$$\|x - \sum_{i=1}^{n} c_i e_i\| \tag{1}$$

is minimized by choosing $c_i = (e_i,x)$. (The coefficients $c_i = (e_i,x)$ are called the *Fourier coefficients* of x with respect to the o.n.s. $\{e_i\}$.)

Remark: This theorem asserts that the best approximation of x (in the Hilbert space norm) by linear combinations of the e_i, is obtained by choosing the c_i to be the Fourier coefficients of x with respect to the e_i. This does *not* make any claims about how good this approximation is. Indeed, it is not in general possible to make the norm of the difference [1] as small as desired by taking n to be large. The theorem asserts only that if the c_i are chosen in any other way, then the approximation will be even worse. See Example 8 and the discussion that follows the proof of this theorem.

Proof: Expanding [1] and omitting the terms $c_i c_j (e_i,e_j) = 0$, when $i \neq j$, we obtain

$$\|x - \sum_{1}^{n} c_i e_i\|^2 = \left(x - \sum_{1}^{n} c_i e_i, x - \sum_{1}^{n} c_i e_i\right) \tag{2}$$

$$= (x,x) + \sum_{1}^{n} c_i^2 - 2\sum_{1}^{n} c_i(e_i,x).$$

If $\sum_{1}^{n} (e_i,x)^2$ is added and subtracted from [2], factorization yields,

$$\|x - \sum_{1}^{n} c_i e_i\|^2 = (x,x) - \sum_{1}^{n} (e_i,x)^2 + \sum_{1}^{n} [(e_i,x) - c_i]^2. \tag{3}$$

The first two terms of [3] do not depend upon how the coefficients c_i are chosen. The third is a sum of nonnegative numbers which is minimized if each $c_i = (e_i,x)$, the Fourier coefficient. ▎

7 Corollary. Let $\{e_i\}$ be an o.n.s. in a Hilbert space H. Then, for any $x \in H$,

$$\|x\|^2 = (x,x) \geq \sum_1^\infty c_i^2; \tag{4}$$

$c_i = (e_i,x)$. Furthermore,

$$\lim_{i\to\infty} (e_i,x) = 0.$$

[4] is called *Bessel's Inequality.*

Proof: [4] is obtained by setting $c_i = (e_i,x)$ in [2].

The convergence of $\sum_1^\infty c_i^2$ implies that the terms $c_i^2 = (e_i,x)^2$ converge to zero. ∎

At this point we return to the remark following the statement of Theorem 6. By means of several examples, we shall try to illuminate the connection between Bessel's inequality and the possibility of approximating an element of H by its *Fourier sums* $\sum_1^n c_ie_i$.

8 Examples. (a) $\{(1,0,0), (0,1,0)\}$ is an orthonormal subset of the Euclidean space R^3. Theorem 6 asserts that the vector $x = (3,-5,6)$ is best approximated by a sum of the type $c_1e_1 + c_2e_2$ if $c_1 = 3$ and $c_2 = -5$. However, this isn't a terribly good approximation, since $\|x - (3e_1 - 5e_2)\| = 6$. It is simply the best we can do with linear combinations of e_1 and e_2. The reason for this is that there are not enough vectors in the orthonormal set. Observe also, that in this case, Bessel's inequality becomes

$$70 = (x,x) > c_1^2 + c_2^2 = 34.$$

By joining $(0,0,1)$ to the given orthonormal set, we obtain

$$\|x - (3e_1 - 5e_2 + 6e_3)\| = 0;$$

Bessel's inequality becomes an equality, $70 = (x,x) = c_1^2 + c_2^2 + c_3^2 = 70.$

(b) Much the same thing happens if $H = l_2$, and $(x,y) = \sum_1^\infty x_iy_i$. The sequence $e_1' = (0,1,0,0,\ldots)$, $e_2' = (0,0,1,0,\ldots)\ldots, e_{n-1}' = (\delta_{n1},\delta_{n2},\ldots,\delta_{nk},\ldots),\ldots$ is orthonormal, but since the vector $(1,0,0,\ldots)$ is not in the sequence, there are not enough vectors to approximate those elements $x = (x_1,x_2,\ldots)$ in H whose first coordinate, $x_1 \neq 0$. Furthermore, Bessel's inequality, for such vectors, is a strict inequality.

These examples suggest the more general statement (Proposition 9) that the Fourier sums $\sum_1^n c_i e_i$ will approximate an element x in H provided that $(x,x) = \sum_1^\infty c_i^2$. This is called *Parseval's identity.*

9 Proposition. Let $\{e_i\}$ be an o.n.s. in a Hilbert space H. Then the elements of H may be approximated (in the Hilbert space norm) by the Fourier sums if and only if Parseval's Identity holds for all $x \in H$. This means that to each $x \in H$ and $\epsilon > 0$, there is an integer $N = N(\epsilon,x)$ for which $\|x - \sum_1^N c_i e_i\| < \epsilon$, $c_i = (e_i,x)$, if and only if $\|x\|^2 = (x,x) = \sum_1^\infty c_i^2$.

Proof: Equation [3], with the c_i taken as the Fourier coefficients, becomes

$$\|x - \sum_1^n c_i e_i\|^2 = (x,x) - \sum_1^n c_i^2,$$

from which the desired result follows. ▮

10 Definition. An o.n.s. of a Hilbert space H is said to be *complete* if to each $x \in H$ and $\epsilon > 0$, there is an integer N such that

$$\|x - \sum_1^N c_i e_i\| < \epsilon, \; c_i = (e_i,x).$$

(Do not confuse the idea of a complete orthonormal sequence with a complete metric space.)

Proposition 9 asserts that an o.n.s. $\{e_i\}$ is complete if and only if Parseval's identity holds for all $x \in H$.

11 Definition. An o.n.s. $\{e_i\}$ of a Hilbert space H is said to be *closed* if, whenever $x \neq y$, there is an integer n, and $(e_n,x) \neq (e_n,y)$.

It is easily seen that by writing $z = x - y$, Definition 11 can be restated as follows:

11′ Definition. An o.n.s. $\{e_i\}$ of a Hilbert space H is said to be *closed* if, whenever $z \neq \overline{0}$, then $(e_n,z) \neq 0$ for some n.

12 Definition. An o.n.s. $\{e_i\}$ of a Hilbert space H is said to be *maximal in H* if $\{e_i\}$ is not a proper subset of another o.n.s. in H. (This means that if $\{e_i'\}$ is an o.n.s. in H and $\{e_i'\} \supset \{e_i\}$, then $\{e_i'\} = \{e_i\}$.)

13 Theorem. Let $\{e_i\}$ be an o.n.s. in a Hilbert space H. The following statements are equivalent·

(a) $\{e_i\}$ is complete.

(b) $\{e_i\}$ is maximal in H.

(c) $\{e_i\}$ is closed.

(d) If $x \in H$, then $(x,x) = \sum_1^\infty c_i^2$; that is, Parseval's identity holds for all x in H.

(e) If $x,y \in H$, then $(x,y) = \sum_1^\infty c_i d_i$; c_i and d_i are the Fourier coefficients of x and y respectively.

Proof: (a) $\Longrightarrow$ (b): If $\{e_i\}$ is not maximal in H, there is a nonzero element e_0 in H, and $\{e_0, e_1, \ldots\}$ is an o.n.s. From $c_i = (e_i,e_0) = 0$, $i = 1,2, \ldots$, it follows that for all n.

$$\| e_0 - \sum_1^n c_i e_i \| = \| e_0 \| = 1,$$

proving that $\{e_i\}$ is not complete.

(b) $\Longrightarrow$ (c): If $\{e_i\}$ is not closed, there is a nonzero $x \in H$, such that $c_i = (e_i,x) = 0, i = 1,2,\ldots$. Let $e_0 = \dfrac{x}{\|x\|}$. Then $\|e_0\| = 1$, and for $i = 1,2,\ldots$,

$$(e_i,e_0) = \left(e_i, \frac{x}{\|x\|}\right) = \frac{1}{\|x\|}(e_i,x) = 0,$$

proving that $\{e_0,e_1,e_2,\ldots\}$ is an o.n.s. that contains the given sequence $\{e_1,e_2,\ldots\}$ as a proper subset. This contradicts the maximality of $\{e_i\}$.

(c) $\Longrightarrow$ (d): Bessel's inequality yields $\|x\|^2 = (x,x) \geq \sum_1^\infty c_i^2$, for all $x \in H$ and $c_i = (e_i,x)$. To reverse the inequality, we prove first that the Fourier sums, $s_n = \sum_1^n c_i e_i$ form a Cauchy sequence in H: By taking $n > m$, we obtain

$$\| s_n - s_m \|^2 = \left(\sum_{m+1}^n c_i e_i, \sum_{m+1}^n c_i e_i \right) = \sum_{m+1}^n c_i^2 \leq \sum_{m+1}^\infty c_i^2;$$

the last term tends to zero as m tends to infinity. Since H is complete, the sequence $\{s_n\}$ converges to an element y in H, and $(e_i,y) = c_i$. Furthermore, lim $\|s_n - y\| = 0$ implies lim $\|s_n\| = \|y\|$, from which it follows that

$$\|y\|^2 = \sum_1^\infty c_i^2 \leq \|x\|^2.$$

This proves that if x is an element of H for which Parseval's identity does *not* hold, then there is a $y \in H$, whose Fourier coefficients are the same as those of x ($(e_i,x) = c_i = (e_i,y)$). However, $\|y\|^2 < \|x\|^2$, which implies that $x \neq y$, contradicting the assumption that $\{e_i\}$ is closed.

$$(d) \Longrightarrow (e): (x - y, x - y) = \|x\|^2 + \|y\|^2 - 2(x,y) \Longleftrightarrow$$

$$\sum_1^\infty (c_i - d_i)^2 = \sum_1^\infty c_i^2 + \sum_1^\infty d_i^2 - 2(x,y) \Longleftrightarrow \sum_1^\infty c_i d_i = (x,y).$$

(c_i and d_i are the Fourier coefficients of x and y respectively.)
Fourier coefficients of x and y respectively.)

(d) $\Longrightarrow$ (a): See Proposition 9. ∎

We conclude this section by showing that every separable Hilbert space is isonormal to l_2. The argument is divided into two parts. First it is shown that every infinite o.n.s. $\{e_i\}$ in H determines a mapping of H onto l_2. This mapping is injective if and only if $\{e_i\}$ satisfies any of the conditions of Theorem 13. The problem is thus reduced to showing that H contains a complete o.n.s. Two proofs are given, one using Zorn's lemma, the second, a constructive proof.

We begin by defining the mapping $\Phi: H \to l_2$. Let $\{e_i\}$ be an o.n.s. in a Hilbert space H. For each x in H, the vector $c = (c_1, c_2, \ldots)$, where $c_i = (e_i, x)$, is in l_2, since $\Sigma c_i^2 = (x,x) < \infty$. We therefore define the mapping Φ by

$$\Phi(x) = (c_1, c_2, \ldots) = c; \qquad c_i = (e_i, x)$$

This mapping preserves the vector space operations (Exercise 6) and is norm preserving if and only if Parseval's identity holds for the o.n.s.:

$$\|x\|^2 = (x,x) = \sum_1^\infty c_i^2 = \|c\|_2^2.$$

It is easily seen that Φ is also surjective; indeed, every $c = (c_1, c_2, \ldots)$ in l_2 determines a Cauchy sequence $s_n = \sum_1^n c_i e_i$ in H which converges to an element x of H (See the proof of Theorem 13, (c) $\Longrightarrow$ (d)). The c_i are the Fourier coefficients of x, and therefore $\Phi(x) = c$.

The mapping Φ is not always injective. For example, consider the o.n.s., $e_1' = (0,1,0,0,\ldots)$, $e_2' = (0,0,1,0,0,\ldots), \ldots$. For all real values of a, the vectors $x_a = (a,0,0,\ldots)$ are mapped into $\bar{0}$ by Φ. More generally, we see that Φ is injective if and only if

$$x \neq y \qquad \text{implies} \qquad (e_i,x) \neq (e_i,y) \text{ for some value of } i;$$

that is, if and only if $\{e_i\}$ is closed (or complete etc., Theorem 13).

This takes us to the final step, which is to prove that if H is separable, it contains an o.n.s. which is closed, or complete, or maximal etc. The first proof of the existence of such a sequence (Theorem 14) uses Zorn's lemma, which gives directly, the desired maximal sequence.

14 Theorem. If H is a separable Hilbert space, then it contains a complete orthonormal sequence.

Proof 1 (nonconstructive): Let $\mathfrak{F}$ be the family of all orthonormal subsets of H. If H contains any nonzero elements, then $\mathfrak{F}$ is not empty; indeed, if $x \geq 0$, then $\{e\} = \dfrac{x}{\|x\|}$ is an o.n.s. that contains a single element.

The separability of H implies that each orthonormal set is countable (Proposition 5).

The inclusion relation "$\subset$" partially orders $\mathfrak{F}$. Suppose that $\{E^\alpha\}$ is a totally ordered subset of $\mathfrak{F}$ indexed by the set $\mathfrak{A}$ (not necessarily countable). The union

$$E = \bigcup_{\alpha \in \mathfrak{A}} E^\alpha = \bigcup_{\alpha \in \mathfrak{A}} \left(\bigcup_{i=1}^{\infty} \{e_i^\alpha\} \right)$$

($E^\alpha = \{e_1^\alpha, e_2^\alpha, \ldots\}$), is countable, and is easily seen to be in $\mathfrak{F}$: For, if e_i^α, $e_j^\beta \in E$, then either $E^\alpha \subset E^\beta$ or vice versa. Assuming the former, there is a vector $e_k^\beta \in E^\beta$ such that $e_k^\beta = e_i^\alpha$. Therefore

$$(e_i^\alpha, e_j^\beta) = (e_k^\beta, e_j^\beta) = \delta_{kj},$$

proving that E is an o.n.s., and is in $\mathfrak{F}$. Moreover, for all $\alpha \in \mathfrak{A}$, $E^\alpha \subset E$, so that E is an upper bound for the totally ordered set $\{E^\alpha\}$. Zorn's Lemma asserts the existence of a maximal element E^* in $\mathfrak{F}$. E^* is itself an o.n.s. and is not contained in a larger one. Theorem 13 yields the desired complete o.n.s. ∎

Proof 2 (constructive): One advantage of this proof over the former, is that it does not rely upon Zorn's Lemma or any of its equivalent statements, which some mathematicians find objectionable. A second is that it provides a procedure for constructing an o.n.s. from any dense sequence in H.

Let $\{x_i\}$ be a dense sequence in H. The first step is to discard "extraneous" elements of the sequence. This is done by selecting a subsequence $\{y_i\}$ as follows: Let y_1 be the first nonzero element that appears in the given sequence $\{x_i\}$. Then choose $y_2 = x_{i(2)}$ so that $\{y_1, y_2\}$ is a linearly independent set of vectors, and

$$j < i(2) \qquad \text{implies} \qquad x_j = cy_1, \quad \text{for some } c \in R.$$

In general, having chosen $y_j = x_{i(j)}$, $j = 1,2,\ldots,n$, let $y_{n+1} = x_{i(n+1)}$ be selected so that $i(n+1) > i(n)$, and the set $\{y_1,\ldots,y_n,y_{n+1}\}$ is linearly independent; but if $j < i(n+1)$ then x_j is a linear combination of $y_1,\ldots,y_n$. The sequence of vectors $\{y_1,y_2,\ldots\}$ is linearly independent (Definition 3), and the collection $\mathfrak{C}$ of finite linear combinations of the y_i contains the original sequence $\{x_i\}$, thus proving that $\mathfrak{C}$ is itself dense in H.

The next step is to "orthogonalize" the dense linearly independent sequence $\{y_n\}$; that is, to obtain from the y_n, a dense orthogonal sequence $\{z_n\}$. Let

$$[6] \qquad z_n = \begin{vmatrix} (y_1,y_1) & (y_1,y_2) & \cdots & \cdots & (y_1,y_{n-1}) & y_1 \\ (y_2,y_1) & (y_2,y_2) & \cdots & \cdots & (y_2,y_{n-1}) & y_2 \\ \vdots & & & & \vdots & \vdots \\ (y_n,y_1) & (y_n,y_2) & \cdots & \cdots & (y_n,y_{n-1}) & y_n \end{vmatrix},$$

where z_n is the element in H obtained by formal expansion of the determinant [6]. The entries in the first $n-1$ columns are real numbers, and those of the last column are the n linearly independent vectors $y_1,y_2, \ldots y_n$. Thus

$$z_n = \sum_{i=1}^{n} a_i^{(n)} y_i;$$

each $a_i^{(n)}$ is the appropriate minor of the determinant. It is easily verified that z_n is orthogonal to $y_1,y_2,\ldots,y_{n-1}$: The inner product (z_n,y_i), $i = 1,2,\ldots,n-1$, is obtained by replacing the last column of [6] by the numbers $(y_1,y_i),\ldots,(y_n,y_i)$. Thus this last column is identical with the ith column, proving that the determinant vanishes. However, $z_1,\ldots,z_{n-1}$ are linear combinations of $y_1,\ldots,y_{n-1}$, from which it follows that for all n,

$$(z_n,z_j) = 0 \quad \text{if} \quad j = 1,2,\ldots,n-1.$$

This proves that $\{z_n\}$ is an orthogonal sequence in H. Again the finite linear combinations of the z_n are dense in H.

An o.n.s. is obtained by setting $e_n = \dfrac{z_n}{\|z_n\|}$. In view of the preceding paragraph, the finite linear combinations of the e_n are dense in H, thus proving that $\{e_n\}$ is a complete o.n.s. in H. ∎

Exercises

1. Determine in each case whether the subset Y is linearly independent, orthogonal, orthonormal.

(a) Let $X = R^3$, $(x,y) = x_1y_1 + x_2y_2 + 2x_3y_3$, $Y = \{(1,0,0),(0,1,0),(0,0,1)\}$.

(b) $X = C[0,1]$, $(f,g) = \int_0^1 fg$. (X is not complete, but *is* an inner product space.) The subset to be tested is the set of monomials, $Y = \{1, x, x^2, \ldots, x^n, \ldots\}$.

(c) Let X and (f,g) be as in part (b). $Y = \{1 - x, 1 + x, x + x^2, 1 + x^2, x^3, x^4, \ldots, x^n, \ldots\}$.

2. Use the procedure suggested in the second proof of Theorem 14 to construct an o.n.s. for each sequence Y of Exercise 1.

*3. Prove that the trigonometric sequence of Example 2(d) is orthonormal.

4. Give an example of a complete o.n.s. in l_2.

5. Show that a Hilbert space whose vector space dimension is n is isonormal to R^n, equipped with the usual inner product that gives rise to the Euclidean norm.

*6. Show that the mapping Φ of the separable Hilbert space H into l_2, defined in the proof of Theorem 14, preserves the vector space operations.

7. Without using [6], find a quadratic polynomial $ax^2 + bx + c$, which is orthogonal to both 1 and x in $C[0,1]$, $(f,g) = \int_0^1 fg$. Is it uniquely determined?

*Exercises which request the proof of a statement in the text, as well as those referred to in the text, are starred.

4 Fourier Series

In Sec. 3 we proved that a separable Hilbert space possesses a complete o.n.s. (orthonormal sequence) (Theorem 10.3.14). This means that for each x in H and $\epsilon > 0$,

$$[1] \qquad \left\| x - \sum_1^n c_i e_i \right\| < \epsilon,$$

for sufficiently large values of n; $\{e_i\}$ is the o.n.s. and $c_i = (e_i, x)$ are the Fourier coefficients. Moreover, the *best* approximation (the one that minimizes [1]) is obtained when the c_i are the Fourier coefficients (Theorem 10.3.6).

In view of the isonormality of all separable Hilbert spaces, we may, without loss of generality, concentrate on L_2, the space of (equivalence classes of) square integrable functions. Suppose that $\{e_i\}$ is a complete o.n.s. in L_2. Although it is possible to approximate a function f of L_2 in the norm [1] by its Fourier sums,

$$[2] \qquad s_n = \sum_1^n c_i e_i, \qquad c_i = (e_i, f) = \int e_i f,$$

we do not as yet know anything about the pointwise convergence of these sums. In this section, we shall restrict ourselves to square integrable functions which vanish outside the fixed interval $[0,2\pi]$, or equivalently, to square integrable functions with *period* 2π, ($f(x) = f(x + 2\pi)$, for all real values of x). The o.n.s. is taken to be the sequence of trigonometric functions (Example 10.3.2(d)), and the partial sums [2] will be the *trigonometric sums*. The questions raised here are roughly the following.

Under what conditions does the (trigonometric) series of an integrable function converge a.e.? If the series belonging to f converges at a point x, does it converge to $f(x)$? Is every a.e. convergent trigonometric series the Fourier series of an integrable function?

We consider the o.n.s. in $L_2[0,2\pi]$,

[3] $$\frac{1}{\sqrt{2\pi}}, \frac{\cos x}{\sqrt{\pi}}, \frac{\sin x}{\sqrt{\pi}}, \ldots, \frac{\cos nx}{\sqrt{\pi}}, \frac{\sin nx}{\sqrt{\pi}}, \ldots$$

[Example 10.3.2(d)]. Since the trigonometric functions of [3] are bounded as well as integrable, it follows that if $f \in L_1[0,2\pi]$, then $f(x) \sin nx$, $f(x) \cos nx \in L_1[0,2\pi]$ for $n = 0, 1, \ldots$, proving that the Fourier series of functions in $L_1[0,2\pi]$ as well as $L_2[0,2\pi]$ are well defined. For convenience, the factor $1/\sqrt{\pi}$ will be omitted in [3]; the resulting sequence is orthogonal, but not orthonormal.

1 Definition. For each $f \in L_1[0,2\pi]$, the series

[4] $$a_0/2 + \sum_1^{\infty} [a_n \cos nx + b_n \sin nx],$$

where

[5] $$a_0 = \frac{1}{\pi} \int_0^{2\pi} f, \qquad \begin{Bmatrix} a_n \\ b_n \end{Bmatrix} = \frac{1}{\pi} \int_0^{2\pi} \begin{Bmatrix} \cos nx \\ \sin nx \end{Bmatrix} f$$

is called the *trigonometric or Fourier series of f.*

Let us suppose that an a.e. (on $[0,2\pi]$) convergent trigonometric series is given. We shall find sufficient conditions for the limit function f to be in $L_1[0,2\pi]$. In this case, it will be shown that the Fourier series [4] of the function f coincides with the given trigonometric series.

2 Theorem. Let

[6] $$a_0/2 + \sum_1^{\infty} [a_n \cos nx + b_n \sin nx]$$

be a trigonometric series that converges a.e. on $[0,2\pi]$ to a function $f(x)$.* If the partial sums,

*Since $\cos n(x + 2\pi) = \cos nx$ and $\sin n(x + 2\pi) = \sin nx$, the series will actually converge a.e. on R to a periodic function f; $f(x + 2\pi) = f(x)$, for all real values of x.

$$s_n = a_0/2 + \sum_1^n [a_j \cos jx + b_j \sin jx]$$

are bounded by a function which is integrable on $[0,2\pi]$, then $f \in L_1[0,2\pi]$, and the trigonometric series [6] is the Fourier series of $f(x)$; that is, the coefficients a_0, b_n, a_n of [6] are given by [5].

Proof: The theorem on dominated convergence gives the integrability of $f(x)$ as well as $f(x) \sin nx$ and $f(x) \cos nx$, $n = 1,2,\ldots.$ The orthogonality of the trigonometric functions (Exercise 10.3.3) and the theorem on dominated convergence yield

$$\int_0^{2\pi} f = \lim_{n\to\infty} \int_0^{2\pi} s_n = \pi\frac{a_0}{2},$$

and

$$\int_0^{2\pi} f \begin{Bmatrix} \cos nx \\ \sin nx \end{Bmatrix} = \lim_{n\to\infty} \int_0^{2\pi} s_n \begin{Bmatrix} \cos nx \\ \sin nx \end{Bmatrix} = \begin{Bmatrix} \pi a_n \\ \pi b_n \end{Bmatrix},$$

thus proving that the given series which converges a.e. to f is indeed its Fourier series. ∎

However, there exist a.e. convergent trigonometric series that are *not* the Fourier series of an integrable function (See Exercise 14, or Gelbaum and Olmstead [1] page 70). In fact, although it will not be proven here, if f is the a.e. limit of a trigonometric series which is *not* a Fourier series, then f is not in $L_1[0,2]$. Taken together with Theorem 2, this yields the following conclusion:

If a trigonometric series converges a.e. to a function f *in* $L_1[0,2\pi]$, *then the Fourier series of* f *coincides with the given trigonometric series.*

The remainder of this section is devoted to finding sufficient conditions for the Fourier series of an integrable function f to converge a.e. to f. The principal results are stated in Theorems 5, 8, 16, and 19, and Corollaries 6 and 9.

The result that is most frequently used in subsequent proofs is the *Riemann-Lebesgue theorem* (Theorem 3). An immediate consequence of this theorem is the assertion that the Fourier coefficients of an integrable function tend to zero as n tends to infinity (Exercise 12).

3 Theorem. If f is integrable, then

[7] $$\lim_{\gamma\to\infty} \int f(x) \begin{Bmatrix} \cos \gamma x \\ \sin \gamma x \end{Bmatrix} = 0.$$

Proof: We prove, in order, that [7] holds when

(i) $f = \chi_{[a,b]}$, the characteristic function of an interval;
(ii) $f = \tau$, a step function;
(iii) $f \in L_1$.

If $f = \chi_{[a,b]}$, then

$$\int f(x) \cos \gamma x = \int_a^b \cos \gamma x = \frac{1}{\gamma} [\sin \gamma b - \sin \gamma a] \to 0,$$

as γ tends to infinity. A similar argument holds for $\int f(x) \sin \gamma x$.

Since every step function is a linear combination of characteristic functions, the linearity of the integral yields the desired result [7] for step functions.

Finally, if $f \in L_1$, there is a Cauchy sequence of step functions $\{\tau_n\}$ which converges (both in norm and a.e.) to f, and

$$\int f(x) \begin{Bmatrix} \cos \gamma x \\ \sin \gamma x \end{Bmatrix} = \lim_{n \to \infty} \int \tau_n \begin{Bmatrix} \cos \gamma x \\ \sin \gamma x \end{Bmatrix};$$

the right-hand side tends to zero as γ tends to infinity. ▌

Theorem 3 will be used to show that the behavior of the Fourier series of an integrable function f at a point z depends only upon the behavior of f in a neighborhood of z. This is stated formally in the *Riemann localization theorem* (4), which is proved below. First, we introduce *Dirichlet's integral*, which provides a very useful representation of the partial sums of the Fourier series of a function in $L_1[0,2\pi]$.

The trigonometric identity

$$[8] \qquad D_n(x) = \frac{\sin (n + 1/2) x}{\sin (x/2)} = \frac{1}{2} + \sum_{k=1}^{n} \cos kx, \quad n = 1, 2, \ldots,$$

may be proved by mathematical induction on n (Exercise 10). The sequence of functions $D_n(x)$ are called the *Dirichlet kernels*.

Substitution of [5] in [4] yields the following expression for the partial sums of [4]:

$$[9] \quad s_n(x) = \frac{1}{2\pi} \int_0^{2\pi} f(u)\, du + \frac{1}{\pi} \sum_{k=1}^{n} \left\{ \left(\int_0^{2\pi} f(u) \cos ku\, du \right) \cos kx + \left(\int_0^{2\pi} f(u) \sin ku\, du \right) \sin kx \right\}.$$

The identity $\cos (A - B) = \cos A \cos B + \sin A \sin B$ and [8] imply

$$[10] \qquad s_n(x) = \frac{1}{\pi} \int_0^{2\pi} f(u) \left[\frac{1}{2} + \sum_{k=1}^{n} \cos k(x - u) \right] du = \frac{1}{2\pi} \int_0^{2\pi} f(u)\, D_n(x - u)\, du.$$

Since the Fourier coefficients of f are given by integrals, they remain unchanged if f is changed at a single point. Therefore, we may assume that

$f(0) = f(2\pi)$, changing the value of $f(2\pi)$ if necessary. Next, we extend $f(x)$ *periodically to* R by defining $f(x) = f(x + 2\pi)$ for all $x \in R$. Then, by setting $u = x + t$ in [10], the periodicity of f and D_n yields

$$[11] \qquad s_n(x) = \frac{1}{2\pi} \int_0^{2\pi} f(x + t)\, D_n(t)\, dt.$$

Expression [11] is called *Dirichlet's integral*. A very useful variation of [11] is obtained by rewriting that part of the integral which is taken over the interval $[\pi, 2\pi]$: By substituting $t = -s$ in this integral, and using the identities $f(x) = f(x + 2\pi)$, $D_n(x) = D_n(x + 2\pi)$, and $D_n(x) = D_n(-x)$, we obtain

$$\int_\pi^{2\pi} f(x + t)\, D_n(t)\, dt = \int_{-2\pi}^{-\pi} f(x - s)\, D_n(-s)\, ds = \int_0^\pi f(x - s)\, D_n(s)\, ds.$$

Thus [11] becomes

$$[12] \qquad s_n(x) = \frac{1}{2\pi} \int_0^\pi [f(x + t) + f(x - t)]\, D_n(t)\, dt.$$

We are now able to state and prove the Riemann localization theorem.

4 Theorem. Let $f, g \in L_1[0, 2\pi]$. If there is a point z and a positive number ϵ such that $f = g$ a.e. in $(z - \epsilon, z + \epsilon) \subset (0, 2\pi)$, then the Fourier series of f and g either both diverge at z or both converge there to the same value.

Proof: Let s_n, $\hat{s}_n$ denote the Fourier sums of f, g respectively. It suffices to show that

$$[13] \qquad \lim_{n \to \infty} [s_n(z) - \hat{s}_n(z)] = 0.$$

From [12] and the assumption that $f = g$ a.e. in $(z - \epsilon, z + \epsilon)$, it follows that

$$\hat{s}_n(z) = \frac{1}{2\pi} \int_0^\epsilon [f(z + t) + f(z - t)]\, D_n(t)\, dt + \int_\epsilon^\pi [g(z + t) + g(z - t)]\, D_n(t)\, dt.$$

Subtracting this partial sum from s_n, we obtain

$$s_n(z) - \hat{s}_n(z) = \frac{1}{2\pi} \int_\epsilon^\pi [f(z + t) + f(z - t) - g(z + t) - g(z - t)]\, D_n(t)\, dt = \frac{1}{2\pi} \int_\epsilon^\pi H(t)\, D_n(t)\, dt;$$

$H(t) \in L_1[\epsilon,\pi]$. By combining the denominator of $D_n(t)$ (See [8]) with $H(t)$, a function $G(t) = [\csc(t/2)]\, H(t)$ is obtained, which is integrable over $[\epsilon,\pi]$. This follows from the integrability of H and the boundedness of $\csc(t/2)$ over that interval. Therefore the Riemann-Lebesgue theorem is applicable to the integral,

$$s_n(z) - \hat{s}_n(z) = \frac{1}{2\pi}\int_\epsilon^\pi \sin((n + 1/2)t)\, G(t)\, dt,$$

and [13] follows. ∎

The Localization Theorem makes no assertion about the convergence of $s_n(z)$ to $f(z)$. Indeed, the theorem says that if $\lim_{n\to\infty} s_n(z)$ exists, its value does *not* depend upon the value $f(z)$.

We turn next to the derivation of several convergence tests for Fourier series. Again, the partial sums s_n of a given function f may be expressed by means of Dirichlet's integral, [11] or [12]. If $g(x) = c$, the constant function, then each of its partial sums, which we denote by $\hat{s}_n$ is equal to c. Thus [12] yields

$$[14] \qquad c = \frac{1}{\pi}\int_0^\pi cD_n(t)\, dt, \qquad \text{for all } c \in R.$$

By setting $c = f(z)$, and subtracting [14] from [12], we obtain

$$[15] \quad s_n(z) - f(z) = \frac{1}{2\pi}\int_0^\pi [f(z + t) + f(z - t) - 2f(z)]\, D_n(t)\, dt.$$

Since z will be fixed throughout the discussion, put

$$F(t) = f(z + t) + f(z - t) - 2f(z).$$

Then, a necessary and sufficient condition that the Fourier series of f converges at z to $f(z)$, is that

$$[16] \qquad \lim_{n\to\infty} \int_0^\pi F(t)\, D_n(t)\, dt = 0.$$

A more useful form of [16] is obtained by applying the Riemann-Lebesgue Theorem (3):

If $0 < \epsilon \leq \pi$, write

$$\int_0^\pi F(t)\, D_n(t)\, dt = \int_0^\epsilon F(t)\, D_n(t)\, dt + \int_\epsilon^\pi F(t)\, D_n(t)\, dt.$$

As in the proof of Theorem 4, the second integral on the right satisfies the conditions of Theorem 3; that is, it is of the type

$$\int_{\epsilon}^{\pi} H(t) \sin\left(n + \frac{1}{2}\right) t \, dt,$$

where $H(t)$ is integrable. Therefore,

$$[17] \qquad \lim_{n\to\infty} \int_0^{\pi} F(t)\, D_n(t)\, dt = \lim_{n\to\infty} \int_0^{\epsilon} F(t)\, D_n(t)\, dt.$$

The identity $\lim_{t\to 0} \frac{\sin t/2}{t/2} = 1$ permits us to rewrite [16] in yet another way: Consider the integral

$$[18] \qquad \int_0^{\epsilon} F(t)\,[\sin(n + 1/2)t]\left[\left(\sin\frac{t}{2}\right)^{-1} - \frac{2}{t}\right] dt.$$

The quantity in brackets is in $L_1[0,\epsilon]$; hence the Riemann-Lebesgue Theorem implies that the integral tends to zero as n tends to infinity. It follows from [7], [17] and [18] that a sufficient condition for the convergence of the Fourier series is given by

$$[19] \qquad \lim_{n\to\infty} \int_0^{\epsilon} \frac{F(t)}{t} \sin(n + 1/2)\, t \, dt = 0$$

This is known as *Dini's test*, and for convenience it is stated formally as a theorem:

5 Theorem (Dini's test). Let z be a point in $(0,2\pi)$ where the function

$$\frac{F(t)}{t} = \frac{f(z + t) + f(z - t) - 2f(z)}{t}$$

$\in L_1[0,\epsilon]$ for some $0 < \epsilon \leq \pi$. Then the Fourier series of $f(x)$ at the point z converges to $f(z)$.

6 Corollary. Let $f \in L_1[0,2\pi]$. If at a point z in $[0,2\pi]$, $f'(z)$ exists, then the Fourier series of $f(x)$ converges to $f(z)$ at $x = z$.

Proof: The existence of the derivative implies that

$$\lim_{t\to 0} \frac{F(t)}{t} = \lim_{t\to 0}\left[\frac{f(x + t) - f(z)}{t}\right] - \lim_{t\to 0}\left[\frac{f(z - t) - f(z)}{-t}\right]$$

$$= f'(z) - f'(z) = 0,$$

from which it follows that $F(t)/t$ is bounded in a neighborhood of z, and therefore integrable in that neighborhood. The convergence of the Fourier series at $x = z$ to $f(z)$ follows from Dini's test. ∎

The convergence test of Jordan (Theorem 8) asserts that if f is integrable over $[0,2\pi]$ and is of bounded variation over some neighborhood of a point z in $(0,2\pi)$, then its Fourier series converges to the "average" value,

[20] $$\frac{f(z+0)+f(z-0)}{2}.$$

The existence of the right and left hand limits $f(z+0)$ and $f(z-0)$ for a function of bounded variation is assured by the possibility of writing $f = g - h$, where g and h are nondecreasing functions (Theorem 9.3.2). Moreover, f is a.e. differentiable in $(z-\epsilon, z+\epsilon)$ (Corollary 9.3.7). If z is a point at which the derivative exists, then the desired result [20] follows. In this case the continuity of f at z reduces [20] to $f(z)$. On the other hand, if the derivative does not exist at z, then Dini's test is not applicable, and it is at these points that Jordan's test is needed. To prove Theorem 8, we need the second mean value theorem for integrals (The reader should compare this with Abel's lemma which appears in Exercise 8).

7 Theorem (Second Mean Value Theorem for Integrals). Let $\varphi \in L_1[a,b]$ and let G be a nonnegative, nondecreasing function on $[a,b]$. Then there is a point $c \in [a,b]$ satisfying

$$\int_a^b G\varphi = G(b-0)\int_c^b \varphi.$$

Proof: The theorem is trivial if G is a constant.

Let us assume that $G(a+0) < G(b-0)$. Then if $\epsilon > 0$ is given, the number

$$a_1 = \inf[y \in [a,b]\colon x < y \quad \text{implies} \quad G(x) - G(a+0) < \epsilon]$$

satisfies the inequality $a < a_1 \le b$. If $a_1 < b$, select a_2 in a similar way:

$$a_2 = \inf[y \in [a,b]\colon x < y \quad \text{implies} \quad G(x) - G(a_1+0) < \epsilon].$$

Continuing in this manner, a finite (why?) number of points, $a = a_0 < a_1 < \cdots < a_n = b$, is chosen satisfying

$$a_k = \inf[y \in [a,b]\colon x < y \quad \text{implies} \quad G(x) - G(a_{k-1}+0) < \epsilon],$$

$k = 1,2,\ldots,n$, where

$$n \le \frac{G(b) - G(a)}{\epsilon}.$$

On the subdivision $\mathcal{S}_\epsilon$ which is determined by the points a_k, $k = 1,2,\ldots,n$, we define a step function

$$\tau(x) = \begin{cases} G(a_k - 0) & \text{if } x \in (a_{k-1}, a_k) \\ 0 & \text{if } x = a_k \end{cases}.$$

It is easily seen that $0 < \tau(x) - G(x) < \epsilon$ except possibly at the points $x = a_k$ (see Figure 30).

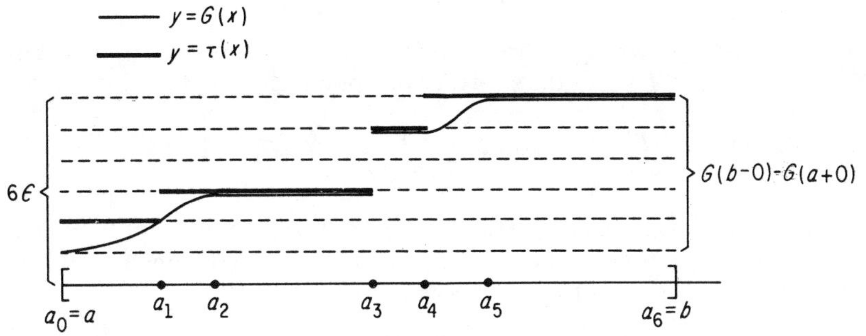

FIGURE 30

From Lemma 9.4.3 we obtain the continuity of the indefinite integral $\Phi(x) = \int_x^b \varphi(t)\,dt$ on the interval $[a,b]$. Since $[a,b]$ is compact, $\Phi(x)$ attains both its maximum and minimum values, M and m, on this interval (Theorem 3.2.14). Let $G(x)$ be extended by setting $G(x) \equiv 0$ whenever $-\infty < x < a$. Then $G(a - 0) = 0$, and since $G(a_k - 0) - G(a_{k-1} - 0) \geq 0$, it follows that

$$\begin{aligned}\int_a^b \tau\varphi &= \sum_1^n G(a_k - 0)\int_{a_{k-1}}^{a_k} \varphi \\ &= \sum_1^n G(a_k - 0)[\Phi(a_k) - \Phi(a_{k-1})] \\ &= \sum_1^n [G(a_k - 0) - G(a_{k-1} - 0)]\,\Phi(a_{k-1}) \\ &\leq M\sum_1^n [G(a_k - 0) - G(a_{k-1})] \\ &= M\cdot G(b - 0).\end{aligned}$$

A similar argument yields the corresponding inequality for m, the minimum value of Φ on $[a,b]$. Therefore

$$m\cdot G(b - 0) \leq \int_a^b \tau\varphi \leq M\cdot G(b - 0). \tag{21}$$

On the other hand,

$$\left|\int_a^b G\varphi - \int_a^b \tau\varphi\right| \leq \int_a^b |\varphi||G - \tau| \leq \epsilon\int_a^b |\varphi|. \tag{22}$$

Since $\epsilon > 0$ was chosen arbitrarily, it follows from [21] and [22] that

$$m \leq \frac{1}{G(b-0)} \int_a^b G\varphi \leq M.$$

The intermediate value theorem (Corollary 3.3.6) applied to the continuous function Φ yields the existence of a point c in $[a,b]$ for which

$$\int_c^b \varphi = \Phi(c) = \frac{1}{G(b-0)} \int_a^b G\varphi. \quad \blacksquare$$

8 Theorem (Jordan's test). If $f \in L_1[0,2\pi]$, and is of bounded variation in an ϵ-neighborhood of a point z which is contained in $(0,2\pi)$, then the Fourier series of f converges at z to

[23] $$\frac{f(z+0)+f(z-0)}{2}.$$

Proof: Let $F(t) = f(z+t) + f(z-t) - f(z+0) - f(z-0)$. If it can be shown that

[24] $$\lim_{n\to\infty} \int_0^\epsilon \frac{\sin(n+1/2)t}{t} F(t)\,dt = 0,$$

then it will follow that the Fourier series converges to the "average" value [23]. (Compare this with [18] and the accompanying discussion.)

Since $F(t)$, as well as $f(t)$, is of bounded variation in the interval $(z-\epsilon, z+e)$, it can be written as the difference $g - h$, of nonnegative, nondecreasing functions. Moreover, $\lim_{t\to 0} F(t) = 0$, and it follows that $\lim_{t\to 0} g(t) = \lim_{t\to 0} h(t) = \alpha \geq 0$. Replacing g and h by functions $G = g - \alpha$ and $H = h - \alpha$ respectively, we obtain

$$F = g - h = (G+\alpha) - (H+\alpha) = G - H,$$

where G and H are nondecreasing, nonnegative functions, and $\lim_{t\to 0} G(t) = \lim_{t\to 0} H(t) = 0$. Thus [24] is verified by showing that the integrals

$$I_n = \int_0^\epsilon \frac{\sin(n+1/2)t}{t} G(t)\,dt, \qquad J_n = \int_0^\epsilon \frac{\sin(n+1/2)t}{t} H(t)\,dt$$

tend to zero as n tends to infinity. We prove only that $\lim_{n\to\infty} I_n = 0$. The convergence of the sequence $\{J_n\}$ is obtained in the same way.

$\lim_{t\to 0} G(t) = 0$ implies that if $\delta > 0$ is given, there is a subinterval $(z-\epsilon', z+\epsilon')$ of $(z-\epsilon, z+\epsilon)$ in which $0 \leq G(t) < \delta$. Then

$$I_n = \int_0^{\epsilon'} \frac{\sin(n+1/2)t}{t} G(t)\,dt + \int_{\epsilon'}^{\epsilon} \frac{\sin(n+1/2)t}{t} G(t)\,dt.$$

The integrability of $G(t)$ implies that the second integral converges to zero as n tends to infinity (Riemann-Lebesgue theorem). The Second Mean Value Theorem for Integrals yields the existence of a point c in $[0,\epsilon']$ for which

$$[25]\quad \int_0^{\epsilon'} \frac{\sin (n + 1/2)t}{t} G(t)\,dt = G(\epsilon' - 0) \int_c^{\epsilon'} \frac{\sin (n + 1/2)t}{t}\,dt.$$

By changing variables, the integral on the right-hand side of [25] becomes

$$[26]\quad \int_{(n+1/2)c}^{(n+1/2)\epsilon'} \frac{\sin u}{u}\,du.$$

Since $\left| \int_a^b \frac{\sin u}{u} \right|$ is uniformly bounded for all intervals $[a,b]$ (Sec. 3 of Chapter 6), [26] is uniformly bounded for all positive integers n and all choices of $0 < c < \epsilon' < \epsilon$. If M is a bound for these integrals, then the absolute values of [25] are bounded by δM. However, δ was an arbitrarily chosen positive number, and it follows that $\lim_{n \to \infty} I_n = 0$. ∎

9 Corollary (Dirichlet Conditions). Let f be a function in $L_1[0,2\pi]$ that satisfies the following conditions:

(i) f has only a finite number of discontinuities in $[0,2\pi]$;

(ii) f has only a finite number of relative maxima and minima. (Points that lie in the *interior* of intervals on which f is constant are excluded.)

Then the Fourier series of $f(x)$ converges everywhere in $(0,2\pi)$ to $[f(x + 0) + f(x - 0)]/2$. At points of continuity, the Fourier series converges to $f(x)$.

Proof: Let $\mathcal{S}$ denote the subdivision whose points

$$0 = a_0 < a_1 < \cdots < a_n = 2\pi,$$

are chosen as follows: For $k = 1, 2, \ldots, n - 1$,

(i) $f(x)$ is discontinuous at $x = a_k$, or

(ii) $f(a_k) > f(x)$ for all x in a neighborhood of a_k (or $f(a_k) < f(x)$ in a neighborhood of a_k), or

(iii) a_k is the left-hand end point of an open interval (a_k,a_{k+1}) over which $f(x)$ is constant.

It is easily seen that $f(x)$ is monotone and continuous over each (a_{k-1},a_k), $k = 1, 2, \ldots, n$, and therefore of bounded variation over each of these intervals. It follows that $f(x)$ is of bounded variation over the union $\bigcup_{k=1}^{n} (a_{k-1},a_k) = [0,2\pi] - \{a_0, \ldots, a_n\}$, and therefore over $[0,2\pi]$ itself. Thus Jordan's test (Theorem 8) is applicable, and at each point x

of $(0,2\pi)$, the Fourier sums converge to $[f(x+0)+f(x-0)]/2$. At points of continuity, $f(x+0)=f(x-0)$, and at such points, $\lim_{n\to\infty} s_n(x) = f(x)$. ∎

Having represented a function as an infinite sum of other functions, or equivalently as the limit of a sequence of functions, it is natural to ask if the integral or derivative of that function (if they exist) may be calculated by integrating or differentiating the terms of the series. It has already been remarked that the pointwise convergence of a series of functions is not sufficient to permit term by term integration or differentiation, even in the case where it is known that the limit function is integrable or differentiable. For example, the validity of term-by-term integration is assured only by making the rather strong assumption that the convergence is uniform. Such strong conditions are not needed in the case of Fourier series. It will be shown that if $\{s_n\}$ is the sequence of Fourier sums of a function f in $L_1[0,2\pi]$, then for all subintervals $[a,b]$ of $(0,2\pi)$,

$$\int_a^b f = \lim_{n\to\infty} \int_a^b s_n;$$

that is, the integral of f is obtained by integrating the terms of the Fourier series. This is true even if $s_n(x)$ is not known to converge anywhere! This fact is proved next.

Let f be a function which is integrable over $[0,2\pi]$. f is extended to R by setting $f(x) = f(x+2\pi)$, for all $x \in R$. If necessary, $f(x)$ is redefined at $x = 2\pi$, so that $f(0) = f(2\pi)$. Then, if a_0 is given by [5], the function

$$F(x) = \int_0^x [f(t) - a_0/2]\,dt$$

is continuous, of bounded variation over any finite interval (Lemmas 9.4.3 and 5), and periodic, since

$$\begin{aligned} F(x+2\pi) &= \int_0^{x+2\pi} [f(t) - a_0/2]\,dt = \int_0^x [f(t-2\pi) - a_0/2]\,dt \\ &= \int_0^x [f(t) - a_0/2]\,dt = F(x). \end{aligned}$$

It follows from Jordan's test (Theorem 8) that $F(x)$ is *everywhere* equal to its Fourier series,

$$\frac{1}{2}A_0 + \sum_1^\infty [A_n \cos nx + B_n \sin nx]. \tag{27}$$

If it can be shown that for all $n \geq 1$,

$$A_n = -\frac{b_n}{n}, \qquad B_n = \frac{a_n}{n},$$

where a_n, b_n are the Fourier coefficients of $f(x)$, then it will follow that [27] is obtained by integrating, term by term, the Fourier series of $f(t) - a_0/2$. (The coefficient A_0 is accounted for below.) Since both equalities are verified in a similar manner, only the case $A_n = -b_n/n$ is considered. Integration by parts yields

$$\begin{aligned} A_n &= \frac{1}{\pi} \int_0^{2\pi} F(x) \cos nx \, dx \\ &= \frac{1}{\pi} \left[F(x) \frac{\sin nx}{n} \right]_0^{2\pi} - \frac{1}{n\pi} \int_0^{2\pi} F'(x) \sin nx \, dx \\ &= -\frac{1}{n\pi} \int_0^{2\pi} f(x) \sin nx \, dx = -\frac{b_n}{n}. \end{aligned}$$

Thus [27] becomes

$$F(x) = \frac{1}{2} A_0 + \sum_1^{\infty} \frac{1}{n} [a_n \sin nx - b_n \cos nx].$$

From $0 = F(0) = \frac{1}{2} A_0 - \sum_1^{n} \frac{b_n}{n}$, it follows that

$$\begin{aligned} F(x) &= \sum_1^{\infty} \frac{1}{n} [a_n \sin nx + b_n(1 - \cos nx)] \\ &= \sum_1^{\infty} \left[a_n \int_0^x \cos nt \, dt + b_n \int_0^x \sin nt \, dt \right] \\ &= \int_0^x \left[f(t) - \frac{a_0}{2} \right] dt. \end{aligned}$$

We state this result as a theorem:

10 Theorem. Let $f(x)$ be periodic with period 2π, and integrable over $[0,2\pi]$. Then the series obtained by integrating [4] term by term, is equal to $\int_0^x f(t)\, dt$, for all $x \in R$.

11 Example. In proving Theorem 10, it was shown that whether or not the Fourier series of f converges, $\sum_1^{\infty} \frac{b_n}{n} = \frac{1}{2} A_0$. The convergence of this constant series enables us to give an example of a convergent trigonometric series which is *not* the Fourier series of a function in $L_1[0,2\pi]$:

Suppose that the series $\sum_{2}^{\infty} \frac{\sin nx}{\log n}$, which converges in $[0,2\pi]$ (Exercise 14) is the Fourier series of a function $f(x)$ in $L_1[0,2\pi]$. Then $\int_0^x f(t)\,dt$ is continuous, and Theorem 10 implies that

$$F(x) = \sum_{2}^{\infty} \frac{1}{\log n} \int_0^x \sin nt\,dt = \sum_{2}^{\infty} \frac{1}{n \log n} [1 - \cos nx]$$

is continuous for all values of x. But it is easily seen that this series diverges for infinitely many values of x; for example, at $x = \pi$,

$$[1 - \cos n\pi] = \begin{cases} 0 & \text{if } n \text{ is even} \\ 2 & \text{if } n \text{ is odd,} \end{cases}$$

and the series

$$\sum_{1}^{\infty} \frac{1}{(2n + 1) \log (2n + 1)}$$

diverges, as can be shown by using the integral test.

The termwise integrability of a Fourier series may also be proved by means of the following theorem:

12 Theorem. Let f be a function in $L_1[0,2\pi]$ which is of bounded variation over a closed subinterval $[a,b]$ of $[0,2\pi]$. Then the Fourier sums of f are uniformly bounded over every closed interval $[c,d]$ contained in (a,b); that is, if $[c,d] \subset (a,b)$, there is a number $M_{[c,d]}$ such that

$$|s_n(x)| \le M_{[c,d]}, \quad n = 1,2,\ldots \quad \text{and} \quad x \in [c,d].$$

Proof: Each of the functions $f(x)$, $s_n(x)$ and $D_n(x) = \dfrac{\sin (n + 1/2)x}{\sin (x/2)}$ has period 2π, so that computations may be done over any interval whose length is equal to 2π. For reasons that will be made clear later, the basic interval will be shifted from $[0,2\pi]$ to $[a,a + 2\pi]$. Using Dirichlet's integral,

$$\begin{aligned} s_n(x) &= \int_a^{a+2\pi} f(t)\, D_n(x - t)\, dt \\ &= \frac{1}{2\pi} \int_a^{a+2\pi} f(t) \sin (n + 1/2)(x - t) \left[\left(\sin \frac{x - 2}{2}\right)^{-1} - \frac{2}{x - t}\right] dt \\ &\quad + \frac{1}{2\pi} \int_a^{a+2\pi} f(t) \sin (n + 1/2)(x - t) \left[\frac{2}{x - t}\right] dt \\ &= I(n,x) + J(n,x). \end{aligned}$$

We must show that the integrals $I(n,x)$ and $J(n,x)$ are uniformly bounded on $[c,d]$, and that the bound is independent of n. To prove the existence of a uniform bound for the sequence of integrals $I(n,x)$, we make use of the continuous function

$$H_n(x,t) = \begin{cases} 0 & \text{if } x = t \\ 1/[\sin{(n + 1/2)(x - t)}] - 2/(x - t) & \text{if } x \neq t, \end{cases}$$

which is defined on the compact subset of R^2,

$$S = \{(x,t): c \leq x \leq d, a \leq t \leq a + 2\pi\}.$$

Clearly, the values of $H_n(x,t)$, for all positive integral n and $x \in S$, are bounded in absolute value by a single number L. It follows that

$$|I(n,x)| \leq \frac{L}{2\pi} \int_a^{a+2\pi} |f(t)|\, dt = M,$$

where M is a number which is independent of both n, and x in S.

To find a similar bound for $|J(n,x)|$, we write $f = g - h$, where g and h are nonnegative and nondecreasing on the interval $[a,b] \subset [a,a + 2\pi]$. Then

$$[28] \quad 2\pi|J(n,x)| \leq \left| \int_a^b g(t) \frac{\sin{(n + 1/2)(x - t)}}{(x - t)/2}\, dt \right| + \left| \int_a^b h(t) \frac{\sin{(n + 1/2)(x - t)}}{(x - t)/2}\, dt \right| + \left| \int_b^{a+2\pi} f(t) \frac{\sin{(n + 1/2)(x - t)}}{(x - t)/2}\, dt \right|.$$

The Second Mean Value Theorem for Integrals (Theorem 7), applied to the first integral of [28] yields

$$\int_a^b g(t) \frac{\sin{(n + 1/2)(x - t)}}{(x - t)/2}\, dt = g(b - 0) \int_{a+\epsilon}^b \frac{\sin{(n + 1/2)(x - t)}}{(x - t)/2}\, dt;$$

$a \leq a + \epsilon \leq b$. The uniform boundedness of this integral follows from the uniform boundedness of $\int_I \frac{\sin u}{u}\, du$ over all bounded intervals I. (See [26].) A similar argument may be given for the second integral of [28].

For the third integral, $x \in [c,d] \subset (a,b)$ implies that $\left| \frac{1}{x - t} \right| \leq$

$\left|\dfrac{1}{b-d}\right|$. Taken together with the inequality $|\sin u| \leq 1$, this yields $\left|\int_a^b f\right|$ as a bound for the third integral. ▮

In Exercise 13, the reader is asked to show that Theorem 12 may be used to justify the termwise integration of a Fourier series.

We have already seen that not every convergent trigonometric series is the Fourier series of an integrable function (Example 11). Let us suppose now that a given series is *known* to be the Fourier series of an integrable function. Can the function be determined by the series? The answer is, obviously not, since every two a.e. equal functions will have the same Fourier series. At best, if the series converges a.e., and the sums remain bounded by a function which is integrable over $[0,2\pi]$, then the a.e. limit of the Fourier sums determines an equivalence class of functions in $L_1[0,2\pi]$, each having the same Fourier series. However, there exist functions in $L_1[0,2\pi]$, in fact continuous functions, whose Fourier series diverge (Titchmarsh [1], p. 416), so that it is not always possible to determine $f(x)$ (or, its equivalence class) by looking at $\lim_{n\to\infty} s_n(x)$. This is sometimes remedied by looking at the sequence of arithmetic means (Definition 13),

$$\sigma_n = \frac{s_1 + s_2 + \cdots + s_n}{n}, \tag{29}$$

which may converge where $\{s_n\}$ diverges (Theorem 16).

13 Definition. Let $\{a_n\}$ be a sequence of real numbers, and let

$$s_n = a_1 + a_2 + \cdots + a_n, \qquad n = 1, 2, \ldots,$$

be the sequence of partial sums associated with the infinite series $\sum_1^\infty a_i$. This series is said to be *(C,1)-summable*, *Cesaro-summable*, or *summable by arithmetic means*, if the sequence [29] of arithmetic means converges.

14 Example. The sequence

$$a_n = \begin{cases} 1 & \text{if } n \text{ is even} \\ -1 & \text{if } n \text{ is odd} \end{cases}$$

determines a divergent sequence of partial sums:

$$s_n = \begin{cases} -1 & \text{if } n \text{ is odd} \\ 0 & \text{if } n \text{ is even.} \end{cases}$$

However, the associated sequence of arithmetic means,

$$\sigma_n = \begin{cases} -\left(\dfrac{n+1}{2n}\right) & \text{if } n \text{ is odd} \\ -\dfrac{n}{2n} = -\dfrac{1}{2} & \text{if } n \text{ is even} \end{cases}$$

converges to $-1/2$, proving that a divergent series may be $(C,1)$-summable.

On the other hand, divergence of [29] implies that $\sum_{1}^{\infty} a_n$ must also diverge. For, if $\lim_{n\to\infty} s_n = \lim_{n\to\infty} \sum_{1}^{n} a_i = s$, then,

$$|s - \sigma_n| = \left|\frac{ns - s_1 - s_2 - \cdots - s_n}{n}\right|$$
$$\leq \frac{1}{n}|(s - s_1) + \cdots + (s - s_k)| + \frac{1}{n}|(s - s_{k+1}) + \cdots + (s - s_n)|,$$

for $k \leq n$. By choosing k so that

$$j > k \qquad \text{implies} \qquad |s_j - s| < \frac{\epsilon}{2},$$

we obtain

$$|s - \sigma_n| \leq \frac{1}{n}\left(\sum_{i=1}^{k} (s - s_i)\right) + \frac{(n - k - 1)\epsilon}{2n}.$$

Keeping k fixed, N may be chosen so that

$$n > N \qquad \text{implies} \qquad \frac{1}{n}\left(\sum_{i=1}^{k} (s - s_i)\right) < \frac{\epsilon}{2}.$$

The second term of $|s - \sigma_n|$ is less than $\epsilon/2$ for all n. This proves:

15 Proposition. If $\sum_{n=1}^{\infty} a_n$ converges to s, then it is $(C,1)$-summable to s.

We turn our attention to deriving an integral representation for the Cesaro sums of a Fourier series. As with Dirichlet's integral, this representation is used to derive conditions for a given Fourier series to be $(C,1)$-summable to an integrable function. Substitution in [11] yields an integral expression for the $(n - 1)$ *st* arithmetic mean of the Fourier sums:

$$[30] \quad \sigma_{n-1}(x) = \frac{1}{2\pi n}\int_0^{\pi} [f(x + t) + f(x - t)]$$
$$\cdot\left[\frac{\sin(t/2) + \sin(3t/2) + \cdots + \sin(n - 1/2)t}{\sin(t/2)}\right] dt$$

By means of the trigonometric identity,

[31] $$\sum_{j=0}^{n-1} \sin\left(j + \frac{1}{2}\right)x = \frac{1 - \cos nx}{2\sin(x/2)} = \frac{\sin^2(nx/2)}{\sin(x/2)}$$

(Exercise 11), [30] may be written in a more convenient form:

[32] $$\sigma_{n-1}(x) = \frac{1}{2\pi n}\int_0^\pi [f(x+t) + f(x-t)]\left[\frac{\sin^2(nt/2)}{\sin^2(t/2)}\right]dt$$
$$= \frac{1}{2\pi n}\int_0^\pi [f(x+t) + f(x-t)]\, F_n(t)\, dt.$$

[32] is called *Fejer's integral.* Comparing [32] with Dirichlet's integral, [10] and [11], we see that the *Fejer Kernel* $F_n(t) \geq 0$, whereas the Dirichlet Kernel $D_n(t)$, oscillates between positive and negative values. It is this property of $F_n(t)$ that makes it easier to verify the $(C,1)$-summability of a Fourier series, or equivalently the convergence of Fejer's integral [32], than it is to verify the convergence of the Fourier sums $s_n(x)$.

Substitution of $f(x) \equiv 1$ into [32] yields

[33] $$1 = \frac{1}{\pi n}\int_0^\pi 2F_n(t)\, dt.$$

To obtain expressions that are similar to [15] and [16], [33] is multiplied by the value $g(x)$, and subtracted from [32]. Thus a necessary and sufficient condition that a given Fourier series be $(C,1)$-summable to $g(x)$ is that

[34] $$\lim_{n\to\infty} [\sigma_n(x) - g(x)]$$
$$= \lim_{n\to\infty} \frac{1}{2\pi n} \cdot \int_0^\pi [f(x+t) + f(x-t) - 2g(x)]\, F_n(t)\, dt = 0.$$

Proceeding as before, (see the discussion that follows [16]), it is easily seen that [34] is true, if and only if for $0 < \epsilon \leq \pi$,

[35] $$\lim_{n\to\infty} \frac{1}{n}\int_0^\epsilon H(t)\,\frac{\sin^2(nt/2)}{\sin^2(t/2)}\, dt = 0;$$

$H(t) = f(x+t) + f(x-t) - 2g(x)$. Finally, the identity $\lim_{u\to 0} \frac{\sin u}{u} = 1$, transforms [35] into

[36] $$\lim_{n\to\infty} \frac{1}{n}\int_0^\epsilon H(t)\,\frac{\sin^2(nt/2)}{t^2}\, dt = 0.$$

(Compare this with Dini's test, [19] and Theorem 5.)

Fejer's theorem, which is proved below, demonstrates the usefulness of $(C,1)$-summability. It asserts that the arithmetic means $\sigma_n(x)$ of the Fourier sums $s_n(x)$ of a function $f(x)$ will converge to

$$(1/2)[f(x+0)+f(x-0)]$$

whenever the right- and left-hand limits exist. In particular, if $\{s_n(x)\}$ is the sequence of Fourier sums of a continuous function on $[0,2\pi]$, then the right- and left-hand limits, $f(x+0)$ and $f(x-0)$, exist and are equal, and $\lim \sigma_n(x) = f(x)$ on $(0,2\pi)$. Thus a continuous function f having a divergent Fourier series, may be written as the limit of the arithmetic means, or Cesaro sums, of the Fourier sums. This assertion is contained in Fejer's theorem:

16 Theorem (Fejer). Let $f \in L_1[0,2\pi]$, and let x be a point of $(0,2\pi)$ at which $f(x+0)$ and $f(x-0)$ exist. Then the Fourier series of f is $(C,1)$-summable to the average value

[37] $$\frac{f(x+0)+f(x-0)}{2}.$$

In particular, if x is a point of continuity of f, then $\lim_{n\to\infty} \sigma_n(x) = f(x)$.

Proof (The proof is similar to that of Jordan's test, Theorem 8):

Let x be a point at which the left- and right-hand limits of f exist. Then since

$$\lim_{t\to 0} H(t) = \lim_{t\to 0} [f(x+t)+f(x-t)-f(x+0)-f(x-0)] = 0,$$

there is an ϵ', $0 < \epsilon' < \epsilon$, and

$$t \in (0,\epsilon') \qquad \text{implies} \qquad |H(t)| < \delta,$$

where δ is any preassigned positive number. Thus the integral of the convergence condition [36] has the following bound:

[38] $$\left|\frac{1}{n}\int_0^{\epsilon} H(t)\,\frac{\sin^2(nt/2)}{t^2}\,dt\right| \le \frac{\delta}{n}\int_0^{\epsilon'} \frac{\sin^2(nt/2)}{t^2}\,dt + \frac{1}{n}\int_{\epsilon'}^{\epsilon} |H(t)|\,\frac{\sin^2(nt/2)}{t^2}\,dt.$$

Setting $(nt/2) = u$ in the first integral on the right yields

$$\frac{\delta}{2}\int_0^{n\epsilon'} \frac{\sin^2 t}{t^2}\,dt < \delta M;$$

Here M is equal to $2\int_{-\infty}^{\infty} \frac{\sin^2 t}{t^2}\,dt$. The second integral is bounded by

$(\epsilon')^{-2}n^{-1}\int_{\epsilon'}^{\epsilon} |H(t)|\,dt$, which tends to zero as n tends to infinity, thus proving that [38] can be made arbitrarily small by taking n to be sufficiently large. This, however, is equivalent to the convergence condition [36], which implies the convergence of the Cesaro sums to the average value. ▌

17 Corollary. If $f(x)$ is continuous on a closed, bounded interval $[a,b]$, then the arithmetic means $\sigma_n(x)$ converge uniformly to $f(x)$ on $[a,b]$.

Proof: The compactness of $[a,b]$ implies that f is uniformly continuous there, and the number ϵ' of Theorem 16 depends only upon δ, and not upon $x \in [a,b]$. Thus [38] is bounded by an expression of the type

[39] $$M\delta + \frac{1}{n}L(\epsilon'); \quad L(\epsilon') = (\epsilon')^{-2}\int_{\epsilon'}^{\epsilon} |H(t)|\,dt.$$

The uniform convergence of $\{\sigma_n\}$ is verified by proving that [39] can be made smaller than any preassigned positive number γ, by choosing n to be sufficiently large. To do this, we first choose $\delta < (\gamma/2M)$, which makes the first term of [39] less than $\gamma/2$. Once δ is so chosen, this determines ϵ', and consequently, $L(\epsilon')$. Then, by requiring that $n > [2L(\epsilon')/\gamma]$, the second term is made less than $\gamma/2$. ▌

The Stone-Weierstrass theorem (Theorem 4.1.13) gave sufficient conditions for a subalgebra of $C(X)$ to be dense in $C(X)$. This means that the functions in the subalgebra may be used to approximate those in $C(X)$. As a special case of this theorem, we have the Weierstrass approximation theorem, which asserts that the functions in $C[a,b]$ may be uniformly approximated by polynomials. The reader is asked to show that a second proof of the Weierstrass theorem (Corollary 18) is obtained by using Corollary 17.

18 Corollary (Weierstrass approximation theorem). If $f \in C[0,2\pi]$, then to each $\epsilon > 0$, there is a polynomial p, such that

$$\|f - p\| = \max_{0\le x\le 2\pi} |f(x) - p(x)| < \epsilon.$$

Although the continuity of $f(x)$ is a sufficient condition for its "reconstruction" by means of Cesaro sums, it is more than we need to determine $f(x)$ almost everywhere (a.e.). Indeed, the *Fejer-Lebesgue theorem* asserts that the Cesaro sums of *any* function in $L_1[0,2\pi]$ converges a.e. to that function:

19 Theorem (Fejer-Lebesgue). If $f \in L_1[0,2\pi]$, then its Fourier series is $(C,1)$-summable to $f(x)$ whenever

$$\lim_{\gamma\to 0}\left[\frac{1}{\gamma}\int_0^\gamma |f(x+t)-f(x-t)|\,dt\right] = 0. \tag{40}$$

Moreover, [40] holds a.e.

Proof: Suppose that x is a point at which [40] is satisfied. The Fourier series of f at x is $(C,1)$-summable to $f(x)$ if and only if [34], which is an integral that represents the difference $\sigma_n(x) - f(x)$, tends to zero as n tends to infinity. This condition is in turn implied by [36], which we shall prove is assured by [40].

Setting $H(t) = f(x+t) + f(x-t) - 2f(x)$ (Cf. [34]), it follows from [40] that if $\alpha > 0$ is given, there is a $\delta > 0$ for which

$$\int_0^\gamma |H(t)|\,dt \le \int_0^\gamma |f(x+t)-f(x)|\,dt + \int_0^\gamma |f(x-t)-f(x)|\,dt < \gamma\alpha, \tag{41}$$

whenever $0 \le \gamma \le \delta$. Let n be an integer chosen so that $1/n < \delta$. Assuming that $\delta < \epsilon$ (see the integral directly below), bounds for [36] may be found as follows:

$$\left|\int_0^\epsilon H(t)\frac{\sin^2(nt/2)}{t^2}\,dt\right| \le \int_0^{1/n} |H(t)|\frac{\sin^2(nt/2)}{t^2}\,dt + \int_{1/n}^\delta \cdots + \int_\delta^\epsilon \cdots = I_1 + I_2 + I_3.$$

(The three dots used in integrals I_2 and I_3 are meant to signify that the integrands are the same as that of I_1.) We must show that for $i = 1, 2, 3$, $\lim_{n\to\infty} I_i/n = 0$.

Since $\lim_{u\to 0}\dfrac{\sin u}{u} = 1$, there is a number $K > 0$ such that

$$\frac{\sin^2(nt/2)}{t^2} = \frac{n^2}{4}\left(\frac{\sin^2(nt/2)}{(nt/2)^2}\right) < Kn^2, \quad \text{if } t \in [0,1/n].$$

From [40] we obtain

$$\frac{I_1}{n} < Kn\int_0^{1/n} |H(t)|\,dt < K\alpha, \tag{42}$$

where α may be chosen to be as small as desired.

We turn now to the second integral. Setting $I(\delta) = \displaystyle\int_{1/n}^\delta |H(t)|\,dt$, integration by parts yields

$$\frac{I_2}{n} \leq \frac{1}{n}\int_{1/n}^{\delta} | H(t) | t^{-2}\,dt = \frac{1}{n}\left[\frac{I(\delta)}{\delta^2} + 2\int_{1/n}^{\delta}\frac{I(t)}{t^3}\,dt\right].$$

From [41], $I(t) < t\alpha$, from which it follows that

$$[43] \qquad \frac{I_2}{n} < \frac{1}{n}\left[\frac{\alpha}{\delta} + 2\alpha\int_{1/n}^{\delta} t^{-2}\,dt\right] < \frac{\alpha}{n\delta} + 2\alpha\left[\frac{1}{n\delta} + 1\right] < 5\alpha.$$

To obtain the third inequality, we used $n\delta > 1$.

And finally, setting $\beta = \max_{0 \leq t \leq \epsilon} | H(t) |$, we obtain

$$[44] \qquad \frac{I_3}{n} = \frac{1}{n}\int_{\delta}^{\epsilon} | H(t) | \frac{\sin^2(nt/2)}{t^2}\,dt < \frac{\beta}{n\delta^2}.$$

Combining [42], [43], and [44] yields

$$[45] \qquad \frac{1}{n}[I_1 + I_2 + I_3] < K\alpha + 5\alpha + \frac{\beta}{n\delta^2}$$

as a bound for [36]. It is easily seen that the right-hand side of [45] can be made as small as desired. For having prescribed α, δ is determined. Thus [44], and hence [45], can be made small by choosing n to be sufficiently large.

This completes the proof that the convergence condition [36] is satisfied, and therefore $\lim_{n\to\infty} \sigma_n(x) = f(x)$.

It remains to be proved that [40] is true a.e. For this, we need the following lemma:

20 Lemma. If $f \in L_1[a,b]$, there is a null subset S of $[a,b]$ for which $x \in [a,b] - S$ and $c \in R$ imply

$$[46] \qquad \lim_{\gamma\to 0}\frac{1}{\gamma}\int_{x}^{x+\gamma} |f(t) - c|\,dt = |f(x) - c|.$$

Proof: Let c be a fixed but arbitrary number. Then [46] is simply the statement that the derivative of an indefinite integral is equal to its integrand. Since this is true a.e. (Theorem 9.4.7) it follows that to each rational number a_n there is a null set $S_n \subset [a,b]$ such that whenever $x \in [a,b] - S_n$, then [46] holds for $c = a_n$. Since $S = \bigcup_{n=1}^{\infty} S_n$ is also a null set, it follows that [46] is true for all rational values of c on the set $[a,b] - S$.

If c is irrational and $\epsilon > 0$ is given, there is a rational number q such that $| c - q | < \epsilon$. Then, the inequality $|f(t) - c| - |f(t) - q| \leq | c - q |$, implies that

[47]
$$\left| \frac{1}{\gamma} \int_x^{x+\gamma} |f(t) - c|\, dt - \frac{1}{\gamma} \int_x^{x+\gamma} |f(t) - q|\, dt \right| \le \frac{1}{\gamma} \int_x^{x+\gamma} |c - q|\, dt = |c - q| < \epsilon.$$

Let $\{q_n\}$ be a sequence of rational numbers that converges to the irrational number c, and let $x \in [a,b] - S$. Then the triangle inequality yields

$$\begin{aligned} &\left| \frac{1}{\gamma} \int_x^{x+\gamma} |f(t) - c|\, dt - |f(x) - c| \right| \\ &\le \left| \frac{1}{\gamma} \int_x^{x+\gamma} |f(t) - c|\, dt - \frac{1}{\gamma} \int_x^{x+\gamma} |f(t) - q_n|\, dt \right| \\ &+ \left| \frac{1}{\gamma} \int_x^{x+\gamma} |f(t) - q_n|\, dt - |f(x) - q_n| \right| \\ &+ \big| |f(x) - q_n| - |f(x) - c| \big| = A_1 + A_2 + A_3. \end{aligned}$$

From [47], it follows that

$$0 \le \lim_{n \to \infty} A_1 \le \lim_{n \to \infty} |c - q_n| = 0.$$

Since the q_n are rational, A_2 can be made as small as desired by taking γ to be sufficiently small; and finally, A_3 is bounded by $|c - q_n|$ which tends to zero as n tends to infinity. This completes the proof of the lemma.

We return to the proof of the Fejer-Lebesgue theorem. By setting $c = f(x)$ in [46], we obtain

[48]
$$\lim_{\gamma \to 0} \frac{1}{\gamma} \int_x^{x+\gamma} |f(t) - f(x)|\, dt = 0,$$

which holds a.e. on $[a,b]$. A simple change of variables transforms [48] into [40], thus proving that [40] is true a.e. ▮

Exercises

*1. A function f is said to be *even* if $f(-x) = f(x)$ for all x and *odd* if $f(-x) = -f(x)$ for all x.
   (a) Let $g(x)$ be any function defined on a *symmetric* interval $[-M,M]$. Show that $[(g(x) + g(-x)]/2$ is an even function, and $[(g(x) - g(-x)]/2$ is an odd function.
   (b) Show that every function defined on a symmetric interval may be written as a sum of an even function and an odd function.
   (c) Let $f \in L_1[-\pi,\pi]$. Prove that f is even if and only if its Fourier series contains no "sin" terms; that is, the coefficients $b_n = 0$ for $n = 1, 2, \ldots$. (See [5].) Prove a similar statement for odd functions.

2. Let

$$f(x) = \begin{cases} 0 & -\pi \leq x \leq 0 \\ 1 & 0 < x \leq \pi. \end{cases}$$

Write the Fourier series of f. Use this series to prove that

$$\frac{\pi}{4} = 1 - \frac{1}{3} + \frac{1}{5} - \frac{1}{7} + \cdots.$$

3. Write the Fourier series of $f(x) = |x|, -\pi \leq x \leq \pi$. At what point does it converge to $|x|$? What is the value of $\displaystyle\sum_{0}^{\infty} \frac{1}{(2n+1)^2}$?

4. Show that if the partial sums of a trigonometric series converge uniformly on the "basic" interval $[-\pi,\pi]$ or $[0,2\pi]$, then the limit function is integrable over that interval, and the given trigonometric series is the Fourier series of the function.

5. Show that

$$\frac{\pi}{8} = \frac{1}{1^2} + \frac{1}{3^2} + \frac{1}{5^2} + \cdots.$$

6. (a) Prove that if a Fourier series converges absolutely at a point x, then it converges absolutely at $-x$.

(b) Prove that if the series converges absolutely at points x and y, then it converges absolutely at $x + y$.

(c) Use (b) to prove that if a Fourier series converges absolutely at all points of an interval, then it converges absolutely everywhere. (Remember, the series is defined everywhere and is periodic.)

7. Let Σa_i be an infinite series each of whose terms is a nonnegative real number. Prove that if the series is $(C,1)$-summable, then it is summable; that is, it converges.

8. Let $\{a_1, \ldots, a_n\}$ and $\{b_1, \ldots, b_n\}$ be sets of real numbers which have the following properties:

(i) There are numbers m and M such that

$$0 \leq m \leq a_1 + \cdots + a_k \leq M, \quad k = 1, \ldots, n;$$

(ii) $b_1 \geq b_2 \geq \cdots \geq b_n \geq 0$.

Show that

$$mb_1 \leq \sum_{1}^{n} b_k a_k.$$

(This is called Abel's Lemma. Compare it with the second mean value theorem for integrals.)

***9.** Use the Second Mean Value Theorem to show that there is a number M such that

$$\left| \int_a^b \frac{\sin x}{x}\, dx \right| < M$$

for all intervals $[a,b]$.

***10.** Prove the trigonometric identity [8].

***11.** Prove the trigonometric identity [31].

***12.** Use Theorem 3 to prove that the Fourier coefficients of an integrable function tend to zero as n tends to infinity.

13. Let $f \in L_1[0,2\pi]$. Prove that $F(x) = \int_a^x f(t)\,dt$, $(a,x) \subset [0,2\pi]$, is obtained by integrating the Fourier series of f term by term. Use Theorem 12.

14. Show that the trigonometric series $\sum_2^\infty \frac{\sin nx}{\log n}$ converges everywhere (Example 11).

15. Prove Corollary 18, the Weierstrass approximation theorem.

16. A function f satisfies a *Lipschitz condition* at a point y if there is a positive number L, and $|f(x) - f(y)| \leq L|x - y|$, for all x in the domain of definition of f.

    Prove that if f is sectionally continuous on $[0,2\pi]$, and satisfies a Lipschitz condition at a point x in $(0,2\pi)$, then the Fourier series of f converges at x to $f(x)$.

*17. Show that if $f = \chi_{[0,2\pi]}$, then its Fourier sums converge to it in the L_2 norm.

    Use this to prove that the trigonometric functions [3], are a complete orthonormal sequence in L_2.

*Exercises which request the proof of a statement in the text, as well as those referred to in the text, are starred.

CHAPTER 11

Linear Functionals

The assignment of the real number $\int f$ to each function f in L_1, defines a mapping of L_1 into R which is linear, homogeneous, monotone and continuous (Proposition 5.4.10). In this chapter we investigate the properties of the class of *all* mappings of a given function space into R, which are linear and homogeneous, and possess a third property called *boundedness.* It will be shown that these mappings, called *bounded, linear functionals*, are given by integrals if the function spaces are L_p or $C[a,b]$. In particular, the bounded linear functions on L_p, if $1 \leq p < \infty$, can be made into a Banach space which is isonormal to the *conjugate* space L_q, $(1/p + 1/q = 1)$, and that each linear functional may be represented by an integral (Theorems 11.2.7, 11.3.1, 11.3.2, and 11.3.3). In Sec. 4, similar results are obtained for $C[a,b]$. In this case, each linear functional is represented by a Stieltjes integral.

1 Bounded Linear Functions; The Hahn-Banach Theorem

1 Definition. A *linear functional* defined on a real vector space X is a mapping $F:X \to R$ that has the following properties:

(i) If $x,y \in X$, then $F(x + y) = F(x) + F(y)$ (linearity);

(ii) If $x \in X$ and $a \in R$, then $F(ax) = aF(x)$ (homogeneity).

In other words, F is a *linear transformation* of the vector space X into the vector space R.

2 Examples. (a) If $X = R^n$, then F is completely determined by what it does to the orthonormal set of basis vectors, $e_i = (\delta_{i1}, \delta_{i2}, \ldots, \delta_{in})$ (δ_{ij} is the Kronecker delta). For, if $F(e_i) = c_i$, where $c_i \in R$, $i = 1,2, \ldots, n$, then since each x in R^n may be represented uniquely by a linear combination

$$x = \sum_1^n x_i e_i, \qquad x_i \in R,$$

the linearity and homogeneity of F imply that

$$F(x) = \sum_{1}^{n} x_i c_i.$$

Thus each linear functional on R^n determines a vector $c = (c_1, \ldots, c_n)$ in R^n, such that for each $x \in X$, $F(x) = (c,x)$; (c,x) denotes the scalar product of c and x (Definition 10.2.1).

Remark. The preceding example illustrates a special case of the more general theorem of linear algebra: Every linear transformation of R^n into R^m may be represented by an $m \times n$ matrix.

(b) It is left the reader to show that every vector in l_2 determines a linear functional on l_2. In Sec. 2 it will be proved that *every* linear functional on l_2 which is bounded (Definition 3), may be represented by a scalar product.

(c) Let $X = L_2$. If $g \in L_2$, then

[1] $$F_g(f) = \int fg = (f,g)$$

is a linear functional on L_2 which may also be written as an inner product. Again, the converse is proved in Sec. 2.

(d) Let $X = L_p$, $1 \leq p \leq \infty$, and let g be a fixed but arbitrary function in the conjugate space L_q. Then

[2] $$F_g(f) = \int fg$$

is a linear functional on L_p. Hölder's inequality assures the existence of this integral. The linearity and homogeneity are inherited from the corresponding properties for integrals. In Sec. 3, we shall show that if $1 \leq p < \infty$, then every linear functional which is also bounded (Definition 3), may be represented by an integral [2].

3 Definition. A linear functional F defined on a normed vector space X is said to be *bounded* if there is a number $M > 0$, such that for all $x \in X$, $|F(x)| \leq M\|x\|$

4 Examples. Each of the linear functions of Example 2 is bounded. In part (a),

$$|F(x)|^2 = (x,c)^2 \leq (x,x)(c,c) = \|x\|^2\|c\|^2,$$

proving that $M = \|c\|$ may be taken as a bound for F.

A similar argument holds for (b) and (c).

In part (d), the boundedness is implied by Hölder's inequality.

5 Proposition. Let F be a linear functional defined on a normed vector space X. Then F is bounded if and only if F is a uniformly continuous mapping of X into R.

Proof: If F is bounded, there is a positive number M, and for all x in X, $|F(x)| \leq M\|x\|$. Let $\epsilon > 0$ be given. Then

$$\|x - y\| < \frac{\epsilon}{M} \text{ implies } |F(x) - F(y)| = |F(x - y)| \leq M\|x - y\| < \epsilon.$$

Thus ϵ/M is a "δ" of uniform continuity corresponding to ϵ.

On the other hand, if F is uniformly continuous, there is a $\delta > 0$ such that

$$[3] \qquad \|x - y\| < \delta \quad \text{implies} \quad |F(x) - F(y)| \leq 1.$$

We shall prove that $M = 2/\delta$ is a bound for $\dfrac{F(x)}{\|x\|}$, $\|x\| \neq 0$. First, observe that for all $x \neq \bar{0}$ in X, and $0 \neq a$ in R,

$$\frac{|F(ax)|}{\|ax\|} = \frac{|a|}{|a|}\frac{|F(x)|}{\|x\|} = \frac{|F(x)|}{\|x\|}$$

If $a = 1/\|x\|$, then $\|ax\| = 1$, proving that

$$\sup_{x \neq \bar{0}} \frac{|F(x)|}{\|x\|} = \sup_{\|x\|=1} \frac{|F(x)|}{\|x\|} = \sup_{\|x\|=1} |F(x)|.$$

It therefore suffices to show that if $\|x\| = 1$, then $|F(x)| < \dfrac{2}{\delta}$. From $F(\bar{0}) = 0$ and $\|\frac{\delta}{2}x - \bar{0}\| = \frac{\delta}{2}\|x\| = \frac{\delta}{2} < \delta$, we obtain

$$|F(x)| = \frac{2}{\delta}\left|F\left(\frac{\delta}{2}x\right)\right| = \frac{2}{\delta}\left|F\left(\frac{\delta}{2}x\right) - F(\bar{0})\right| \leq \frac{2}{\delta} = M.$$

The inequality follows from [3]. ▮

We show next that if X is a normed vector space, then the collection X^* of bounded linear functionals on X can be made into a normed linear space. The vector space operations on X^* coincide with the addition of real valued functions and multiplication by a real number; that is, if $F,G \in X^*$ and $a \in R$,

$$[4] \qquad (F + G)(x) = F(x) + G(x), \qquad (aF)(x) = a(F(x)).$$

The verification of the vector space properties is left to the reader. Since the elements of X^* are bounded, we may assign to each functional F in X^*, the real number,

$$[5] \qquad \|F\|^* = \sup_{x \neq \bar{0}} \frac{|F(x)|}{\|x\|} = \sup_{\|x\|=1} |F(x)|.$$

It is easily verified that [5] defines a norm on X^*:

(i) $\|F\|^* \geq 0$, and $\|F\|^* = 0$ only if $F(x) = 0$ for all $x \in X$. This however means that $F = 0^*$, the additive identity of X^*.

(ii) $\|aF\|^* = \sup_{\|x\|=1} |aF(x)| = |a| \|F\|^*$.

(iii) $$\|F + G\|^* = \sup_{\|x\|=1} |F(x) + G(x)| \leq \sup_{\|x\|=1} |F(x)| + \sup_{\|x\|=1} |G(x)| = \|F\|^* + \|G\|^*.$$

Thus X^* *is a normed vector space.*

We prove next that X^* *is complete:*

As the proof of the completeness of X^* will demonstrate, this property in no way depends upon the completeness of X, but only upon the completeness of R.

Suppose that $\{F_n\}$ is a Cauchy sequence of linear functionals on X. Then for each $x \in X$,

$$|F_n(x) - F_m(x)| = |(F_n - F_m)(x)| \leq \|F_n - F_m\|^* \|x\| \to 0$$

as n,m tend to infinity, proving that $\{F_n(x)\}$ is a Cauchy sequence of real numbers for every x in X. Since R is complete, each sequence converges to a real number which we denote by $F(x)$. It remains to be proved that this mapping F, of X into R, is in X^*, and that $\lim_{n \to 0} \|F - F_n\|^* = 0$.

The linearity and homogeneity of F follow easily from the linearity and homogeneity properties of the F_n. Indeed,

$$F(x + y) = \lim_{n \to \infty} F_n(x + y) = \lim_{n \to \infty} [F_n(x) + F_n(y)] = F(x) + F(y).$$

Similarly, $F(ax) = aF(x)$.

To verify the boundedness of F, we observe first that if $\{F_n\}$ is a Cauchy sequence, than the sequence of norms, $\{\|F_n\|^*\}$ converges, and is therefore bounded. Thus,

$$\sup_{\|x\|=1} |F(x)| = \sup_{\|x\|=1} \lim_{n \to \infty} |F_n(x)| = \lim_{n \to \infty} \|F_n\|^*.$$

Finally, for all x satisfying $\|x\| = 1$,

$$|F_n(x) - F(x)| = \lim_{m \to \infty} |F_n(x) - F_m(x)| \leq \lim_{m \to \infty} \|F_n - F_m\|^*$$

Since $\{F_n\}$ is a Cauchy sequence in X^*, the last expression tends to zero as n tends to ∞. Thus for sufficiently large n,

$$\sup_{\|x\|=1} |F_n(x) - F(x)| = \|F_n - F\|^* < \epsilon,$$

where ϵ is any preassigned positive number, proving that $\{F_n\}$ converges to F in the $*$-norm and that X^* is complete.

Gathering together these results, we have:

6 Theorem. If X is a real normed vector space, the set X^*, of bounded linear functionals on X, is a Banach space. The vector space operations in X^* and the norm are defined by [4] and [5].

We prove next the *Uniform Boundedness Theorem* for linear functions, and give one of its many applications.

7 Theorem. Let X be a Banach space and let $\{F_n\}$ be a sequence in X^* satisfying:

[6] $$\forall x \in X, \exists M_x \in R, \text{ and } |F_n(x)| \leq M_x, \quad n = 1,2,\ldots.$$

Then the sequence $\{\|F_n\|^*\}$ of real numbers is bounded.

Proof: For each positive integer k, let

$$S_k = \{x: |F_n(x)| \leq k, n = 1,2,\ldots\}$$
$$= [|F_1| \leq k] \cap [|F_2| \leq k] \cap \ldots = \bigcap_{n=1}^{\infty} [|F_n| \leq k];$$

(The set $[|F_i| \leq k] = \{x: |F_i(x)| \leq k\}$.) From [6], it follows $X = \bigcup_{k=1}^{\infty} S_k$. The completeness of X and the Baire category theorem (4.3.4) imply that at least one S_k is *not* nowhere dense. Thus the closed set S_k must contain a closed neighborhood $\overline{S_\epsilon(\theta)}$. Letting $\bar{0}$ denote the additive identity of X, we obtain

$$y \in S_\epsilon(\bar{0}) \text{ if and only if } x = (\theta + y) \in \overline{S_\epsilon(\theta)};$$

and so, whatever n may be, $y \in S_\epsilon(\bar{0})$ implies

$$|F_n(y)| = |F_n(x - \theta)| \leq |F_n(x)| + |F_n(\theta)| \leq 2k.$$

If $\|z\| = 1$, then $\epsilon z = y$ is in $S_\epsilon(\bar{0})$, and

$$|F_n(z)| = \frac{1}{\epsilon} |F_n(y)| \leq \frac{2k}{\epsilon}, \quad n = 1,2,\ldots,$$

from which it follows that

$$\|F_n\|^* = \sup_{\|z\|=1} |F(z)| \leq \frac{2k}{\epsilon}. \quad \blacksquare$$

An application follows.

8 Theorem. There is a continuous function $f(x)$ with period 2π $(f(x) = f(x + 2\pi))$, whose Fourier series diverges at the points $x = 2n\pi$, $n = 0, \pm 1, \pm 2, \ldots$.

Proof: Let X be the subset of $C[0,2\pi]$ which consists of those functions satisfying, $f(0) = f(2\pi)$. Clearly, X is a complete, normed, linear subspace of $C[0,2\pi]$ (Exercise 3).

The identity [11] of Sec. 4, Chapter 10, yields the following expression for $s_n(0)$, the nth partial sum of the Fourier series of $f(x)$ evaluated at $x = 0$:

$$[7] \qquad s_n(0) = \frac{1}{2\pi}\int_0^{2\pi} f(t)\, D_n(t)\, dt;$$

$D_n(t)$ denotes the Dirichlet kernel ([8] of Chapter 10, Sec. 4). For each n, $F_n(f) = s_n(0)$ is a linear functional defined on X. Moreover,

$$|F_n(f)| \le \|f\|\left\{\frac{1}{2\pi}\int_0^{2\pi}\left|\frac{1}{2} + \cos t + \cdots + \cos nt\right| dt\right\} \le (n+1)\|f\|,$$

where $\|f\| = \max\limits_{0\le x\le 2\pi} |f(x)|$, proving that the F_n are also bounded.

We wish to show that there is a function f in X for which the sequence $s_n(0) = F_n(f)$ diverges; that is, the Fourier series will diverge at $x = 0$. The periodic extension of such a function to R, $(f(x) = f(x + 2\pi))$, is the desired function whose Fourier series diverges at the points $x = 2n\pi$. Suppose that there is no such function. Then for every f in X, $\{F_n(f)\}$ is a convergent sequence of real numbers, and therefore bounded. Let M_f denote a bound for such a sequence. Theorem 7 asserts the existence of a number M such that

$$\|F_n\|^* \le M, n = 1,2,\ldots;$$

or equivalently, $|F_n(f)| \le M\|f\|$. In particular, the sequence of functions $f_n(x) = \sin(n + 1/2)x$ is in X and

$$[8] \qquad |F_n(f_n)| \le M\|f_n\| = M.$$

The trigonometric identity $2 \sin A \cos B = \sin(A + B) + \sin(A - B)$ yields

$$[9] \qquad \int_0^{2\pi}\left[\sin\left(n + \frac{1}{2}\right)t\right]\cos kt\, dt = \frac{1}{2}\left\{\int_0^{2\pi}\sin(n + 1/2 + k)t\, dt + \int_0^{2\pi}\sin(n + 1/2 - k)t\, dt\right\}.$$

A change of variables transforms the right side of [9] into

$$\frac{1}{2}\left\{\int_0^{2\pi[n+1/2+k]}\sin u\, du + \int_0^{2\pi[n+1/2-k]}\sin u\, du\right\} = \frac{1}{n + 1/2 + k} + \frac{1}{n + 1/2 + k}.$$

Combining this with [7], we obtain

$$\begin{aligned}
F_n(f_n) &= \frac{1}{2\pi}\int_0^{2\pi} \sin(n + 1/2)t\, D_n(t)\, dt \\
&= \frac{1}{2\pi}\int_0^{2\pi} \sin(n + 1/2)t[1/2 + \cos t + \cdots + \cos nt]\, dt \\
&= c\sum_{k=0}^{n}\left(\frac{1}{n + 1/2 + k} + \frac{1}{n + 1/2 - k}\right) \\
&\qquad\qquad > c\sum_{k=0}^{n}\frac{1}{n + 1/2 - k},
\end{aligned}$$

where c is a real constant. By setting $n - k = j$,

$$F_n(f_n) > c\sum_{j=0}^{n}\frac{1}{j + 1/2},$$

from which it follows that $\{F_n(f_n)\}$ is unbounded, thus contradicting [8], and the assumption that $\{s_n(0)\}$ converges for all $f \in X$. ∎

We conclude this section with a brief discussion of the Hahn-Banach theorem, which asserts conditions under which a linear functional defined on a subspace may be extended to a linear functional on the full space which shares certain boundedness conditions with the given functional. This theorem has many interesting and important applications, several of which are included here.

The proof requires Zorn's lemma (Chapter 0.2.10) which, the reader will recall, is equivalent to the Axiom of Choice.

9 Theorem (Hahn-Banach). Let f be a linear functional defined on a subspace S of a real vector space X, and let p be a real-valued function defined on X which satisfies the following conditions: (i) for each pair of elements $x,y \in X$, $p(x + y) \leq p(x) + p(y)$, (subadditivity), and (ii) for all $x \in X$ and $a \geq 0$, $p(ax) = ap(x)$. If $f(x) \leq p(x)$ whenever $x \in S$, then there is a linear functional $F(x)$ defined on X satisfying (i′) $F(x) = f(x)$ for $x \in S$, and (ii′) $F(x) \leq p(x)$ for all $x \in X$. That is, there is an extension F of f to the full space X which is also bounded from above by p.

Proof: Let $\mathfrak{F}$ denote the class of linear functionals each of which is an extension of f to some subspace of X, and each $g \in \mathfrak{F}$ is bounded from above by p wherever it is defined. The relation $\prec$ given by:

$$g \prec h \Leftrightarrow h \text{ is an extension of } g \text{ (to a larger subspace)},$$

partially orders $\mathfrak{F}$. If $\mathfrak{S}$ is a totally ordered subset of $\mathfrak{F}$, each $g_\alpha \in \mathfrak{S}$ is a linear functional defined on a subspace S_α of X; and if $g_\alpha, g_\beta \in \mathfrak{S}$, then either $g_\alpha < g_\beta$ or $g_\beta < g_\alpha$. In the first case, this means that $S_\alpha \subset \dot{S}_\beta$ and $g_\alpha = g_\beta$ on S_α. It is easily seen that the function $g(x) = g_\alpha(x)$ if $x \in S_\alpha$ is a linear functional on the subspace $\hat{S} = \bigcup S_\alpha$ of X and that $g_\alpha < g$ (Exercise 11). Thus every totally ordered subset of $\mathfrak{F}$ has an upper bound, and therefore $\mathfrak{F}$ has a maximal element (Zorn's Lemma) which shall be denoted by F. This maximal element F is a linear functional defined on a subspace Y of X satisfying (i′) $F(x) = f(x)$ for $x \in S$, and (ii″) $F(x) \leq p(x)$ for $x \in Y$.

To complete the proof, it must be shown that $Y = X$, or that the known property (ii″) of F is really (ii′). This is done by showing that $X - Y \neq \emptyset$ implies the existence of a suitably bounded extension of F to an even larger subspace, which contradicts the maximality of F.

Suppose that $X - Y$ contains a point u. Then

$$Z = \{z \in X : z = y + au; y \in Y, a \in R\}$$

is a subspace of X which contains Y as a proper subspace. Any extension $\tilde{F}$ of F to Z must satisfy

$$[10] \qquad \tilde{F}(z) = \tilde{F}(y + au) = \tilde{F}(y) + a\tilde{F}(u) = F(y) + a\tilde{F}(u).$$

We must show that it is possible to define $\tilde{F}(u)$ so that $\tilde{F}(z) \leq p(z)$ on Z. Then $\tilde{F}$ will be a proper of extension of F, which contradicts the latter's maximality.

If $v, w \in Y$, then

$$F(v) + F(w) = F(v + w) \leq p(v + w) \leq p(v - u) + p(w + u),$$

implying that

$$[11] \qquad -p(v - u) + F(v) \leq p(w + u) - F(w).$$

Since [11] holds for all $v, w \in Y$, it follows that

$$[12] \qquad \sup_{y \in Y} [-p(y - u) + F(y)] \leq \inf_{y \in Y} [p(y + u) - F(y)].$$

Let γ be a real number which is not less than the left-hand side of [12] and does not exceed the right side. We show next that $\tilde{F}(u) = \gamma$ gives the desired extension. First let us consider $z = y + au$ for $a > 0$. [10] and [12] yield

$$\begin{aligned}\tilde{F}(z) = F(y) + a\gamma &= a[F(y/a) + \gamma] \leq a[F(y/a) + p(y/a + u) - F(y/a)] \\ &= a[p(y/a + u)] = p(y + au) = p(z).\end{aligned}$$

If on the other hand $a = -b < 0$, then

$$\begin{aligned}&\tilde{F}(z) = F(y) - b\gamma = b[F(y/b) - \gamma] \\ &\leq b[F(y/b) - \{-p(y/b - u) + F(y/b)\}] = p(y - bu) = p(z).\end{aligned}$$

Thus F is not maximal unless $Y = X$. ∎

10 Corollary. Let S be a subspace of a real vector space X. If $z \in X - S$ and if $d(z,S) = \delta > 0$, then there is a bounded linear functional F on X satisfying (i) $\|F\|^* \leq 1$, (ii) $F(z) = \delta$, and (iii) $F(x) = 0$ whenever $x \in S$.

Proof: Let $T = \{t \in X: t = s + az, s \in S \text{ and } a \in R\}$. Set $f(t) = f(s + az) = a\delta$. Clearly f is a linear functional on T and (ii) and (iii) are satisfied. Moreover, $\|s + az\| = |a| \cdot \|s/a + z\| \geq |a|\delta$, from which $|f(t)| \leq \|t\|$ follows for all $t \in T$. The conditions of Theorem 9 are satisfied if we set $p(t) = \|t\|$, and therefore an extension F of f exists which satisfies $|F(x)| \leq p(x) = \|x\|$. This last inequality is however simply (i). Conditions (ii) and (iii) are satisfied by the restriction f of F to T. ∎

11 Definition. A subset C of a real vector space X is called *convex* if $x,y \in C$ implies that $ax + (1 - a)y \in C$ for $0 \leq a \leq 1$. That is, C contains the segment that joins x to y.

A neighborhood $S_\delta(x)$ or a closed ball $\overline{S_\delta(x)}$ is a convex set. So is a subspace. In R^2, rectangles, ovals (that is, the areas bounded by these curves), and lines (or line segments) are convex.

The verification of the following simple properties of convex sets is left to the reader.

12 Proposition. Let C_1 and C_2 be compact subsets of a real vector space X. Then each of the following sets is convex: (i) $C_1 \cap C_2$; (ii) $aC_1 = \{x \in X: x = ac, c \in C\}$, where a is a real number; (iii) $C_1 + C_2 = \{x \in X: x = c_1 + c_2, c_i \in C_i\}$; (iv) $C_{1,2} = \{x \in X: x = c_1 - c_2, c_i \in C_i\}$.

13 Definition. A point s is said to be an *internal* point of a subset S of a real vector space X, if the intersection of S and every line through s contains an open interval which contains s.

Remark. An interior point of a set is necessarily an internal point. The converse is easily seen to be false. (See Exercise 14).

14 Definition. A pair of convex sets C_1 and C_2 are said to be *separated* by a linear functional F if a real number γ may be found such that $F(x) \leq \gamma$ whenever $x \in C_1$ and $F(x) \geq \gamma$ on C_2.

15 Example. $C_1 = \{(x,y): x^2 + y^2 \leq 1\}$ and $C_2 = \{(x,y): (x - 2)^2 + y^2 \leq 1\}$ are convex subsets of R^2, and are separated by the functional $F(x,y) = x$. For, if $(x,y) \in C_1$, then $F(x,y) \leq 1$, and if $x \in C_2$, $F(x,y) \geq 1$.

In more geometrical language, we can say that the compact sets C_1 and C_2 are separated by the hyperplane $F = \gamma$. In the case of Example 15, the hyperplane is the line $x = 1$.

In the application that follows, we shall show that usually a pair of disjoint convex sets may be separated by a linear functional. In this case, the function $p(x)$ which is required to apply the Hahn-Banach theorem has an interesting geometric interpretation.

Let C be a convex set which contains $\bar{0}$ an internal point. The function $p(x) = \inf [\alpha:\alpha^{-1}x \in C, \alpha > 0]$ is called the *support function* of C with respect to $\bar{0}$. It follows readily that $p(x)$ is homogeneous with respect to nonnegative real numbers; that is, $p(\alpha x) = \alpha p(x)$ if $\alpha \geq 0$, and that

[13] $$[p < 1] \subset C \subset [p \leq 1].$$

It is left to the reader to show that $p(x)$ is subadditive.

16 Theorem. If C_1 and C_2 are disjoint convex sets in a real vector space X, one of which contains an internal point, then there is a linear function F which separates them.

Proof: Let us suppose that the set C_1 contains an internal point u_1. Then the set $C_{1,2}$ [see Proposition 12(iv)] is convex, and if u_2 is any point of C_2, then $u = u_1 - u_2$ is an internal point of $C_{1,2}$ (why?). In order to construct a support function, the set $C_{1,2}$ is "translated" so that u is carried into 0. That is, we consider instead the set

$$C = \{x:x = x_1 - x_2 - u, x_i \in X_i\},$$

which is convex and contains $\bar{0}$ as an internal point. The disjointness of C_1 and C_2 implies that $\bar{0} \in C_{1,2}$ and that therefore $-u \notin C$. If p is the *support function of* C with respect to $\bar{0}$, then [13] implies that $p(-u) \geq 1$. We begin by defining a linear functional on the subspace S of all multiples of the vector u. Set $f(au) = -a$, for all $a \in R$. It is easily verified that on S, $f(x) \leq p(x)$ and that p satisfies the conditions of the Hahn-Banach theorem. Thus f may be extended to a linear functional F which is defined on the full space X and which is everywhere bounded from above by p. To complete the proof, we must show that F separates C_1 and C_2. The definition of the support function implies that if $x \in C$, then $F(x) \leq 1$. It follows therefore that if $x_i \in C_i$, then $x_1 - x_2 - u \in C$ and

$$F(x_1) - F(x_2) - F(u) = F(x_1 - x_2 - u) \leq 1.$$

However, $F(u) = -1$, from which it follows that $F(x_1) \leq F(x_2)$ for all $x_1 \in C_1$ and $x_2 \in C_2$. Thus

[14] $$\sup_{x_1 \in C_1} F(x_1) \leq \inf_{x_2 \in C_2} F(x_2).$$

The number γ required by Definition 14 may be taken to be any value which does not exceed the right side of [14] and which is not exceeded by the left-hand side. ∎

Exercises

***1.** Show that every vector of l_2 determines a bounded linear functional on l_2 [see Example 2(b)].

***2.** Prove that [1] and [2] define bounded linear functionals.

3. Show that the subset X of $C[0, 2\pi]$ which consists of the functions satisfying $f(0) = f(2\pi)$ is a complete normed subspace.

The *null space* N_F of a linear functional defined on a vector space X is the subset

$$N_F = \{x : F(x) = 0\}.$$

A subspace S of a vector space X is called a *hyperspace* if there is a vector $e \in X - S$ such that

$$X = \{x : x = s + ae,\ s \in S \text{ and } a \in R\}.$$

(Example: Any plane passing through the origin in R^3 is a hyperspace.)

4. Prove that the null space of a linear functional is a vector subspace, and unless $F = 0^*$, N_F is a hyperspace. [*Hint:* Choose a vector $e \in X$ such that $F(e) \neq 0$. Show that if $x \in X$, then a constant c can be determined, and an element $y \in N_F$, such that $x = y + ce$. Do this by applying F to $y + ce$.]

5. Prove that if S is a hyperspace of X, then there is a linear functional F, and $S = N_F$. [*Hint:* Again, choose $e \in X - S$, and write each vector x of X as a sum $s + ce$, where $s \in S$ and $c \in R$. Show that $F(x) = c$ is the desired functional.]

6. Describe the null functionals for the spaces of Example 2.

7. Show that if F is a bounded linear functional on X, then N_F is a closed subset of X. (*Hint:* Use Proposition 5.)

8. Prove that if the null space of a linear functional is closed, then the functional is bounded.

9. Show that a hyperspace of a normed vector space X is either closed in X or dense in X. In the latter case, the functional is not continuous.

***10.** How do we know that the class $\mathfrak{F}$ of Theorem 9 contains any other elements besides the given linear functional f? That is, are you sure that proper extensions, also bounded by p, exist? (Such an extension is actually constructed in the course of the proof.)

***11.** Prove that the relation "$\prec$" defined at the beginning of the proof of Theorem 9 is a partial ordering of $\mathfrak{F}$. Show also that $g = g_\alpha$ on S_α is a well defined linear functional on $\bigcup S_\alpha$, and is an upper bound for $\mathcal{S}$.

12. Prove that the set $C = \{(x,y) \in R^2 : 0 \le x \le 1, 0 \le y \le 2\}$ is convex.

***13.** Prove Proposition 12.

***14.** Show that every point of an open set is internal to that set, but that not every internal point is in the interior of that set. In particular, give an example of a planar set which has an internal point that is not a point of the interior of the set.

15. Prove that the image of a convex set under a linear transformation is convex.

Show by giving an example that a nonconvex set may have a convex image under a linear transformation.

16. Show that the convex sets $C_1 = \{(x,y) : x^2 + y^2 < 1\}$ and $C_2 = \{(x,y) : (x - 2)^2 + (y - 2)^2 < 1\}$ may be separated by a linear functional.

***17.** Prove that the support function $p(x) = \inf [\alpha : \alpha^{-1}x \in C,\ \alpha \in R]$ of the convex set C with respect to $\bar{0}$ satisfies the following properties: (i)

$p(\alpha x) = \alpha p(x)$ if $\alpha \geq 0$; (ii) $p(x + y) \leq p(x) + p(y)$; (iii) $[p < 1] \subset C \subset [p \leq 1]$.

18. Let P be a subset of a real vector space X, which has the following properties: (i) if $x,y \in P$ and $a,b \geq 0$, then $ax + by \in P$; (ii) if both x and $-x \in P$, then $x = \bar{0}$. The set P which we shall call the set of *positive* elements of X, is sometimes called a *convex cone.* Show first that the relation "$\prec$" defined below is a partial ordering on X:

$$x \prec y \iff y - x \in P.$$

Let S be a subspace of X for which the following is true: if $u \in X$, then the set $(S + \{u\}) \cap P$, $(S + \{u\}$ is the set of points of the form $s + u$, where $s \in S)$ is not empty if and only if the set $(S + \{-u\}) \cap P$ is not empty. Show that a linear functional f on S which satisfies: $x \in S$ and $x \succ \bar{0}$ imply $f(x) \geq 0$, may be extended to a linear functional F on X which satisfies the similar condition: $x \in X$ and $x \succ \bar{0}$ implies $F(x) \geq 0$. [*Hint:* Let $p(x) = \inf_{\substack{y \in S \\ y \succ x}} [f(y)]$, for all x in the smallest subspace which contains both P and S. Then use the Hahn-Banach theorem. This theorem may also be proved directly by using Zorn's lemma.]

*Exercises which request the proof of a statement made in the text, as well as those referred to in the text, are starred.

2 Linear Functionals on a Hilbert Space

In Sec. 1 of this chapter, several examples of linear functionals on inner product spaces were given. In each case (Example 11.1.2) the linear functional was represented by a scalar product. It is easily seen that if X is *any* inner product space, then each vector in X determines a linear functional which can be represented by a scalar product. For, if $z \in X$, then

$$[1] \qquad F_z(x) = (z,x), \qquad x \in X,$$

is a bounded, linear functional on X. The main result of this section is the *Riesz Representation Theorem* (7), which asserts that *every* bounded, linear functional defined on a Hilbert space can be represented by a scalar product. For finite-dimensional Hilbert spaces (R^n equipped with the Euclidean metric and scalar product) this has already been proved [Example 11.1.2(a)]. To prove the more general theorem, it is necessary first to derive some additional geometric properties of a Hilbert space.

1 Definition. Let S be a subset of a vector space X. The smallest (vector) subspace of X that contains S, is denoted by $\langle S \rangle$, or by $\langle x_\alpha, x_\beta, \ldots \rangle$ if $S = \{x_\alpha, x_\beta, \ldots\}$.

Recalling that a subset Y of a vector space X is a subspace if and only if

$$x,y \in Y \text{ and } a,b \in R \qquad \text{imply} \qquad ax + by \in Y,$$

it follows that $\langle S \rangle$ may be described also as the totality of (finite) linear combinations of elements in S.

2 Definition. The *orthogonal complement* of a subset S of an inner product space X, written $S^\perp$, is the set of elements in X which are orthogonal to all vectors in S; that is,

$$S^\perp = \{x \in X:(x,y) = 0, \forall\, y \in S\}.$$

3 Proposition. (i) $S^\perp$ is a subspace of X; (ii) $S^\perp = \langle S \rangle^\perp$; (iii) If $S \neq \phi$, then $S^\perp \cap \langle S \rangle = \{\bar{0}\}$.

Proof: If $x,x' \in S^\perp$ and $y \in S$, then $(x,y) = (x',y) = 0$, and it follows that for real numbers a and a',

$$(ax + a'x',y) = a(x,y) + a'(x',y) = 0,$$

which implies that $ax + a'x' \in S^\perp$, thus proving (i).

The inclusion $S \subset \langle S \rangle$ implies that $S^\perp \supset \langle S \rangle^\perp$; for, if z is orthogonal to all vectors in $\langle S \rangle$, it is orthogonal to all vectors in S. Conversely, if $x \in S^\perp$ and $y \in \langle S \rangle$, then by writing $y = \sum_{i=1}^{n} a_i y_i$, where the $y_i \in S$,

$$(x,y) = (x, \Sigma\, a_i y_i) = \Sigma\, a_i(x,y_i) = 0,$$

since each $(x,y_i) = 0$. Thus $S^\perp = \langle S \rangle^\perp$.

Since both $S^\perp$ and $\langle S \rangle$ are subspaces, their intersection is a subspace and must contain the additive identity $\bar{0}$. To show that there are no other vectors in the intersection, let us assume that $x \in \langle S \rangle \cap S^\perp$. Definition 2 implies that x is orthogonal to itself; that is, $(x,x) = 0$. However, it follows from Definition 10.2.1 that $x = \bar{0}$. ∎

4 Example. (a) Let $\Pi[-1,1]$ denote the vector space of polynomial functions defined on $[-1,1]$. The integral

$$(f,g) = \int_{-1}^{1} f(x)g(x)\,dx$$

defines an inner product on $\Pi[-1,1]$. Let $S_E = \{1,x^2,\ldots,x^{2n},\ldots\}$ be the subset of monomials of even degree. Then the subspace $\Pi_E = \langle S_E \rangle$ consists of polynomials which have only terms of even degree. It is left to the reader to verify that

$$\Pi_E^\perp = \Pi_0 = \langle x,x^3,\ldots,x^{2n-1},\ldots \rangle,$$

the subspace of polynomials that contain only terms of odd degree.

(b) The preceding example can be generalized in the following way: Let $L_2[-\gamma,\gamma]$ be the space of functions whose squares are integrable over the interval $[-\gamma,\gamma]$. $L_2[-\gamma,\gamma]$ is a Hilbert space whose inner product is given by

$$(f,g) = \int_{-\gamma}^{\gamma} f(x)g(x)\,dx.$$

Let S_E denote the subset of (a.e.) even functions; that is,

$$S_E = \{f \in L_2[-\gamma,\gamma]: f(x) = f(-x) \text{ a.e.}\}.$$

It is easily verified that $\langle S_E \rangle$ is a vector subspace, and therefore $\langle S_E \rangle = S_E$. The orthogonal space $S_E^{\perp}$ contains the space

$$S_0 = \{f \in L_2[-\gamma,\gamma]: f(x) = -f(-x) \text{ a.e.}\},$$

of (a.e.) odd functions. For, if $f_E \in S_E$ and $f_0 \in S_0$, then the product $g = f_E f_0 \in S_0$ and

$$[2] \quad \int_{-\gamma}^{\gamma} g(x)\,dx = \int_{-\gamma}^{0} g(x)\,dx + \int_{0}^{\gamma} g(x)\,dx = \int_{\gamma}^{0} g(-x)\,(-dx) + \int_{0}^{\gamma} g(x)\,dx = -\int_{0}^{\gamma} -g(x)\,(-dx) + \int_{0}^{\gamma} g(x)\,dx = 0.$$

Thus $S_E^{\perp} \supset S_0$. To reverse the inclusion, let $f(x) \in S_E^{\perp}$. Since

$$f(x) = \frac{f(x) + f(-x)}{2} + \frac{f(x) - f(-x)}{2} = f_E(x) + f_0(x),$$

where f_E is an a.e. even function and f_0 is a.e. odd, it suffices to show that,

$$f(x) \in S_E^{\perp} \quad \text{implies} \quad f(x) = f_0(x).$$

For any even function g,

$$0 = (f,g) = (f_E,g) + (f_0,g) = (f_E,g),$$

if $f \in S_E^{\perp}$. Setting $g = f_E$ yields

$$\int_{-\gamma}^{\gamma} f_E^2 = 0,$$

which is true only if $f_E(x) = 0$ a.e.

(c) Let $H = R^3$ be equipped with the Euclidean inner product, and let S be the set consisting of the two vectors, $x = (1,1,0)$ and $y = (0,1,0)$. Then

$$\langle S \rangle = \{z \in R^3 : z = ax + by, \quad a,b \in R\},$$

which is the plane determined by the two vectors x and y. The orthogonal subspace consists of all vectors which are orthogonal (or normal) to the plane. To describe $S^{\perp}$ analytically, we note that

$$w = (w_1, w_2, w_3) \in S^{\perp} \Leftrightarrow (w,x) = (w,y) = 0 \Leftrightarrow$$
$$w_1 + w_2 = 0 \text{ and } w_2 = 0 \Leftrightarrow w = (0,0,w_3).$$

That is, $S^{\perp}$ is the 1-dimensional subspace of multiples of the vector $e_3 = (0,0,1)$.

To anticipate the *Projection Theorem*, which follows, observe that the vectors $x = (1,1,0)$ and $y = (0,1,0)$, which span $\langle S \rangle$, are linearly independent, from which it follows that $\{x,y,e_3\}$ is also a linearly independent set and must therefore span R^3. That is, every vector in R^3 can be expressed uniquely as a linear combination $ax + by + ce_3$. Since $(ax + by) \in \langle S \rangle$, this may be restated as follows: Every vector in R^3 can be expressed uniquely as a sum $u + v$, where $u \in \langle S \rangle$ and $v \in \langle S \rangle^{\perp}$. The general theorem follows.

5 Theorem (Projection Theorem). Let S be a (vector) subspace of the Hilbert space H. If S is a closed subset of H, in the metric induced by the inner product, then

[3] $$H = \{z : z = x + y, x \in S \text{ and } y \in S^{\perp}\}.$$

That is, every element of H may be written as the sum of a vector in S and a vector in $S^{\perp}$. Moreover, this decomposition is unique.

Remark. Before proving this theorem, it should be noted that if H is finite dimensional, as in Example 4(c), *every* subspace is closed (Exercise 1). That this may not be the case if H does not have finite dimension is illustrated by Example 6, which follows the proof of this theorem (see also Exercise 11.1.9).

Proof: The accompanying diagram illustrates the two-dimensional case. Its value rests in our ability to translate a simple heuristic geometric proof in R^2 into a rigorous proof in H.

Let S be the given subspace and let z be an arbitrary but fixed element of H. If $S = H$ or $S = \phi$, then there is nothing to prove, and so we may assume that $\phi \subsetneq S \subsetneq H$. Consider the set of vectors $\{v : v = z - u, u \in S\}$. Since the set of numbers $\{\|z - u\|\}$ is bounded from below, the number

$$0 \le \delta = \inf_{u \in S} \{\|z - u\|\} = d(z,S),$$

is well defined. If $\delta = 0$, then z itself must be in S since S is closed. In this case $x = z$ and $y = 0$ is the desired decomposition of z.

On the other hand, if $\delta > 0$, then $z \in H - S$, and there exists a sequence $\{x_n\} \subset S$ such that

[4] $$0 < \delta \le \|z - x_n\| < \delta + 1/n.$$

The parallellogram identity (Theorem 10.2.5) yields

[5] $$\begin{aligned}\|x_n - x_m\|^2 &= \|(x_n - z) + (z - x_m)\|^2 \\ &= 2\|x_n - z\|^2 + 2\|x_m - z\|^2 - \|(x_n - z) + (x_m - z)\|^2.\end{aligned}$$

The inequality [4] implies that

$$\lim_{n,m\to\infty}\left[2\|x_n - z\|^2 + 2\|x_m - z\|^2\right] = 4\delta^2.$$

By rewriting the third term of [5], we obtain

$$[6] \qquad \|(x_n - z) + (x_m - z)\|^2 = 4\left\|\frac{(x_n + x_m)}{2} - z\right\|^2.$$

It is left to the reader to verify that [4] implies that [6] also tends to $4\delta^2$ as $n,m \to \infty$ (Exercise 2). Thus the right side of [5] tends to $4\delta^2 - 4\delta^2 = 0$, proving that $\{x_n\}$ is a Cauchy sequence in S. But S is a closed subset of a complete space, and is therefore itself complete. Thus $\{x_n\}$ converges in S to some element x. Since the inner product is a continuous function of two variables (Exercise 10.1.8), it follows that

$$\|z - x\| = \lim_{n\to\infty}\|z - x_n\| = \delta.$$

(See Figure 31.)

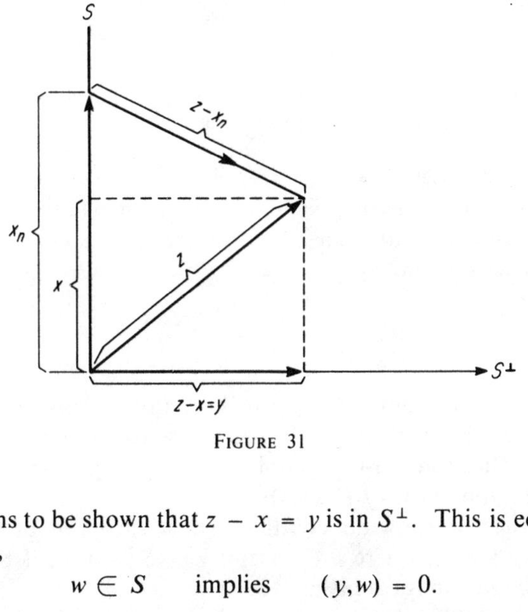

FIGURE 31

It remains to be shown that $z - x = y$ is in $S^\perp$. This is equivalent to proving that,

$$w \in S \qquad \text{implies} \qquad (y,w) = 0.$$

If $w \in S$, then for every real number a, $(x - aw) \in S$. Since $\delta = d(z,S)$, it follows that

$$\begin{aligned}\delta^2 \le \|z - (x - aw)\|^2 &= \|(z - x) + aw\|^2 \\ &= \|z - x\|^2 + a^2\|w\|^2 + 2a(z - x, w).\end{aligned}$$

The identity $\delta^2 = \| z - x \|^2$ yields

$$0 \leq a^2 \| w \|^2 + 2a(z - x, w). \tag{7}$$

If $(z - x, w) = (y, w)$ does not vanish, then the right side of [7] is a quadratic polynomial in a:

$$P(a) = a[\| w \|^2 a + 2(z - x, w)],$$

which has two real roots, $a = 0$ and $a = \dfrac{-2(z - x, w)}{\| w \|^2}$. Since w is an arbitrary element of S, we may assume that it is not the additive identity $\bar{0}$. Moreover, $(z - x, w) \neq 0$ implies that the two roots of $P(a)$ are different, from which it follows that for some real values of a, $P(a) < 0$. But this contradicts [7], which asserts that $P(a) \geq 0$. Thus it must follow that $(z - x, w) = 0$ for all $w \in S$, proving that $y = z - x \in S^{\perp}$. This completes the proof of the existence of at *least* one decomposition of z into a sum $x + y$, where $x \in S$ and $y \in S^{\perp}$. If there were another such decomposition, $x' + y'$, then it would follow that $x - x' = y' - y$. Since both S and $S^{\perp}$ are vector spaces, we obtain

$$x - x' \in S \qquad \text{and} \qquad y' - y \in S^{\perp},$$

implying that $S \cap S^{\perp} \supset \{x - x'\}$. Proposition 3 yields $x - x' = y' - y = \bar{0}$, thus proving that $x = x'$ and $y = y'$. ▌

6 Example. Let $S = \langle x, x^3, \ldots, x^{2n-1}, \ldots \rangle$ be the subspace of $L_2[-\gamma, \gamma]$ which consists of polynomials each of whose terms has odd degree [Example 4]. Although S is a subspace of a Hilbert space, it is not closed, since the function

$$f(x) = \begin{cases} \sin x & \text{if } -\gamma \leq x \leq \gamma \\ 0 & \text{if } \gamma < |x| \end{cases}$$

is in the closure of S, but not in S itself (Exercise 3). Another consequence of the fact that S is not closed, is the proper inclusion of S in $(S^{\perp})^{\perp}$. If S were closed, Theorem 5 would imply that $S = (S^{\perp})^{\perp}$ (Exercise 4). A simple computation yields $f \in (S^{\perp})^{\perp}$.

It is also easy to show directly that $f(x) = \sin x$ cannot be written as the sum $u(x) + v(x)$, where $u \in S$ and $v \in S^{\perp}$. For, if this could be done, a contradiction is readily arrived at:

(i) $f(x)$ is odd and $v(x)$ is even imply that $(f, v) = 0$;

(ii) $f(x) \notin S$ and $u(x) \in S$ imply that

$$0 = (v, f) = (v, u) + (v, v) = (v, v),$$

since $(v, u) = 0$.

However, $(v, v) = 0$ implies that $v = 0$ a.e., from which it follows that $f = u \in S$.

7 Theorem (Riesz Representation Theorem for Hilbert Spaces). Let F be a bounded linear functional defined on a real Hilbert space H. Then there is a unique element $\alpha \in H$ such that $F(x) = (\alpha,x)$ for all $x \in H$, and $\|F\|^* = \|\alpha\|$; $\|\alpha\| = \sqrt{(\alpha,\alpha)}$ and $\|F\|^* = \sup_{x \neq \bar{0}} \dfrac{|F(x)|}{\|x\|}$.

Proof: Since every bounded linear functional is a continuous mapping of H into R, the set $S = \{x \in H : F(x) = 0\}$ is a closed subset of H. Moreover, the linearity and homogeneity of F imply that S is a subspace (Exercises 11.1.4 and 11.1.7). Thus S satisfies the hypotheses of the Projection Theorem, and every vector z in H can be represented uniquely as a sum $z = x + y$, where $x \in S$ and $y \in S^\perp$.

Case (i). If $S^\perp = \{\bar{0}\}$, then $S = H$ and $F(z) = 0 = (\bar{0},z)$ for all $z \in H$. Thus $\alpha = \bar{0}$.

Case (ii). If $S^\perp$ contains a nonzero element, then it contains at least two distinct, nonzero elements u and y, since it is a subspace. Thus neither $F(u)$ nor $F(y)$ vanishes, and therefore $a = \dfrac{F(y)}{F(u)} \neq 0$. It follows that

$$F(y - au) = F(y) - aF(u) = 0,$$

and therefore $y - au \in S$. However, it was assumed that both y and u are in $S^\perp$. Thus we obtain $y - au \in S \cap S^\perp$. Proposition 3 implies that $y = au$.

We have just proved that any pair of elements in $S^\perp$ are linearly dependent, which means that $\dim S^\perp = 1$. Let e be a vector in $S^\perp$ whose norm is equal to 1. Since every element of $S^\perp$ is a real multiple of e, it follows from the Projection Theorem that, if $z \in H$, then there is a vector $x \in S$, and a real number c, such that $z = x + ce$. Setting $F(e) = A$, we obtain

$$\begin{aligned} F(z) &= F(x) + cA = cA = A(e,ce) \\ &= A[(e,x) + (e,ce)] = (Ae,x + ce) = (Ae,z). \end{aligned}$$

(Remember, $x \in S$ if and only if $F(x) = 0$, which is in turn equivalent to $(e,x) = 0$.) Thus $\alpha = Ae$ is the desired vector.

The uniqueness is easily verified. Suppose that there is a second vector β satisfying $(\alpha,z) = (\beta,z)$ for all $z \in H$. Then by taking $z = \alpha - \beta$, we obtain $(\alpha - \beta, \alpha - \beta) = 0$, which implies that $\alpha = \beta$.

The Cauchy-Schwarz inequality yields

$$|F(z)| = |(\alpha,z)| \leq \|\alpha\| \, \|z\|,$$

thus implying that $\|F\|^* \leq \|\alpha\|$. Setting $z = \alpha$, we obtain

$$|F(\alpha)| = (\alpha,\alpha) = \|\alpha\|^2,$$

or $\|F\|^* \geq \|\alpha\|$. ∎

Exercises

1. Prove that a finite-dimensional subspace of a Hilbert space is closed.
*2. Verify that the expression [6] converges to $4\delta^2$.
3. Show that $S = \langle x, x^3, \ldots, x^{2n-1}, \ldots \rangle$ is *not* a closed subspace of $L_2[-\pi, \pi]$ by verifying that

$$f(x) = \begin{cases} \sin x & \text{if } -\pi \leq x \leq \pi \\ 0 & \text{otherwise} \end{cases}$$

is an accumulation point of S, but is not itself in S. (Be sure to use the L_2-norm.)
4. Show that if S is a closed subspace of H, then $(S^{\perp})^{\perp} = S$.

Let X be an inner-product space that contains a subset S. A vector x in S is called the *projection* of a second vector z in $X - S$ onto S, if for all $y \in S, (z - x, y) = 0$. (Draw a picture.)

5. Prove the following "projection" theorem: If S is a closed subspace of a Hilbert space H, then every vector in H has a projection in S. Show also that this is true if S is a complete (in the metric sense) subspace of an inner product space.
6. Prove that the "projection" theorem of Exercise 5 remains true if the condition that S is a *subspace* is replaced by the following assumption: S is *convex*, which means that if $x,y \in S$ then the points $z = ax + (1 - a)y, 0 \leq a \leq 1$, are also in S. (Geometrically, this says that if two points belong to a convex set, then so do all the points which lie on the segment that joins these points.)

*Exercises which request the proof of a statement made in the text, as well as those referred to in the text, are starred.

3 Linear Functionals on L_p Spaces

Earlier [Example 11.1.2(d)] it was shown that if p and q are conjugate indices, then $L_q \subset L_p^*$. More precisely, if $\alpha \in L_q$, then

$$F_\alpha(f) = \int f\alpha$$

is a bounded, linear functional on L_p. The Riesz Representation Theorem (3) for L_p spaces asserts that if $1 \leq p < \infty$, then $L_p^* = L_q$. It can be shown, however, that L_1 is a *proper* subset of L_∞^*; for, there exist bounded, linear functionals on the space L_∞ of a.e. bounded, measurable functions, which can not be expressed in integral form (See Taylor [2]).

1 Theorem. Let L_p and L_q be conjugate spaces ($1 \leq p,q \leq \infty$ and $1/p + 1/q = 1$). Then to each $\alpha \in L_q$, there is a functional F_α in L_p^*, and

$$\| F_\alpha \|_p^* = \| \alpha \|_q .$$

($\| \, \|_p$, $\| \, \|_q$ denote the norms in L_p, L_q respectively, and $\| \, \|_p^*$, $\| \, \|_q^*$ are the norms in the corresponding conjugate spaces of bounded, linear functionals.)

Proof: It has already been verified that

[1] $$F_\alpha(f) = \int f\alpha$$

is a bounded linear functional on L_p [Example 11.1.2(d)]: Indeed, the Hölder inequality yields

$$|F_\alpha(f)| \le \|\alpha\|_q \|f\|_p,$$

from which $\|F_\alpha\|_p^* \le \|\alpha\|_q$ follows. We prove next that these two norms are equal by exhibiting a function f in L_p for which $|F_\alpha(f)| = \|\alpha\|_q \|f\|_p$:

If $1 < p < \infty$, set $f = (\operatorname{sgn} \alpha) |\alpha|^{q/p}$, where

$$\operatorname{sgn} \alpha(x) = \begin{cases} +1 & \text{if } \alpha(x) \ge 0 \\ -1 & \text{if } \alpha(x) < 0. \end{cases}$$

From $\alpha \in L_q$, we obtain the integrability of $|f|^p = |\alpha|^q$, and

$$|F_\alpha(f)| = \int f\alpha = \int |\alpha|^{1+q/p} = \int |\alpha|^q = \|\alpha\|_q^q.$$

Moreover,

$$\|\alpha\|_q^q = \|\alpha\|_q \|\alpha\|_q^{q-1} = \|\alpha\|_q \left(\int |\alpha|^q\right)^{(q-1)/q}$$

$$= \|\alpha\|_q \left(\int |f|^p\right)^{1/p} = \|\alpha\|_q \|f\|_p,$$

from which it follows that $\|F_\alpha\|_p^* = \|\alpha\|_q$.

For the case $p = 1$ and $q = \infty$, it will be shown that if $\epsilon > 0$, then $\|F_\alpha\|_1^* \ge \|\alpha\|_\infty - \epsilon$. (The opposite inequality follows again from Hölder's inequality.) Let

[3] $$S_\epsilon = \{x : |\alpha(x)| \ge \|\alpha\|_\infty - \epsilon\},$$

and

$$f_\epsilon(x) = \begin{cases} \operatorname{sgn} \alpha(x) & \text{if } x \in S_\epsilon \\ 0 & \text{if } x \notin S_\epsilon. \end{cases}$$

If $m([f_\epsilon \ne 0]) = 0$, then S_ϵ is a null set and $|\alpha(x)| < \|\alpha\|_\infty - \epsilon$ a.e., from which it follows that $\|\alpha\|_\infty$ is not the a.e. least upper bound of $\alpha(x)$. Thus $m([f_\epsilon \ne 0]) > 0$. Since $|f_\epsilon(x)| = 1$ on S_ϵ, and vanishes outside S_ϵ, it follows that $[f_\epsilon \ne 0] = S_\epsilon$, and

[4] $$|F_\alpha(f_\epsilon)| = \int f_\epsilon \alpha = \int_{S_\epsilon} |\alpha| \ge m(S_\epsilon)[\|\alpha\|_\infty - \epsilon].$$

Moreover,

[5] $$\|f_\epsilon\|_1 = \int |f_\epsilon| = \int \chi_{S_\epsilon} = m(S_\epsilon) = m([f_\epsilon \ne 0]) > 0.$$

Substituting [5] into [4] yields

$$|F_\alpha(f_\epsilon)| \ge m(S_\epsilon)(\|\alpha\|_\infty - \epsilon) = \|f_\epsilon\|_1(\|\alpha\|_\infty - \epsilon).$$

Since ϵ is an arbitrary positive number, it follows that

$$\| F_\alpha \|_1^* = \| \alpha \|_\infty .$$

Finally, if $p = \infty$, $q = 1$ and $\alpha \in L_q$, set $f = \operatorname{sgn} \alpha(x)$. Then $f \in L_\infty$, $f\alpha \in L_1$ and unless $\alpha = 0$ a.e., $\|f\|_\infty = 1$ and

$$| F_\alpha(f) | = \int | \alpha | = \| \alpha \|_1 \|f\|_\infty . \quad \blacksquare$$

The converse of Theorem 1 is proved next for functions that vanish outside a closed bounded interval $[a,b]$.

2 Theorem (Riesz Representation Theorem). To each $F \in L_p^*[a,b]$, $1 \le p < \infty$, there is a function $\alpha \in L_q[a,b]$, $1 < q \le \infty$ and $1/p + 1/q = 1$, such that for all $f \in L_p[a,b]$,

$$F(f) = F_\alpha(f) = \int f\alpha. \tag{6}$$

Moreover, α is a.e. uniquely determined.

Proof: As in some earlier proofs, we shall begin with the simplest class of functions in $L_p[a,b]$, the characteristic functions of subintervals I of $[a,b]$. The action of F on this subclass will determine the function α that satisfies [6]. Following this, the functional is extended by linearity to the class of step functions that vanish outside the interval $[a,b]$, and finally to L_p by taking limits and truncations.

Case 1. Let $f = \chi_I$, the characteristic function of an interval $I \subset [a,b]$. Since these functions are in $L_p[a,b]$, it follows that $F(\chi_{[a,t]})$ is a well defined function $\varphi(t)$ on $[a,b]$. We shall show that $\varphi(t)$ *is absolutely continuous and that its a.e. derivative is the function* $\alpha(t)$ *that we seek.*

Let $I_i = (x_i, y_i)$, $i = 1, 2, \ldots, n$, be a disjoint set of subintervals of $[a,b]$, whose total length

$$m\left(\bigcup_{i=1}^{n} I_i\right) = \sum_1^n (y_i - x_i) < \delta.$$

(If some $x_i = a$, we shall assume, for convenience, that $I_i = [a, y_i]$.) Taking the variation of $\varphi(t)$ over $\bigcup_1^n I_i$, we obtain

$$\sum_1^n | \varphi(y_i) - \varphi(x_i) | = \sum_1^n | F(\chi_{[a,y_i]}) - F(\chi_{[a,x_i]}) | .$$

The linearity of F, and the identity $\chi_{[a,y_i]} - \chi_{[a,x_i]} = \chi_{I_i}$ yield

$$\sum_1^n | F(\chi_{I_i}) | = \sum_1^n (\operatorname{sgn} F)(F(\chi_{I_i})) = F\left\{\sum_1^n (\operatorname{sgn} F)\chi_{I_i}\right\}$$

$$\le \| F \|_p^* \left\| \sum_1^n (\operatorname{sgn} F)\chi_{I_i} \right\|_p \le \| F \|_p^* \delta^{1/p},$$

proving that the variation of $\varphi(t)$ over a system of intervals tends to zero as the total length of the system tends to zero. Thus $\varphi(t)$ is absolutely continuous on $[a,b]$. Theorem 9.4.12 asserts that there is a function $\alpha \in L_1[a,b]$ satisfying

[7] $$\varphi(t) = \int_a^t \alpha; \quad \varphi'(t) = \alpha(t) \text{ a.e.}$$

The identity $\varphi(t) = F(\chi_{[a,t]})$, and [7] yield

[8] $$F(\chi_{[a,t]}) = \int_a^t \alpha = \int_a^b \alpha\chi_{[a,t]} = \int \alpha\chi_{[a,t]},$$

which is the desired identity [6] for the case $f = \chi_{[a,t]}$, $a \leq t \leq b$. If $I = [c,d]$ is contained in $[a,b]$, then

$$\chi_{[c,d]} = \chi_{[a,d]} - \chi_{[a,c]} + \chi_{\{c\}} = \chi_{[a,d]} - \chi_{[a,c]} \text{ a.e.},$$

and the linearity of the integral implies [6] for $f = \chi_{[c,d]}$.

Case 2. Suppose that $f = \sigma$, a step function which vanishes outside $[a,b]$. Then there is a subdivision $a = x_0 < x_1 < \cdots < x_n = b$ of $[a,b]$, and

$$\sigma(x) = \begin{cases} c_i & \text{if } x_{i-1} < x < x_i;\ i = 1,2,\ldots,n. \\ \lambda_i & \text{if } x = x_i;\ i = 0,1,\ldots,n. \\ 0 & \text{otherwise.} \end{cases}$$

Thus, except possibly at the points $x_1, x_2, \ldots, x_n$, (that is, a.e.),

$$\sigma = \sum_1^n c_i\chi_{(x_{i-1},x_i)} = c_n\chi_{[a,x_n]} + (c_{n-1} - c_n)\chi_{[a,x_{n-1}]} + \cdots + (c_1 - c_2)\chi_{[a,x_1]}.$$

The linearity of F and [8] yield

[9] $$F(\sigma) = \int \alpha\sigma,$$

for all step functions σ that vanish outside $[a,b]$.

Case 3. Let g be a bounded measurable function that vanishes outside $[a,b]$. It follows from the theorem on dominated convergence that $g \in L_p[a,b]$, $1 \leq p < \infty$. Obviously, g is also in L_∞. The integrability of g implies the existence of a sequence $\{\tau_n\}$ of step functions which is a Cauchy sequence in $L_1[a,b]$, converges a.e. to g, and converges in norm to g; that is, $\lim \|\tau_n - g\|_1 = 0$. It is left to the reader as an exercise to show that for $p > 1$, $\lim \|\tau_n - g\|_p = 0$ also.

If M is a bound for $|g|$, then the truncated sequence of step functions

$$\sigma_n(x) = (\tau_n(x))_M = \begin{cases} \tau_n(x) & \text{if } |\tau_n(x)| \leq M \\ M & \text{if } \tau_n(x) > M \\ -M & \text{if } \tau_n(x) < -M, \end{cases}$$

is a Cauchy sequence in all p-norms (Exercise 2), and

$$\| \sigma_n - g \|_p \le \| \tau_n - g \|_p, \quad n = 1, 2, \dots,$$
$$| \sigma(x) - g(x) | \le | \tau_n(x) - g(x) |, \quad a \le x \le b.$$

Thus $\lim \| \sigma_n - g \|_p = 0$ and $\lim | \sigma_n - g | = 0$ a.e., and it follows that $\lim | \alpha\sigma_n - \alpha g | = 0$ a.e. Moreover, $| \alpha\sigma_n - \alpha g | \le 2M | \alpha |$, where $2M | \alpha | \in L_1[a,b]$, so that the theorem on dominated convergence is applicable. Therefore

$$\lim_{n\to\infty} | F(\sigma_n) - F(g) | = \lim_{n\to\infty} | \int (\alpha\sigma_n - \alpha g) | = 0.$$

This proves that for bounded, measurable functions g,

[10] $$F(g) = \lim_{n\to\infty} F(\sigma_n) = \lim_{n\to\infty} \int \alpha\sigma_n = \int \alpha g.$$

Case 4. Suppose now that f is any function in $L_p[a,b]$ (unbounded or bounded), $1 \le p < \infty$. To prove the existence of the integral $\int f\alpha$ of [6], we must show first that $\alpha \in L_q[a,b]$ for $1 < q \le \infty$. For the moment, let us assume that this has already been proved, and continue with the verification of the identity [6]:

The sequence $\{f_n\}$ of truncations of f at n, are bounded measurable functions, and $\lim \alpha f_n = \alpha f$. Therefore, αf is integrable, and

[11] $$\lim_{n\to\infty} F(f_n) = \lim_{n\to\infty} \int \alpha f_n = \int \alpha f.$$

Furthermore, the continuity of F implies that

[12] $$| F(f) - F(f_n) | \le \| F \|_p^* \, \| f - f_n \|_p \to 0$$

as n tends to infinity. Combining [11] and [12], we obtain

$$F(f) = \lim_{n\to\infty} F(f_n) = \int \alpha f,$$

which is the desired result [6].

If β is a second function in $L_q[a,b]$ for which $F(f) = \int \beta f$ whenever $f \in L_p[a,b]$, then it follows that for all f in $L_p[a,b]$, $\int (\alpha - \beta) f = 0$, in particular for $f = \operatorname{sgn}(\alpha - \beta) | \alpha - \beta |^{q-1}$. This yields $\int | \alpha - \beta |^q = 0$, from which it follows that $\alpha = \beta$ a.e. on $[a,b]$.

It remains to be proved that $\alpha \in L_q[a,b]$, $1 < q \le \infty$. First, assume that $q < \infty$. We shall show that α is the monotone limit of a sequence in $L_q[a,b]$ whose norms are uniformly bounded, and then use the theorem on monotone convergence (6.1.5). Let $\{\alpha_n\}$ denote the sequence of truncations of α. From $| \alpha_n | \le n$, it follows that $\alpha_n \in L_q[a,b]$, for all $q \ge 1$ and $n = 1, 2, \dots$. Furthermore, $\alpha_n \uparrow \alpha$. It remains to be shown that for fixed but arbitrary $q < \infty$, the set of q-norms $\{ \| \alpha_n \|_q \}$ is bounded. Indeed, this bound turns out to be $\| F \|_p^*$: Let $\beta_n = \operatorname{sgn} \alpha | \alpha_n |^{q/p}$, where p and

q are conjugate indices, $1 < p < \infty$. Clearly each β_n is a bounded measurable function, and so, substitution in [10] yields $F(\beta_n) = \int \alpha\beta_n$. Furthermore,

$$\|\beta_n\|_p = \left(\int |\beta_n|^p\right)^{1/p} = \left(\int |\alpha_n|^q\right)^{1/p} = \|\alpha_n\|_p^{q/p}.$$

Combining these last two expressions and using the identity $1/p + 1/q = 1$, we obtain

[13] $$\|\alpha_n\|_q^q = \int |\alpha_n|^{1+q/p} \leq \int |\alpha| \, |\alpha_n|^{q/p} = \int \alpha\beta_n$$
$$= |F(\beta_n)| \leq \|F\|_p^* \, \|\beta_n\|_p = \|F\|_p^* \, \|\alpha_n\|_q^{q/p}.$$

(The existence of the integrals in [13] follows from the integrability of α (See [7]) and the fact that $|\alpha_n|^{q/p}$ is bounded and measurable.) Dividing [13] by $\|\alpha_n\|_q^{q/p}$, we obtain the uniform bound for the q-norms of the α_n. The theorem on monotone convergence yields the integrability of the monotone limit $|\alpha|^q$, and $\|\alpha\|_q \leq \|F\|_p^*$.

We prove next that $\alpha \in L_\infty[a,b]$: Choose numbers c and M satisfying $a \leq c < c + 1/M \leq b$. The function

$$\beta(x) = \begin{cases} M & \text{if } c \leq x \leq c + 1/M \\ 0 & \text{otherwise,} \end{cases}$$

is bounded and integrable. Therefore

[14] $$|F(\beta)| = \left|\int \alpha\beta\right| = \left|M \int_c^{c+1/M} \alpha\right| \leq \|F\|_1^* \, \|\beta\|_1 = \|F\|_1^*.$$

On the other hand,

[15] $$M \left|\int_c^{c+1/M} \alpha\right| = \frac{|\varphi(c + 1/M) - \varphi(c)|}{1/M} \to |\alpha(c)| \text{ a.e.}$$

as $M \to \infty$. Combining [14] and [15], we obtain the a.e. bound $|\alpha(c)| \leq \|F\|_1^*$, proving that $\alpha \in L_\infty[a,b]$. ∎

We conclude with the extension of the Riesz Representation Theorem to the full L_p spaces; that is, we remove the restriction that the functions vanish outside a bounded interval. The idea is to use Theorem 2 to obtain a function α_n which corresponds to the restriction of a given functional F (in L_p) to the interval $I_n = [-n,n]$. It should come as no great surprise that we shall be able to prove that $F(f) = \int \alpha f$, where $\alpha = \lim \alpha_n$ a.e.

Throughout the proof of this theorem, it will be necessary to restrict certain functions in L_p to finite subintervals, or conversely, to extend functions in $L_p[-n,n]$ by defining them on $R - I_n$. To avoid an unwieldy excess of superscripts and subscripts, we shall freely abuse the functional notation as follows: a function f in $L_p[-n,n]$ is, strictly speaking, not defined outside the interval I_n, and is not the same as the "extended" function

$$f_e(x) = \begin{cases} f(x) & \text{if } -n \le x \le n \\ 0 & \text{otherwise.} \end{cases}$$

Obviously, $f_e \in L_p$ whenever $f \in L_p[-n,n]$, and vice versa, and

$$\int f_e = \int f_e \chi_{[-n,n]} = \int_{-n}^{n} f.$$

For this reason, the subscript e will generally be omitted, and f and f_e will be considered to be the same function. This "abuse of notation" permits us to write, whenever $n \le m$,

$$L_p[-n,n] \subset L_p[-m,m] \subset L_p;$$

that is, the functions in $L_p[-n,n]$ may be described as those p-integrable functions that vanish outside $[-n,n]$. Conversely, every $f \in L_p$ may be restricted to I_n. The *restriction* of f to I_n is denoted by

[16]
$$f_{(n)}(x) = \begin{cases} f(x) & \text{if } x \in I_n \\ 0 & \text{otherwise.} \end{cases}$$

Clearly, $f_{(n)} \in L_p[-n,n]$.

3 Theorem (Riesz Representation Theorem). If $F \in L_p^*$, $1 \le p < \infty$, there is a function $\alpha \in L_q$, $1 < q \le \infty$ and $1/p + 1/q = 1$, such that for all $f \in L_p$,

$$F(f) = \int \alpha f \quad \text{and} \quad \|F\|_p^* = \|\alpha\|_q.$$

Moreover, α is a.e. uniquely determined by F.

Proof: Each $F \in L_p^*$ determines a sequence $F_n \in L_p^*[-n,n]$ as follows: The inclusion $L_p[-n,n] \subset L_p$ suggests that the F_n may be defined as the restriction of F to $L_p[-n,n]$:

[17]
$$F_n(f) = F(f), \quad f \in L_p[-n,n].$$

To each F_n, there is a function α_n in $L_q[-n,n]$, determined uniquely a.e., satisfying

$$F_n(f) = \int_{I_n} \alpha_n f, \quad f \in L_p[-n,n],$$

and $\|F_n\|_p^* = \|\alpha_n\|_q$ (Theorem 2). From [17] and the inclusion relation, it follows that if $f \in L_p[-n,n]$, then $F_n(f) = F_{n+1}(f)$. Since $f(x) \equiv 0$ outside I_n, this becomes

$$F_n(f) = \int_{I_n} \alpha_n f = F_{n+1}(f) = \int_{I_{n+1}} \alpha_{n+1} f = \int_{I_n} \alpha_{n+1} f.$$

Thus $\alpha_{n+1} = \alpha_n$ a.e. on I_n. This suggests that the function α we seek may be defined as follows:

$$[18] \qquad \alpha(x) = \begin{cases} \alpha_n(x) & \text{if } x \in I_n - I_{n-1}, \quad n = 2, 3, \ldots. \\ \alpha_1(x) & \text{if } x \in I_1 \end{cases}$$

Thus [18] and the identity $\alpha_{n+1} = \alpha_n$ a.e. on I_n imply

$$[19] \qquad \alpha(x) = \alpha_n(x) \text{ a.e.} \quad \text{on } I_n, \quad i = 1, 2, \ldots.$$

To show that α, as defined by [19], is the desired function, requires the verification of

(i) $\alpha \in L_q$ and (ii) $F(f) = \int \alpha f$, for all $f \in L_p$.

The case $p = 1$, $q = \infty$, is left to the reader.

If $p > 1$, and p,q are conjugate indices, then

$$[20] \quad \left(\int_{I_n} |\alpha|^q\right)^{1/q} = \left(\int_{I_n} |\alpha_n|^q\right)^{1/q} = \|F_n\|_p^* = \sup_{\substack{\|f\| \neq 0 \\ f \in L_p[-n,n]}} \frac{|F_n(f)|}{\|f\|_p}$$

$$\leq \sup_{\substack{\|f\| \neq 0 \\ f \in L_p}} \frac{|F(f)|}{\|f\|_p} = \|F\|_p^*.$$

Thus $\{|\alpha_n|^q\}$ is a nondecreasing sequence of integrable functions whose integrals are uniformly bounded (by $\|F\|_p^*$). The theorem on monotone convergence and the definition of α ([19]), yield the integrability of $|\alpha|^q$, thus proving (i).

To verify (ii), let $f \in L_p$ be given, and let $f_{(n)}$ denote the restriction of f to I_n ([16]). Theorem 2 implies

$$[21] \qquad F(f_{(n)}) = F_n(f_{(n)}) = \int_{I_n} \alpha_n f_{(n)}.$$

Furthermore, since $f - f_{(n)}$ is in L_p, and vanishes on I_n, it follows that

$$[22] \qquad \lim_{n\to\infty} \|f_{(n)} - f\|_p^p = \lim_{n\to\infty} \int |f_{(n)} - f|^p$$

$$= \lim_{n\to\infty} \int_{R-I_n} |f|^p = 0.$$

The continuity of F and [22] yield

$$[23] \qquad |F(f) - F(f_n)| \leq \|F\|_p^* \, \|f - f_{(n)}\|_p,$$

which tends to zero as n tends to infinity. Combining [21] and [23] we obtain,

$$[24] \qquad F(f) = \lim_{n\to\infty} F(f_{(n)}) = \lim_{n\to\infty} \int_{I_n} \alpha_n f_{(n)} = \lim_{n\to\infty} \int \alpha_n f_{(n)}.$$

Hölder's inequality yields the integrability of αf, and since lim $| \alpha_n f_{(n)} - \alpha f | = 0$, it follows from [23] and [24] that

$$F(f) = \int \alpha f.$$

It is left to the reader to verify the a.e. uniqueness of α, and that $\| F \|_p^* = \| \alpha \|_q$. ∎

Exercises

*1. Prove (in case 3 of Theorem 2), that lim $\| \tau_n - g \|_1 = 0$ implies lim $\| \tau_n - g \|_p = 0$ for $p > 1$.
*2. Prove (in case 3 of Theorem 2), that the truncated sequence $\{\sigma_n\}$ is a Cauchy sequence in all p-norms.
*3. Verify (i) and (ii) for the case $p = 1$ and $q = \infty$ ((i) and (ii) follow [19]).
*4. Complete the proof of Theorem 3 by proving that $\| F \|_p^* = \| \alpha \|_q$ and that $\alpha(x)$ is a.e. uniquely defined.

*Exercises which request the proof of a statement made in the text, as well as those referred to in the text, are starred.

4 The Stieltjes Integral and Linear Functionals on $C[a,b]$

The proof of the Representation Theorem for linear functionals defined on the Banach space $C[a,b]$ (Example 2.2.3) occupies the major part of this section. Again, the linear functional is represented by an integral, in this case the *Stieltjes integral*, discussed below.

The Lebesgue integral may be thought of as a generalization of the Riemann integral which is obtained by enlarging both the class of integrable functions, and measurable or integrable sets. The advantages of this particular generalization have already been discussed at great length. One fact that bears comparison with the subject of this section is that every Riemann integrable function is Lebesgue integrable, and for such a function the two integrals coincide. Similarly, for Riemann and Lebesgue measurable sets: in enlarging the class of measurable sets, the measure assigned to the Riemann measurable sets remain unchanged.

In the case of the Stieltjes integral, however, the generalization proceeds along different lines. Again, the Riemann integral appears as a special case of this new integral, but here the idea of the generalization is not to enlarge the class of integrable functions, but to consider new measures on R. For example, the Riemann measure (or length) of the interval [1,2] is given by

$$\int \chi_{[1,2]} = \int_1^2 1\,dx = 1.$$

If the positive x axis were to be deformed so that each point with coordinate x was mapped into the point $\alpha = x^2$, then the line segment given above would be "stretched" to length three. This may be expressed as follows:

$$\int_1^4 1\, d\alpha = \int_1^2 1\, \frac{d\alpha}{dx}\, dx;$$

$d\alpha = (d\alpha/dx)\, dx$ denotes the "stretched" differential arc length in terms of x and dx. In a similar manner, if the Riemann integral

$$\int_1^2 f(x)\, dx$$

is interpreted geometrically as an area, then as a result of stretching the right half-plane so that (x,y) is mapped into (x^2,y), a new figure is obtained whose area is given by

$$\int_1^2 f(x^2)\, dx^2.$$

By varying the stretching factor (in this case $\alpha = x^2$), a whole class of integrals is obtained. Roughly speaking, these are the Stieltjes integrals, and they are described in detail below.

1 Definition. Let $f(x)$ and $\alpha(x)$ be functions defined on the interval $[a,b]$, and let $\mathcal{S}$ be a subdivision of $[a,b]$:

$$a = x_0 < x_1 < \cdots < x_n = b.$$

If points c_i are chosen from the intervals $[x_{i-1},x_i]$, then the sum

$$[1] \qquad S(f,\alpha,\mathcal{S},c_i) = \sum_1^n f(c_i)[\alpha(x_i) - \alpha(x_{i-1})] = \sum_1^n f(c_i)\, \Delta\alpha_i$$

is called the Stieltjes approximating sum for f and α with respect to $\mathcal{S}$ and $\{c_i\}$.

Let S be a real number with the following property: To each $\epsilon > 0$, there is a $\delta > 0$ such that

$$|\mathcal{S}| = \max_{1 \le i \le n} \{(x_i - x_{i-1})\} < \delta \qquad \text{implies} \qquad |S - S(f,\alpha,\mathcal{S},c_i)| < \epsilon,$$

for all choices of c_i. The number S is called the *Stieltjes integral of f with respect to α* over the interval $[a,b]$, and is denoted by

$$[2] \qquad \int_a^b f\, d\alpha = S = \lim_{|\mathcal{S}| \to 0} S(f,\alpha,\mathcal{S},c_i).$$

If $\alpha(x) \equiv x$, [2] is simply the Riemann integral.

2 Proposition. Let f,g be (Stieltjes) integrable over $[a,b]$ with respect to both α and β, and let $s,t \in R$. Then

(i) $$\int_a^b [sf + tg]\,d\alpha = s\int_a^b f\,d\alpha + t\int_a^b g\,d\alpha;$$

(The integral is a linear, homogeneous functional defined on the class containing both f and g.)

(ii) $$\int_a^b fd(s\alpha + t\beta) = s\int_a^b f\,d\alpha + t\int_a^b f\,d\beta.$$

(The integral is linear and homogeneous with respect to α and β.)

The proofs are similar to those given for the corresponding properties of Riemann integrals, and are therefore left to the reader.

3 Proposition. If $\int_a^b f\,d\alpha$ exists, then for any $c \in [a,b]$, the integrals $\int_a^c f\,d\alpha$ and $\int_c^b f\,d\alpha$ exist, and their sum is equal to the integral from a to b.

Again, the proof is left to the reader.

4 Theorem. If $f \in C[a,b]$ and α is of bounded variation over $[a,b]$, then $\int_a^b f\,d\alpha$ exists.

Proof: In view of the linearity of the integral (Proposition 2), and the fact that every function of bounded variation may be written as the difference of monotone functions (Theorem 9.3.2), it may be assumed, without loss of generality, that α is a nondecreasing function on $[a,b]$.

Let $\mathcal{S}$ be a subdivision given by $a = x_0 < x_1 < \cdots < x_n = b$. The compactness of the intervals $[x_{i-1},x_i\}$ and the continuity of f imply the existence of points $\lambda_i,\mu_i \in [x_{i-1},x_i]$ at which

[3] $$f(\lambda_i) = \min_{x_{i-1}\le x\le x_i} f(x) \le \max_{x_{i-1}\le x\le x_i} f(x) = f(\mu_i).$$

Since f and α shall remain fixed throughout this proof, we shall write $\underline{S}(\mathcal{S})$, $\bar{S}(\mathcal{S})$, and $S(\mathcal{S},c_i)$ in place of $S(f,\alpha,\mathcal{S},\lambda_i)$, $S(f,\alpha,\mathcal{S},\mu_i)$, and $S(f,\alpha,\mathcal{S},c_i)$ respectively. From [3] it follows that

[4] $$\underline{S}(\mathcal{S}) \le S(\mathcal{S},c_i) \le \bar{S}(\mathcal{S}).$$

The desired conclusion, that $\lim_{|\mathcal{S}|\to 0} S(\mathcal{S},c_i)$ exists, is reached in two steps:

(i) We show first that $\lim_{|\mathcal{S}|\to 0} [\bar{S}(\mathcal{S}) - \underline{S}(\mathcal{S})] = 0$, and then

(ii) prove that $\lim_{|\mathcal{S}|\to 0} \bar{S}(\mathcal{S})$ exists.

It is readily seen that (i) and (ii), taken together with [4], yield the existence of the integral, $\int f\,d\alpha$.

The uniform continuity of f implies that a $\delta > 0$ may be chosen so that if $|\mathcal{S}| < \delta$ then

$$|\lambda_i - \mu_i| < |\mathcal{S}| < \delta \qquad \text{and} \qquad (f(\mu_i) - f(\lambda_i)) < \epsilon.$$

Furthermore, $\alpha(x)\uparrow$ on $[a,b]$, and therefore its total variation,

$$V(\alpha) = \sup \Sigma \Delta\alpha_i = \alpha(b) - \alpha(a) < \infty$$

(Definition 9.3.1). Hence it follows that

$$0 \leq \bar{S}(\mathcal{S}) - \underline{S}(\mathcal{S}) = \Sigma[f(\mu_i) - f(\lambda_i)]\,\Delta\alpha_i < \epsilon V(\alpha),$$

which proves (i).

Suppose now that $\epsilon > 0$ is given. Let $\xi > 0$ be chosen so that if $x,y \in [a,b]$, then

$$|x - y| < \xi \qquad \text{implies} \qquad |f(x) - f(y)| < \epsilon/2V(\alpha)$$

(again, uniform continuity). If $\mathcal{S}, \mathcal{S}'$, are subdivisions whose norms are less than ξ, then their common refinement $\mathcal{S}''$, is a subdivision, $a = y_0 < y_1 < \cdots < y_m = b$, obtained by taking the union of the points of $\mathcal{S}$ and $\mathcal{S}'$, discarding duplicates and ordering them, and $|\mathcal{S}''| = \max\limits_{1\leq i\leq m} (y_i - y_{i-1}) \leq \min(|\mathcal{S}|, |\mathcal{S}'|) < \xi$. The triangle inequality yields

$$[5] \quad |\bar{S}(\mathcal{S}) - \bar{S}(\mathcal{S}')| \leq |\bar{S}(\mathcal{S}) - \bar{S}(\mathcal{S}'')| + |\bar{S}(\mathcal{S}'') - \bar{S}(\mathcal{S}')| < \frac{\epsilon}{2V(\alpha)} V(\alpha) + \frac{\epsilon}{2V(\alpha)} V(\alpha) = \epsilon.$$

This is very much like the Cauchy Property for sequences. Indeed, by constructing a sequence of subdivisions $\mathcal{S}_n$ for which $\{\bar{S}(\mathcal{S}_n)\}$ is a Cauchy sequence, the convergence of the upper sums is obtained:

Let $\mathcal{S}_n$ denote the subdivision obtained by dividing $[a,b]$ into n equal parts; that is, $(x_i - x_{i-1}) = (b - a)/n$, $i = 1,2,\ldots,n$. The corresponding sequence of upper sums $\bar{S}_n = \bar{S}(\mathcal{S}_n)$, is easily seen to be a Cauchy sequence. For if $\epsilon > 0$ is given, and ξ is chosen as described above, then [5] implies that if $(b - a)/n$, $(b - a)/m$ are less than ξ, then it follows that $|\bar{S}_n - \bar{S}_m| < \epsilon$. Let S denote the limit of this sequence of real numbers.

We shall prove that $S = \int_a^b f\,d\alpha$: Again, if $\epsilon > 0$ is given, choose $\xi > 0$ so that [5] holds whenever both $|\mathcal{S}|$ and $|\mathcal{S}'|$ are less than ξ. Then choose an integer n that satisfies $(b - a)/n < \xi$. $\lim \bar{S}_n = S$ and [5] imply that if $|\mathcal{S}| < \xi$, then

$$|\bar{S}(\mathcal{S}) - S| \leq |\bar{S}(\mathcal{S}) - \bar{S}_n| + |\bar{S}_n - S| < 2\epsilon.$$

Thus $\int f\,d\alpha$ exists if $f \in C[a,b]$ and α is monotone on $[a,b]$. The corresponding result for the case that α is of bounded variation follows from the linearity of the integral. ∎

Unlike the case of the Riemann integral, the following example illustrates that the existence of the Stieltjes integrals $\int_a^c f\,d\alpha$ and $\int_c^b f\,d\alpha$ does not imply the existence of $\int_a^b f\,d\alpha$:

5 Example. $a = 0, c = 1$, and $b = 2$; $f(x) = 0$ on the interval $[0,1)$ and $f(x) = 1$ on $[1,2]$; $\alpha(x) = 0$ on $[0,1]$ and $\alpha(x) = 1$ on $(1,2]$. Theorem 4 implies that the integrals of f with respect to α exist over both $[0,1]$ and $[1,2]$. It is left to the reader to verify that the approximating sums over $[0,2]$ do not approach the sum of these integrals. Indeed, $\int_0^2 f\,d\alpha$ does not exist.

Example 5 illustrates that Theorem 4 can not be extended further to include sectionally continuous functions $f(x)$. However, careful study of this simple example reveals that it is not simply the discontinuity of $f(x)$ at $x = 1$ that is responsible for the nonexistence of the integral. For, $\alpha(x)$ is also discontinuous at this point. It can be shown that a necessary condition for the existence of $\int_a^b f\,d\alpha$ is that at every point of $[a,b]$, *either* $f(x)$ or $\alpha(x)$ be continuous (Exercise 18).

Remark. A generalized Stieltjes integral $\int_a^b f\,d\alpha$ may be defined for which f and α may be the functions of Example 5. (See either Exercise 19 or Burrill and Knudsen [1].)

Turning now to the main results of this section, we observe first that Proposition 2 implies that the expression

$$F_\alpha(f) = \int_a^b f\,d\alpha$$

is a linear functional defined on the normed vector space $C[a,b]$. Moreover, F_α is bounded:

$$\|F_\alpha\|^* = \sup_{\|f\| \neq 0} \frac{|F_\alpha(f)|}{\|f\|} \leq V(\alpha).$$

(Remember, $\|f\| = \max_{a \leq x \leq b} |f(x)|$.) We conclude by proving that every bounded linear functional on this space can be represented by a Stieltjes integral.

6 Theorem (Riesz). If $F \in C^*[a,b]$, (the space of bounded linear functionals on $C[a,b]$, with $\|F\|^* = \sup_{\|f\| \neq 0} \frac{|F(f)|}{\|f\|}$), then there is a function $\alpha(x)$ of bounded variation over $[a,b]$ such that for all $f \in C[a,b]$,

$$F(f) = \int_a^b f\, d\alpha,$$

and $\|F\|^* = V(\alpha)$.

Moreover, α is uniquely determined by F up to an additive constant ($d\alpha = d(\alpha + c)$, $c \in R$) except possibly at points of discontinuity.

Proof: It is required that $C[a,b]$ be temporarily enlarged to a space which contains step functions. Then, as in the proof of the Representation Theorem for L_p spaces, α is determined by the action of the linear functional F on the step functions (Compare step 7 of this proof with case 1 of Theorem 11.3.2).[1]

Step 1. Let Γ be the set of functions obtained by taking monotone (nondecreasing) limits of sequences in $C[a,b]$ with uniformly bounded norms.

That is, if f is in Γ, there is a sequence $\{f_n\} \subset C[a,b]$ and a number M such that $f_n(x) \uparrow f(x)$ on $[a,b]$ and $\|f_n\| \leq M$, $n = 1, 2, \ldots$.

It is easily seen that Γ is not a vector space. The only closure condition that is satisfied is

$$f,g \in \Gamma \quad \text{and} \quad a,b \geq 0 \qquad \text{imply} \qquad af + bg \in \Gamma.$$

However, the space $\hat{\Gamma}$ of differences of functions in Γ, is a real vector space. (Cf. Sec. 1 of Chapter 6. There, the space Γ is the set of monotone limits of step functions with uniformly bounded integrals, and $\hat{\Gamma} = L_1$.)

The boundedness of the functions in Γ and $\hat{\Gamma}$ permits us to extend the norm by defining $\|f\| = \sup_{a \leq x \leq b} |f(x)|$, for f in Γ or $\hat{\Gamma}$ (Example 2.2.3 and Exercise 13).

Step 2. If $f \in \Gamma$ and if $\{f_n\}$ is a uniformly bounded sequence in $C[a,b]$ which converges monotonically (up) to f, then $\lim_{n \to \infty} F(f_n)$ exists.

The linearity of F implies that

$$F(f_{n+1}) = F(f_1) + \sum_1^n [F(f_{i+1}) - F(f_i)].$$

Thus $\lim F(f_n) = \lim F(f_{n+1})$ exists if and only if the series of real numbers

$$\sum_1^\infty [F(f_{i+1}) - F(f_i)] \tag{6}$$

[1] The process of extending the space and the linear functional can be avoided by using the Hahn-Banach theorem. This eliminates most of the first six steps of the proof. The details of this method may be found in Taylor [2], pp. 195–200.

converges. We prove next that [6] converges absolutely. Setting

$$\epsilon_i = \operatorname{sgn}[F(f_{i+1}) - F(f_i)] = \begin{cases} 1 & \text{if } [F(f_{i+1}) - F(f_i)] \geq 0 \\ -1 & \text{if } [F(f_{i+1}) - F(f_i)] < 0, \end{cases}$$

we obtain

$$\sum_1^n |F(f_{i+1}) - F(f_i)| = \sum_1^n \epsilon_i F(f_{i+1} - f_i)$$

$$\leq \|F\|^* \|\sum_1^n \epsilon_i (f_{i+1} - f_i)\|.$$

Since $f_i \uparrow$ and $\|f_i\| \leq M$, it follows that

$$\|F\|^* \|\sum_1^n \epsilon_i (f_{i+1} - f_i)\| \leq 2M \|F\|^*,$$

which proves the absolute convergence of [6], and therefore of $F(f_n)$.

Step 3. If $\{f_n\}$ and $\{g_n\}$ are uniformly bounded sequences in $C[a,b]$ which converge monotonically to f, then

[7] $$\lim_{n\to\infty} F(f_n) = \lim_{n\to\infty} F(g_n).$$

Remark. Once [7] is verified, F may be extended to Γ by choosing *any* sequence $\{f_n\}$ in $C[a,b]$ which converges monotonically to f, and setting $F(f) = \lim F(f_n)$.

We begin by replacing the sequences $\{f_n\}$ and $\{g_n\}$ by *strictly* increasing sequences which also converge to f. Let

$$f'_n = f_n - 1/n, \quad g'_n = g_n - 1/n.$$

Then $f_{n+1} \geq f_n$ if and only if $f'_{n+1} = f_{n+1} - 1/(n+1) > f_n - 1/n = f'_n$, and

[8] $$0 < f - f'_n = (f - f_n) + 1/n \to 0$$

as n tends to infinity. Similarly for $\{g'_n\}$. We shall show next that there is an increasing sequence of integers $m_1 < m_2 < \cdots$, for which the following inequalities hold on $[a,b]$:

[9] $$f'_{m_1} < g'_{m_2} < f'_{m_3} < g'_{m_4} < \cdots < f'_{m_{2n-1}} < g'_{m_{2n}} < \cdots.$$

Set $m_1 = 1$. If there is no integer n for which $f'_1 < g'_n$ on $[a,b]$, then the sets $S_n = [f'_1 \geq g'_n]$ are closed, nested, nonempty subsets of $[a,b]$, and the intersection contains at least one point (Theorem 1.3.19). Suppose ζ is in the intersection. Then $f(\zeta) > f'_1(\zeta) \geq g'_n(\zeta)$ for all values of n, which contradicts $g'_n \uparrow f$ on $[a,b]$. Thus, an integer m_2 exists for which the first inequality of [9] is satisfied. The principle of mathematical induction yields the desired sequence [9]. For convenience, the terms of [9] are designated by $h_1 = f'_{m_1}$, $h_2 = g'_{m_2}, \ldots$. $\{h_n\}$ is an increasing sequence of continuous functions whose norms are bounded, and $h_n \uparrow f$. From step 2, $\lim F(h_n)$ exists. This implies that

$$\lim_{n\to\infty} F(h_n) = \lim_{n\to\infty} F(h_{2n-1}) = \lim_{n\to\infty} F(f'_{m_{2n-1}})$$
$$= \lim_{n\to\infty} \left[F(f_{m_{2n-1}}) - F\left(\frac{\chi_{[a,b]}}{2n-1}\right)\right] = \lim_{n\to\infty} F(f_n),$$

since $F\left(\dfrac{\chi_{[a,b]}}{2n-1}\right) \leq \dfrac{1}{2n-1}\|F\|^* \to 0$ as n tends to infinity. A similar argument yields $\lim\limits_{n\to\infty} F(h_n) = \lim\limits_{n\to\infty} F(g_n)$.

Thus $F(f)$ is well defined on Γ. Furthermore, if $a,b \geq 0$ and if $f,g \in \Gamma$, then $af + bg \in \Gamma$ and

[10] $$F(af + bg) = \lim_{n\to\infty} F(af_n + bg_n)$$
$$= a \lim_{n\to\infty} F(f_n) + b \lim_{n\to\infty} F(g_n)$$
$$= a\,F(f) + b\,F(g),$$

proving that F is linear on Γ and homogeneous with respect to nonnegative numbers.

Step 4. There is an extension $\hat{F}$ of F to $\hat{\Gamma}$, which is linear and homogeneous.

If $h \in \hat{\Gamma}$, then there are functions $f,g \in \Gamma$, and $h = f - g$. Set

[11] $$\hat{F}(h) = F(f) - F(g).$$

To prove that $\hat{F}$ is well defined, let f',g' denote a second pair of functions in Γ whose difference is also equal to h. Then $f + g' = f' + g$, and [10] implies

$$F(f) + F(g') = F(f + g') = F(f' + g) = F(f') + F(g).$$

Subtracting, we obtain $F(f) - F(g) = F(f') - F(g')$, and $\hat{F}(f - g) = \hat{F}(f' - g')$ follows from [11]. The verification of the linearity and homogeneity is left to the reader.

Step 5. $\hat{F}$ is bounded, and its norm coincides with the norm of F on $C[a,b]$; that is, $\|\hat{F}\|^*_{\hat{\Gamma}} = \|F\|^*_C$.

(Temporarily, the subscripts "$\hat{\Gamma}$" and "C" will be used to distinguish between the norms taken in $\hat{\Gamma}$ and those taken in $C[a,b]$. $\hat{F}$ denotes the extension of F to $\hat{\Gamma}$ ([7] and [11]).)

Since $C[a,b] \subset \hat{\Gamma}$,

$$\|\hat{F}\|^*_{\hat{\Gamma}} = \sup_{\bar{0}\neq h\in\Gamma} \frac{|\hat{F}(h)|}{\|h\|_{\hat{\Gamma}}} \geq \sup_{\bar{0}\neq h\in C[a,b]} \frac{|F(h)|}{\|h\|_C} = \|F\|^*_C.$$

To reverse this inequality, and thereby prove that the norm of the extended functional coincides with that of the given functional, let $h = f - g$ be a decomposition of a function $h \in \hat{\Gamma}$ into the difference of functions in Γ. Also let $\{f_n\}$ and $\{g_n\}$ be "defining" sequences for f and g respectively (step 1). Then [11] yields

$$[12] \quad |\hat{F}(h)| = |F(f) - F(g)| \leq |\hat{F}(f - f_n)| + |\hat{F}(f_n - g_n)| + |\hat{F}(g_n - g)| \leq |F(f) - F(f_n)| + \|F\|_C^* \|f_n - g_n\|_C + |F(g_n) - F(g)|.$$

From Step 3, the first and third terms on the right tend to zero as n tends to infinity, and since

$$\begin{aligned}\lim_{n\to\infty} \|f_n - g_n\|_C &= \lim_{n\to\infty} \max_{a\leq x\leq b} |f_n(x) - g_n(x)| \\ &= \sup_{a\leq x\leq b} |f(x) - g(x)| \\ &= \|f - g\|_{\hat{\Gamma}} = \|h\|_{\hat{\Gamma}},\end{aligned}$$

it follows that $|\hat{F}(h)| \leq \gamma_n + \|F\|_C \|h\|_{\hat{\Gamma}}$, where $\{\gamma_n\}$ is a sequence of real numbers that converges to zero, thus proving that $\|\hat{F}\|_\Gamma^* = \|F\|_C^*$.

Having shown that $\hat{F}$ is indeed an extension of F in the sense that it coincides with F on $C[a,b]$ and that $\|\hat{F}\|_{\hat{\Gamma}}^* - \|F\|_C^*$, we may hereafter omit the subscripts "C" and "Γ," and the circumflex, and simply write F for $\hat{F}$.

Step 6. *$\hat{\Gamma}$ contains all the step functions which vanish outside* $[a,b]$. Each step function is the monotone limit of piecewise linear functions (Exercise 15).

Step 7. *The function $\alpha(t) = F(\chi_{[a,t]})$, $a \leq t \leq b$, is absolutely continuous, and*

$$[13] \qquad F(\chi_{[a,t]}) = \int_a^b \chi_{[a,t]}\, d\alpha.$$

The proof of the absolute continuity is identical with the one given in Theorem 11.3.2, and will not be repeated. The absolute continuity of α implies that it is of bounded variation and therefore the integral [13] exists (Exercise 3). Writing this integral as the limit of approximating sums, [1], and considering only subdivisions $\mathcal{S}$ having t as one of their points, say $t = x_{m(\mathcal{S})}$, we obtain

$$\begin{aligned}\int_a^b \chi_{[a,b]}\, d\alpha &= \lim_{|\mathcal{S}|\to 0} \sum_1^{m(\mathcal{S})} 1[\alpha(x_i) - \alpha(x_{i-1})] \\ &= \lim_{|\mathcal{S}|\to 0} \sum_1^{m(\mathcal{S})} F(\chi_{[a,x_i]-[a,x_{i-1}]})\end{aligned}$$

$$= \lim_{|\mathcal{S}| \to 0} F \sum_{1}^{m(\mathcal{S})} \chi_{(x_{i-1}, x_i]} = F(\chi_{[a,t]}).$$

Step 8. If σ is a step function which vanishes outside $[a,b]$, then $\sigma \in \hat{\Gamma}$ and

$$F(\sigma) = \int_a^b \sigma d\alpha.$$

The proof is left to the reader.

Step 9. If $f \in C[a,b]$,

$$F(f) = \int_a^b f\, d\alpha.$$

Let $\{\tau_n\}$ be a sequence of step functions which converges uniformly to f. The uniform continuity of the functional F on $\hat{\Gamma}$ implies that

$$F(f) = \lim_{n \to \infty} F(\tau_n) = \lim_{n \to \infty} \int_a^b \tau_n\, d\alpha.$$

However,

$$\left| \int_a^b (f - \tau_n)\, d\alpha \right| \leq \| f - \tau_n \|\, V(\alpha) \to 0,$$

($V(\alpha)$ is the variation of α), from which it follows that

[14] $$F(f) = \lim_{n \to \infty} \int_a^b \tau_n\, d\alpha = \int_a^b f\, d\alpha.$$

This completes the proof of the assertion that if *F is a bounded, linear, functional defined on $C[a,b]$, then there is a function of bounded variation α, and $F(f)$ is the Stieltjes integral of f with respect to α.*

Step 10. $\| F \|^* = V(\alpha)$.

It is easily seen that $| F(f) | = \left| \int_a^b f\, d\alpha \right| \leq \|f\|\, V(\alpha)$, or equivalently, $\| F \|^* \leq V(\alpha)$. Recalling that the norm of the extended functional is also equal to $\| F \|^*$ (Step 5), it suffices to exhibit a function $f \in \hat{\Gamma}$ for which $| F(f) | \geq \|f\|\, V(\alpha)$.

Let $a = x_0 < x_1 < \cdots < x_n = b$ be the points of a subdivision $\mathcal{S}$, and let $\epsilon_i = \text{sgn}\,[\alpha(x_i) - \alpha(x_{i-1})]$. The function

$$f(x) = \sum_{1}^{n} \epsilon_i [\alpha(x_i) - \alpha(x_{i-1})] = \sum_{1}^{n} \epsilon_i \chi_{[x_{i-1}, x_i]}(x)$$

is a step function and therefore in $\hat{\Gamma}$. Furthermore, $\epsilon_i = \pm 1$ implies that $f(x) = \pm 1$, and so $\|f\| = 1$. Hence

[15] $$\| F \|^* \geq | F(f) | = \left| \int_a^b f\, d\alpha \right| = \sum_1^n | \alpha(x_i) - \alpha(x_{i-1}) |$$

holds for all subdivisions. Taking the least upper bound of the right side yields $\| F \|^* \geq V(\alpha)$.

Step 11. (*uniqueness*) *If α and β are of bounded variation over $[a,b]$ and satisfy*

[16] $$\int_a^b f d\alpha = \int_a^b f\, d\beta, \qquad \text{for all } f \in C[a,b],$$

then there is a real number c, and $\alpha(x) - \beta(x) = c$ whenever both $\alpha(x)$ and $\beta(x)$ are continuous.

The function $\alpha - \beta$ is of bounded variation, and therefore its set of discontinuities is countable (Theorem 9.3.3). Let $y < z$ be points of con-

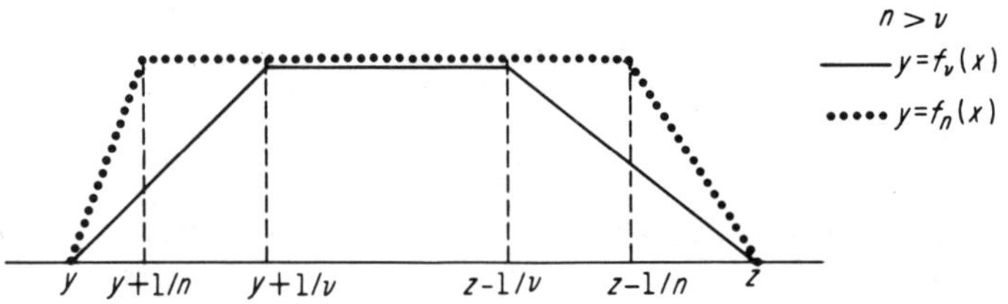

FIGURE 32

tinuity of both functions. Choose an integer ν so that $y + 1/\nu < z - 1/\nu$. Then for $n \geq \nu$, the sequence of polygonal functions (see Figure 32)

$$f_n(x) = \begin{cases} 0 & \text{if } a \leq x \leq y \\ nx - ny & \text{if } y < x \leq y + 1/n \\ 1 & \text{if } y + 1/n < x \leq z - 1/n \\ -nx + nz & \text{if } z - 1/n \leq x < z \\ 0 & \text{if } z < x \leq b \end{cases}$$

converges monotonically (up) to $\chi_{[y,z]}$. Step 3 implies that

[17] $$\lim_{n \to \infty} \int_a^b f_n[d\alpha - d\beta] = \int_a^b \chi_{[y,z]}(d\alpha - d\beta).$$

However, each $f_n \in C[a,b]$, and it follows from [16] that the integrals [17] vanish. Since $y < z$ were chosen to be *any* pair of points of continuity, we have

[18] $$0 = \alpha(z) - \beta(z) - \alpha(y) + \beta(y) \Leftrightarrow a(z) - \beta(z)$$
$$= \alpha(y) - \beta(y) = c;$$

that is, there is a real number c and $\alpha(x) - \beta(x) = c$ whenever x is a point of discontinuity.

We conclude by showing that every point of $[a,b]$ must be either a point of continuity for *both* α and β, or for neither. Suppose that α is discontinuous at a point $x = \eta$, $a < \eta < b$. Since the set of discontinuities is denumerable, every "half" neighborhood $(\eta, \eta + \epsilon)$ contains points of continuity of both functions. Let $\{x_n\}$ be a sequence of points, $\eta < x_n < b$, which converges to η, and at which both α and β are continuous. From [18] we obtain

$$\alpha(\eta + 0) - \beta(\eta + 0) = \lim\,[\alpha(x_n) - \beta(x_n)] = c.$$

A similar calculation yields $\alpha(\eta - 0) - \beta(\eta - 0) = c$. Hence

$$\alpha(\eta + 0) - \alpha(\eta - 0) = \beta(\eta + 0) - \beta(\eta - 0),$$

and the discontinuity of α at η implies that the left-hand side does not vanish, which in turn implies that the right-hand side is not equal to zero, or that β is also discontinuous at η. ∎

Exercises

***1.** Prove Proposition 2.

***2.** Prove Proposition 3.

***3.** Prove that $\int_a^b \chi_{[a,t]}\,d\alpha$ exists if $a \leq t \leq b$ and if α is of bounded variation over $[a,b]$.

4. Compute the total, positive and negative variations of each of the following functions:

(a) $\alpha(x) = [x] - x$, $0 \leq x \leq 3$. (The function $[x]$ denotes the greatest integer which does not exceed x.)

(b) $\alpha(x) = \begin{cases} x^2 - 3 & \text{if } 0 \leq x < 1 \\ 2x + 1 & \text{if } 1 \leq x \leq 3. \end{cases}$

(c) $\alpha(x) = \begin{cases} 2 & \text{if } x = -1 \\ 3 & \text{if } -1 < x \leq 0 \\ -4 & \text{if } 0 < x \leq 1. \end{cases}$

(d) $\alpha(x)$ is the Cantor function (Example 9.4.2).

5. Evaluate the following Stieltjes integrals, $\int_a^b f\,d\alpha$:

(a) $a = 0$, $b = 3$, $a(x) = [x] - x$, and $f(x) = x$;

(b) Let a, b, and $f(x)$ be as in part (a); $\alpha(x)$ is the step function of Exercise 4(c);

(c) a, b, and $f(x)$ are the same and $\alpha(x)$ is the Cantor function.

6. (*First Mean Value Theorem.*) Let $f \in C[a,b]$, $\alpha(x)\uparrow$ on $[a,b]$. Show that there is a point ξ, $a \leq \xi \leq b$, satisfying

$$\int_a^b f\,d\alpha = f(\xi)[\alpha(b) - \alpha(a)].$$

7. A second version of the *First Mean Value Theorem* is as follows: Let f and α be as in Exercise 6, and let g be a nonnegative function which is integrable with respect to α. Show that there is a point ξ in the interval $[a,b]$ satisfying

$$\int_a^b fg\,d\alpha = f(\xi)\int_a^b g\,d\alpha.$$

8. (*Second Mean Value Theorem.*) If $f(x)$ and $\alpha(x)$ are monotonic and continuous on $[a,b]$, then there is a point ξ in $[a,b]$ for which

$$\int_a^b f\,d\alpha = f(a)[\alpha(\xi) - \alpha(a)] + f(b)[\alpha(b) - \alpha(\xi)].$$

9. State and prove a second version of the second mean value theorem which corresponds to Exercise 7.
10. (*Change of Variables.*) If $f \in C[a,b]$ and α is continuous and strictly increasing on $[a,b]$, then denoting the inverse of α by β, we have

$$\int_a^b f(x)\,dx = \int_{\alpha(a)}^{\alpha(b)} f(\beta(x))\,d\beta(x).$$

The integral on the left may be thought of as either a Riemann or a Stieltjes integral; the one on the right is the Stieltjes integral of the function $g(x) = f(\beta(x))$ with respect to $\beta(x)$ over the interval $[\alpha(a),\alpha(b)]$.
11. What is the relationship between $V(\alpha)$ and arc length?
12. Let $f(x)$ be continuous and $\alpha(x)$ be of bounded variation over $[a,b]$. Show that the function $\psi(x) = \int_a^x f\,d\alpha$ is of bounded variation over $[a,b]$, and that if $g(x) \in C[a,b]$, then

$$\int_a^b g\,d\psi = \int_a^b gf\,d\alpha.$$

*13. Verify that $\hat{\Gamma}$ (step 1 of Theorem 6) is a normed vector space, with $\|f\|_\Gamma = \sup_{a \le x \le b} |f(x)|$. Is $\hat{\Gamma}$ complete? (See the corresponding discussion of Σ, Γ, and L_1.)
*14. Show that the sets $S_n = [f_1' \ge g_n']$ are closed, nested, nonempty subsets of $[a,b]$ (step 3 of Theorem 6).
*15. Prove that every step function is the monotone limit of a sequence of piecewise linear functions (step 6 of Theorem 6).
16. Prove that if $f \in C[a,b]$ and α is of bounded variation, then the Lebesgue integral

$$\int f\alpha'\chi_{[a,b]}$$

exists and is equal to the Stieltjes integral of f with respect to α. (*Hint:* Use the results of Sec. 4 of Chapter 9.)

*17. Compute the integrals of Example 5, and show that $\int_0^2 f\,d\alpha$ does not exist by proving that the approximating sums do not converge to

$$\int_0^1 f\,d\alpha + \int_1^2 f\,d\alpha.$$

18. Prove that a necessary condition for the existence of $\int_a^b f\,d\alpha$ is that at each point of $[a,b]$ either $f(x)$ or $\alpha(x)$ be continuous.

19. The number $S = \mathfrak{G}\!\!\int_a^b f\,d\alpha$ is said to be the *generalized Stieltjes* integral of f with respect to α over $[a,b]$, if to each $\epsilon > 0$ there is a subdivision $\mathcal{S}_\epsilon$ having the property that if $\mathcal{S}$ is any refinement of $\mathcal{S}_\epsilon$, and $c_1, \ldots, c_n$ are points of the intervals of $\mathcal{S}$, then $|\,S(f,\alpha,\mathcal{S},c_i) - S\,| < \epsilon$.

(a) Show that if the Stieltjes integral of f with respect to α exists, then so does the generalized Stieltjes integral and the two are the same.

(b) Show that if the generalized integrals of f with respect to α over both $[a,c]$ and $[c,b]$ exist, then the integral over $[a,b]$ exists and is equal to their sum. (Cf. Example 5.)

(c) Prove that the generalized Stieltjes integral exists for the functions of Example 5.

(d) Prove that if $\mathfrak{G}\!\!\int_a^b f\alpha$ exists, then at each point $x \in [a,b]$, either $f(x)$ or $\alpha(x)$ must be equal to the value of their respective right or left limits.

*Exercises which request the proof of a statement made in the text, as well as those referred to in the text, are starred.

PART IV

The Daniell Integral and Measure Spaces

CHAPTER **12**

The Daniell Integral and Measure Spaces (I)

It was remarked earlier, that once the four "basic" properties (linearity, homogeneity, monotonicity and continuity) of the Lebesgue integral were established, along with the vector space and lattice properties of L_1, it was no longer necessary to return to the original definitions and constructions of the Lebesgue integral and the space of integrable functions, which required Cauchy sequences of step functions. Indeed, these basic properties sufficed to derive all of the major convergence theorems of Chapter 6, as well as to establish the properties of measurable functions and sets. In this chapter, these "basic" properties of functions in L_1 and their (Lebesgue) integrals will be taken as the *defining properties*, or *axioms*, of a more general integral. The development of the abstract theory proceeds along the lines of the special case of the Lebesgue integral, and the principal theorems are reproved in a more general setting. Included in this discussion is the multiple (Lebesgue) integral on R^n, and integrals over abstract measure spaces.

Again, there are essentially two approaches. The first (Secs. 1–4), is modeled after the construction of L_1 by completing the normed space of step functions. Here, instead of using Cauchy sequences (which may be done), monotone limits are taken in a class of *elementary functions.* (Cf. Sec. 1 of Chapter 6.) The second approach (Sec. 5) takes certain properties of Theorem 7.2.2 as the defining properties of the measurable subsets of a given nonempty set Ω. As in Sec. 2 of Chapter 8, the discussion of measurable sets precedes that of measurable and integrable functions.

The chapter concludes with a brief discussion of the relationship between the two approaches.

1 Elementary Functions and Their Integrals

We begin by defining the abstract integral on a class of functions that corresponds to the real step functions (Cf. Sec. 1 of Chapter 5).

1 Definition. A collection C_0 of bounded real-valued functions defined on a nonempty set Ω is said to be a class of *elementary functions* if the following conditions are satisfied:

(i) $\sigma, \tau \in C_0$ and $a, b \in R$ imply $a\sigma + b\tau \in C_0$.

(ii) $\sigma, \tau \in C_0$ implies $\max(\sigma, \tau) \in C_0$.

The reader should have no difficulty in proving that (i) and (ii) are equivalent to asserting that C_0 is a real vector space and a lattice (Exercise 1).

If $\Omega = R$ (or $[a,b]$), then the classes of continuous functions, step functions, polygonal functions, and Riemann integrable functions are classes of elementary functions.

2 Definition. A mapping of C_0 into R, denoted by

[1] $$\int \sigma\, d\mu,$$

is called a *Daniell integral*, a *μ-integral*, or simply an *integral* if it has the following properties, ($\sigma, \tau, \sigma_n \in C_0$ and $a, b \in R$):

(i) $\int (a\sigma + b\tau)\, d\mu = a\int \sigma\, d\mu + b\int \tau\, d\mu$ (The integral is a *linear, homogeneous functional* on C_0);

(ii) $\sigma \geq 0$ implies $\int \sigma\, d\mu \geq 0$ (*positivity*);

(iii) $\sigma_n \downarrow 0$ on Ω implies $\int \sigma_n\, d\mu \downarrow 0$ (*continuity*).

Remark. The continuity of the integral is not in general equivalent to the boundedness or (uniform) continuity of a linear functional (Cf. Definitions 11.1.1 and 11.1.3 and Proposition 11.1.5). For, if C_0 is made into a normed space by setting

$$\| \sigma \| = \sup_{x \in \Omega} |\sigma(x)|,$$

then the integral will be a bounded linear functional provided that χ_Ω, the characteristic function of the space Ω, is in C_0. That is, if $\sigma \in C_0$, then

$$\left| \int \sigma\, d\mu \right| \leq \| \sigma \| \; \left| \int \chi_\Omega\, d\mu \right|.$$

This will follow from parts (c) and (e) of Proposition 3 which is proved below. It is not difficult to show that if χ_Ω is *not* in C_0, then the integral may not be a bounded functional. The reader is asked (Exercise 2) to give an example of such a case. To avoid confusion, the word "continuity" will be reserved for property (ii) of Definition 2.

Before giving examples of a Daniell integral, we list some properties, several of which have already been noted above.

3 Proposition. Let $\Omega \neq \phi$; C_0, σ, τ, σ_n and $\int \sigma\, d\mu$ are as defined before. Then

(a) C_0 is a normed, linear space with $\| \sigma \| = \sup_{x \in \Omega} | \sigma(x) |$;

(b) C_0 is a lattice;

(c) $\sigma \geq \tau$ implies $\int \sigma\, d\mu \geq \int \tau\, d\mu$ (monotonicity);

(d) $\int \bar{0}\, d\mu = 0$ ($\bar{0}$ is the additive identity of C_0);

(e) $| \int \sigma\, d\mu | \leq \int | \sigma |\, d\mu$ for all $\sigma \in C_0$;

(f) $\sigma_n \uparrow 0$ on Ω implies $\int \sigma_n\, d\mu \uparrow 0$; and $\sigma_n \uparrow \tau$ (or $\sigma_n \downarrow \tau$), where $\tau \in C_0$, implies $\int \sigma_n\, d\mu \uparrow \int \tau\, d\mu$ (or $\downarrow \int \tau\, d\mu$).

Proof: The vector space and lattice properties have already been discussed, and the verification of the norm axioms does not differ from Example 2.2.3.

Definition 2 yields

$$\sigma \geq \tau \Leftrightarrow \sigma - \tau \geq 0 \Rightarrow \int (\sigma - \tau)\, d\mu \geq 0 \Leftrightarrow \int \sigma\, d\mu \geq \int \tau\, d\mu.$$

(d) follows from the homogeneity of the integral:

$$\int \bar{0}\, d\mu = \int 0\bar{0}\, d\mu = 0 \int \bar{0}\, d\mu = 0.$$

The identity $\pm\sigma = \pm(\sigma^+ - \sigma^-) \leq \sigma^+ + \sigma^- = | \sigma |$, implies $\pm \int \sigma\, d\mu \leq \int | \sigma |\, d\mu$, which is (e). •

(f) follows easily from the continuity property, and is left to the reader as an exercise. ▌

4 Examples. (a) Let $\Omega = R$, $C_0 = \Sigma$, the vector lattice of real step functions. The (Lebesgue) integral of a step function (Definition 5.1.1) satisfies the conditions of Definition 2 (Proposition 5.1.2 and Theorem 6.1.1).

(b) $\Omega = R$ and $C_0 = C[a,b]$, the class of real valued, continuous functions defined on the interval $[a,b]$. It is left to the reader to verify that C_0 is a class of elementary functions, and that the Riemann integral is an example of a Daniell integral. [Dini's Theorem yields the continuity property, although in this case, the Riemann integrability of the function $f(x) \equiv 0$ permits us to use the theorem on monotone convergence. See Sec. 3 of Chapter 6.]

It is not conceptually difficult to extend the preceding examples to $\Omega = R^n$. The major problems are notational. However, since n-dimensional step functions appear in Sec. 5 of Chapter 13 (Fubini's theorem), the necessary definitions are included below. During the discussion that follows, the reader is advised to provide the appropriate diagrams for the special case $n = 2$.

5 Definition. A real-valued function $\sigma(x)$, $x = (x_1, \ldots, x_n) \in R^n$, is called a *step function* if the following conditions are satisfied:

•See special note on page 447.

(i) There is a number $L > 0$ such that $\sigma(x) = 0$ whenever some coordinate $|x_i| > L$, $i = 1,2,\ldots,n$; that is, $\sigma(x)$ vanishes outside a closed cube.

(ii) There is a doubly indexed set of numbers a_j^i, $i = 1,\ldots,n$,

$$-L = a_0^i < a_1^i < \cdots < a_{m(i)}^i = L,$$

which determines a *grid*, or *subdivision* of the cube into rectangular parallelopipeds on whose interiors $\sigma(x)$ is constant. That is, there are numbers $c_{k_1 k_2 \ldots k_n}$, $0 < k_i \leq m(i)$, and

$$\sigma(x) = c_{k_1 k_2 \ldots k_n} \qquad \text{if} \qquad a_{k_i - 1}^i < x_i < a_{k_i}^i.$$

(iii) $\sigma(x)$ may take on any values on the hyperplanes $x_i = a_j^i$, provided the set of values of $|\sigma(x)|$ is bounded on the cube.

To simplify the computations, it may be assumed that $m(i) = m(j)$ for all pairs $i,j = 1,\ldots,n$. This is done by setting $m = \max_{i \leq n} [m(i)]$, and inserting additional hyperplanes $x_i = a_j^i$, $j = m(i), m(i) + 1,\ldots,m$, whenever necessary. Although this increases the total number of parallelopipeds, the values of $\sigma(x)$ remain unchanged. For example, consider the step function defined on R^2:

$$\sigma(x) = \begin{cases} 1 & \text{if } 0 < x_1 < 1 \quad \text{and} \quad 0 < x_2 < 1 \\ 2 & \text{if } 1 < x_1 < 2 \quad \text{and} \quad 0 < x_2 < 1 \\ 6 & \text{if } x_1 = 0,\ x_1 = 1,\ x_1 = 2,\ x_2 = 0 \text{ or } x_2 = 1. \end{cases}$$

(See Figure 33.)

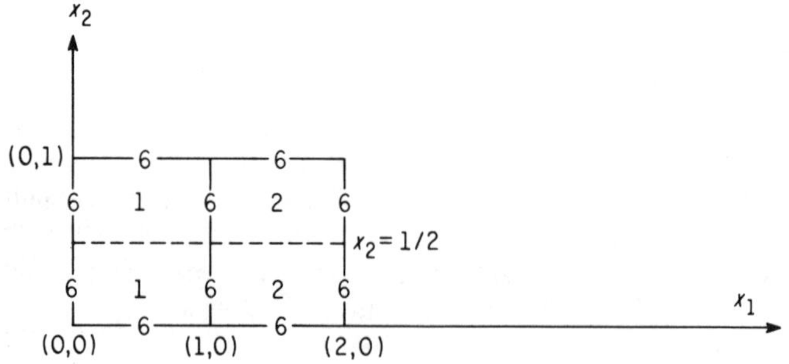

FIGURE 33

Here $m(1) = 2$ and $m(2) = 1$. By adding the (dotted) line $x_2 = 1/2$, a grid composed of four rectangles is obtained. The function remains unchanged.

The reader should have no difficulty in verifying that the n-dimensional step functions satisfy (i) and (ii) of Definition 1, and are therefore a class of elementary functions.

The definition of the (n-dimensional) Lebesgue integral of an n-dimensional step function should come as no great surprise. For $n = 1$, the integral gives the sum of the (signed) areas under the steps (See Sec. 1 of Chapter 5). If $n = 2$, the integral is the sum of the (signed) volumes under the constant steps. In the general case, the integral of an n-dimensional step function is the sum of the $(n + 1)$-dimensional (signed) volumes of the parallelopipeds whose "heights" are given by the constants $c_{k_1 \dots k_n}$, and whose n-dimensional bases are the parallelopipeds $a^i_{k_i - 1} \leq x_i \leq a^i_{k_i}$, $1 \leq k_i \leq m$ and $i = 1, \dots, n$. The formal definition follows:

6 Definition. Let $\sigma(x)$ be a step function defined on R^n. The *integral* (Lebesgue or Riemann) of $\sigma(x)$, written $\int \sigma$, is the sum

[1] $$\sum_{i=1}^{n} \sum_{k_i = 1}^{m} c_{k_1 \dots k_i \dots k_n} \Delta A_{k_1 \dots k_i \dots k_n},$$

where

$$\Delta A_{k_1 \dots k_i \dots k_n} = (a^l_{k_1} - a^1_{k_1 - 1})(a^2_{k_2} - a^2_{k_2 - 1}) \dots (a^n_{k_n} - a^n_{k_n - 1}).$$

The linearity, homogeneity, and positivity of [1] are readily verified. Again, the continuity is proved by using Dini's theorem.

7 Example. Let $\Omega = R^n$ and let C_0 be the class of continuous functions each of which vanishes outside a compact subset of R^n. It is left to the reader to verify that C_0 is a class of elementary functions and that the (multiple) Riemann integral is a Daniell integral.

8 Example. Let $\Omega = \{a_1, a_2, \dots, a_k\}$ be a nonempty, finite set of points, and let C_0 be the family of real valued functions defined on Ω. The expression

$$\int_{[a_1, \dots, a_k]} f = \sum_{1}^{k} f(a_i)$$

is a Daniell integral on the class C_0 of elementary functions.

9 Example. Let Ω be any nonempty set and let C_0 be the family of real valued functions on Ω each of vanishes except at finitely points. Again, it is easily verified that C_0 is a class of elementary functions and that

[2] $$\int f = \sum_{x \in \Omega} f(x)$$

defines a Daniell integral on C_0. We shall verify only the continuity property: Suppose that $\{f_n\}$ is a sequence of functions in C_0 which converges monotonically (down) to 0 on Ω. If $x_1, x_2, \ldots, x_m$ are the points at which $f_1(x)$ does *not* vanish, then the monotonicity of the sequence implies that for *all* values of n, $f_n(x) = 0$ except possibly at $x = x_i$, $i = 1, 2, \ldots, m$. Moreover, for each i, $f_n(x_i) \downarrow 0$ as n tends to infinity. Therefore, if $\epsilon > 0$ is given, an integer N may be chosen so that $n > N$ implies $0 \le f_n(x_i) \le f_N(x_i) < \epsilon/m$, $i = 1, 2, \ldots, m$, from which it follows that

$$\int f_n \le \int f_N = \sum_{x \in \Omega} f_N(x) = \sum_{i=1}^{m} f_N(x_i) < \epsilon.$$

Exercises

*1. Show that (i) and (ii) of Definition 1 are equivalent to the assertion that C_0 is a real vector space and a lattice.

2. Let $\Omega = R$ and let C_0 be the class of Riemann integrable functions on R. Clearly, χ_R is not in C_0. Using Dini's Theorem, it can be shown that the Riemann integral has the continuity property and is therefore a Daniell integral. However, the integral, which is a linear, homogeneous functional is not bounded. Prove this.

*3. Prove (f) of Proposition 3.

*4. Show that the n-dimensional step functions are a family of elementary functions.

*5. Prove the assertions made in Example 8.

*6. In each of the following cases, determine whether C_0 is a class of elementary functions. Is the "integral" defined a Daniell integral?

(a) $\Omega = R$ and C_0 is the collection of polygonal functions which vanish outside a bounded interval. The integral is defined as the Riemann integral. Now let C_0 be the collection of functions which vanish outside a bounded interval and on that interval are equal to polynomials. The integral is the same. •

(b) $\Omega = R$ and C_0 is the class of functions which vanish outside a fixed interval $[a, b]$ and are continuous on that interval. Let $\alpha(x)$ be a function of bounded variation over $[a, b]$. The integral is the Stieltjes integral $\int f\,d\alpha$. Is the problem affected if it is assumed that $\alpha(x)$ is nondecreasing?

(c) Let Ω and C_0 be as in Example 9, and let $\int f = \sum_{x \in \Omega} [f(x)]^2$.

(d) Let Ω and C_0 be as defined in Example 9, and let $H(x)$ be any real valued function defined on Ω. $\int^{(H)} f = \sum_{x \in \Omega} f(x) H(x)$.

(e) Let $\Omega = R$ and let C_0 be the family of real valued functions which vanish except on a countable set and that have the additional property that $\sum_{x \in \Omega} |f(x)|$ converges. We define $\int f = \sum_{x \in \Omega} f(x)$. What happens if it assumed only that $\sum_{x \in \Omega} f(x)$ converges?

*Exercises which request the proof of a statement made in the text, as well as those referred to in the text, are starred.

•See special note on page 447.

2 μ-Null Sets

Let Ω, C_0 and $\int \sigma\, d\mu$ have the meanings assigned to them in Sec. 1. Since C_0 is a lattice, $|\sigma| \in C_0$ whenever σ does, and therefore the expression

$$\|\sigma\|_\mu = \int |\sigma|\, d\mu$$

is well defined. Definition 12.1.2 and Proposition 12.1.3 yield

for all $\sigma, \tau \in C_0$ and $c \in R$,

$$\|\sigma\|_\mu \geq 0, \qquad \|c\sigma\|_\mu = |c|\ \|\sigma\|_\mu,$$

and

$$\|\sigma + \tau\|_\mu \leq \|\sigma\|_\mu + \|\tau\|_\mu.$$

However, it is not in general true that

$$\|\sigma\|_\mu = 0 \qquad \text{implies} \qquad \sigma(x) = 0 \text{ for all } x \in \Omega,$$

so that $\|\ \ \|_\mu$ may not be taken as a norm on C_0. To overcome this difficulty for the space $C_0 = \Sigma$ of real step functions, Σ was partitioned into equivalence classes, each class containing step functions which differ only on a finite set (Sec. 1 and 2 of Chapter 5). Thus equivalent step functions determine identical areas and hence have the same integral. This is expressed symbolically as follows:

$$\sigma \sim \tau \Leftrightarrow \int |\sigma - \tau| = \|\sigma - \tau\| = 0$$
$$\Leftrightarrow \sigma(x) = \tau(x) \text{ for } x \in R - \{x_1, x_2, \ldots, x_n\}.$$

In this way the collection of equivalence classes of step functions is made into a normed vector space. The elements of the completion L_1 of Σ are equivalence classes of "integrable" functions, two functions being equivalent if and only if they differ at most on a null set. These ideas may be carried over to the Daniell integral, with the elementary functions replacing the step functions. To partition C_0 into equivalence classes of "a.e. equal functions" it is necessary first to define the *null sets of* Ω. It will be recalled that a null subset of R was defined initially as a set that could be covered by a countable union of open intervals whose length was arbitrarily small. Since it is not assumed that Ω has any topological or metrical properties, it is not in general possible to define null sets by using arbitrarily "small" covers. The reader will recall however that Theorem 5.3.8 gave a second characterization of null sets, describing them in terms of nondecreasing sequences of step functions and their integrals. This theorem about real null sets is taken as the definition of μ-null sets.

Throughout the remainder of this section, Ω, C_0, $\int \sigma\, d\mu$ etc. will have the meanings assigned to them in Sec. 1.

1 Definition. A subset $E \subset \Omega$ is said to be μ-null if there is a nondecreasing sequence $\{\sigma_n\}$ in C_0, and a real number $M > 0$, such that

(i) if $x \in E$, then $\sigma_n(x) \uparrow \infty$ as n tends to infinity, and
(ii) $\int \sigma_n d\mu < M$, $n = 1, 2, \ldots$.

It is left to the reader to show that μ-null sets may also be characterized as follows:

1′ Definition. A subset $E \subset \Omega$ is said to be *μ-null* if to each $A > 0$, there is a nonnegative, nondecreasing sequence $\{\sigma_n\}$ in C_0 satisfying

(i′) if $x \in E$, then $\sigma_n(x) \uparrow \infty$ as n tends to infinity, and
(ii′) $\int \sigma_n d\mu \uparrow A$.

Before deriving some of the familiar properties of μ-null sets, we shall determine the μ-null sets for the examples discussed in Sec. 1.

2 Examples. (a) The μ-null sets of Example 12.1.4(a) are the Lebesgue null sets.

(b) The elementary functions of Example 12.1.4(b) are the real continuous functions defined on the interval $[a,b]$. It is left to the reader to show, using Definition 1 and part (a) of this example, that the μ-null sets defined by sequences of continuous functions coincide with those defined by sequences of step functions.

(c) If $\Omega = R^n$ and C_0 is either the space of n-dimensional step functions or the space of continuous functions which vanish outside a compact set, then the μ-null sets are the same.

Remark. In each of the preceding cases the μ-null sets may also be described as those which have arbitrarily "small" open coverings; that is, a subset E of R^n is μ-null (using either step functions or continuous functions) if to each $\epsilon > 0$ there is a countable set $\{S_k\}$ of open parallelopipeds in R^n whose union contains E, and whose n-dimensional volume does not exceed ϵ. The verification of the equivalence of these definitions is left to the reader.

3 Examples. (a) The finite sum $\sum_{i=1}^{k} f(a_i)$ defines a Daniell integral on the class of real valued functions whose domain of definition is the finite set $\{a_1, \ldots, a_k\}$ (See Example 12.1.8). It is easily seen that only the empty set is null. For if $E = \{a_{i_1}, \ldots, a_{i_m}\}$ is any nonempty subset of the given set, and if $\{\sigma_n\}$ is a nondecreasing sequence of elementary functions which diverges on E, then the sums $\sum_{j=1}^{m} \sigma_n(a_{i_j})$ diverge as n tends to infinity, since each term $\sigma_n(a_{i_j}) \uparrow \infty$ as n tends to infinity.

(b) A similar discussion is applicable to Example 12.1.9.

(c) In Exercise 12.1.6(d), Ω is taken to be any nonempty set; C_0 is the class of real valued functions defined on this set, each of which vanishes except on a finite subset of Ω. If $H(x)$ is a fixed but arbitrary nonnegative function defined on Ω, then

$$\int^{(H)} f = \sum_{x \in \Omega} f(x)H(x)$$

is a Daniell integral. The subsets of

$$S_H = \{x \in \Omega : H(x) = 0\}$$

are the null sets. For, if $E \subset S_H$, the sequence

$$\sigma_n(x) = \begin{cases} n \text{ if } x \in E \\ 0 \text{ if } x \in \Omega - E \end{cases}$$

possesses the desired properties. Conversely, if E contains a single point y which is not in S_H, then $\sigma_n(y) \uparrow \infty$ implies

$$\int^{(H)} \sigma_n \geq \sigma_n(y)H(y) \uparrow \infty,$$

thus proving that the sequence of integrals of the σ_n can not remain bounded; that is, E cannot be a null set if it contains any points that are not in S_H.

4 Proposition. (a) Every subset of a μ-null set is μ-null. (b) If $\{E_\alpha\}$ is a collection of μ-null sets, then $\bigcap E_\alpha$ is μ-null. (c) If $\{E_n\}$ is a sequence of μ-null sets, then $\bigcup E_n$ is μ-null.

Proof: If E is μ-null, there is a sequence $\{\sigma_n\}$ of nondecreasing, nonnegative elementary functions such that $\sigma_n \uparrow \infty$ on E and $\int \sigma_n \, d\mu \uparrow A$. Let E' be any subset of E. Clearly, the sequence $\{\sigma_n\}$ diverges also on E' and its integrals remain unchanged. This proves (a).

Since $\bigcap E_\alpha \subset E_\alpha$ for all α, (b) follows from (a).

Suppose now that $\{E_n\}$ is a sequence of μ-null sets. For each n, there is a nondecreasing sequence $\sigma_{n,1} \leq \sigma_{n,2} \leq \cdots \leq \sigma_{n,k} \leq \cdots$ of elementary functions which diverges on E_n and satisfies $\int \sigma_{n,k} < 2^{-n}$ for all values of k. To prove that $E = \bigcup E_n$ is μ-null requires a sequence $\{\sigma_k\}$ satisfying the conditions of Definition 1. Let

$$\sigma_k = \sum_{n=1}^{k} \sigma_{n,k}.$$

The nonnegativeness of each sequence $\{\sigma_{n,k}\}$ and the monotonicity with respect to k yield

$$\sigma_k = \sum_{n=1}^{k} \sigma_{n,k} \leq \sum_{n=1}^{k+1} \sigma_{n,k+1} = \sigma_{k+1}.$$

Suppose that x is in E. Then x is in some μ-null set E_m, and if $M > 0$ is given, there is an integer N, which may be assumed to be greater than

m, for which the following is true:

$$k \geq N > m \qquad \text{implies} \qquad \sigma_{m,k}(x) > M.$$

Thus

$$\sigma_k(x) = \sigma_{1,k}(x) + \sigma_{2,k}(x) + \cdots + \sigma_{m,k}(x) + \cdots + \sigma_{k,k}(x) \geq \sigma_{m,k}(x) > M,$$

proving that $\{\sigma_k\}$ diverges on E. Finally, for all values of k,

$$\int \sigma_k \, d\mu = \sum_{n=1}^{k} \int \sigma_{n,k} \, d\mu < \sum_{n=1}^{k} 2^{-n} < 1. \quad \blacksquare$$

Hereafter, the sentence

"P is true a.e. (μ)"

("a.e." reads "almost everywhere") will stand for

"P is true except on a μ-null set."

The monotonicity of the Daniell integral guarantees only (so far) that the integral is nonnegative if the integrand is *everywhere* nonnegative. Similarly, the continuity and the related properties of Proposition 12.1.3 are stated only for sequences which converge *everywhere* on Ω. Propositions 5, 6, 7, and 8, (below) demonstrate that the nonnegativity of σ [Definition 12.1.2(ii)] and the convergence of $\{\sigma_n\}$ (Definition 12.1.2(iii) and Proposition 12.1.3) need only be assumed a.e. (μ).

5 Proposition. If $\sigma \in C_0$ and $\sigma \geq 0$ a.e. (μ), then

[1] $$\int \sigma \, d\mu \geq 0.$$

Proof: The inequality [1] is equivalent to proving that if $\epsilon > 0$ is given, then

[2] $$\int \sigma \, d\mu \geq -2\epsilon.$$

Let E be the null set on which $\sigma(x) < 0$, and let $\{\tau_n\}$ be a nonnegative, nondecreasing sequence of elementary functions which diverges on E, and whose μ-integrals are bounded by 1 (Definition 1′).

Form the sequence $\omega_n = (\sigma^- - \epsilon\tau_n)^+$ of elementary functions. Since $\epsilon\tau_n \geq 0$ on Ω and $\sigma^- = 0$ on $\Omega - E$, it follows that, for $n = 1, 2, \ldots,$

[3] $$\omega_n(x) = 0 \qquad \text{if } x \in \Omega - E.$$

If $x \in E$, the sequence $-\epsilon\tau_n(x) \downarrow -\infty$, from which it follows that $[\sigma^-(x) - \epsilon\tau_n(x)] \downarrow -\infty$, and therefore

[4] $$\omega_n(x) = [\sigma^-(x) - \epsilon\tau_n(x)]^+ \downarrow 0 \qquad \text{if } x \in E.$$

From [3] and [4] and the continuity of the integral, we obtain

[5] $$\int \omega_n \, d\mu \downarrow 0,$$

and so

$$\int \sigma \, d\mu = \int \sigma^+ \, d\mu - \int \sigma^- \, d\mu = -\epsilon \int \tau_n \, d\mu - \int (\sigma^- - \epsilon \tau_n) \, d\mu.$$

However, $\int \tau_n \, d\mu \leq 1$, and therefore

$$\int \sigma \, d\mu \geq -\epsilon - \int (\sigma^- - \epsilon \tau_n) \, d\mu \geq -\epsilon - \int \omega_n \, d\mu > -2\epsilon$$

for sufficiently large values of n. ∎

6 Proposition. If $\sigma = \tau$ a.e. (μ), then

$$\int \sigma \, d\mu = \int \tau \, d\mu.$$

Proof: $\sigma = \tau$ a.e. (μ) implies that $\sigma \geq \tau$ and $\tau \geq \sigma$ a.e. (μ), which in turn implies $\sigma - \tau \geq 0$ and $\tau - \sigma \geq 0$ a.e. (μ). Proposition 5 yields

$$\int (\sigma - \tau) \, d\mu \geq 0 \quad \text{and} \quad \int (\tau - \sigma) \, d\mu \geq 0.$$

The desired result follows from the linearity of the integral. ∎

7 Proposition. If $\sigma \in C_0$, $\sigma \geq 0$ and $\int \sigma \, d\mu = 0$, then $\sigma = 0$ a.e. (μ).

Proof: Let E be the set on which σ is positive. The sequence $\sigma_n = n\sigma$ of functions in C_0 is nondecreasing, diverges on E, and $\int \sigma_n \, d\mu = \int n\sigma \, d\mu = n \int \sigma \, d\mu = 0$. Definition 1 implies that E is μ-null. ∎

8 Proposition. If $\{\sigma_n\}$ is a nonincreasing (or nondecreasing) sequence of elementary functions, and if $\lim_{n \to \infty} \sigma_n(x) = 0$ a.e. (μ), then

$$\lim_{n \to \infty} \int \sigma_n \, \mu = 0.$$

Proof: Let E be the μ-null set on which $\{\sigma_n\}$ does not converge monotonically to zero, and let $\{\tau_n\}$ be the sequence described in the proof of Proposition 5. Then it is easily verified that for $\epsilon > 0$, the sequence $\omega_n = (\sigma_n - \epsilon \tau_n)^+ \downarrow 0$ on Ω, from which it follows that $\int \omega_n \, d\mu \downarrow 0$. Since the σ_n are a.e. (μ) nonnegative, Proposition 5 implies that

$$0 \leq \int \sigma_n \, d\mu = \epsilon \int \tau_n \, d\mu + \int (\sigma_n - \epsilon \tau_n) \, d\mu \leq \epsilon + \int \omega_n \, d\mu < 2\epsilon$$

for sufficiently large values of n, proving that

$$\lim_{n \to \infty} \int \sigma_n \, d\mu = 0.$$

(The monotonicity of the sequence $\{\sigma_n\}$ implies further that the convergence of the sequence of integrals is also monotonic.) ∎

Proposition 8 remains true if it is assumed only that the sequence $\{\sigma_n\}$ is a.e.(μ) monotonic. The verification of this result is left to the reader.

9 Proposition. If $\{\sigma_n\}$ is a sequence of elementary functions which is a.e.(μ) nonincreasing, and if $\lim_{n\to\infty} \sigma_n = 0$ a.e.(μ), then

$$\int \sigma_n \, d\mu \downarrow 0.$$

Exercises

***1.** Demonstrate the equivalence of Definitions 1 and 1′.

***2.** Prove that the μ-null sets of parts (a) and (b) of Example 12.1.4 (or Example 2) are the same.

***3.** Discuss the μ-null sets of Example 12.1.9.

***4.** Let $f(x)$ be a continuous function defined on the cube

$$S_L = \{x \in R^n : |x_1|, |x_2|, \ldots, |x_n| \leq L\}.$$

Show that if $\epsilon > 0$ is given, there is an integer N having the property that if S_L is divided into cubes whose edges are equal to $2L/N$, then there are two step functions σ and τ, defined on S_L which satisfy the following conditions: (i) $\sigma(x) \leq f(x) \leq \tau(x)$ for all $x \in S_L$, (ii) $(\tau(x) - \sigma(x)) < \epsilon$ and (iii) $\tau(x)$ and $\sigma(x)$ are constant on the interiors of the small cubes.

5. Use Exercise 4 to prove that if a subset of R^n is null with respect to the continuous elementary functions, then it is null with respect to the class of step functions. (See Examples 2(a) and (b).) [*Hint:* The proof requires showing that the existence of a sequence of continuous functions that satisfy the conditions of Definition 1 implies the existence of a sequence of step functions which also satisfy these conditions.]

6. Let $\sigma(x)$ be the step function displayed in Figure 34. The numbers that appear in the small squares are the constant values that $\sigma(x)$ takes on in the interiors of those squares. It is assumed that $\sigma(x) = 0$ on the grid lines and outside the large square. Show that there is a continuous function f defined on the large square satisfying $|f(x) - \sigma(x)| < \epsilon$, where ϵ is some preassigned positive number. [*Hint:* Cover the edges of the small rectangles by strips of width δ, where δ is assumed to be less than 1/3. Define $f(x)$ so that it

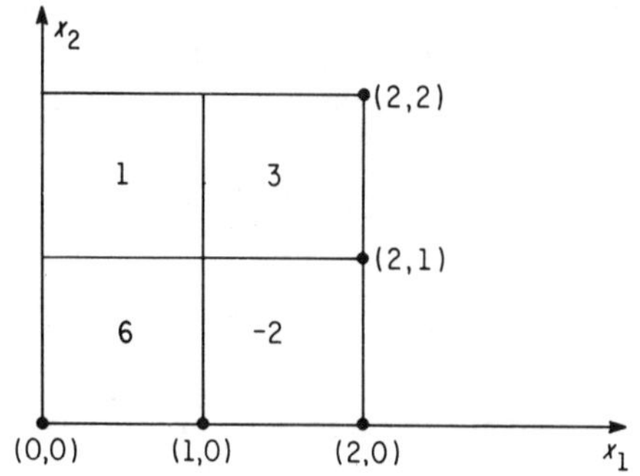

FIGURE 34

coincides with $\sigma(x)$ in the "reduced" rectangles (dotted lines) and is extended continuously across the strips.]

7. Generalize the result of Exercise 6.

8. Use Exercise 7 to reverse the conclusion of Exercise 5.

***9.** Prove Proposition 9.

*Exercises which request the proof of a statement made in the text, as well as those referred to in the text, are starred.

3 Integrable Functions and Convergence Theorems

The space L_1^μ of μ-integrable functions is obtained by extending the class C_0 of elementary functions in three stages (cf. Sec. 1 of Chapter 6). At each point the μ-integral is defined on the enlarged class of functions.

Let $\hat{C}_0^\mu$ be the class of real valued functions defined on Ω, each of which differs from an elementary function on at most a μ-null set, that is,

$$\hat{C}_0^\mu = \{\hat{\sigma} : \exists\, \sigma \in C_0 \quad \text{and} \quad \hat{\sigma} = \sigma \text{ a.e. } (\mu)\}.$$

The integral is extended to $\hat{C}_0^\mu$ in the natural way:

$$[1] \qquad \int \hat{\sigma}\, d\mu = \int \sigma\, d\mu, \qquad \text{if } \hat{\sigma} = \sigma \text{ a.e. } (\mu).$$

It must be verified that [1] is well defined and that the conditions of Definition 12.1.2 are satisfied:

If σ and τ are elementary functions which are a.e. (μ) equal, then Proposition 12.2.6 implies that their integrals are the same, proving that the integral of $\hat{\sigma}$ does not depend upon the choice of the function in C_0 to which it is a.e. (μ) equal. The positivity and monotonicity of the extended integral follow from Proposition 12.2.5, and the linearity and homogeneity follow immediately from the known properties of the integral defined on C_0. The reader should have no difficulty in verifying these statements, as well as the assertion that $\hat{C}_0^\mu$ is a lattice (Exercise 2). The only condition that remains to be verified is the continuity of the extended integral.

1 Proposition. If $\{\hat{\sigma}_n\}$ is a nonincreasing sequence of functions in $\hat{C}_0^\mu$ that converges a.e. (μ) to zero, then

$$\lim_{n\to\infty} \int \hat{\sigma}_n\, d\mu = 0.$$

Proof: Let E_0 be the μ-null set on which $\{\hat{\sigma}_n\}$ does not converge to zero, and let $\{\sigma_n\}$ be a sequence of elementary functions (in C_0) satisfying $\hat{\sigma}_n(x) = \sigma_n(x)$ except on a μ-null set E_n. Then $\{\sigma_n\}$ is an a.e. (μ) nonincreasing sequence of elementary functions that converges to zero except on the μ-null set $E = \bigcup_{i=0}^{\infty} E_i$ (Proposition 12.2.4(c)). Proposition 12.2.8

and [1] yield

$$\lim_{n\to\infty} \int \hat{\sigma}_n \, d\mu = \lim_{n\to\infty} \int \sigma_n \, d\mu = 0. \quad \blacksquare$$

For notational convenience, we shall hereafter omit the circumflex "$\hat{}$" and assume that the given class of elementary functions has the following property: If $\sigma \in C_0$ and if $\tau = \sigma$ a.e. (μ), then $\tau \in C_0$.

The next step is to extend the integral by taking monotone limits of functions in C_0. This is simply Method II of constructing the class of Lebesgue integrable functions (see Sec. 1 of Chapter 6 and the epilogue to Part II).

2 Definition. Let $\{\sigma_n\}$ be an a.e. (μ) nondecreasing sequence of functions in C_0 whose integrals are uniformly bounded. If the sequence converges a.e. (μ) to a function f (See Remark that follows), then the *μ-integral of f* is given by

$$[2] \qquad \int f \, d\mu = \lim_{n\to\infty} \int \sigma_n \, d\mu.$$

Remark. The boundedness and the monotonicity of the sequence $\left\{ \int \sigma_n \, d\mu \right\}$ of real numbers assures the existence of the limit [2]. Definition 12.2.1 implies that the sequence $\{\sigma_n\}$ converges a.e. (μ).

Let Γ^μ denote the class of real-valued functions defined on Ω which are the a.e. (μ) limits of nondecreasing sequences of elementary functions whose integrals are uniformly bounded. In order that [2] be an extension of the Daniell integral to Γ^μ which coincides with [1] on the subset C_0 of Γ^μ, it must be verified that [2] does not depend on the choice of the sequence $\{\sigma_n\}$ that converges monotonically to f. This and the related result,

$$f, g \in \Gamma^\mu \quad \text{and} \quad f = g \text{ a.e. } (\mu) \qquad \text{imply} \qquad \int f \, d\mu = \int g \, d\mu,$$

will follow from Proposition 3.

3 Proposition. If $\{\sigma_n\}$ and $\{\tau_n\}$ are nondecreasing sequences of elementary functions that converge a.e. (μ) to f and g respectively, and if $f \geq g$ a.e. (μ), then

$$[3] \qquad \int f \, d\mu = \lim_{n\to\infty} \int \sigma_n \, d\mu \geq \lim_{n\to\infty} \int \tau_n \, d\mu = \int g \, d\mu,$$

provided that the limit on the left exists.

Proof: Keeping the integer m fixed, set $\omega_n = \min(\sigma_n, \tau_m)$. Since C_0 is a lattice $\omega_n \in C_0$. The monotonicity of the sequence $\{\sigma_n\}$ and the a.e. inequality, $\lim_{n\to\infty} \sigma_n = f \geq g \geq \tau_m$, imply that

(i) $\omega_n \leq \sigma_n$, $n = 1, 2, \ldots$, and

(ii) $\omega_n \uparrow \tau_m$ a.e. (μ) as n tends to infinity.

Only (ii) requires verification: Let E be the set on which $\{\omega_n\}$ does not converge to τ_m. If $x \in E$, then for $n = 1, 2, \ldots, \omega_n(x) \leq \sigma_n(x) \leq \tau_m(x)$.

Thus $E \subset E_1 \cup E_2 \cup E_3$, where

$$E_1 = \{x : \lim \sigma_n(x) \neq f(x)\}, \quad E_2 = \{x : \lim \tau_n(x) \neq g(x)\}$$

and

$$E_3 = \{x : f(x) < g(x)\}$$

are μ-null sets. Thus E is μ-null, (ii) follows, and the sequence $\varphi_n = \tau_m - \omega_n$ (remember, m is kept fixed) of elementary functions tends monotonically (down) to zero except on the μ-null set E. Proposition 1 yields

$$0 = \lim_{n\to\infty} \int (\tau_m - \omega_n)\, d\mu = \int \tau_m\, d\mu - \lim_{n\to\infty} \int \omega_n\, d\mu,$$

from which it follows that for all values of m,

$$\int \tau_m\, d\mu = \lim \int \omega_n\, d\mu \leq \lim_{n\to\infty} \int \sigma_n\, d\mu.$$

If the right-hand limit exists, then it is equal to $\int f\, d\mu$, and [3] is obtained by letting m tend to infinity. ∎

Returning to the claim made before proving Proposition 3, let us suppose that $\{\sigma_n\}$ and $\{\tau_n\}$ are sequences in C_0 which converge monotonically a.e. (μ) to f. Taking $f = g$ in Proposition 3, the desired result follows; that is, the integral of f does not depend upon the choice of the sequence.

It is readily seen that Γ^μ is not a vector space, for it does not follow from $f \in \Gamma^\mu$ that $-f \in \Gamma^\mu$ (cf. Example 6.1.3). However,

(i) $f,g \in \Gamma^\mu$ and $a,b \geq 0$ imply $af + bg \in \Gamma^\mu$, and

(ii) $f,g \in \Gamma^\mu$ implies $\max(f,g)$, $\min(f,g)$, f^+, f^- and $|f| \in \Gamma^\mu$.

We prove only that $\max(f,g) \in \Gamma^\mu$. The reader should have no difficulty in verifying (i) and the remaining portions of (ii).

Let $\{\sigma_n\}$, $\{\tau_n\}$ be a.e. (μ) nondecreasing sequences of elementary functions whose integrals are uniformly bounded, and which converge a.e. (μ) to f and g respectively. Since C_0 is a lattice, $\max(\sigma_n, \tau_n) \in C_0$, and except on a μ-null set, $f(x) < g(x)$ implies that for sufficiently large n, $\sigma_n(x) < \tau_n(x)$, in which case the sequence $\max[\sigma_n(x), \tau_n(x)] = \tau_n(x)$ converges monotonically to $g(x)$. Similarly, except on a μ-null set, $g(x) < f(x)$ implies $\max[\sigma_n(x), \tau_n(x)] \uparrow f(x)$. And finally, except on a μ-null set, $f(x) = g(x)$ implies $\max[\sigma_n(x), \tau_n(x)] \uparrow f(x) = g(x)$. Thus $\{\max(\sigma_n, \tau_n)\}$ is an a.e. (μ) nondecreasing sequence of elementary functions that converges a.e. (μ) to $\max(f,g)$. Moreover, the sequence of integrals is bounded by $\sup_n \left\{\int \sigma_n\, d\mu, \int \tau_n\, d\mu\right\}$, and therefore it follows that the a.e. (μ) limit of $\{\max(\sigma_n, \tau_n)\}$ is in Γ^μ.

In developing the Lebesgue theory, the continuity property appeared as a special case of the theorem on monotone convergence. The same assertion can be made here. The proof is essentially the same. (See Theorems 6, 7, 8, and 9.)

Since Γ^μ is not in general closed under taking differences, it must be further extended (cf., again, Method II, Sec. 1, Chapter 6):

4 Definition. The class of real-valued functions

$$L_1^\mu = \{f : f = g - h,\ g,h \in \Gamma^\mu\},$$

is called the class of *μ-integrable functions* defined on Ω. The *μ-integral* of a function f in L_1^μ is given by

[3] $$\int f\, d\mu = \int g\, d\mu - \int h\, d\mu;$$

$g,h \in \Gamma^\mu$ and $f = g - h$.

5 Theorem. (a) The μ-integral [3] is well defined; that is, if $f = g - h = g' - h'$, where $g,h,g',h' \in \Gamma^\mu$, then

$$\int g\, d\mu - \int h\, d\mu = \int g'\, d\mu - h'\, d\mu;$$

(b) L_1^μ is a real vector space;
(c) L_1^μ is a lattice;
(d) the integral [3] is linear, homogeneous, monotonic and continuous.

Corollary 9 (below) yields the continuity, and the remainder of the properties appear in the exercises.

As stated earlier, the principle convergence theorems of Chapter 6 hold also for L_1^μ. The arguments are virtually the same, the spaces C_0, Γ^μ, and L_1^μ replacing Σ, Γ, and L_1 respectively. For convenience, these theorems are restated for μ-integrals, along with the major preliminary steps.

First, Γ^μ must be shown to be closed under monotone convergence:

6 Theorem. If $\{f_n\}$ is a nondecreasing sequence of functions in Γ^μ whose μ-integrals are uniformly bounded, then there is a function $f \in \Gamma^\mu$ such that

(i) $\lim_{n\to\infty} f_n = f$ a.e. (μ), and

(ii) $\lim_{n\to\infty} \int f_n\, a\mu = \int f\, d\mu$.

See the proof of Theorem 6.1.2.

The statement of Theorem 7 (to follow) requires the extension of the norm (See Sec. 2) to L_1^μ. The space of μ-integrable functions, (more precisely, the space of equivalence classes of μ-integrable functions) can be made into a normed vector space by setting

[4] $$\|f\|_\mu = \int |f|\, d\mu, \quad f \in L_1^\mu.$$

The verification of the norm axioms is left to the reader.

7 Theorem. If $f \in L_1^\mu$, then for any $\epsilon > 0$, there are functions $G,H \in \Gamma^\mu$ satisfying, (i) $f = G - H$, and (ii) $\| H \|_\mu < \epsilon$.

Proof: If it can be shown that f is the a.e. (μ) limit of a Cauchy sequence of elementary functions, then the proof of Theorem 6.1.4 may be used here. To obtain such a sequence, we first write $f = g - h$, where $g,h \in \Gamma^\mu$ (Definition 4). Let $\{\sigma_n\}$, $\{\tau_n\}$, be nondecreasing sequences of elementary functions that converge a.e. (μ) to g,h respectively. Then $\gamma_n = \sigma_n - \tau_n$ is a sequence of elementary functions which converges a.e. (μ) to f. The triangle inequality yields

$$[5] \quad \| \gamma_n - \gamma_m \|_\mu = \int | \gamma_n - \gamma_m | \, d\mu \le \int | \sigma_n - g | \, d\mu + \int | g - \sigma_m | \, d\mu + \int | \tau_m - h | \, d\mu + \int | h - \tau_n | \, d\mu.$$

However, since $\sigma_i \uparrow g$ and $\tau_i \uparrow h$ a.e.(μ), it follows that each of the four integrals on the right-hand side of [5] tends to zero as n,m tend to infinity, thus proving that $\{\gamma_n\}$ is a Cauchy sequence. ∎

8 Theorem (Monotone Convergence). Let $\{f_n\}$ be a nondecreasing sequence of μ-integrable functions whose μ-integrals are uniformly bounded. Then the sequence converges a.e.(μ) to a function f in L_1^μ, and

$$\int f \, d\mu = \lim_{n \to \infty} \int f_n \, d\mu.$$

See the proof of Theorem 6.1.5.

Theorem 8 implies that the μ-integral is continuous:

9 Corollary. The μ-integral defined on L_1^μ is continuous (Definition 12.1.2 (iii)).

With no alteration of the proofs, both Fatou's lemma and the Theorem on Dominated Convergence follow. They are stated below.

10 Theorem (Fatou). If $\{f_n\}$ is a sequence of nonnegative, μ-integrable functions, and if $\underline{\lim}_n \int f_n \, d\mu$ is finite, then the function $f(x) = \underline{\lim}_n f_n(x)$ is μ-integrable, and

$$\int \underline{\lim}_n f_n \, d\mu \le \underline{\lim}_n \int f_n \, d\mu.$$

11 Theorem (Dominated Convergence). Let $\{f_n\}$ be a sequence of μ-integrable functions converging a.e.(μ) to a function f. If there is a function $g \in L_1^\mu$ and $|f_n| \le g$ for all n, then $f \in L_1^\mu$ and

$$\int f \, d\mu = \lim_{n \to \infty} \int f_n \, d\mu.$$

Exercises

*1. Prove that the integral [1] defined on $\hat{C}_0$ is linear, homogeneous and monotone.
*2. Prove that $\hat{C}_0$ is a lattice.
*3. Verify property (i) of Γ^μ and the remaining portions of (ii).
*4. Prove Theorem 5.
*5. Prove Theorem 6.
*6. Verify the norm axioms for $\|f\|_\mu$ (see [4]). You must prove that, $\|f\|_\mu = 0$ implies $f = 0$ a.e. (μ). (Cf. Proposition 12.2.7.)
*7. Prove Theorems 8, 10, and 11.
*8. Prove that L_1^μ is complete; that is, show that every Cauchy sequence of μ-integrable functions converges to a μ-integrable function.

*Exercises which request the proof of a statement made in the text, as well as those referred to in the text, are starred.

4 Measurable Functions and Measurable Sets

In the preceding sections of this chapter it was shown that a Daniell integral defined on a class of elementary functions (Definitions 12.1.1 and 12.1.2) could be extended, first to the class Γ^μ of monotone limits of elementary functions with uniformly bounded integrals, and then to space L_1^μ of differences in Γ^μ. In the course of proving Theorem 12.3.7, we showed that each function in L_1^μ (that is, each μ-integrable function) could also be represented as the a.e. (μ) limit of a Cauchy sequence of elementary functions. The converse, that every Cauchy sequence in L_1^μ converges, was left as an exercise. Section 3 concluded with the statements of the three principal convergence theorems.

We turn next to a discussion of μ-measurable functions and μ-measurable sets. (In Sec. 5 we shall see that in the abstract case too, it is possible to develop the theory of Integration and Measure by defining first, measurable sets, and then measurable and integrable functions.)

The *Lebesgue* measurable functions were defined as a.e. limits of sequences of integrable functions or step functions, or equivalently, as functions whose truncations are integrable (Sec. 1 of Chapter 7). A variation of the truncation method is used in the abstract case. However, it is first necessary to alter the definition of a truncated function. For, if $\Omega \neq R$, there may not be subsets of Ω which can serve the same purpose as the "truncating" intervals $[-n,n]$. Thus, instead of truncating a function at an *integral* value, and setting it equal to zero outside the corresponding interval, the *function is truncated with respect to a second function*, which is assumed to be nonnegative and μ-integrable: The function *f truncated at g* is a function $f_g(x)$, which takes on, at each point $x \in \Omega$, the middle value of the three numbers $f(x)$, $g(x)$ and $-g(x)$. This is denoted by

$$f_g = \text{mid}(-g,f,g).$$

This definition yields the following equivalent expressions for f_g.

$$\begin{aligned}\text{mid}(-g,f,g) &= \max[\min(f,g), \min(f,-g), \min(-g,g)]\\ &= \min[\max(f,g), \max(f,-g), \max(-g,g)].\end{aligned}$$

Also,

$$[\text{mid}(-g,f,g)]^+ = \min(f^+,g)$$

and

$$[\text{mid}(-g,f,g)]^- = \min(f^-,g).$$

(See Exercise 1.)

1 Definition. A real valued function $f(x)$ defined on Ω is said to be *μ-measurable*, if for each nonnegative elementary function σ, $f_\sigma = \text{mid}(-\sigma,f,\sigma)$ is μ-integrable. The class of μ-measurable functions is denoted by M^μ.

2 Proposition. (i) M^μ is a real vector space that contains L_1^μ. (ii) M^μ is a lattice. (This means that if $g,h \in M^\mu$, then so do $\max(g,h)$, $\min(g,h)$, g^+, g^-, and $|g|$.) (iii) If f is the a.e.(μ) limit of μ-measurable functions, then f is μ-measurable.

Proof: The verification of (i) and (ii) is left to the reader.

Suppose that $\{f_n\}$ is a sequence of μ-measurable functions converging a.e.(μ) to a function f. We shall prove that if σ is a nonnegative, elementary function, then f_σ is μ-integrable: From Definition 1 it follows that for $n = 1,2,\ldots$, the functions $\sigma_n = \text{mid}(-\sigma,f_n,\sigma)$ are μ-integrable, and $\lim_{n\to\infty} \sigma_n = \text{mid}(-\sigma,f,\sigma)$ a.e.(μ). However, since $|\sigma_n| \leq \sigma$ for all n, it follows from the theorem on dominated convergence (Theorem 12.3.11), that the limit function $\text{mid}(-\sigma,f,\sigma)$ is μ-integrable, proving that f is μ-measurable. ∎

3 Proposition. f is μ-measurable if and only if f_g is μ-integrable whenever g is a nonnegative, μ-integrable function.

Proof: In view of the inclusion $C_0 \subset L_1^\mu$, the "if" part is trivial.

Suppose now that f is μ-measurable. If g is a nonnegative μ-integrable function, then $g \in \Gamma^\mu$ (Why?), and there is a nondecreasing sequence $\{\sigma_n\}$ of nonnegative, elementary functions which converges a.e.(μ) to g. Then for each n, $f_n = \text{mid}(-\sigma_n,f,\sigma_n)$ is μ-integrable (Definition 1) and $|f_n| \leq g$. Therefore, the theorem on dominated convergence implies that the function $(-g,f,g) = \lim_{n\to\infty} f_n$ is μ-integrable. ∎

The reader should have no difficulty in verifying the following additional properties of μ-measurable functions (cf. the corresponding properties for the Lebesgue measurable functions, Sec. 1 of Chapter 7).

4 Proposition. (i) If f and g are a.e.(μ) equal functions, then f is μ-measurable if and only if g is; (ii) If $f \in M^{\mu}$, $g \in L_1^{\mu}$, and $|f| \leq g$ a.e.(μ), then $f \in L_1^{\mu}$; (iii) f is μ-measurable if and only if, whenever g and h are μ-integrable functions satisfying $g \leq 0 \leq h$, the function

$$\text{mid}(g,f,h) = \begin{cases} f(x) & \text{if } g(x) \leq f(x) \leq h(x) \\ h(x) & \text{if } g(x) \leq h(x) \leq f(x) \\ g(x) & \text{if } f(x) \leq g(x) \leq h(x) \end{cases}$$

is μ-integrable (This is a more general type of truncation). (iv) Let $\{f_n\}$ be a sequence of μ-measurable functions. Then the functions $\sup[f_1,f_2,\ldots]$, $\inf[f_1,f_2,\ldots]$, $\overline{\lim}\, f_n$ and $\underline{\lim}\, f_n$ are μ-measurable provided they are a.e.(μ) finite.

5 Definition. A subset S of Ω is called *μ-measurable* if the characteristic function of S, χ_S, is a μ-measurable function. If moreover, χ_S is μ-integrable, then S is said to be *μ-integrable* or *finitely μ-measurable*, and the *μ-measure of S* is given by the integral

[1] $$\mu(S) = \int \chi_S \, d\mu.$$

The class of μ-measurable sets is denoted by $\mathfrak{M}^{\mu}$.

The assumptions made about Ω, C_0 and $\int \sigma\, d\mu$ do not imply the measurability of either the set Ω or its characteristic function χ_Ω. This is the case in Example 12.1.9 if Ω is assumed to be noncountable (See also Exercise 6). To assure the measurability of Ω and χ_Ω, the following assumption must be made about the space of μ-integrable functions:

6 Stone's Axiom. If $f \in L_1^{\mu}$ then $\min(\chi_\Omega, f) \in L_1^{\mu}$.

The verification of Proposition 7 is left to the reader.

7 Proposition. Stone's Axiom is satisfied if and only if χ_Ω is μ-measurable.

8 Theorem (Properties of μ-measurable sets). Let $S, S_1, S_2, \ldots,$ be μ-measurable sets. (If a set T is μ-measurable but not μ-integrable, then we write $\mu(T) = \infty$.)

(i) $S_1 \subset S_2$ implies $\mu(S_1) \leq \mu(S_2)$.

(ii) S is a μ-null set if and only if $\mu(S) = 0$.

(iii) $S_1 \subset S_2$ implies $S_2 - S_1$ is μ-measurable and $\mu(S_2 - S_1) = \mu(S_2) - \mu(S_1)$.

(iv) $\bigcup S_n$ is μ-measurable, and if $\Sigma \mu(S_n) < \infty$, then $\bigcup S_n$ is μ-integrable, and $\mu(\bigcup S_n) \leq \Sigma \mu(S_n)$. (The union and summation are either finite or countably infinite.)

(v) $\bigcap S_n$ is μ-measurable, and if at least one of the S_n is finitely μ-

measurable, then so is $\bigcap S_n$, and $\mu(\bigcap S_n) \leq \inf[\mu(S_1), \mu(S_2), \ldots]$.

(vi) If Stone's axiom is satisfied, Ω is μ-measurable, and if S is μ-measurable, then $\Omega - S$ is μ-measurable.

Proof: (i) If $S_1 \subset S_2$, then $\chi_{S_1} \leq \chi_{S_2}$, and the monotonicity of integral yields (if both sets are finitely μ-measurable),

$$\mu(S_1) = \int \chi_{S_1} d\mu \leq \int \chi_{S_2} d\mu = \mu(S_2).$$

If either set has infinite μ-measure, then there is nothing to prove.

(ii) From Definition 5, it follows that if the μ-measure of S is zero, then $0 = \int \chi_S d\mu = \int n\chi_S d\mu$, implying that $\sigma_n = n\chi_S$ is a sequence of elementary functions satisfying the conditions of Definition 12.2.1. Thus S is μ-null.

Conversely, if S is μ-null, then $\chi_S = 0$ a.e.(μ), and it follows from Proposition 12.3.3 that $\mu(S) = \int \chi_S d\mu = 0$.

(iii) $S_1 \subset S_2$ implies $\chi_{S_2 - S_1} = \chi_{S_2} - \chi_{S_1}$. The linearity of the integral yields

$$\mu(S_2 - S_1) = \int (\chi_{S_2} - \chi_{S_1}) d\mu = \mu(S_2) - \mu(S_1).$$

(iv) The μ-measurability of the functions χ_{S_n} implies the μ-measurability of the function $\chi_{\cup S_n} = \sup[\chi_{S_1}, \chi_{S_2}, \ldots]$. Thus Definition 5 yields the μ-measurability of the union $\bigcup S_n$. If moreover, $\Sigma \mu(S_n) < \infty$, then $\sigma_k = \chi_{\bigcup_1^k S_n}$ is a nondecreasing sequence of μ-integrable functions, and

$$\int \sigma_k d\mu \leq \sum_{n=1}^{k} \mu(S_n) \leq \sum_{n=1}^{\infty} \mu(S_n) < \infty.$$

From the theorem on monotone convergence (Theorem 12.3.8) we obtain the integrability of the monotone limit $\chi_{\cup S_n}$ of the σ_k, and

$$\mu\left(\bigcup_{n=1}^{\infty} S_n\right) = \int \chi_{\cup S_n} d\mu \leq \sum_{n=1}^{\infty} \mu(S_n).$$

The verification of the remaining properties is left to the reader. ▮

9 Theorem (Countable Additivity of μ-measure). If $\{S_n\}$ is a sequence of disjoint μ-integrable sets, then

$$\mu\left(\bigcup_{n=1}^{\infty}\right) S_n = \sum_{n=1}^{\infty} \mu(S_n).$$

The value ∞ for the sum of the μ-measures on the right, is permitted.

See the proof of Theorem 7.2.4.

We conclude this section by stating the analog of Theorem 7.3.8. This theorem, which characterizes the μ-measurable functions as those for which the sets $[f < A]$, (or equivalently, $[f > A], [f \leq A], \ldots, [B > f > A]$, ...) are μ-measurable, will serve as the definition of a μ-measurable function in Sec. 5.

10 Theorem. A real valued function f defined on Ω is μ-measurable if and only if the sets $[f < A]$ are μ-measurable for all real values of A.

The proof is the same as the one given for Theorem 7.3.8, and it is also true that the sets $[f < A]$ may be replaced by the sets of Lemma 7.3.7.

Exercises

***1.** Show that if x, y and z are real numbers, then

$$\begin{aligned} \text{mid}(x,y,z) &= \max[\min(y,z), \min(y,x), \min(x,z)] \\ &= \min[\max(y,z), \max(y,x), \max(x,z)]. \end{aligned}$$

Also, if $z \geq 0$.

$$[\text{mid}(-z,y,z)]^+ = \min(y^+,z) \quad \text{and} \quad [\text{mid}(-z,y,z)]^- = \min(y^-,z).$$

($y^+ = y$ if $y \geq 0$, and $y^+ = 0$ if $y < 0$. Also $y^- = -y$ if $y \leq 0$, and $y^- = 0$ if $y > 0$.)

***2.** Prove (i) and (ii) of Proposition 2.

***3.** Show that the $\{\sigma_n\}$ of Proposition 2 converge to $\text{mid}(-\sigma, f, \sigma)$.

***4.** Prove Proposition 4.

***5.** Prove that if f and g are μ-measurable, then f^2, $f \cdot g$, and $\sqrt{|f|}$ are μ-measurable.

***6.** Prove that the spaces of Examples 12.1.4, 12.1.7, and 12.1.8 satisfy Stone's axiom, but that of Example 12.1.9 does not if Ω is assumed to be uncountable.

***7.** Prove Proposition 7.

8. Let $\Omega = (0,1]$ and let C be the space of real valued functions defined on $(0,1]$ which can be written in the form $f(x) = ax$, $a \in R$. Show that C is a real vector space, but not a lattice. Prove that $F(f) = \int_0^1 f\,dx$ is a linear, homogeneous, monotone, continuous, functional. Describe the class Γ^μ of monotone limits of functions in C whose integrals are uniformly bounded. Let the measurable functions be those which are obtained by taking monotone limits. Describe these functions. Why is the empty set the only measurable set? Show that Stone's axiom is not satisfied.

***9.** Prove (v) and (vi) of Theorem 8.

***10.** Prove Theorem 9.

*Exercises which request the proof of a statement made in the text, as well as those referred to in the text, are starred.

5 Measure Spaces and the Integral

After constructing the Lebesgue integrable functions by means of Cauchy sequences of step functions, several alternate definitions of (Lebesgue) measurable functions and sets were given in Chapter 8. Each of these suggests a corresponding approach to the abstract theory of integration and measure. For example, it is possible to define first an "outer" or "exterior" measure on the totality of subsets of the given space, and then by suitably restricting this function to a subclass (the measurable sets), a nonnegative, monotone, countably additive set function is obtained. (Cf. Sec. 1 of Chapter 8. See Taylor [1], Sec. 9 of Chapter 4 for a discussion of the abstract theory.)

Here we shall start with a given class of sets on which a *measure function* (Definition 2) has been defined. Theorem 12.4.10 is used to define measurable functions. An integral is first defined on a special class of measurable functions (the generalized step functions), which take on only finitely many values, and then extended to a larger class of functions. This extension may take place in two ways: First, by observing that the integral defined on the class of generalized step functions is a Daniell integral, the extension described in the first three sections of this chapter is possible. This method, however, leads to a (possibly new) class of measurable functions and sets (See Sec. 4), and so it must be determined under what circumstances these functions and sets coincide with the given functions and sets (See the discussion that precedes the statement of Lemma 17). The second method, outlined here, may be compared with the procedure discussed in Chapter 8.

1 Definition. A collection $\mathfrak{A}$ of subsets of a given set X is said to be a *σ-ring* if

(i) $S,T \in \mathfrak{A}$ implies $S - T \in \mathfrak{A}$, and

(ii) $S_1, S_2, \ldots \in \mathfrak{A}$ implies $\bigcup_{i=1}^{\infty} S_i \in \mathfrak{A}$.

If, in addition to satisfying (i) and (ii), $\mathfrak{A}$ contains the set X, then $\mathfrak{A}$ is said to be a *σ-algebra*.

Setting $S = \bigcup S_n$ (finite or countable union), De Morgan's laws yield,

$$\bigcap S_n = S - \bigcup (S - S_n).$$

Thus a σ-ring contains also the countable intersections.

Theorem 12.4.8 asserts that $\mathfrak{M}^{\mu}$ is a σ-ring. If Stone's Axiom is satisfied, then the μ-measurable sets are a σ-algebra.

2 Definition. Let $\mathfrak{C}$ be a family of subsets of a nonempty set Ω, and let ν denote a mapping of $\mathfrak{C}$ into the nonnegative, extended real numbers. The triple $(\Omega,\mathfrak{C},\nu)$ is called a *measure space* if the following conditions are satisfied:

(i) $\mathfrak{C}$ is a σ-algebra;

(ii) $\nu(\phi) = 0$;

(iii) If $\{S_n\}$ is a disjoint sequence of sets in $\mathfrak{C}$, then

$$\nu(\bigcup S_n) = \Sigma \nu(S_n)$$

(Countable additivity of measure).

The mapping ν is called a *ν-measure*, or simply a *measure*. The sets of $\mathcal{C}$ are the *ν-measurable*, or simply *measurable*, subsets of Ω.

3 Examples. (a) $(R,\mathcal{B},m)$, where $\mathcal{B}$ is the class of Borel sets and m denotes the Lebesgue measure, is a measure space.

(b) If $\mathfrak{M}$ denotes the class of Lebesgue measurable sets, then $(R,\mathfrak{M},m)$ is a measure space.

(c) Let $\mathfrak{M}^{\mu}$ be the σ-ring of μ-measurable sets that arises from a Daniell integral. Then $(\Omega,\mathfrak{M}^{\mu},\mu)$ is a measure space if Stone's axiom is satisfied.

The following properties are easily verified.

4 Theorem. Let S, T, S_1, S_2, ... denote measurable subsets of a measure space $(\Omega,\mathcal{C},\nu)$. Then

(i) $S \subset T$ implies $T - S$ is measurable, and $\nu(T - S) = \nu(T) - \nu(S)$.

(ii) $S \subset T$ implies $\nu(S) \leq \nu(T)$.

(iii) $\bigcup S_n$ is measurable and $\nu(\bigcup S_n) \leq \Sigma \nu(S_n)$, provided this sum is finite.

(iv) $\bigcap S_n$ is measurable, and $\nu(\bigcap S_n) \leq \inf [\nu(S_1), \nu(S_2) \ldots]$. If moreover, $S_n \supset S_{n+1}$ for all n, then $\nu(\bigcap S_n) = \lim \nu(S_n)$.

5 Definition. A real valued function f defined on Ω, or on a measurable subset of Ω, is said to be *measurable* (or *ν-measurable*) if the sets $[f < A]$ are measurable for all real values of A. The class of measurable functions is denoted by N^{ν}.

It is easily verified that N^{ν} is a real vector space which contains, for all real values of c, the constant functions $c\chi_{\Omega}$.

Two additional conditions, both satisfied by $(R,\mathfrak{M},m)$, must be imposed on $(\Omega,\mathcal{C},\nu)$ to insure that the measure space and the class of measurable functions has several desirable properties. These are the *σ-finiteness of* Ω and the *completeness of* ν.

The extension of the definition of the Lebesgue integral to functions which do not vanish outside a bounded interval, requires taking the limit (as n tends to infinity) of the sequence of integrals over the finite intervals $[-n,n]$. To carry out a similar extension in the abstract case, it must be assumed that Ω may be "approximated" by an increasing sequence of sets each having finite measure. This is guaranteed by the property which is described below:

6 Definition. The measure space $(\Omega,\mathcal{C},\nu)$ is said to be *σ-finite* if Ω can be written as a countable union $\bigcup S_n$, where each S_n has finite measure.

The following is an equivalent characterization of a σ-finite space:

6′ Definition. $(\Omega,\mathcal{C},\nu)$ is called *σ-finite*, if there is a nested sequence $\Omega_1 \subset \Omega_2 \subset \cdots \subset \Omega_n \subset \Omega_{n+1} \subset \cdots$ of finitely measurable sets whose union is Ω.

The verification of the equivalence is left to the reader.

7 Examples. (a) $(R,\mathfrak{M},m)$ is σ-finite, since $R = \bigcup_{\infty -}^{\infty} [-n,n]$; each of these sets has measure equal to $2n$.

(b) If Ω is any nonempty set, then a measure space is obtained by setting $\mathcal{C} = \{\Omega,\emptyset\}$, and $\nu(\Omega) = \infty$, $\nu(\emptyset) = 0$. This space is not σ-finite.

In addition to σ-finiteness, it must be assumed that if a set has measure zero, then each of its subsets is measurable, in which case, the measure of each subset must also be zero (Theorem 4). (Cf. the extension of the space C_0 of elementary functions to the the space $\hat{C}_0$, of those functions, each of which is a.e. (μ) to a function in C_0.) Without this property, it is not possible to give a meaning to "a.e. (ν)." For, let us suppose that $S \supset E \supset E'$, where both S and E are measurable, and $\nu(E) = 0$. If some statement can be made for all points of $S - E$, then we should like to say that this statement is true a.e. (ν) on S. Certainly, if the statement is known to be true on the (possibly) larger set $S - E'$, we should like to continue to say that it is true a.e. (ν) on S. That is, if the set E is "negligible" from the point of view of measure and integration, then so should each of its subsets be. (See Theorem 11.)

8 Definition. A measure space $(\Omega,\mathcal{C},\nu)$ is called *complete* if $\nu(S) = 0$ and $T \subset S$, imply $\nu(T) = 0$.

9 Examples. (a) In view of Proposition 12.2.4, the measure which arises from a Daniell integral is complete.

(b) If $\mathfrak{M}$ is replaced by $\mathcal{B}$, the class of Borel sets, then $(R,\mathcal{B},m)$ is not complete (See Secs. 3 and 4 of Chapter 7, in particular Exercise 7.4.5). The reader is asked to supply the details (Exercise 7).

We show next that every measure space may be completed by adding all subsets of sets having measure zero:

10 Theorem. If $(\Omega,\mathcal{C},\nu)$ is a measure space, then there is a complete measure space $(\Omega,\hat{\mathcal{C}},\hat{\nu})$ satisfying the following conditions:

(i) $\mathcal{C} \subset \hat{\mathcal{C}}$,

(ii) If $S \in \mathcal{C}$, then $\nu(S) = \hat{\nu}(S)$, and

(iii) $\hat{S} \in \hat{\mathcal{C}}$ if and only if $\hat{S} = S \cup E$, $S \in \mathcal{C}$, and E is a set whose $\hat{\nu}$-measure is zero.

Proof: The third condition suggests the definitions of $\hat{\mathcal{C}}$ and $\hat{\nu}$: $\hat{\mathcal{C}}$ is obtained by adding to $\mathcal{C}$ all subsets of ν-null sets, and taking complements and countable unions and intersections. This insures that if $\hat{S} \in \hat{\mathcal{C}}$, then $\hat{S} = S \cup E$, where $S \in \mathcal{C}$ and $\hat{\nu}(E) = 0$. The extended measure is defined in the natural way by setting $\hat{\nu}(\hat{S}) = \nu(S)$. It left to the reader to verify that $(\Omega, \hat{\mathcal{C}}, \hat{\nu})$ is complete and that (i) and (ii) are satisfied. ❙

The following theorem illustrates the importance of the completeness property.

11 Theorem. Let f be a measurable function defined on a complete measure space $(\Omega, \mathcal{C}, \nu)$. If g is a second function which is a.e.(ν) equal to f, then g is measurable.

Proof: For all real values of A, the sets $[f < A]$ are measurable (Definition 5). Since $[g < A]$ differs from $[f < A]$ by a set whose measure is zero, it follows from the completeness of the measure space that the set $[g < A]$ is measurable, and therefore g is a measurable function. ❙

The procedure for defining an integral on a subspace of the measurable functions arising from a given measure space, does not differ greatly from the special case of the Lebesgue integral: First, an integral is defined for a class of functions which are similar to the real step functions. Following this, the integral is extended to a class of nonnegative, bounded measurable functions, and finally to the full class of integrable functions.

12 Definition. Let $S_1, S_2, \ldots, S_n$ be disjoint measurable subsets of Ω, and let $\alpha_1, \alpha_2, \ldots, \alpha_n$ be real numbers. The function

$$\sigma(x) = \Sigma \alpha_i \chi_{S_i}$$

is called a *generalized step function*. The class of these functions is denoted by C^ν. The subclass of C^ν which consists of those functions for which the S_i have finite measure is denoted by C_0^ν.

The reader should have no difficulty in verifying that C_0^ν (as well as C^ν) is both a real vector space and a lattice, and therefore a class of elementary functions (Definition 12.1.1).

13 Proposition. The generalized step functions are measurable.

Proof: Definition 5 implies that if S is a measurable set, then χ_S is a measurable function. Since N^ν is a vector space, it contains all linear combinations of characteristic functions, and hence all generalized step functions. ❙

14 Definition. The *ν-integral*, or simply *integral* of a generalized step function σ in C_0^ν is given by the following sum:

[1] $$\int \sigma\, d\nu = \sum_{1}^{n} \alpha_i \nu(S_i);$$

$\sigma(x)$ is given by the sum in Definition 12.

It follows easily that [1] is linear, homogeneous and monotone (Exercise 10). The continuity property is a corollary to the special form of Egoroff's theorem which is proved next.

15 Theorem. If $\{\sigma_n\}$ is a nonnegative, monotone sequence of generalized step functions which converges to 0 at every point of a finitely measurable set S, then for any $\epsilon > 0$, there is a measurable subset S' of S satisfying, (i) $\nu(S') < \epsilon$, and (ii) $\{\sigma_n\}$ converges uniformly to zero on $S - S'$.

Proof: It may be assumed that the σ_n vanish outside S.

To each x in S and $\delta > 0$, there is a smallest integer $n(x,\delta)$ such that

$$m \geq n(x,\delta) \qquad \text{implies} \qquad \sigma_m(x) < \delta.$$

If δ remains fixed, the sets $S_j(\delta) = \{x \in S : n(x,\delta) = j\}$ are disjoint, and

$$S = S_1(\delta) \cup S_2(\delta) \cup \cdots \cup S_n(\delta) \cup \cdots.$$

The measurability of the σ_j, $\nu(S) < \infty$, and Definition 5 imply that the sets $S \cap [\sigma_j < \delta]$ are finitely measurable. Hence, for $j = 1, 2, \ldots,$

$$S_j(\delta) = S \cap [\sigma_{j-1} \geq \delta] \cap [\sigma_j < \delta] \cap [\sigma_{j+1} < \delta] \cap \cdots$$

has finite measure, and from Definition 2 it follows that

$$\sum_{1}^{\infty} \nu(S_j(\delta)) = \nu(S) < \infty.$$

The convergence of this series implies that for each choice of $\delta = 1/k$, $k = 1, 2, \ldots,$ there is an integer N_k such that

[2] $$\nu\left[\bigcup_{j=N_k+1}^{\infty} S_j(1/k)\right] = \sum_{j=N_k+1}^{\infty} \nu(S_j(1/k)) < \epsilon/2^k,$$

where ϵ is a preassigned positive number. By setting

$$\Sigma_k = \bigcup_{j=N_k+1}^{\infty} S_j(1/k),$$

we obtain,

$$x \in (S - \Sigma_k) \quad \text{and} \quad n > N_k \qquad \text{imply} \qquad \sigma_n(x) < 1/k.$$

We shall show now that $S' = \bigcup_{k=1}^{\infty} \Sigma_k$ is the desired set: The inequality [2] yields

$$\nu(S') \leq \sum_1^\infty \nu(\Sigma_k) < \sum_1^\infty \epsilon/2^k = \epsilon.$$

To prove that the sequence converges uniformly on $S - S'$, let $1 > \gamma > 0$ be given. There is a unique integer k satisfying $1/k < \gamma \leq 1/(k-1)$. Let N_k be the integer determined by taking $\mathrm{d} = 1/k$. Then for $x \in S - S'$,

$$n \geq N_k \quad \text{implies} \quad \sigma_k(x) < 1/k < \gamma,$$

which proves that the convergence is uniform. ∎

16 Corollary. (Continuity of the integral for generalized step functions.) Let $\{\sigma_n\}$ be a sequence of nonnegative functions in C_0^ν which converge monotonically to zero. Then

$$\lim_{n\to\infty} \int \sigma_n \, d\nu = 0.$$

Proof: The monotonicity of the sequence implies that for all values of n, $[\sigma_n \neq 0] \subset [\sigma_1 \neq 0] = S$, where S is a finitely measurable set. Therefore, Egoroff's theorem (15) implies that for any positive number ϵ, there is a subset S' of S whose measure is less than ϵ, and $\{\sigma_n\}$ converges uniformly on $S - S'$. That is, given $\gamma > 0$, $\sigma_n < \gamma$ on $S - S'$ for sufficiently large values of n. It follows that

$$0 \leq \int \sigma_n \, d\nu = \int \sigma_n \chi_{S-S'} \, d\nu + \int \sigma_n \chi_{S'} \, d\nu$$
$$< \gamma \cdot \nu(S - S') + \epsilon \cdot \max_{x \in S} [\sigma_n(x)] < \gamma \cdot \nu(S) + K\epsilon,$$

where $K = \max_{x \in S} [\sigma_1(x)] \geq \max_{x \in S} [\sigma_n(x)]$. Since γ and ϵ are arbitrary, this implies that the integral tends to zero. ∎

Having just proved that [1] is a Daniell integral on the class of generalized step functions, each of which vanishes outside a set of finite measure, we may extend this integral in the manner described in Sec. 1-4. This leads, in order, to the classes L_1^ν (integrable functions, Definition 12.3.4), M^ν (measurable functions, Definition 12.4.1), and $\mathfrak{M}^\nu$ (measurable sets, Definition 12.4.5). The question that naturally arises is: Does the *given* class $\mathfrak{C} = \mathfrak{C}^\nu$, of measurable sets (Definition 2) coincide with the class $\mathfrak{M}^\nu$ that arises from the Daniell integral? Certainly, the latter includes the former, for, if anything, additional measurable sets will arise from the Daniell integral (Exercise 11). Also, does $N^\nu = M^\nu$? (N^ν is described in Definition 5.) Again, N^ν is a subset of M^ν, and, Definition 4 and Theorem 12.4.10 imply

$$\mathfrak{C}^\nu = \mathfrak{M}^\nu \quad \text{if and only if} \quad N^\nu = M^\nu.$$

We shall prove that if the measure space $(\Omega, \mathfrak{C}^\nu, \nu)$ is σ-finite and complete, then $\mathfrak{C}^\nu = \mathfrak{M}^\nu$ and $N^\nu = M^\nu$. We begin by proving this for the null sets:

17 Lemma. If $(\Omega,\mathcal{C}'',\nu)$ is a complete measure space, then the subclass of sets in $\mathcal{C}''$ having measure zero coincides with the subclass of sets in $\mathfrak{M}''$ which have measure zero.

Proof: As observed above, $\mathcal{C}'' \subset \mathfrak{M}''$, so that it suffices to prove that

$$E \in \mathfrak{M}'' \quad \text{and} \quad \nu(E) = 0 \qquad \text{imply} \qquad E \in \mathcal{C}''.$$

If E is a null set of $\mathfrak{M}''$, there is a nondecreasing sequence $\{\sigma_n\}$ of generalized step functions which diverges on E (and possibly on a larger set) and whose integrals remain uniformly bounded.

Let E' be the set on which the sequence $\{\sigma_n\}$ diverges. Then E' contains E, and if it can be shown that E' is in $\mathcal{C}''$, then $\nu(E') = 0$ follows, for otherwise the boundedness of the sequence of integrals is contradicted. The completeness of the measure space would then yield both $E \in \mathcal{C}''$, and the measure of E (as a set in $\mathcal{C}''$) is zero.

For $k = 1, 2, \ldots$, $E' \subset \bigcup_{n=1}^{\infty} [\sigma_n > k] = X_k$; $X_k \supset X_{k+1}$. Also, $E' = \bigcap_{k=1}^{\infty} X_k$. Since $\mathcal{C}''$ is a σ-algebra, and $[\sigma_n > k]$ is in $\mathcal{C}''$ for all positive integral values of n and k, it follows that E' is in $\mathcal{C}''$. ▌

Having proved that both $\mathcal{C}''$ and $\mathfrak{M}''$ have the same null sets, the expression "a.e. (ν)" may be hereafter used unambiguously.

We show next that if the measure space is also σ-finite, then $\mathcal{C}'' = \mathfrak{M}''$. This requires Proposition 18.

18 Proposition. Let $(\Omega,\mathcal{C}'',\nu)$ be a complete, σ-finite measure space. If f is a nonnegative function in M'', then there is a nonnegative, nondecreasing sequence of generalized step functions $\{\sigma_n\}$ which converge a.e. (ν) to f.

The proof is left to the reader.

Let us suppose now that S is a set in $\mathfrak{M}''$, that is, a measurable set arising from the Daniell integral. Definition 12.4.5 implies that χ_S is in M''. From Proposition 18, it follows that there is a nonnegative, nondecreasing sequence $\{\sigma_n\}$ of generalized step functions which converge to χ_S a.e. (ν). Thus, the set $S' = [\sigma_1 > 0] \cup [\sigma_2 > 0] \cup \cdots \cup [\sigma_n > 0] \cup \cdots$, is in $\mathcal{C}''$, and differs from S by at most a null set, which in view of Lemma 17 implies that S is in $\mathcal{C}''$.

This completes the proof that if $(\Omega,\mathcal{C}'',\nu)$ is a given complete, σ-finite measure space, then the Daniell integral defined on the class of generalized step functions, each of which vanishes outside a set of finite measure, leads back to (via Definitions 12.3.2, 12.3.4, 12.4.1, and 12.4.5) the original measurable sets and functions.

A second method of defining the integral, once a measure space is given, is outlined below. The details are left to the reader, and it should be observed that this approach bears comparison with the procedure described in Sec. 2 of Chapter 8.

Let $(\Omega,\mathcal{C}^{\nu},\nu)$ be a complete, σ-finite measure space for which generalized step functions and a Daniell integral have already been defined. Let us assume further that Theorem 4 has been proved and that the measurable functions (Definition 5) have been defined. The next step is to extend the integral to the class of nonnegative, bounded, measurable functions, each of which vanishes outside a set whose measure is finite (cf. Definition 8.2.1). This is done by using monotone sequences of generalized step functions, $\sigma_n \uparrow f$, and $\tau_n \downarrow f$, which vanish outside the set $S = [f \neq 0]$ (Proposition 18 and Exercises 13 and 14). The existence of these sequences makes the following definition possible:

19 Definition. Let f be a bounded, nonnegative measurable function which vanishes outside a set S which has finite measure. The *ν-integral*, or simply the *integral*, of f is the least upper bound of the set of values $\int \sigma \, d\nu$, where the σ ranges over the generalized step functions which satisfy the inequality, $0 \le \sigma \le f$.

The finiteness of the least upper bound taken in Definition 19 is assured by the inequality

$$\sigma \le \sup_{x \in S} \{f(x)\} \cdot \chi_S.$$

Thus the set of integrals $\left\{\int \sigma \, d\nu\right\}$ is bounded from above by the number $\nu(S) \cdot \{\sup_{x \in S} f(x)\}$.

For bounded, nonnegative, measurable functions f which vanish outside sets having finite measure, the integral may also be defined by taking the greatest lower bound of the set of integrals $\int \tau \, d\nu$, where τ ranges over the functions in C_0^{ν} that satisfy $f \le \tau$. This requires proving that there is a nonincreasing sequence of such functions which converges to f (Exercise 13). It is left to the reader to show that these definitions are identical, and that the integral so obtained is an extension of the Daniell integral on C_0^{ν}.

The next step is to extend the integral to a wider class of nonnegative, measurable functions, which are neither assumed to be bounded nor to vanish outside a set having finite measure. Again, such a function f is an upper bound for some nonnegative, generalized step functions. If

$$\sup_{0 \le \sigma \le f} \int \sigma \, d\nu \tag{3}$$

is finite (σ is assumed to be in C_0^{ν}), then f is said to be integrable, and [3] is called the integral of f. This clearly is an extension of the definition of the integral of a nonnegative, bounded, measurable function which vanishes outside a set of finite measure (Definition 19). The verification of the basic properties (linearity, homogeneity, positivity, and continuity) is left to the reader, along with the proof of the following theorem:

20 Theorem. Let f be a nonnegative, measurable function. If $\int f\,d\nu$ exists, then the set $[f > 0]$ is σ-finite. Conversely, if $[f > 0]$ is σ-finite, and if

[4] $$\sup_{0 \le g \le f} \int g\,d\nu$$

is finite, g ranging over the set of bounded, nonnegative, measurable functions which vanish outside a set whose measure is finite, then f is integrable, and its integral is equal to the number [4].

The final extension of the integral is made as follows: Writing $f = f^+ - f^-$, the measurable function f is said to be integrable if both f^+ and f^- are integrable, and

$$\int f\,d\nu = \int f^+\,d\nu - \int f^-\,d\nu.$$

(Cf. Sec. 2 of Chapter 8.)

Exercises

1. Prove that if $\mathcal{A}$ is a σ-ring, then $S_1, S_2, \ldots, \in \mathcal{A}$, implies $\bigcap_{i=1}^{\infty} S_i \in \mathcal{A}$.
2. In each of the following cases, decide whether or not the family of sets is a σ-ring, a σ-algebra. Give reasons. (a) the bounded subsets of R; (b) the countable subsets of R; (c) if X is any set, then $\mathcal{A}$ is the collection of sets whose complement in X is finite; (d) alter (c) by letting $\mathcal{A}$ be the collection of subsets of X, each of whose complements in X is countable; (e) the Borel subsets of R.
3. Extend Example 3(a) and (b) to R^n.

*4. Prove Theorem 4.

*5. Verify that N^ν as given by Definition 5, is a real vector space that contains the constant functions.

*6. Prove that the two characterizations of a σ-finite measure space (Definitions 6 and 6′) are equivalent.

*7. Why is the measure space with $\mathcal{C}$ taken to be the family of Borel sets incomplete? (See Example 9.)

*8. Complete the proof of Theorem 10.

*9. Show that both C^ν and C_0^ν (Definition 12) are vector spaces and lattices.

*10. Show that [1] is linear, homogeneous and monotone.

*11. Prove that $\mathcal{C}^\nu \subset \mathcal{M}^\nu$. (See the discussion that precedes Lemma 17.)

*12. Prove Proposition 18.

*13. Prove that if f is a bounded, measurable function which vanishes outside a set whose measure is finite, then there is a nonincreasing sequence $\{\tau_n\}$ in C_0^ν which converges to f.

14. Prove that if f is a bounded, measurable, nonnegative function which vanishes outside a set whose measure is finite, then,

$$\int f\,d\nu = \sup_{\sigma \le f} \int \sigma\,d\nu = \inf_{f \le \tau} \int \tau\,d\nu;$$

the functions σ and τ are generalized step functions, each vanishing outside a set whose measure is finite.

***15.** Prove that each of the extended integrals satisfies the basic properties: linearity, homogeneity, positivity and continuity.

*Exercises which request the proof of a statement made in the text, as well as those referred to in the text, are starred.

CHAPTER **13**

Measure Spaces II: Outer Measures, Signed Measures, Radon-Nikodym Theorem, and Product Measures (Fubini Theorem)

Introduction

In Sec. 1 of this chapter we shall show how a measure function defined on a σ-algebra of sets is generated by a nonnegative, subadditive function defined on the family of *all* subsets of a fixed set. The analogy to be made here is to the creation of the Lebesgue measure from the *outer* or *exterior* measure which is defined on the family of all subsets of the real numbers. The section concludes with a short description of how outer measures are generated by certain set functions defined on a collection of sets that is not quite a σ-algebra. The measures induced by these outer measures are extensions of the given set functions. As a particular example of measures constructed in this manner, we have the Lebesgue-Stieltjes measures, which are discussed in Sec. 2.

The construction of the Lebesgue-Stieltjes measures and integrals leads naturally to a question discussed earlier in a less general setting: When can a given function be represented by an integral? In Sec. 2, the *distribution functions*, in particular those which are absolutely continuous, will play the role of the primitives or indefinite integrals in the earlier discussions. Here it will be shown that certain Lebesgue-Stieltjes measures (the absolutely continuous ones) may be represented by Lebesgue integrals. Again, in Sec. 4, this question is posed in an even more general setting: If μ and ν are measures defined on the same class $\mathcal{C}$ of sets, when is it possible to write, for all $S \in \mathcal{C}$, $\nu(S) = \int_S f_0 \, d\mu$, where f_0 is a μ-integrable function? This is similar to the question posed in Chapter 9: When is a function $F(x)$ the indefinite integral of a second function $f(x)$? The answer to the question about measures bears analogy to the more familiar question about real functions, and indeed, we shall even be able to say that if one measure is given by an integral with respect to another, then its abstract derivative with respect to the second measure exists, and is equal to the integrand f_0.

Before settling the question of the integral representation of one measure in terms of another, certain preliminary results are necessary. These appear in Sec. 3, as part of a brief discussion of *signed measures*, which are, roughly speaking, the set functions that arise by taking linear combinations of measures. It will be proved that each signed measure may be decomposed into the difference of two measures, in much the same way that a function of bounded variation may be written as the difference of monotone functions. Moreover, each signed measure induces a partition of the space into a "positive" and "negative" part, much as a charge distribution on a surface partitions it into positively charged and negatively charged parts.

In the final section of the chapter, we shall see how a measure is generated on a product space by the measures given on the constituent spaces. Here we shall show that an integral on the product space is identical with the iterated integral—that is, an n-fold integral taken, in order, over each of the spaces of the product.

1 Outer Measures and the Generation of Measures

The reader will recall that the first step in defining the Lebesgue measure is to assign the length of an interval to be its measure (Intervals of infinite length, such as $(2, \infty)$ are permitted). After proving that every open subset of R is the countable disjoint union of open intervals, the measure function is extended to the open sets simply by summing the measures of the disjoint intervals of the union. This of course assures the countable additivity of the measure function. At this point in the development it is not possible to extend the measure function to the class of all measurable subsets of R without first knowing which subsets are measurable. Therefore an extension of this "measure function" is first defined for the family of *all* subsets of R by setting its value on a subset S of R to be the greatest lower bound of the set of values of the measures of all *open* sets which contain S (cf. Definition 8.1.1). This extended function (the exterior measure) retains the non-negativity and monotonicity of the restricted measure function, but as a result of having extended it to too large a class of sets, it is no longer countably additive (see Proposition 8.1.2). The more general case, which follows, is very similar to the special case described in Sec. 1 of Chapter 8. Therefore, many of the proofs will be omitted, and those which are not completely straightforward are outlined in the exercises.

1 Definition. An extended real-valued function μ^* defined on the totality of subsets of a nonempty set Ω is called an *outer measure* if it has the following properties:

(i) $\mu^*(\emptyset) = 0$;

(ii) If $S \subset T$ then $\mu^*(S) \leq \mu^*(T)$;

(iii) $\mu^*\left(\bigcup_{i=1}^{\infty} S_i\right) \leq \sum_{i=1}^{\infty} \mu^*(S_i)$.

The nonnegativity of μ^* follows from (i) and (ii).

2 Definition. A subset S of Ω is said to be *μ-measurable* if for every subset T of Ω,

$$\mu^*(T) = \mu^*(T \cap S) + \mu^*(T \cap (\Omega - S)). \tag{1}$$

The proof of the following theorem is left as an exercise.

3 Theorem. The class of μ-measurable sets is a σ-algebra, and μ is a complete (Definition 12.5.8) measure defined on this class.

Theorem 3 asserts that a given outer measure defined on the family of all subsets of a fixed set Ω may be suitably restricted (See Definition 2) so that this restriction is a complete measure on a σ-algebra of subsets of Ω. The next question, again in analogy with the real case, is: How are outer measures generated? Let us suppose that a function μ is defined on an algebra[1] $\mathfrak{A}$ of subsets of Ω (For example, think of an algebra that contains the open intervals of R with μ taken to be the sum of the lengths of the intervals that comprise a set in this algebra). It is required only that $\mu(\emptyset) = 0$, and if $\{S_i\}$ is a disjoint countable collection of sets in $\mathfrak{A}$ whose union S is also in $\mathfrak{A}$, then $\mu(S) = \Sigma\mu(S_i)$ (Think of S as an open subset of R which is expressed as a disjoint union of open intervals.) If $\mathfrak{A}$ were a σ-algebra, that is if it were closed under complementation and *countable* union and intersection, then μ would be a measure. The problem is to extend μ to a σ-algebra which contains $\mathfrak{A}$. For convenience, we shall call μ a *measure function defined on the algebra* $\mathfrak{A}$. Keeping in mind the special case in which $\Omega = R$ and $\mathfrak{A}$ is generated by the collection of all open subsets of R (that is, $\mathfrak{A}$ is the smallest algebra of subsets of R which contains the open sets), we define an outer measure on the family of *all* subsets of Ω: For each subset S of Ω, put

[1]A family $\mathfrak{A}$ of subsets of a fixed set Ω is said to be an *algebra* if it is closed under *finite* intersection and union, and contains along with each of its members A, the complement $\Omega - A$. In particular, it contains Ω and $\emptyset$.

[2] $$\mu^*(S) = \inf_{\{S_i\}} \sum_{i=1}^{\infty} \mu(S_i);$$

the greatest lower bound is taken over the countable collections $\{S_i\}$ of sets in $\mathfrak{A}$ for which $S \subset \bigcup_{i=1}^{\infty} S_i$ is true (cf. Definition 8.1.1). It is left to the reader in the exercises to prove that μ^* is an outer measure (Definition 1), and that if $S \in \mathfrak{A}$, then $\mu(S) = \mu^*(S)$ (cf. Definition 8.1.3 and Theorem 8.1.4). Letting $\hat{\mu}$ denote the measure induced by μ^* (Definition 2 and [1]), it is easily verified that the class $\mathfrak{M}$ of μ-measurable sets contains $\mathfrak{A}$, and that if $S \in \mathfrak{A}$, then $\hat{\mu}(S) = \mu(S)$ (Exercises 2 and 3). The measure $\hat{\mu}$ is said to be *induced* or *generated* by the set function μ (If μ is the "length" function described above, which is defined on the open subsets of R, then $\hat{\mu}$ is the Lebesgue measure and $\mathfrak{M}$ the class of Lebesgue measurable sets). Since the induced measure $\hat{\mu}$ is an extension of the given measure function μ which is defined on $\mathfrak{A}$, we are justified in omitting the circumflex from the former, and henceforth we shall write simply μ to denote both the given set function μ (on $\mathfrak{A}$) and its extension to the σ-algebra of μ-measurable sets.

In some cases, in particular the case of the Lebesgue-Stieltjes measures which are discussed in Sec. 2, it is convenient to start with a set function which is defined on a family of sets that has some but not all the properties of an algebra. Such a family of sets is a *semialgebra*, which like an algebra is closed under finite intersection, contains both $\emptyset$ and Ω, and for each pair of sets S and T in the semialgebra $\mathcal{S}$, the complement $S - T$ is the disjoint finite union of sets in $\mathcal{S}$. The semialgebra $\mathcal{S}$ which appears in Sec. 2 contains $\emptyset$, R and all intervals of the type $(a,b]$ as well as the unbounded intervals $(a,\infty]$ and $(-\infty,b]$. Let $\mathfrak{A}$ be the algebra generated by $\mathcal{S}$, that is, the smallest algebra which contains $\mathcal{S}$. $\mathfrak{A}$ is formed by joining to $\mathcal{S}$ those sets which are obtained by taking disjoint finite unions of sets in $\mathcal{S}$. The reader is asked to show that a set function μ defined on $\mathcal{S}$ may be extended to a measure function on $\mathfrak{A}$ provided that it has the following properties: μ is nonnegative and $\mu(\emptyset) = 0$, $\mu(S) = \sum_{i=1}^{n} \mu(S_i)$ if S is the disjoint finite union of sets S_i in $\mathcal{S}$ (finite additivity), and if a set S is the disjoint *countable* union $\bigcup S_i$ of sets in $\mathcal{S}$, then

$$\mu(S) \le \sum_{i=1}^{\infty} \mu(S_i) \quad \text{(countable subadditivity).}$$

We prove next the analogy of Theorem 7.3.3 and its corollaries. The reader will recall that although Lebesgue measurable sets exist that are not Borel sets, every Lebesgue measurable set differs only by a null set from a Borel set. This property is suggested by the definition of the outer

(Lebesgue) measure of a set S as the greatest lower bound of the measures of all open sets which contain S. This is simply a special case of [2], where $\bigcup S_i$ represents a countable disjoint union of open intervals. The intersection of an appropriate sequence of these open sets containing S is a G_δ set whose measure coincides with that of S. In the more general case considered here, the open sets are replaced by sets in an algebra $\mathfrak{A}$ of subsets of Ω. The classes $\mathfrak{A}_\sigma$ and $\mathfrak{A}_\delta$ are the families of sets obtained by taking countable unions and intersections respectively of sets in $\mathfrak{A}$. $\mathfrak{A}_{\sigma\delta}$ is the family of sets that arises by taking countable intersections of sets in $\mathfrak{A}_\sigma$, etc. (cf. Sec. 3 of Chapter 7).

A measure function μ defined on an algebra $\mathfrak{A}$ is said to be *finite* if $\mu(\Omega) < \infty$, and σ-finite if Ω can be written as the countable union $\bigcup \Omega_i$, where each Ω_i is in $\mathfrak{A}$ and $\mu(\Omega_i) < \infty$ (cf. Definition 12.5.6). It follows from the definition of the outer measure of a set as the greatest lower bound of the measures of sets in $\mathfrak{A}$ (see [2]), and from the fact that the measure $\hat{\mu}$ (or μ) is simply a restriction of μ^* to a σ-algebra containing $\mathfrak{A}$, that the finiteness or σ-finiteness of the measure function on $\mathfrak{A}$ implies the finiteness or σ-finiteness of the induced measure $\hat{\mu}$ (or μ).

Preliminary to the main theory which asserts that a set S is μ-measurable if and only if it is contained in an $\mathfrak{A}_{\sigma\delta}$ set T' and $\mu^*(T' - S) = \mu(T' - S) = 0$, we shall prove two similar properties about the outer measure of sets.

If S is any subset of Ω whose outer measure is finite, and if $\epsilon > 0$ is given, then there is a set $T \in \mathfrak{A}_\sigma$ (or simply an $\mathfrak{A}_\sigma$ set) which contains S and $\mu^*(T) \le \mu^*(S) + \epsilon$. This follows easily from [2]. If $\{T_n\}$ is taken to be a sequence of $\mathfrak{A}_\sigma$ sets corresponding to the values $\epsilon = 1/n$, it follows that the intersection $T = \bigcap T_n$ is an $\mathfrak{A}_{\sigma\delta}$ set which contains S, and the monotonicity of the outer measure implies that

$$\mu^*(S) \le \mu^*(T) \le \mu^*(S) + 1/n,$$

for all n, so that $\mu^*(S) = \mu^*(T)$ follows. We summarize these results:

4 Proposition. Let μ^* be the outer measure generated by a measure function μ defined on an algebra $\mathfrak{A}$ of subsets of Ω. If S is any subset of Ω whose outer measure is finite, then

(i) to each $\epsilon > 0$, there is an $\mathfrak{A}_\sigma$ set T which contains S and $\mu^*(T) \le \mu^*(S) + \epsilon$;

(ii) there is an $\mathfrak{A}_{\sigma\delta}$ set T' which contains S and $\mu^*(T') = \mu^*(S)$.

With one important difference, this looks very much like the theorem we have set out to prove for *measurable* sets S. However, if S is *not* measurable, then neither is $T' - S$ (since T' is measurable) and it would not follow from $\mu^*(T') = \mu^*(S)$ that $\mu^*(T' - S) = 0$, but only that $\mu^*(T' - S) \ge 0$ (remember, μ^* is subadditive). In fact, as Theorem 5 implies, $\mu^*(T' - S) > 0$ unless S is measurable.

5 Theorem. Let μ^* be the outer measure induced by a σ-finite measure function μ defined on an algebra $\mathfrak{A}$ of subsets of Ω. A subset S of Ω is measurable (Definition 2) if and only if there is a set T in $\mathfrak{A}_{\sigma\delta}$ which contains S and $\mu^*(T - S) = 0$.

Proof: Since the measure is σ-finite, the space Ω can be written as the disjoint union of countably many finitely measurable sets Ω_i. Let S be any measurable subset of Ω. Then the sets $S_i = \Omega_i \cap S$ are finitely measurable, mutually disjoint, and $S = \bigcup S_i$. Proposition 4(i) implies that to each pair of positive integers n and i, there is an $\mathfrak{A}_{\sigma\delta}$ set $T_{n,i}$ which contains S_i and

$$\mu(T_{n,i}) \leq \mu(S_i) + 1/2^i n, \tag{3}$$

or equivalently $\mu^*(T_{n,i} - S_i) \leq 1/2^i n$ (the measurability of the sets $T_{n,i}$ and S_i permit the substitution of μ for μ^*). For each n, S is contained in the union $T_n = \bigcup_{i=1}^{\infty} T_{n,i}$, and from [3] it follows that $\mu(T_n - S) \leq 1/n$. The intersection $T = \bigcap_{n=1}^{\infty} T_n$ is an $\mathfrak{A}_{\sigma\delta}$ set which contains S, and $(T - S) \subset (T_n - S)$ for all n. Thus $\mu(T - S) \leq \mu(T_n - S) \leq 1/n$ for all n, from which it follows that $\mu(T - S) = 0$.

The converse is left to the reader (Exercise 5). ∎

The generation of outer measures and measures is more fully discussed in Munroe [1], Taylor [1], Halmos [2], and Titchmarsh [1].

Exercises

***1.** Prove Theorem 3. (Outline of proof: Show first that Ω and $\emptyset$ satisfy [1]. In general, the symmetry of [1] with respect to a set and its complement guarantees that the complement of a measurable set is also measurable. To show that the union of two measurable sets S_1 and S_2 is measurable, use [1] twice, first taking $S = S_2$, and then repeating this procedure, obtain for any set T in Ω,

$$\mu^*(T) = \mu^*(T \cap S_2) + \mu^*(T \cap (\Omega - S_2) \cap S_1) + \mu^*(T \cap (\Omega - S_2) \cap (\Omega - S_1)).$$

The subadditivity of μ^* and DeMorgan's laws yield the measurability of the union. The general finite case is proved by mathematical induction. Extend the result to the countable union $S = \bigcup_{i=1}^{\infty} S_i$ by applying the finite case to the sequence of sets $T_n = \bigcup_{i=1}^{n} S_i$. The verification of the completeness of the measure is straightforward.)

***2.** Show that μ^* as defined by [2] is an outer measure and that if $S \in \mathfrak{A}$, then $\mu(S) = \mu^*(S)$.

***3.** Letting $\hat{\mu}$ denote the measure generated by the outer measure μ^*, show that the sets in the algebra $\mathfrak{A}$ are measurable and $\hat{\mu}(S) = \mu(S)$ for all $S \in \mathfrak{A}$.

***4.** Let μ be an extended real function defined on a semialgebra $\mathcal{S}$ of subsets of a fixed set Ω. Show that this function may be extended to a measure function on the smallest algebra of subsets of Ω that contains $\mathcal{S}$ provided that the following conditions are satisfied: μ is nonnegative, $\mu(\emptyset) = 0$, μ is finitely additive and countably subadditive. The latter two conditions mean that if S is a set in the semialgebra which can be expressed as the disjoint finite union of sets $S_1, S_2, \ldots, S_n$ which are also in the semialgebra, then $\mu(S) = \sum_{i=1}^{n} \mu(S_i)$; and if S is the countable union of sets S_i in the semialgebra, then $\mu(S) \leq \sum_{i=1}^{\infty} \mu(S_i)$.

***5.** Complete the proof of Theorem 5; that is, show that a subset S is measurable if it is contained in an $\mathcal{A}_{\sigma\delta}$ set T such that $\mu^*(T - S) = 0$.

***6.** Let μ^* be the outer measure induced by a measure function μ which is defined on an algebra $\mathcal{A}$ of subsets of Ω. Let $\mathcal{B}$ be the smallest σ-algebra which contains $\mathcal{A}$. If $\tilde{\mu}$ is any extension of μ to $\mathcal{B}$ which is a measure on $\mathcal{B}$, prove that $\tilde{\mu}(S) = \mu^*(S)$ whenever S is a set in $\mathcal{B}$ which has finite outer measure. (The particular extension $\hat{\mu}$ is defined by [1]. The reader is asked to prove that any measure on $\mathcal{B}$ agreeing with the given measure function which is defined on $\mathcal{A}$, gives the same value to sets in $\mathcal{B}$ as does the outer measure μ^*.) Prove also that if μ is σ-finite, then there is exactly one extension to $\mathcal{B}$.

In the remaining exercises, $\mathcal{F}$ will denote the family of all subsets of a fixed set Ω.

7. Show that the set function $\mu^*(\emptyset) = 0$, and $\mu^*(S) = 1$ if S is a nonempty subset of Ω, is an outer measure. Prove that μ^* is a measure if and only if Ω consists of a single point. Describe the μ-measurable subsets if Ω consists of at least two points.

8. Show that the set function $\mu^*(\emptyset) = 0$, $\mu^*(\Omega) = 2$, and $\mu^*(S) = 1$ if $S \neq \emptyset$ or Ω, is an outer measure. What are the measurable sets?

9. Let $\mu^*(S)$ be the number of points in S if S is finite, and let $\mu^*(S) = \infty$ otherwise. Show that μ^* is an outer measure. Is it also a measure?

10. Let Ω be the set of positive integers, and let $\mu^*(S) = \sum_{n \in S} 1/n^2$ whenever S is finite, but not empty, $\mu^*(\emptyset) = 0$ and $\mu^*(S) = \infty$ otherwise. Why isn't μ^* a measure? How can it be altered so that it becomes a measure?

11. Show that if $\{\mu_n^*\}$ is a sequence of outer measures, and if $\{a_n\}$ is a sequence of nonnegative numbers, then $\Sigma a_n \mu_n^*$ is an outer measure.

In Exercises 12-16 another method of generating outer measures is introduced. A collection $\mathcal{S}$ of subsets of Ω is said to be a *countable covering class* of Ω if $\emptyset \in \mathcal{S}$ and if every subset S of Ω is contained in a countable union of sets in $\mathcal{S}$. Let λ denote a nonnegative mapping of $\mathcal{S}$ into the extended real numbers, which maps $\emptyset$ into zero. A set function is defined on $\mathcal{F}$ by

$$\mu^*(S) = \inf_{\{S_i\}} \Sigma\lambda(S_i);$$

the greatest lower bound is taken over all sequences $\{S_i\}$ in $\mathcal{S}$ for which $\bigcup_{i=1}^{\infty} S_i \supset S.$

***12.** Show that μ^* is an outer measure.

13. Let $\Omega = R$. Show that the collection of open intervals is a countable covering class, and that if λ is taken to be the length of an interval, then μ^* is the Lebesgue exterior measure.

14. Let $\Omega = R$ and let $\mathcal{S}$ be the collection of open intervals (including the empty set). What is the outer measure that is induced by the set function $\lambda((0,1)) = 2$, $\lambda((0,2)) = 1$ and $\lambda(S) = 0$ for all other S in $\mathcal{S}$? Show that if S is an open interval, then $\mu^*(S) \neq \lambda(S)$.

15. Let Ω and $\mathcal{S}$ be as in Exercise 14. Define a set function λ which generates a class of measurable sets that does *not* contain the class $\mathcal{S}$.

***16.** Let $\Omega = R$; $\mathcal{S}$ consists of ϕ and all bounded intervals $(a,b]$ which are "closed" on the right. Let $F(x)$ be a nondecreasing function defined on R which is everywhere right continuous; that is,

$$F(b) = F(b + 0) = \lim_{\substack{x \to b \\ x > b}} F(x).$$

Show that $\mathcal{S}$ is a countable covering class. Define $\lambda((a,b]) = F(b) - F(a)$. Show that this gives rise to an outer measure μ^* which satisfies $\mu^*((a,b]) = \lambda((a,b])$. What are $\mu^*((a,b))$, $\mu^*([a,b))$ and $\mu^*([a,b])$?

The measure induced by λ is called a *Lebesgue-Stieltjes measure* (see Sec. 2, in particular [3]).

*Exercises which request the proof of a statement in the text, as well as those referred to in the text, are starred.

2 Lebesgue-Stieltjes Measures and Integrals

It has already been remarked that the Lebesgue measure is a generalization of the Riemann measure, in the sense that it assigns a measure to sets in a family which includes the Riemann measurable sets as a proper subfamily. It is an extension of the "natural" measure on the real line, which assigns to each interval its length as its measure. Subsequently, new measures were assigned to intervals by means of a variable "weighting" function (Stieltjes measures). These measures may be given the following physical interpretations. If

[1]
$$f(x) = \begin{cases} 0 & \text{if } x < -2 \\ x + 2 & \text{if } -2 \leq x \leq 4 \\ 0 & \text{if } x > 4, \end{cases}$$

then for $a < b$, $\mu((a,b)) = \int_a^b f(x)\, dx$ is a nonnegative, countably additive function defined on the class of all bounded intervals. It may be extended to a measure defined on the class of Borel sets, the family of Riemann measurable sets, or even the family of Lebesgue measurable sets by letting $\mu(S) = \int_S f$; the integral is the Lebesgue integral, or in the case of the Riemann measurable sets, it may be interpreted as a Riemann integral. The function $f(x)$, or more properly one of its indefinite integrals $F(x)$, may be thought of as describing a mass distribution along the x-axis. Here, no mass is located in either of the unbounded intervals $(-\infty,2)$ and $(4,\infty)$. The distribution on the interval $[-2,4]$ is continuous, very "sparse" near the end point $x = -2$ and increasing in density as the end point $x = 4$ is approached. Indeed, the respective measures of the intervals $[-2,0]$ and $[2,4]$, both having identical lengths, are 2 and 12 respectively. In Sec. 4 of Chapter 11 we considered such generalizations of Riemann measure which did *not* enlarge the class of measurable sets, but changed the *values* of the measure function defined on this class in such a way that the measure of an interval was no longer necessarily its length. The Lebesgue-Stieltjes measures both enlarge the class of (Riemann) measurable sets (as does the Lebesgue measure but not the Stieltjes measure), and change the values assigned to these sets (as does the Stieltjes but not the Lebesgue measure).

1 Definition. A nondecreasing function $F(x)$, defined on R, which is continuous from the right at all points, that is

$$F(b) = \lim_{x \to b+0} F(x) = F(b+0),$$

is called a *distribution function.*

An indefinite integral of a nonnegative function (such as $f(x)$, defined by [1]), is a distribution function.

Remark. There is often a finiteness condition imposed upon the distribution function. For example, whenever these functions appear in probability theory (see Munroe [1]), it is assumed that $\lim_{x \to -\infty} F(x) = 0$ and $\lim_{x \to +\infty} F(x) = 1$. Such a function describes an "event" in the space R, by saying that the probability that it occurs in the interval $(a,b]$ is $F(b) - F(a)$. The boundedness conditions insure that the probability of the event in the entire space R is equal to one.

More generally, it is sometimes assumed only that $F(x)$ is bounded. For our purposes this assumption is not necessary.

The relationship between distribution functions and measures is discussed below. It will be shown first that a measure defined on a σ-algebra containing the class $\mathcal{B}$ of Borel sets, which assigns finite values to all bounded sets, determines a distribution function, or more precisely a family of distribution functions, the difference between any pair of them being a constant. (This suggests the corresponding well known result for primitives or indefinite integrals of a given function. Just how far this analogy may be carried will be seen at the end of the section, where absolutely continuous distribution functions are discussed; in Sec. 4 a more general notion of absolute continuity defined for abstract measures is introduced, and the question of representing one measure in terms of another is settled.) Following this, it will be shown that every distribution function gives rise to a complete measure defined on a σ-algebra that contains $\mathcal{B}$.

Let μ be a measure defined on a σ-algebra which contains $\mathcal{B}$, and which assigns a finite value to each bounded set. For each real number c the real function

[2] $$F_c(x) = \begin{cases} -\mu((x,c]) & \text{if } x < c \\ 0 & \text{if } x = c \\ \mu((c,x]) & \text{if } c < x \end{cases}$$

is nondecreasing and continuous from the right (Exercise 6), and therefore a distribution function. If $a < b$ then the difference $F_b(x) - F_a(x) = \mu((a,b])$ is constant for all x in R, thus proving that a measure function determines a distribution function up to an additive constant. For convenience, a fixed but arbitrary point c may be chosen, and henceforth the distribution function $F_c(\mathrm{x})$, which shall be simply denoted by $F(x)$, may be regarded as *the* distribution function that corresponds to the given measure μ.

Let us suppose now that a distribution function $F(x)$ is given. The question that naturally arises is: Does there exist a measure μ, defined on a σ-algebra containing $\mathcal{B}$, that gives rise to $F(x)$, as in [2]? The answer is yes, and the construction is a special case of the general method described in Sec. 1. We shall begin by defining μ on the family of intervals of the type $(a,b]$, (a,∞) and $(-\infty,b]$. (If left continuity of $F(x)$ were required instead of right continuity, the intervals would be of the type $[a,b)$, $[a,\infty)$ and $(-\infty,b)$. See Exercise 1.) Since the family of these intervals is a semialgebra, the results of Sec. 1 will yield an extension of this function to a complete measure defined on a σ-algebra containing $\mathcal{B}$, provided that the function defined on the semialgebra is finitely additive and countably subadditive (see Exercise 13.1.4).

The proper definition of μ on this class of intervals is suggested by the relation $\mu((a,b]) = F(b) - F(a)$, which is known to hold if $F(x)$ is the distribution function determined by a given measure (see [2]). Thus, if $F(x)$ is given, we *define*

[3] $$\mu((a,b]) = F(b) - F(a)$$

for all bounded intervals $(a,b]$. Setting $F(-\infty) = \lim_{x\to-\infty} F(x)$ and $F(\infty) = \lim_{x\to+\infty} F(x)$, μ is extended to the unbounded intervals by defining

$$\mu((-\infty,b]) = F(b) - F(-\infty), \quad \mu((a,\infty)) = F(\infty) - F(a).$$

In order to be able to extend μ to a measure function on the algebra $\mathcal{A}$

which is generated by the semialgebra of intervals, the conditions of Exercise 13.1.4 must be verified: If $(a,b]$ is the disjoint union $\bigcup_{i=1}^{n}(a_i, b_i]$, where it is assumed that the interval $(a_i,b_i]$ lies to the left of $(a_j,b_j]$ whenever $i < j$, then it is easily seen that $a_1 = a$, $a_2 = b_1$, $a_3 = b_2, \ldots,$ $a_n = b_{n-1}$, $b_n = b$; and [3] implies that

$$\sum_{i=1}^{n} \mu((a_i,b_i]) = [F(b_n) - F(a_n)] + [F(b_{n-1}) - F(a_{n-1})] + \cdots + [F(b_1) - F(a_1)] = F(b) - F(a) = \mu((a,b]).$$

A similar argument holds for $(-\infty,b]$ and (a,∞).

It is left to the reader to show that if $(a,b]$ is the disjoint *countable* union $\bigcup(a_i,b_i]$, then $\mu((a,b]) \leq \Sigma\mu((a_i,b_i])$ (Exercise 7). Thus the conditions of Exercise 13.1.4 are satisfied and μ may be extended to the algebra generated by the family of intervals, and subsequently to an outer measure. The complete measure (also denoted by μ) induced by the outer measure (see Sec. 1), is called the *Lebesgue-Stieltjes measure corresponding to* or *induced by the distribution function* $F(x)$. Several examples appear in the exercises, and there the reader is asked to give physical interpretations of the following distribution functions:

(a) $F(x)$ is everywhere constant.

(b) $F(x)$ is continuous on an interval $[a,b]$ and constant on $(-\infty,a)$ and (b,∞).

(c) There is a pair of points $a < b$ such that $\mu((a,b)) < F(b) - F(a) < \mu([a,b])$.

(d) $F(x)$ has a single discontinuity at the point $x = a$. (In particular, what is $\mu(\{a\})$?)

(e) There are points $a < b$ and $c < d$ such that $\mu([a,b)) < F(b) - F(a)$ and $\mu([c,d) > F(d) - F(c)$.

The relationship between the Borel sets and the sets which are measurable with respect to a Lebesgue Stieltjes measure μ induced by a distribution function, is similar to the relationship between the Borel sets and the Lebesgue measurable sets. Theorem 7.3.3 asserts that every bounded (Lebesgue) measurable set S is contained in a Borel set, in particular a G_δ set (a countable intersection of open sets) S', and $m(S) = m(S')$. An immediate corollary to this theorem is the assertion that S contains an F_σ set (a countable union of closed sets) whose measure also coincides with that of S. A similar result may be obtained for Lebesgue-Stieltjes measurable sets (Theorem 2) as a special case of the more general theorem proved in the preceding section (Theorem 13.1.5).

2 Theorem. Let μ be the Lebesgue-Stieltjes measure determined by the distribution function $F(x)$. If S is μ-measurable, then there is a Borel set B containing S and $\mu(S) = \mu(B)$. Moreover, S contains a Borel set B' and $\mu(S) = \mu(B')$.

The reader is asked in the exercises to give a direct proof of this theorem, that is, one which is independent of the more general result, Theorem 13.1.5.

Let μ be the complete Lebesgue-Stieltjes measure that corresponds to a distribution function $F(x)$. According to Sec. 5 of Chapter 12, this complete measure determines classes of measurable and integrable functions and a μ-integral $\int f\,d\mu$, which is called the *Lebesgue-Stieltjes integral* with respect to μ (or F). It is also proved there that if S is a finitely measurable set, then $\mu(S) = \int \chi_S\,d\mu$. To indicate the dependency upon the distribution function, the Lebesgue-Stieltjes integral is generally denoted by $\int f\,dF$, or as we shall write here in order to distinguish it from the Stieltjes integral (Definition 11.4.1), by $\mathbb{L}\!\!\int f\,dF$.

A slight modification of Definition 12.5.19 yields the following definition of the Lebesgue-Stieltjes integral. It is left to the reader to show that Definition 12.5.9 and Definition 3 yield the same class of μ-integrable functions and that their intergrals are the same.

3 Definition. Let μ be the Lebesgue-Stieltjes measure corresponding to a distribution function $F(x)$; T is a finitely μ-measurable set and f is a bounded function. The collection $\{T_1, T_2, \ldots, T_n\}$ of measurable sets is said to be a *partition* of T if the T_i are mutually disjoint and their union is T. Setting

$$M_i = \sup_{x \in T_i} f(x), \qquad m_i = \inf_{x \in T_i} f(x),$$

the sums

$$\overline{S}(f,F,T_i) = \sum_{i=1}^{n} M_i \mu(T_i), \qquad \underline{S}(f,F,T_i) = \sum_{i=1}^{n} m_i \mu(T_i)$$

are called the *upper and lower Lebesgue-Stieltjes sums of f with respect to F and the partition $\{T_i\}$. The upper and lower Lebesgue-Stieltjes integrals of f with respect to F over the set T* are respectively $\inf_{\{T_i\}} \overline{S}(f,F,T)$ and $\sup_{\{T_i\}} \underline{S}(f,F,T)$, the least upper bound and greatest lower bound being taken over all partitions of T. The function f is said to be *Lebesgue-Stieltjes integrable with respect to F over T* if the upper and lower integrals are equal, and in this case their common value is called the *Lebesgue-Stieltjes integral of f with respect to F over T.*

It is easily seen that if f is a bounded μ-measurable function then the integral exists (cf. Sec. 2 of Chapter 8, in particular Theorem 8.1.12). It can be extended to unbounded functions and to sets having infinite measure in the usual way, by taking truncations.

Drawing the analogy from earlier remarks about the Lebesgue integral being an extension of the Riemann integral, we should like to make a similar statement about the Lebesgue-Stieltjes integrals and the Stieltjes integrals. Indeed it is easy to see that if f is continuous and F any distribution function, then the Stieltjes integral exists and is equal to the Lebesgue-Stieltjes integral. However, it has been proved that the Stieltjes integral of a continuous function exists with respect to *any* monotone function—actually with respect to any function of bounded variation. For this reason we shall extend the Lebesgue-Stieltjes integral in such a way that it is defined for all functions of bounded variation. Since each function of bounded variation is the difference of two monotone functions, it suffices to extend the integral for monotone functions. The general result is obtained by using the linearity of the integral.

If $F(x)$ is any nondecreasing function (not necessarily right continuous), then the distribution function

$$\hat{F}(x) = \begin{cases} F(x) & \text{if } x \text{ is a point of continuity of } F(x) \\ F(x+0) & \text{if } x \text{ is a point of discontinuity of } F(x) \end{cases}$$

differs from $F(x)$ only at points of discontinuity (a countable set according to Theorem 9.3.3). We therefore define the *Lebesgue-Stieltjes integral of f over T with respect to F*, written $\mathcal{L}\int_T f\,dF$, to be simply $\mathcal{L}\int_T f\,d\hat{F}$, where $\hat{F}$ is the distribution function that is equal to F except possibly at points of discontinuity.

Remark. If $F = G - H$ is a function of bounded variation written as the difference of monotone functions, then the set function $\mu(S) = \mathcal{L}\int \chi_S\,dF$ is not generally a measure function since there may be sets on which it takes on negative values. Such set functions, called *signed measures*, are discussed in Sec. 3.

We conclude this section with a brief discussion of absolutely continuous distribution functions (Definition 9.4.8). The reader will recall that if F is a function of bounded variation, then the derivative F' exists a.e. and $\int_a^b F' \leq F(b) - F(a)$. This becomes an equality if F is absolutely continuous (Theorem 9.4.12). The following analogous theorem will be proved here:

4 Theorem. Let F be an absolutely continuous distribution function and S a Lebesgue measurable set. If the function f is either

Lebesgue-Stieltjes integrable over S with respect to F, or if fF' is Lebesgue integrable over S, then both integrals exist and are equal.

Remark. Once Theorem 4 is proved, the linearity of the integral implies that it holds for all absolutely continuous (not necessarily monotone) functions.

The following lemmas are needed to prove Theorem 4.

5 Lemma. If F is an absolutely continuous distribution function and S is a set which can be written as the disjoint countable union $\bigcup I_n$ of intervals, then

[4] $$L\!\!\int_S dF = L\!\!\int \chi_S\, dF = \int_S F'.$$

Proof: If S is itself a bounded interval, then Theorem 9.4.12 yields $\int_a^b F' = F(b) - F(a)$. The continuity of F and the definition of μ (see [3]) imply that $F(b) - F(a)$ is equal to $\mu((a,b])$, $\mu((a,b))$, $\mu([a,b))$ and $\mu([a,b])$, which in turn is simply $\int \chi_S\, dF$. The countable additivity of the integral yields the more general case. ∎

6 Lemma. Let μ denote the Lebesgue-Stieltjes measure that corresponds to the absolutely continuous distribution function F. If S is Lebesgue measurable and $m(S) = 0$, (m denotes the Lebesgue measure), then S is Lebesgue-Stieltjes measurable and $\mu(S) = 0$. Conversely if $\mu(S) = 0$, then $F'(x) = 0$ a.e. (in the Lebesgue measure) on S.

Proof: Assume first that S is a bounded set contained in the open interval (a,b). The absolute continuity of F yields the integrability of F' over (a,b) and hence over all (Lebesgue) measurable subsets of (a,b). The condition $m(S) = 0$ implies that there exists a sequence of open subsets T_n of (a,b) such that $S \subset T_n$ and $m(T_n) \leq 1/n$. The monotonicity of the integral and Lemma 5 imply that

$$\mu(S) = L\!\!\int \chi_S\, dF \leq L\!\!\int \chi_{T_n} dF = \int_{T_n} F' \to 0.$$

Conversely, if $\mu(S) = 0$, Theorem 2 implies that there is a sequence of open sets T_n each of which contains S and $\mu(T_n) \leq 1/n$. Setting $T = \bigcap_{n=1}^{\infty} T_n$, the monotonicity of the integral and Lemma 5 yield

$$\int_T F' \leq \int_{T_n} F' = L\!\!\int \chi_{T_n} dF = \mu(T_n) \leq 1/n$$

for all positive integral values of n, from which it follows that $\int_T F' = 0$.

Therefore $F'(x) = 0$ a.e. on T, and since S is a subset of T, $F'(x) = 0$ a.e. on S. ∎

7 Lemma. If F is an absolutely continuous distribution function and S a Lebesgue measurable set, then $\mathcal{L}\!\!\!\int_S dF = \int_S F'$. (This extends Lemma 5 to general measurable sets.)

Proof: It is easily verified that every Lebesgue measurable set is also μ-measurable (Exercise 9). Let us assume that S is contained in the bounded interval (a,b). For unbounded sets, a similar result is obtained by taking truncations (Exercise 10).

Let $\{T_n\}$ be a sequence of open subsets of (a,b) which contain S and $m(T_n - S) \leq 1/n$. The intersection $T = \bigcap T_n$ contains S, and Lemma 6 implies that $\mu(T - S) = m(T - S) = 0$. Since Lemma 5 is applicable to the sets T_n, taking limits yields

$$\mathcal{L}\!\!\!\int_S dF = \mu(S) = \mu(T) = \lim_{n\to\infty} \mathcal{L}\!\!\!\int_{T_n} dF$$

$$= \lim_{n\to\infty} \int_{T_n} F' = \int_T F' = \int_S F'. \quad ∎$$

Remark. Lemma 7 and Exercise 9 imply that if S is a Lebesgue measurable set and F is an absolutely continuous distribution function, then S is μ-measurable and its μ-measure may be expressed as a Lebesgue integral; that is, $\mu(S) = \int_S F'$, where F' is the a.e. derivative of the distribution function that corresponds to μ. (Why is this false if it is assumed only that F is of bounded variation? See Example 9.4.2.) The integral representation of the Lebesgue-Stieltjes measure by a Lebesgue integral is a special case of the more general Radon-Nikodym theorem, which is proved in Sec. 4. There, a relation called *absolute continuity* is defined for two measures having the same class of measurable sets, and it is shown that this condition is both necessary and sufficient to insure that one of these measures be represented as an integral with respect to the other. The abstract version of Theorem 4, also known as the Radon-Nikodym theorem, is proved there too (Theorem 13.4.6).

Although every Lebesgue measurable set (or function) is also Lebesgue-Stieltjes measurable (Exercise 9), the converse is easily seen to be false. A trivial example is provided by the constant distribution function which yields a measure on the class of all subsets of R. This rather uninteresting measure assigns the value 0 to each subset of R. However, it suggests less trivial examples. Since $\mu(S) = \int_S F'$ whenever the set S is also Lebesgue measurable, the idea is to define a distribution function having the property that the (Lebesgue measurable) set $E = [F' = 0]$ contains a nonmeasurable (Lebesgue) set C. Then since $\mu(E) = 0$, it follows that C is μ-measurable and $\mu(C) = 0$. The reader is asked to give examples of such measures (Exercise 11). The following lemma asserts that in general, μ-measurable sets differ from Lebesgue measurable sets by subsets of E:

8 Lemma. Let F be an absolutely continuous distribution function with corresponding measure μ. If S is any μ-measurable set, then $S - E$, where $E = [F' = 0]$, is Lebesgue measurable.

The proof is left to the reader (Exercise 12).

The final lemma needed for the proof of Theorem 4 asserts the equivalence of the Lebesgue measurability of fF' with the μ-measurability of f:

9 Lemma. Let f be a function defined on a Lebesgue measurable set S. If F is an absolutely continuous distribution function with corresponding μ-measure, then f is μ-measurable if and only if fF' is Lebesgue measurable.

Proof: Let us assume first that f is μ-measurable. Since S is μ-measurable, Lemma 8 implies that the set $S' = S - E$, where $E = [F' = 0]$, is both Lebesgue measurable and μ-measurable. The Lebesgue measurability of F' is known (Theorem 9.3.6) and therefore it suffices to prove the Lebesgue measurability of f. This is done by showing that the sets $S_A = [f < A]$ are Lebesgue measurable for all real values of A. Let $S'_A = S_A \cap S' = S_A - E$. The Lebesgue measurability of E reduces the problem to showing that S'_A is Lebesgue measurable. This follows easily from the μ-measurability of f, which implies the μ-measurability of the sets S_A and S'_A. From Lemma 8 it follows that $S'_A = S'_A - E$ is Lebesgue measurable.

The proof of the converse is left to the reader (Exercise 13). ∎

We return to the proof of Theorem 4. It suffices to prove the theorem for nonnegative f; the linearity of the integral implies the more general result.

Lemma 9 asserts that the integrability (Lebesgue or μ) of either fF' or f respectively, implies the measurability (μ or Lebesgue) of the other. The monotone convergence theorem for integrals will yield the desired integrability.

Let us assume that f is μ-integrable. The converse is proved in a similar way and is therefore left to the reader in the exercises. Since f is nonnegative, there is a nondecreasing sequence $\{\sigma_n\}$ of generalized step functions which converges monotonically to f a.e. (μ); that is,

$$\sigma_n = \sum_{i=1}^{m(n)} a_{i,n} \chi_{S_{i,n}},$$

where

$$S = \bigcup_{i=1}^{m(n)} S_{i,n},$$

the disjoint union of μ-measurable sets. Then the sequence $\{\sigma_n F'\}$ also converges a.e. (μ) to fF'. Lemma 7, the linearity of the integral, the Lebesgue measurability of the sets $S - E$ and $S_{i,n}$, and $\mu(E) = 0$ imply

$$[5] \qquad \mathcal{L}\!\!\int_S \sigma_n \, dF = \mathcal{L}\!\!\int_{S-E} \sigma_n \, dF = \sum_{i=1}^{m(n)} a_{i,n} \, \mathcal{L}\!\!\int \chi_{S_{i,n}} \, dF$$

$$= \sum_{i=1}^{m(n)} a_{i,n} \int \chi_{S_{i,n}-E} F' = \sum_{i=1}^{m(n)} a_{i,n} \int \chi_{S_{i,n}} F' = \int_S \sigma_n F'.$$

The known convergence of the monotone sequence of integrals on the left implies that the monotone sequence of integrals $\int_S \sigma_n F'$ (on the right) converges also to $\mathcal{L}\!\!\int_S f \, dF$, and it follows from the theorem on monotone convergence (for Lebesgue integrals) that the limit function fF' is Lebesgue integrable, and therefore

$$\mathcal{L}\!\!\int_S f \, dF = \lim_{n \to \infty} \int_S \sigma_n F' = \int_S fF'. \quad \blacksquare$$

Exercises

1. Show that a distribution function may be defined as a *left* continuous non-decreasing function, and that in this case the semialgebra generating the measurable sets is the family of intervals of the type $[a,b)$, $[a,\infty)$ and $(-\infty,b)$. Why must either right or left continuity be assumed? Show that if neither is assumed, then the quantity $F(b) - F(a)$ can not be said to be the measure of an interval.
2. Use the idea of a countable covering class to construct a Lebesgue-Stieltjes measure (See Exercise 13.1.16).

*3. Show that if μ is the measure that corresponds to a distribution function F, then $\mu((a,b)) = F(b-0) - F(a)$, $\mu([a,b)) = F(b-0) - F(a+0)$, $\mu([a,b]) = F(b) - F(a-0)$ and $\mu(\{a\}) = \lim_{x \to a} \mu((x,a]) = F(a) - F(a-0)$.

4. Describe mass distributions that correspond to the following distribution functions:
   (a) $F(x) \equiv 0$.
   (b) The set of discontinuities of $F(x)$ is the finite set $\{a_1, a_2, \ldots, a_n\}$. In particular, what can you say about $\mu(\{a_i\})$?
   (c) $$F(x) = \begin{cases} 0 \text{ on the interval } (-\infty,a) \\ \text{continuous on the interval } [a,b] \\ \text{constant on the interval } (b,\infty) \end{cases}$$
   Discuss the two cases, $F(a) = 0$ and $F(a) \neq 0$.
5. Find distribution functions that have the following properties:
   (a) There are real numbers $a < b$ satisfying

$$\mu((a,b)) < F(b) - F(a) < \mu([a,b]).$$

(b) There are real numbers $a < b$ and $c < d$ satisfying

$$\mu([a,b)) < F(b) - F(a) \quad \text{and} \quad \mu([c,d)) > F(d) - F(c)$$

6. Prove that the function $F(x)$ defined by [2] is a distribution function.
7. Show that the set function μ defined by [3] is countably additive; that is, if an interval $S = (a,b]$ in the semialgebra can be expressed as the countable union of intervals $S_i = (a_i,b_i]$, then $\mu(S) \le \Sigma\mu(S_i)$. This is equivalent to proving that if $F(x)$ is a nondecreasing function that is continuous from the right, then $F(b) - F(a) \le \Sigma[F(b_i) - F(a_i)]$. (See the remarks following [3] for definitions of $F(\infty)$ and $F(-\infty)$.)
8. Give a proof of Theorem 2 that is independent of Theorem 13.1.5.
*9. Let F be an absolutely continuous distribution function whose corresponding measure is denoted by μ. Prove that if S is a Lebesgue measurable set (f is a Lebesgue measurable function) then S is μ-measurable (f is μ-measurable).
*10. Prove Lemma 7 for unbounded sets.
*11. Let μ be the Lebesgue-Stieltjes measure of an absolutely continuous distribution function. Show that the μ-measurability of a set (or a function) does not imply the Lebesgue measurability of the set (or function). See the discussion that precedes Lemma 8.
*12. Prove Lemma 8.
*13. Complete the proof of Lemma 9.
*14. Complete the proof of Theorem 4 by showing that the Lebesgue integrability of fF' implies the μ-integrability of f.

*Exercises which request the proof of a statement made in the text, as well as those referred to in the text, are starred.

3 Signed Measures and the Hahn Decomposition Theorem

One of the conditions imposed upon a measure function is nonnegativity. The justification for this requirement is that measures are generalizations of magnitudes, such as length, area, and volume, each of which is a nonnegative set function. If this condition is removed, and if we consider instead all countably additive functions, bounded either from below or above, which are defined on a σ-algebra $\mathcal{C}$ of subsets of a fixed set, and which map $\emptyset$ into zero, we obtain a wider class of set functions called the *signed measures* (Definition 1). They share many properties with the measure functions, and may in fact be represented as the difference of two measures. To illustrate these remarks, suppose that f is an a.e. (μ) nonnegative, integrable function. Then

$$\nu(S) = \int_S f\,d\mu = \int f\chi_S\,d\mu$$

defines a new measure on $\mathcal{C}$ (Exercise 1). However if f is negative on a set T whose μ-measure is positive, then $\nu(T) < 0$, and therefore it is not a measure. On the other hand, ν is represented by an integral, and therefore inherits its countable additivity. Moreover, it maps the empty set into zero. Thus the integral in this case does not represent a measure on $\mathcal{C}$, but is an example of a signed measure provided that it does not take on both the values $+\infty$ and $-\infty$. By writing $f = f^+ - f^-$, we obtain

$$[1] \qquad \nu(S) = \int_S f^+ \, d\mu - \int_S f^- \, d\mu = \nu^+(S) - \nu^-(S),$$

where both ν^+ and ν^- are readily seen to be measures defined on $\mathcal{C}$. This decomposition of ν into the difference of two measures suggests the following physical interpretation of a signed measure: If f is thought of as a function that describes a charge distribution on the space Ω, then $\nu(S)$, as given by [1], is the total charge on the set S. A negative value for $\nu(S)$ may be interpreted to mean that the negative charge on S exceeds the positive charge by the quantity $|\nu(S)|$. We shall show that if a signed measure ν is given, the space Ω may be decomposed with respect to ν into "positive" and "negative" parts, P and N. That is, Ω may be written as a disjoint union $P \cup N$, where P and N have the following properties: $S \subset P$ implies $\nu(S) \geq 0$, and $S \subset N$ implies $\nu(S) \leq 0$ (Hahn Decomposition Theorem). This decomposition will permit us to prove further that *every* signed measure, including those which do not have an integral representation, may be written as the difference of two measures, in much the same way that every function of bounded variation can be written as the difference of two nonnegative monotone functions.

The Hahn Decomposition Theorem, just described, is the principal result of this section, and is needed in the proof of the Radon-Nikodym Theorem (13.4.6). The Radon-Nikodym Theorem is a generalization to abstract spaces of Theorem 13.2.4 which states conditions under which a Lebesgue-Stieltjes measure and its integrals can be represented by Lebesgue integrals.

1 Definition. Let $\mathcal{C}$ be a σ-algebra of subsets of a nonempty set Ω. An extended real valued function μ defined on $\mathcal{C}$, is said to be a *signed measure* if $\mu(\emptyset) = 0$, μ is countably additive, and μ does not take on both the values $+\infty$ and $-\infty$.

2 Examples. (a) The set function ν defined by the integral [1] is a signed measure if either f^+ or f^- is integrable over Ω. $\mathcal{C}$ is the class of μ-measurable sets.

(b) If μ and ν are measures defined on the same class of "measurable" sets $\mathcal{C}$, then $\mu - \nu$ is a signed measure if either $\mu(\Omega) < \infty$ or $\nu(\Omega) < \infty$.

(c) See Exercise 2.

3 Definition. A measurable set P is said to be *positive with respect to a signed measure* μ, if whenever S is a measurable subset of P, then $\mu(S) \geq 0$.

A measurable set N is called *negative with respect to* μ if each of its measurable subsets has nonpositive signed measure.

A set which is both positive and negative is said to be *μ-null*, or simply null.

It should be emphasized that a set which has positive signed measure is not necessarily a positive set, for it may contain subsets having negative signed measure. Similarly, a set whose signed measure is zero is not necessarily a null set, for it may contain subsets which have positive signed measure as well as subsets having negative signed measure. For example, let us consider the following signed measure defined on the class of Borel sets: $\mu(S) = \int_S f$, where $f(x) = x^3$ if $-12 \leq x < \infty$ and $f(x) \equiv 0$ otherwise. (The integral may be interpreted as the Lebesgue integral.) Let $S = (-1,2)$ and $T = (-1,1)$. Then $\mu(S) > 0$ and $\mu(T) = 0$. However, it is easily seen that S contains the negative subset $(-1,0)$, and T contains the positive subset $(0,1)$ as well as the negative subset $(-1,0)$.

As the preceding example suggests, sets having positive (negative) signed measure must contain positive (negative) subsets (Proposition 4(d)).

4 Proposition. (a) Every subset of a positive (negative) set is positive (negative).

(b) Countable unions and intersections of positive (negative) sets are positive (negative).

(c) S is a null set if and only if each of its measurable subsets has zero signed measure.

(d) If $0 < \mu(S)$, then S contains a positive subset which is not null.

Proof: The proofs of (a) and (c) are left to the reader.

Proof of (*b*)*:* The closure under intersection follows from (a).

Suppose now that $P = \bigcup_{i=1}^{\infty} P_i$, where the P_i are positive. The sets $P'_n = P_n - \left(\bigcup_{i=1}^{n-1} P_i\right)$ are positive since $P'_n \subset P_n$. Furthermore they are disjoint and $\bigcup_{i=1}^{n} P_i = \bigcup_{i=1}^{n} P'_i$ for all n, from which it follows that $P = \bigcup_{i=1}^{\infty} P'_i$. If S is a subset of P, then $S = \bigcup_{i=1}^{\infty} (S \cap P'_i)$, where each $S \cap P'_i$ is a positive set. Since this union is disjoint, the countable additivity of the signed measure yields

$$\mu(S) = \sum_{i=1}^{\infty} \mu(S \cap P'_i) > 0,$$

thus proving that the union P of countably many positive sets is positive.

Proof of (d): If S itself is not positive, then it contains sets having negative (signed) measure. Thus there is a smallest positive integer n_1 for which there is a measurable set S_1 satisfying the inequality, $\mu(S_1) < -1/n_1 < 0$. If $S - S_1$ is a positive set, then this is the set we seek. If not, repeat the argument, removing from $S - S_1$ a set S_2 which satisfies the following inequality: $\mu(S_2) < -1/n_2$, n_2 being the smallest positive integer for which there exists a subset of $S - S_1$ having signed measure less than $-1/n_2$. Proceeding inductively, a set $P = S - \bigcup S_j$ is obtained, the union being finite if for some integer m, the set $S - \bigcup_{j=1}^{m} S_j$ is positive. Let us assume that there is no such integer m, for otherwise the theorem is proved. Then for each j, $\mu(S_j) < -1/n_j$, and since the S_j are mutually disjoint as well as disjoint from P, it follows that

$$\mu(S) = \mu(P \cup (\cup S_j)) = \mu(P) + \Sigma\mu(S_j).$$

The inequalities $\mu(S) > 0$ and $\mu(S_j) < 0$ imply that $\mu(P) > 0$. To prove that P is a positive set, we must show that if T is a subset of P, then $\mu(T) > 0$. This is equivalent to proving that for every $\epsilon > 0$, $\mu(T) \geq -\epsilon$. In order to prove this, we need to show first that $\lim_{k\to\infty} 1/n_k = 0$. This however is implied by the convergence of the series $\Sigma 1/n_k$, which is dominated by the series $\Sigma\,|\,\mu(S_k)\,|$. To verify the convergence of the latter series, two cases must be considered. First, let us suppose that $\mu(S) < \infty$. If the sum diverges to $-\infty$, then it follows that $\mu(P) = \infty$, so that μ is a signed measure which takes on both the values $+\infty$ and $-\infty$, which cannot be (Definition 1). Thus the series converges if $\mu(S) < \infty$ is assumed. On the other hand, if $\mu(S) = \infty$, then a similar argument again yields the convergence of the series of negative terms $\mu(S_k)$, from which it follows that $\lim_{k\to\infty} 1/n_k = 0$. Choose k so that $1/(n_k - 1) < \epsilon$. Then

$$P = S - \bigcup_{j=1}^{\infty} S_j \subset S - \bigcup_{j=1}^{k} S_j = \Sigma_k.$$

If T is any subset of Σ_k, then $\mu(T) > -1/n_k > -\epsilon$. Since every subset of P is a subset of Σ_k, the desired inequality $\mu(T) > -\epsilon$ follows. ▎

5 Lemma. If μ is a signed measure which omits the value $+\infty$, then the set of values of $\mu(S)$, $S \in \mathfrak{C}$, is bounded from above. If the value $-\infty$ is omitted, then this set is bounded from below.

Proof: If $+\infty$ is omitted and the set of values is not bounded from above, then there is a sequence of measurable sets S_n and $\mu(S_n) > n$. It is easily seen that the measurable set $S = \bigcup_{n=1}^{\infty} S_n$ has signed measure equal to ∞.

A similar argument may be given for the remaining case. ▎

6 Theorem (Hahn Decomposition). Let μ be a signed measure defined on a σ-algebra $\mathcal{C}$ of subsets of Ω. Then Ω may be written as a disjoint union $P \cup N$, where P is a positive set with respect to μ, and N is negative.

Proof: We may assume that the signed measure omits the value $+\infty$. Let α be the least upper bound of the set of values $\mu(S)$ as S ranges over the positive subsets of $\mathcal{C}$ (Lemma 5). If $\alpha = 0$, then every measurable set is negative, and the desired decomposition is obtained by setting $P = \emptyset$ and $N = \Omega$.

If $\alpha > 0$, let $\{S_n\}$ be a sequence of positive sets whose signed measures converge to α. Since their union P is positive (Proposition 4(b)), $\mu(P) \leq \alpha$. On the other hand, the positivity of the sets $P - S_n$, and the additivity of the measure, yield $\mu(P) = \mu(P - S_n) + \mu(S_n) \geq \mu(S_n)$ for all n. Letting n tend to infinity gives the inequality $\mu(P) \geq \alpha$, from which $\mu(P) = \alpha$ follows.

The desired decomposition is easily seen to be given by $\Omega = P \cup (\Omega - P) = P \cup N$. Clearly the union is disjoint, and it has already been shown that P is positive. If N is not negative, then it contains a subset whose signed measure is positive, and by virtue of Proposition 4(d) it must contain a positive set P' whose measure is positive. Then $P \cup P'$ is a positive set, and the additivity yields $\mu(P \cup P') = \mu(P) + \mu(P') > \mu(P) = \alpha$, which cannot be since α is the least upper bound of the set of values $\mu(S)$, for positive sets S. ∎

The Hahn decomposition of Ω with respect to a given signed measure is not unique. However, it is readily verified that two such decompositions may differ only by a null set (Exercise 5).

If $P \cup N$ is a Hahn decomposition of Ω with respect to a signed measure μ, then $\mu^+(S) = \mu(S \cap P)$ and $\mu^-(S) = -\mu(S \cap N)$ are measures on the σ-algebra $\mathcal{C}$ (Exercise 7). It is easily seen that $\mu = \mu^+ - \mu^-$, which proves the assertion made earlier that *every signed measure may be written as the difference of two measures*. Although there are in general many ways to write a signed measure as the difference of two measures, this particular representation has a minimality property which is stated below in Proposition 7.

The decomposition of a signed measure into the difference of positive and negative parts suggests the decomposition of a function of bounded variation, in particular an absolutely continuous function, into the difference of two nonnegative monotone functions. The sum of the positive and negative parts of the signed measure, which is denoted by $|\mu|$, is called the *total variation of* μ. Again, interpreting μ as a function which describes a charge distribution on Ω, we consider the related distribution which is obtained by making all charges positive. The measure associated with this distribution is $|\mu|$, since $|\mu|(S) = \mu(S \cap P) + \mu(S \cap N) =$

$\mu^+(S) + \mu^-(S)$. The reader should have no difficulty verifying that

[2] $$|\mu|(S) = \sup_{\{S_i\}} \sum_{j=1}^{n} |\mu(S_j)|,$$

the supremum being taken over all finite, disjoint collections $\{S_1, S_2, \ldots, S_n\}$ of measurable sets whose union is S. This is of course a generalization of the total variation of a real function over an interval (cf. Definition 9.3.1). The following identities may be easily verified:

[3] $$\mu^+(S) = \sup_{\{S_i\}} \sum_{j=1}^{n} \mu^+(S_j), \qquad \mu^-(S) = \sup_{\{S_i\}} \sum_{j=1}^{n} \mu^-(S_j).$$

7 Proposition. Let μ be a signed measure on a σ-algebra $\mathcal{C}$ of subsets of Ω.

(i) If $P \cup N$ is a Hahn decomposition of Ω with respect to μ, then $\mu^+(S)$ and $\mu^-(S)$ are measures on $\mathcal{C}$, and $\mu^+(N) = \mu^-(P) = 0$.

(ii) If μ_1 and μ_2 are a pair of measures on $\mathcal{C}$, and if $\mu = \mu_1 - \mu_2$, then for all $S \in \mathcal{C}$, $\mu^+(S) \leq \mu_1(S)$ and $\mu^-(S) \leq \mu_2(S)$.

The proof of Proposition 7 is left to the reader.

Exercises

***1.** Let $(\Omega, \mathcal{C}, \mu)$ be a measure space, and let f be an a.e. (μ) nonnegative integrable function. Show that $\nu(S) = \int_S f\, d\mu$ is a measure on $\mathcal{C}$.

***2.** (a) How does an absolutely continuous function, bounded either from above or below, give rise to a signed measure?
(b) Why is it necessary to assume that a signed measure does not take on both the values $+\infty$ and $-\infty$?

***3.** Prove (a) and (c) of Proposition 4.

4. Let ν be a signed measure on $\mathcal{C}$, and let S be the disjoint countable union $\bigcup_{n=1}^{\infty} S_n$ of sets in $\mathcal{C}$. Prove that if $|\nu(S)| < \infty$, then $\Sigma |\nu(S_n)| < \infty$. [*Hint:* Use the condition that ν cannot take on both the values $+\infty$ and $-\infty$.]

5. Give an example of a signed measure for which there exist several Hahn decompositions.

Show that if $P \cup N$ and $P' \cup N'$ are Hahn decompositions of a space Ω with respect to the same signed measure, then $P \cap N'$ and $P' \cap N$ are null sets. This says that P and P' differ by at most a null set. Similarly for N and N'.

6. Discuss the relation between the decomposition of a signed measure into the difference of its positive and negative parts, and the decomposition of a function of bounded variation into the difference of two nonnegative, monotone functions.

***7.** (a) Prove that if μ is a signed measure, then $\mu^+(S) = \mu(S \cap P)$ and $\mu^-(S) = -\mu(S \cap N)$ are measures.
(b) Show that $\mu^+(S) = \sup \mu(T)$, where T ranges over the positive subsets of S. State and prove a similar identity for $\mu^-(S)$.

***8.** Verify [2] and [3].

9. Prove Proposition 7. (To verify (ii), use $|\mu(S)| \leq \mu_1(S) + \mu_2(S)$, and [2] and [3].)

10. Let $\mathcal{C}$ be a σ-algebra of subsets of a nonempty set Ω, and let $\mathfrak{F}$ denote the family of finite signed measures on $\mathcal{C}$. (A *finite signed measure* takes on *neither* the value $+\infty$ or $-\infty$.)

(a) Prove that $\mathfrak{F}$ is a real vector space under the operations:

$$(\mu + \nu)(S) = \mu(S) + \nu(S), \quad (a \cdot \mu)(S) = a\mu(S), \quad a \in R.$$

Why is this not true for the class of all signed measures on $\mathcal{C}$?

(b) Show that if μ and ν are in $\mathfrak{F}$, then for all measurable sets S,

$$|\mu + \nu|(S) \leq |\mu|(S) + |\nu|(S).$$

(c) Show that $\|\mu\| = |\mu|(\Omega)$ defines a norm on $\mathfrak{F}$.

(d) Let $\mu \prec \nu$ mean that $\mu(S) \leq \nu(S)$ for all $S \in \mathcal{C}$. Show that "$\prec$" is a partial ordering of $\mathfrak{F}$.

(e) Prove that μ^+ is the least upper bound of μ and $\bar{0}$ ($\bar{0}$ is the additive identity of $\mathfrak{F}$). We shall denote the least upper bound of two elements μ and ν by $\mu \vee \nu$.

What is the greatest lower bound of μ and $\bar{0}$?

(f) Show that $\mathfrak{F}$ is a lattice. [*Hint:* Try $\mu \vee \nu = \mu + (\nu - \mu) \vee \bar{0}$.]

*Exercises which request the proof of a statement made in the text, as well as those referred to in the text, are starred.

4 The Radon-Nikodym Theorem

In Chapter 9 it was proved that a real function $F(x)$ could be written in the form $\int_a^x f(t)\,dt$, if and only if it is absolutely continuous on a closed interval that contains both a and x. A related result appeared in Sec. 2 of this chapter. There the relation between Lebesgue-Stieltjes measures and Lebesgue measure was investigated, and it was shown that a Lebesgue-Stieltjes measure corresponding to an absolutely continuous distribution function $F(x)$ has an integral representation

$$\mu(S) = \mathbb{L}\!\!\int_S dF = \int_S F',$$

whenever S is a Lebesgue measurable set (and therefore also Lebesgue-Stieltjes measurable). More generally, Theorem 13.2.5 asserts that if F is an absolutely continuous distribution function, then for any measurable set S, $\mathbb{L}\!\!\int_S f\,dF = \int_S fF'$; the left side is a Lebesgue-Stieltjes integral, the the right-hand side a Lebesgue integral. In this section, these results are generalized for the abstract case. We shall define the notion of absolute continuity for a pair of measures given on the same σ-algebra of sets (Definition 1). One form of the Radon-Nikodym Theorem asserts that a measure ν has an integral representation in terms of a second measure μ, if ν is absolutely continuous with respect to μ. As in Sec. 2, a corresponding theorem is also proved for μ and ν-integrals (Theorem 6 or 7). Even the idea of a derivative of one measure with respect to second can be introduced here (Exercise 1).

Two proofs of the Radon-Nikodym theorem will be given. The first relies upon the results obtained in Sec. 3, in particular the Hahn decomposition theorem. The second proof uses Hilbert space theory.

At the end of this section, the Radon-Nikodym theorem is used to prove the *Lebesgue Decomposition Theorem* for measures: if μ is a fixed but arbitrary σ-finite measure on a σ-algebra $\mathfrak{C}$, then any other σ-finite measure on $\mathfrak{C}$ may be represented uniquely as the sum of a measure that is absolutely continuous with respect to μ and another which is said to be *singular* with respect to μ. (This decomposition theorem is reminiscent of the Projection Theorem for Hilbert spaces, which asserts that an element of a Hilbert space can be written as the sum of a vector in a fixed closed subspace and one which is orthogonal to that space.) The reader will observe, upon reading Definitions 1 and 9 that absolute continuity describes a dependency of one measure upon another, and mutual singularity, as the name suggests, asserts the independence of two measures from each other.

1 Definition. Let μ and ν be measures on the same σ-algebra of sets. *ν is said to be absolutely continuous with respect to μ*, written $\nu \ll \mu$, if

$$\mu(S) = 0 \qquad \text{implies} \qquad \nu(S) = 0.$$

If μ and ν are signed measures defined on the same σ-algebra of sets, then ν is said to be absolutely continuous with respect to μ, again written $\nu \ll \mu$, if $|\nu| \ll |\mu|$.

2 Example. Let $(\Omega,\mathfrak{C},\mu)$ be a measure space and let f be an a.e. (μ) nonnegative integrable function. Then the integral $\int_S f\,d\mu$ defines a measure $\nu(S)$ on $\mathfrak{C}$, which is absolutely continuous with respect to μ.

The converse of Example 2 is the Radon-Nikodym theorem, which asserts that absolute continuity is a sufficient condition for one measure to have an integral representation with respect to another. Its proof requires the following two lemmas.

3 Lemma. Let $(\Omega,\mathfrak{C},\mu)$ be a measure space and let ν be a signed measure on $\mathfrak{C}$. If both μ and ν are finite, and if $\mathfrak{J}$ denotes the collection of μ-integrable functions f which satisfy

$$\nu(S) \geq \int_S f \, d\mu, \quad \text{for all } S \in \mathfrak{C},$$

then there is a function $f_0 \in \mathfrak{J}$ such that

$$\int_\Omega f_0 \, d\mu = \sup_{f \in \mathfrak{J}} \int_\Omega f \, d\mu.$$

Proof: Hereafter, we shall write simply $\int f \, d\mu$ to denote the integral taken over the entire space Ω.

The finiteness of the signed measure ν implies that $\sup_{f \in \mathfrak{J}} \int f \, d\mu \leq \nu(\Omega) < \infty$. Let α denote this number. Then there is a sequence $\{g_n\}$ of functions in $\mathfrak{J}$ whose integrals converge to α. The desired function f_0 is obtained as the limit of a monotone sequence which is defined as follows: Let

$$f_n = \max\,[g_1, g_2, \ldots, g_n].$$

The sequence is nondecreasing, and the lattice properties of L_1^μ guarantee that the f_n are μ-integrable. The reader should have no difficulty in verifying that the f_n are in $\mathfrak{J}$. This requires showing that for $S \in \mathfrak{C}$ and all n, $\int_S f_n \, d\mu \leq \nu(S)$. The finiteness of the measure ν implies that the integrals of the f_n are uniformly bounded, and the theorem on monotone convergence asserts the existence of an integrable limit function f_0 which satisfies the inequality

$$\int_S f_0 \, d\mu = \lim_{n \to \infty} \int_S f_n \, d\mu \leq \nu(S)$$

for all measurable sets S, including $S = \Omega$. The definition of the sequence $\{f_n\}$ implies that

$$\lim_{n \to \infty} \int f_n \, d\mu \geq \lim_{n \to \infty} \int g_n \, d\mu = \alpha.$$

However, the f_n are in $\mathfrak{J}$, and therefore α is also an upper bound on the integrals of the f_n. Thus, integrating over Ω, we obtain

$$\int f_0 \, d\mu = \lim_{n \to \infty} \int f_n \, d\mu = \alpha = \sup_{f \in \mathfrak{J}} \int f \, d\mu,$$

which is what we set out to prove. ▌

4 Lemma. If ν is a nonzero measure which is absolutely continuous with respect to μ, then there is a positive number ϵ and a measurable set T whose measure is positive which has the following property: If S is a measurable subset of T then $\nu(S) \geq \epsilon\mu(S)$.

Proof: To each nonzero rational number q, $\lambda_q = \nu - (1/q)\mu$ is a signed measure. Let $\Omega = P_q \cup N_q$ be a Hahn decomposition with respect to λ_q. The sets $P_0 = \bigcup_{q\in Q} P_q$ and $N_0 = \bigcap_{q\in Q} N_q$ are disjoint and their union is Ω. Since N_0 is contained in each of the N_q, it follows that for every rational number q,

$$0 \geq \lambda_q(N_0) = \nu(N_0) - (1/q)\mu(N_0).$$

However, ν is also a measure and is therefore nonnegative, and so

$$0 \leq \nu(N_0) \leq (1/q)\mu(N_0),$$

for all rational values of q. Therefore $\nu(N_0) = 0$. The assumption that ν is not the zero measure yields $\nu(P_0) > 0$, from which it follows that for some rational number q, $\nu(P_q) > 0$. This in turn yields $\mu(P_q) > 0$. For, if $\mu(P_q) = 0$, then the absolute continuity would imply $\nu(P_q) = 0$. It is easily seen that $\epsilon = 1/q$ and $T = P_q$ satisfy the conditions stated in the lemma. ∎

5 Theorem (Radon-Nikodym I). Let $(\Omega,\mathfrak{C},\mu)$ and $(\Omega,\mathfrak{C},\nu)$ be σ-finite measure spaces, $\nu \ll \mu$. Then there is an a.e. (μ) nonnegative, measurable function f_0, and

$$\nu(S) = \int_S f_0\, d\mu,$$

for all $S \in \mathfrak{C}$. Moreover, f_0 is a.e. (μ) uniquely determined by μ and ν.

Proof—Case (*i*): First, let us assume that both μ and ν are finite. Choose a function f_0 as described in Lemma 4. Then

[1] $$\nu(S) \geq \int_S f_0\, d\mu, \qquad S \in \mathfrak{C},$$

and if $f \in \mathfrak{I}$ (see the statement of Lemma 3), $\int f\, d\mu \leq \int f_0\, d\mu$. We shall show that [1] is an equality, and is therefore the desired integral representation of $\nu(S)$. Form the new measure

$$\lambda(S) = \nu(S) - \int_S f_0\, d\mu.$$

(Why is λ a measure?) From $\nu \ll \mu$ it follows that $\lambda \ll \mu$. If $\lambda \neq \overline{0}$, then Lemma 4 yields the existence of a set T having positive μ-measure, and a positive number ϵ, such that $\lambda(S) \geq \epsilon\mu(S)$ for all measurable subsets of T. (The finiteness of the measures is used here.) We show next that this leads to a contradiction: Let $h = f_0 + \epsilon\chi_T$. If S is measurable, then

$$\begin{aligned}\int_S h\,d\mu &= \int_S f_0\,d\mu + \epsilon\int_{S\cap T}\chi_T\,d\mu\\ &= \int_S f_0\,d\mu + \epsilon\mu(S\cap T) \leq \int_S f_0\,d\mu + \lambda(S\cap T)\\ &= \int_S f_0\,d\mu + \nu(S\cap T) - \int_{S\cap T} f_0\,d\mu\\ &= \int_{S-(S\cap T)} f_0\,d\mu + \nu(S\cap T)\\ &\leq \nu(S-(S\cap T)) + \nu(S\cap T) = \nu(S),\end{aligned}$$

proving that h is in the family $\mathfrak{J}$ described in Lemma 3. On the other hand,

$$\int h\,d\mu = \int f_0\,d\mu + \epsilon\mu(T) > \int f_0\,d\mu,$$

contradicting the fact that the integral on the right is an upper bound for all integrals of functions in $\mathfrak{J}$. Thus it follows that λ is the zero measure, and that $\nu(S) = \int_S f_0\,d\mu$ for all measurable set S, which completes the proof for finite measures.

Case (ii): If μ and ν are σ-finite measures, then there are nested sequences of measurable sets, $S_n \subset S_{n+1}$ and $T_n \subset T_{n+1}$, satisfying $\bigcup_{n=1}^{\infty} S_n = \bigcup_{n=1}^{\infty} T_n = \Omega$, and $\mu(S_n)$, $\nu(T_n) < \infty$. The sequence of intersections $V_n = S_n \cap T_n$ is also a nested sequence, and its union is Ω. Moreover, both μ and ν are finite on the V_n. Case (i) asserts the existence of a sequence $\{f_n\}$ of a.e. (μ) nonnegative functions with the following properties: each f_n is μ-integrable over V_n, vanishes outside V_n, and $\nu(S) = \int_S f_n\,d\mu$ for all measurable subsets S of V_n. If $n < m$, $V_n \subset V_m$, and and therefore if S is a measurable subset of V_n,

$$\nu(S) = \int_S f_n\,d\mu = \int_S f_m\,d\mu,$$

from which it follows that $f_n = f_m$ a.e. (μ) on V_n. Let

$$g_n = \max\,[f_1, f_2, \ldots, f_n].$$

$\{g_n\}$ is a nondecreasing sequence of measurable, a.e. (μ) nonnegative functions, each g_n vanishing outside V_n. Since $g_n = f_n$ a.e. (μ) (why?), it follows that for measurable subsets S of V_n, $\nu(S) = \int_S g_n\,d\mu$. Let $f_0(x) =$

$\lim_{n\to\infty} g_n(x)$ for all $x \in \Omega$ (the value $f_0(x) = \infty$ is permitted). If S is any finitely ν-measurable set, then

$$\nu(S \cap V_n) = \int_S g_n \, d\mu \uparrow \nu(S),$$

which by the monotone convergence theorem is equal to $\int_S f_0 \, d\mu$. If $\nu(S) = \infty$, then the sequence of integrals diverges to ∞.

The uniqueness of the representation is a simple consequence of integration theory: Suppose that there are two functions f_0 and g_0 having the properties stated in the Theorem. Let $T = [f_0 > g_0]$. Unless $\mu(T) = 0$, we have

$$\int_T (f_0 - g_0) \, d\mu > 0,$$

contradicting the assumption that the integrals of both f_0 and g_0 give the value $\nu(T)$. Similarly for the set $[g_0 > f_0]$, thus proving that $f_0 = g_0$ a.e. (μ). ∎

The Radon-Nikodym theorem remains true if it is assumed only that μ is σ-finite (see Exercises 10 and 11, or Kuller [1] and Royden [1]).

The theorem is also true if ν is a signed measure. This follows easily by writing the signed measure as the difference of two measures, applying Theorem 5 to each of the two measures, and using the linearity of the integral.

The following more general version of the Radon-Nikodym theorem may be proved without any great difficulty as a corollary to Theorem 5, and is left to the reader as an exercise.

6 Theorem (Radon-Nikodym II). Let μ and ν be σ-finite measures defined on a σ-algebra $\mathfrak{C}$ of subsets of Ω, and let $\nu \ll \mu$. Then there is an a.e. (μ) nonnegative, μ-measurable function f_0, for which the following statement is true: If $f \in L_1^\nu$, then $ff_0 \in L_1^\mu$ and

$$\int f \, d\nu = \int f f_0 \, d\mu.$$

We shall give a second proof of this theorem which is independent of Theorem 5 and the results of Sec. 3, in particular the Hahn decomposition theorem. This proof of the Radon-Nikodym theorem uses the Riesz representation theorem for bounded linear functionals on a Hilbert space (Theorem 11.2.7). The Hilbert space which plays an important role in the proof is L_2^λ, the space of functions whose squares are λ-integrable, $\lambda = \mu + \nu$. These three measures appear in the proof, along with their respective L_1 and L_2 spaces. The reader should keep in mind that the measure λ and the various L_2 spaces are introduced solely to make possible the use of the Riesz theorem, which (almost) gives us the desired integral representation of the measure ν.

In the next paragraph we shall outline the procedure for defining the L_p spaces for the abstract case. Everything is exactly like the real case, and only some notational changes are required.

Let $(\Omega,\mathcal{C},\lambda)$ be a measure space, and let C_0^λ, Γ^λ, and L_1^λ denote respectively the class of generalized step functions each of which vanishes outside a finitely measurable set, the monotone limits of sequences in C_0^λ whose integrals are uniformly bounded, and the class of integrable functions (see Chapter 12). Although they were not discussed in Chapter 12, it is of course possible to define the remaining L_p spaces, $1 < p < \infty$ (The space L_∞^λ will not interest us here): A measurable function f is said to be in L_p^λ if $|f|^p$ is integrable. The inequalities of Hölder and Minkowski are proved in the same way as for the special case $\Omega = R$ and $\lambda = m$, and if the measure space is complete, which we shall assume throughout, L_p^λ can be partitioned into equivalence classes of a.e. (λ) equal functions, and then made into a Banach space by setting

$$\|f\|_p = \left(\int |f|^p \, d\lambda\right)^{1/p}.$$

(Again, L_p^λ is used to denote both the space of functions and the space of equivalence classes.) If $p = 2$, we have a Hilbert space whose inner product is given by

$$(f,g) = \int fg \, d\lambda.$$

For this case, the Riesz representation theorem asserts that if F is a bounded linear functional defined on L_2^λ, then there is function α in L_2^λ such that for all $f \in L_2^\lambda$, $F(f) = \int \alpha f \, d\lambda = (\alpha,f)$.

The Radon-Nikodym theorem, which follows, is proved here only for finite measures. The extension to σ-finite measures can be made by means of the procedures described in Theorem 5 and Exercises 10 and 11.

7 Theorem (Radon-Nikodym II). Let μ and ν be finite measures defined on a σ-algebra $\mathcal{C}$ of subsets of Ω. If $\nu \ll \mu$, then there is a nonnegative, μ-integrable function f_0 having the property that if $f \in L_1^\nu$, then $ff_0 \in L_1^\mu$ and

$$\int f \, d\nu = \int ff_0 \, d\mu. \tag{2}$$

The function f_0 is uniquely determined a.e. (μ) by ν.

Proof: Let $\lambda = \mu + \nu$. The class L_1^λ of λ-integrable functions is equal to $L_1^\mu \cap L_1^\nu$.

For the sake of clarity, the statements to be proved in each step below appear in italics.

Step (i): If $f \in L_2^\lambda$, then $f \in L_1^\lambda$, L_1^μ and L_1^ν.

Since all three measures are finite, χ_Ω, the characteristic function of the space, is in L_1^λ, and $\lambda(\Omega) = \int \chi_\Omega \, d\lambda < \infty$. Therefore the Cauchy-Schwarz inequality (or the Hölder inequality for $p = q = 2$) yields the integrability of the function $f = f\chi_\Omega = f(\chi_\Omega)^2$ with respect to the measure λ. The desired result follows from the inclusion $L_1^\lambda \subset L_1^\mu, L_1^\nu$.

Step (ii): The integral $\int f \, d\nu$ defines a bounded, linear functional on L_2^λ.

Step (i) insures that this integral is defined on the space L_2^λ. The linearity and homogeneity are properties of the ν-integral. The boundedness follows from the Cauchy-Schwarz inequality:

$$\left| \int f \, d\nu \right| \leq \|f\|_2 \sqrt{\nu(\Omega)}. \tag{3}$$

The Riesz representation theorem for bounded, linear functionals on a Hilbert space asserts the existence of a function g in L_2^λ which has the following property: For all $f \in L_2^\lambda$

$$F(f) = \int f \, d\nu = \int fg \, d\lambda \tag{4}$$

The function g is a.e. (λ) uniquely determined.

Step (iii): $g \geq 0$ a.e. (λ).

Let S be the (finitely) λ-measurable set on which $g < 0$. Since $\nu(S)$ is also finite, it follows from [4] that

$$0 \leq \nu(S) = \int \chi_S \, d\nu = \int \chi_S \, g \, d\lambda.$$

The definition of S implies that the function $\chi_S g$ is nowhere positive, and so the integral on the right hand side is less than or equal to zero. The nonnegativity of the measure function implies that this integral must vanish, proving that $\chi_S g = 0$ a.e. (λ). However, both χ_S and g are nonzero on the set $S = [g < 0]$, and it follows therefore that S is a λ-null set; that is, $g \geq 0$ a.e. (λ). We shall assume that $g \geq 0$ everywhere.

Step (iv): [4] holds for all functions in L_1^λ.

Since L_2^λ contains the characteristic functions of all measurable sets (sets in $\mathfrak{C}$), it follows that [4] holds for these functions and hence for all generalized step functions. Every λ-integrable function is the difference of functions in Γ^λ, and therefore if [4] is verified for functions in Γ^λ, it may be extended to L_1^λ by using the linearity of the integral. Let f be a function in Γ^λ. Then there is a nondecreasing sequence $\{\sigma_n\}$ of generalized step functions with the following properties: $\sigma_n \uparrow f$ a.e.(λ), and $\int \sigma_n \, d\lambda < A$ for some number A and all positive integers n. The nonnegativity of g implies that $g\sigma_n \uparrow gf$ a.e. (λ), and the inclusion $L_2^\lambda \subset L_1^\lambda = L_1^\mu \cap L_1^\nu$ yields the μ-integrability of the $g\sigma_n$. Moreover, each σ_n is bounded, and therefore $g\sigma_n$ is in L_2^λ. Thus $g\sigma_n$ may be substituded for f in [4]. The monotonicity of the sequence, and the bound on the sequence $\left\{ \int \sigma_n \, d\lambda \right\}$ yield

$$\int g\sigma_n \, d\lambda = \int \sigma_n \, d\nu = \int \sigma_n \, d\lambda - \int \sigma_n \, d\mu < A - \int \sigma_1 \, d\mu. \tag{5}$$

Applying the theorem on monotone convergence to the sequences $\{\sigma_n\}$ and $\{g\sigma_n\}$, we obtain

$$\lim_{n\to\infty} \int g\sigma_n \, d\lambda = \int gf \, d\lambda \qquad \text{and} \qquad \lim_{n\to\infty} \int \sigma_n \, d\nu = \int f \, d\nu,$$

and it follows from [5] that $\int gf \, d\lambda = \int f \, d\nu$. Hence [4] holds for functions in Γ^λ. The linearity of the integral implies that [4] is also true for all λ-integrable functions.

The desired function f_0 is obtained by expanding [4] and iterating: If $f \in L_1^\lambda$, then $gf \in L_1^\lambda$ [step (iv)], and

$$[6] \qquad \int f \, d\nu = \int gf \, d(\mu + \nu) = \int gf \, d\mu + \int gf \, d\nu.$$

The second integral on the right may be further expanded by using [4] again. Replacing f in [4] by gf, we obtain

$$\int gf \, d\mu + \int g^2 f \, d(\mu + \nu) = \int f(g + g^2) \, d\mu + \int g^2 f \, d\nu.$$

The principle of mathematical induction yields, for $n = 1,2,\ldots,$

$$[7] \qquad \int f \, d\nu = \int f(g + g^2 + \cdots + g^n) \, d\mu + \int g^n f \, d\nu.$$

If it can be shown that $0 \leq g < 1$, then it is reasonable to hope that upon allowing n to tend to infinity, the first integral on the right-hand side of [7] will tend to $\int f[g/(1 - g)] \, d\mu$, and the second to zero. Thus $f_0 = g/(1 - g)$ is the function we seek. This heuristic argument is justified below.

Step (v): $o \leq g < 1$ *a.e.* (λ)

Let $S = [g \geq 1]$. Then [4], [6], and [7], with $f = \chi_S$, imply

$$\nu(S) = \int \chi_S \, d\nu = \int (g + g^2 + \cdots + g^n) \, d\mu + \int g^n \, d\nu \geq n\mu(S) + \nu(S),$$

which leads to a contradiction unless $\mu(S) = 0$. However, $\nu \ll \mu$, and therefore $\nu(S) = 0$, from which it follows that $\lambda(S) = 0$ too. For convenience, we shall assume that $0 \leq g < 1$ everywhere.

Step (vi): *The function* $f_0 = g/(1 - g)$ *satisfies* [2].

$f_0 \geq 0$, and since $[f_0 \geq A] = [g \geq (A/A + 1)]$, the λ-measurability of g implies the λ-measurability of f_0 (Theorem 12.4.10 and/or Definition 12.5.5). We must verify [2] for all ν-integrable functions. Suppose first that f is a nonnegative function. Then $\{g^n f\}$ is a nonincreasing sequence of ν-integrable functions which converges everywhere to zero. The continuity of the integral implies that $\lim_{n\to\infty} \int g^n f d\nu = 0$ (see [7]). It remains to be shown that

$$[8] \qquad \lim_{n\to\infty} \int f(g + g^2 + \cdots + g^n) \, d\mu = \int f f_0 \, d\mu.$$

We first verify [8] for nonnegative functions f in L_2^λ: Since f and $(g + g^2 + \cdots + g^n)$ are in L_2^λ, their product is λ-integrable, and therefore

μ-integrable, and the sequence $\varphi_n = f(g + g^2 + \cdots + g^n)$ of μ-integrable functions is nondecreasing and $\int \varphi_n \, d\mu \le \int f \, d\nu$. The theorem on monotone convergence implies that the monotone limit $ff_0 = f[g/(1 - g)]$, of the φ_n, is μ-integrable, and

$$\int \varphi_n \, d\mu \uparrow \int ff_0 \, d\mu.$$

Thus [8] holds for nonnegative functions in L_2^λ. The extension to all functions in L_2^λ is made in the usual way.

To show that [2] holds for all ν-integrable functions, we repeat an argument that appeared earlier in the proof: Since the generalized step functions are in L_2^λ, [2] holds for these functions. By taking monotone limits [cf. step (iv)], [2] may be extended to Γ^ν, and then by linearity to L_1^ν. The details are left to the reader. ▌

Let $(\Omega, \mathcal{C}, \mu)$ be a fixed measure space. Until now, we have focussed our attention only upon measures closely related to, in fact dependent upon μ. These measures, which are absolutely continuous with respect to the given measure, are characterized by the property that they possess an integral representation with respect to the given measure. In particular, the vanishing of the μ-measure of a set S implies the vanishing of the ν-measure of S whenever the latter is absolutely continuous with respect to the former. In the case of the Lebesgue-Stieltjes measures, this situation could be described as follows: If ν is an absolutely continuous measure (with respect to the Lebesgue measure m), then the mass distribution to which it corresponds is given by an integral of a "mass density" function f, which is a.e. equal to the derivative of the absolutely continuous distribution function F which generates ν. From $\nu \ll m$ it follows in particular that the ν-measure of a (Lebesgue) null set is zero if ν is an absolutely continuous Lebesgue-Stieltjes measure. However, measures (on R) which assign non-zero values to null sets are of interest. More generally, we shall want to consider ν-measures on $\mathcal{C}$ which assign non-zero values to μ-null sets. This leads quite naturally to the concept on *mutually singular measures*. Roughly speaking, mutually singular measures act upon disjoint subsets of the space Ω; that is, the null sets of one contain the sets whose measures with respect to the other are not zero. An example of such a pair of measures appeared in Sec. 3: Let $\Omega = P \cup N$ be a Hahn decomposition with respect to a signed measure λ (Theorem 13.3.6). Then $\lambda = \lambda^+ - \lambda^-$, where $\lambda^+(S) = \lambda(S \cap P)$ and $\lambda^-(S) = -\lambda(S \cap N)$. It is easily seen that λ^+ and λ^- are mutually singular measures since λ^+ vanishes on the set N and all its subsets, and λ^- vanishes on P and all its subsets. The formal definition and discussion of mutually singular measures follow:

8 Definition. Two measures μ and ν defined on the same class $\mathcal{C}$ of measurable subsets of Ω are said to be *mutually singular* if there are disjoint measurable sets S and T whose union is Ω, and $\mu(S) = \nu(T) = 0$. The relation is denoted by $\mu \perp \nu$.

Although singularity is a symmetric relation, it is sometimes said that *μ is singular with respect to ν*, or vice versa.

9 Examples. (a) If λ is a signed measure, then the measures λ^+ and λ^- are mutually singular as observed above. The reader is asked to show that they are the only pair of mutually singular measures whose difference is the given signed measure.

(b) If μ is a Lebesgue-Stieltjes measure that corresponds to a distribution function F, then μ is singular with respect to m if and only if $F' = 0$ a.e. (Exercise 17).

(c) To illustrate part (b), let $F(x)$ be the Cantor function (Example 9.4.2) on the interval $[0,1]$, $F(x) \equiv 0$ on $(-\infty,0)$ and $F(x) \equiv 1$ on $(1,\infty)$. Then $F'(x) = 0$ except on the Cantor set, which is a (Lebesgue) null set. Letting μ denote the measure generated by this distribution function, the only sets having μ-measure not equal to zero, are (certain) subsets of the Cantor set.

An important consequence of the Radon-Nikodyn theorem is the Lebesgue decomposition theorem for measures, whose proof follows. It asserts that if ν is a second σ-finite measure defined on a σ-finite measure space $(\Omega,\mathcal{C},\mu)$, then it can be represented by the sum of a measure that is absolutely continuous with respect to μ and one which is singular with respect to μ. As observed in the introduction, this suggests the Projection Theorem for Hilbert spaces, which asserts the decomposition of any element into the sum of a vector in a fixed closed subspace, and one which is orthogonal to that space. Even more closely related, is the decomposition of a given function of bounded variation into the difference $g - h$, where g is absolutely continuous and $h' = 0$ a.e. (Exercise 9.4.11).

10 Theorem (Lebesgue decomposition of measures). Let μ and ν be σ-finite measures on $\mathcal{C}$. Then there are measures ξ and ζ such that

(i) $\nu = \xi + \zeta$, (ii) $\xi \perp \mu$, and (iii) $\zeta \ll \mu$.

Moreover, these measures are uniquely determined by μ and ν.

Proof: The sum $\lambda = \mu + \nu$ is a σ-finite measure, and $\lambda(S) = 0$ implies that $\mu(S) = \nu(S) = 0$, so that both μ and ν are absolutely continuous with respect to λ. The Radon-Nikodym theorem yields the existence of two a.e. (λ) nonnegative, λ-measurable functions f and g, and

$$\mu(S) = \int_S f\,d\lambda, \qquad \nu(S = \int_S g\,d\lambda,$$

for all measurable sets S. We may assume that f is nowhere negative. Then

$$\Omega = [f = 0] \cup [f > 0] = U \cup V,$$

where U and V are disjoint measurable sets. The desired measures are defined as follows:

$$\xi(S) = \nu(S \cap U), \qquad \zeta(S) = \nu(S \cap V).$$

From

$$\xi(V) = \xi(V \cap U) = \xi(\phi) = 0$$

and

$$\mu(U) = \int_{[f=0]} f\,d\lambda = 0$$

it follows that $\xi \perp \mu$, which is (ii). Furthermore, if S is any measurable set, then $S = (S \cap U) \cup (S \cap V)$, and $\nu(S) = \nu(S \cap U) + \nu(S \cap V) = \xi(S) + \zeta(S)$, proving (i). It remains to be proved that $\zeta \ll \mu$: If $\mu(S) = 0$, then $\int_S f\,d\lambda = 0$, and it follows that $f = 0$ a.e. (λ) on S. The definition of the set V yields $\lambda(S \cap V) = 0$, and $\nu \ll \lambda$ implies that $\nu(S \cap V) = 0$. However, $\zeta(S) = \nu(S \cap V) = 0$ and (iii) is proved.

The proof of the uniqueness is left to the reader. ∎

Exercises

1. It is assumed that each of the measures that appear in these exercises is defined on the same class $\mathcal{C}$ of measurable sets. The function f_0 of Theorems 5, 6, and 7 is called the Radon-Nikodym derivative of ν with respect to μ, and is denoted by $d\nu/d\mu$.

(a) Show that if a and b are positive numbers, then

$$\frac{d(a\nu + b\lambda)}{d\mu} = a\frac{d\nu}{d\mu} + b\frac{d\lambda}{d\mu}.$$

(It is assumed that $\nu \ll \mu$ and $\lambda \ll \mu$.)

(b) Prove that $\nu \ll \mu$ and $\mu \ll \lambda$ imply $\nu \ll \lambda$, and that

$$\frac{d\nu}{d\lambda} = \left(\frac{d\nu}{d\mu}\right)\left(\frac{d\mu}{d\lambda}\right) \qquad \text{(Chain rule).}$$

(c) Prove that if $\nu \ll \mu$ and if $\mu \ll \nu$, then

$$\left(\frac{d\nu}{d\mu}\right)^{-1} = \left(\frac{d\mu}{d\nu}\right).$$

2. Let μ be a fixed signed measure.

(a) Show that the collection $\mathcal{S}$ of signed measures which are singular with respect to μ is a real vector space.

(b) Show that the collection $\overline{\mathcal{S}}$ of signed measures which are absolutely continuous with respect to μ is a real vector space.

(c) Show that $\mathcal{S} \cap \overline{\mathcal{S}}$ contains only the zero measure.

***3.** Prove that if λ is a signed measure which can be written as the difference $\mu - \nu$ of mutually singular measures, then $\lambda^+ = \mu$ and $\lambda^- = \nu$.

4. Let μ, ν, be signed measures. Show that $\nu \ll \mu$ if and only if $\nu^+, \nu^- \ll \mu$.

***5.** Show that "$\ll$" is not a symmetric relation by giving an example of a pair of measures μ and ν satisfying $\mu \ll \nu$, but $\nu \ll \mu$ is false.

***6.** For signed measures, prove that $\nu \perp \mu$ implies $\nu^+, \nu^-, |\nu| \perp \mu$.

***7.** Complete the proof of Lemma 4 by showing that $\epsilon = 1/q$ and T_q are the sought after positive number and measurable set.

***8.** Prove Theorem 6.

***9.** Prove the uniqueness of the Lebesgue decomposition.

10. μ is a finite measure and $\nu \ll \mu$; $P_n \cup N_n$ is a Hahn decomposition for the signed measure $\nu - n\mu$, $n = 1, 2, \ldots$. Set $P = \bigcap P_n$ and $N = \bigcup N_n$. Show that N is σ-finite with respect to the measure ν, and if S is a measurable subset of P, then either $\nu(S) = 0$ or $+\infty$.

***11.** Use Exercise 10 to prove that the Radon-Nikodym theorem is true if it is assumed only that μ is σ-finite and $\nu \ll \mu$.

12. Let $\mathcal{C}$ be the family of subsets of R which have the following properties: $S \in \mathcal{C}$ if and only if either S or its complement $R - S$ is countable.

(a) Prove that $\mathcal{C}$ is a σ-algebra.

(b) Prove that the two set functions defined below are measures on the σ-algebra $\mathcal{C}$:

$$\mu(S) = \begin{cases} \text{number of elements in } S \text{ if } S \text{ is a finite set,} \\ \infty \quad \text{otherwise.} \end{cases}$$

$$\nu(S) = \begin{cases} 0 & \text{if } S \text{ is countable,} \\ \infty & \text{if } S \text{ is not countable.} \end{cases}$$

(c) Prove that $\nu \ll \mu$, but that the Radon-Nikodym theorem is not applicable.

13. Let $(\Omega, \mathcal{C}, \mu)$ be a σ-finite measure space, and let $1 \le p < \infty$. Prove that if F is a bounded linear functional on L_p^μ, then there is an a.e. (μ) uniquely determined function α in L_q^μ, and

$$F(f) = \int \alpha f \, d\mu, \qquad \text{for all } f \in L_p^\mu.$$

[q is conjugate to p. This is the Riesz Representation theorem, which has already been proved for the real case. The proof of this theorem (11.3.2) may be considerably shortened if the Radon-Nikodym theorem is used to obtain the function α. Show first that $\nu(S) = F(\chi_S)$ is a signed measure and $\nu \ll \mu$.]

14. Complete the last step of the proof of Theorem 7 by extending [2] from the generalized step functions to L_1^ν.

***15.** Let $(\Omega, \mathcal{C}, \mu)$ be given, and let ν be a signed measure defined on this space. (a) Prove that $\nu \ll \mu$, if to each $\epsilon > 0$ there is a $\delta > 0$ such that $\mu(S) < \delta$ implies $\nu(S) < \epsilon$. (b) Prove that if $\mu(S) < \infty$ implies that $\nu(S) < \infty$, then the converse of (a) is true (cf. Proposition 9.4.4).

***16.** Let μ be the Lebesgue-Stieltjes measure that corresponds to the distribution function $F(x)$. Prove that if S is a finitely μ-measurable set, then $\int_S F'$ exists and is not greater than $\mu(S)$. When are these two quantities equal?

***17.** Prove the statement made in Example 9(b).

*Exercises which request the proof of a statement in the text, as well as those referred to in the text, are starred.

5 Product Spaces and Fubini's Theorem

By completing the space of n-dimensional real step functions in the manner described in Secs. 1 and 3 of Chapter 12, we obtain the complete space of integrable functions of n real variables. Since the space R^n may be thought of as the n-fold product of R with itself, it is natural to inquire into the possibility of representing the "n-dimensional" integral as an n-fold iterated real integral. The latter is simply the value obtained by integrating a quantity n times, once with respect to each of its coordinates. For example, the function $f(x,y) = 3x^2 + 2xy$ is continuous on the rectangle $S = \{(x,y): 0 \leq x \leq 1, 0 \leq y \leq 2\}$, and its integral (Riemann or Lebesgue) $\int_S f(x,y)\, dA$ is equal to the iterated integral

$$\int_0^2 \left\{ \int_0^1 (3x^2 + 2xy)\, dx \right\} dy = \int_0^2 (1 + y)\, dy = 4.$$

It is clear that the iterated integral has great computational advantages. Even for a relatively simple function such as $3x^2 + 2xy$, the calculation of the multiple integral requires the summation of several infinite series, and is at best tedious. By reducing a double integral to an iterated integral it is often possible to avoid taking the limit, which is required whenever an integral is calculated directly from the definition. The problem is thus reduced to two successive real integrations. Here we have at our disposal the Fundamental Theorem of Calculus, which is a great computational advantage whenever antiderivatives of the integrands are known.

There are several ways to go about defining the product measure and the integral over subsets of the product space. One very natural method, outlined in the exercises (Exercise 1) makes use of the construction of measures which is discussed in Sec. 1. Here a measure function is first defined on the *rectangular subsets* of the product space (Definition 1). This collection of sets is easily seen to be a semialgebra, and therefore the measure function has an extension to a measure defined on a σ-algebra of subsets of the product space provided that the conditions of Exercise 13.1.4 are satisfied. We shall instead make use of the Daniell theory of the abstract integral and prove the Fubini theorem first for a class of elementary functions (the generalized step functions). The theorem is then extended to Γ^μ by taking monotone limits, and finally by linearity to L_1^μ.

Let $(\Omega_i,\mathcal{C}_i,\mu_i)$, $i = 1,2,\ldots,n$, be a collection of complete, σ-finite, measure spaces, and let C^i, Γ^i, L_1^i, and M^i denote the corresponding spaces of generalized step functions, monotone limits of sequences of generalized step functions with uniformly bounded integrals, μ_i-integrable functions, and μ_i-measurable functions. The familiar case of the n-dimensional Lebesgue integral suggests the proper choice of elementary functions to be defined on the product space $\Omega = \Omega_1 \otimes \Omega_2 \otimes \cdots \otimes \Omega_n$, whose elements are the n-tuples $x = (x_1, x_2, \ldots, x_n)$, $x_i \in \Omega_i$. These are easily seen to be the natural generalization of the n-dimensional step functions. Preliminary to the general Fubini Theorem, we shall prove first that the integral of an elementary function may be computed by successive iterations over the Ω_i.

1 Definition. A subset S of Ω that can be written as a product $S_1 \otimes S_2 \otimes \cdots \otimes S_n$ of μ_i-measurable subsets of the Ω_i is called a *rectangular set*. If, moreover, the S_i are finitely measurable, then S is said to be a *finite rectangule*. (See Exercise 1.)

2 Definition. By a *step function* on Ω is meant a (finite) linear combination of characteristic functions of finite rectangular sets. That is, $\sigma(x)$, where $x = (x_1, x_2, \ldots, x_n)$, $x_i \in \Omega_i$, is a step function if there are finite rectangular sets $S^j = S_1^j \otimes S_2^j \otimes \cdots \otimes S_n^j$, $j = 1, 2, \ldots, m$, in Ω, and numbers $\alpha_1, \alpha_2, \ldots, \alpha_m$ such that

$$\sigma(x) = \begin{cases} \alpha_j & \text{if } x \in S^j \\ 0 & \text{otherwise.} \end{cases} \tag{1}$$

We may also write $\sigma(x) = \sum_{j=1}^{m} \alpha_j \chi_{S^j}$

3 Definition. The *μ-integral of a step function* is given by

$$\int \sigma \, d\mu = \sum_{j=1}^{m} \alpha_j \mu_1(S_1^j) \cdot \mu_2(S_2^j) \cdot \cdots \cdot \mu(S_n^j). \tag{2}$$

(Cf. Sec. 1 of Chapter 12 and Definition 5.1.1.)

The reader should have no difficulty in verifying that the collection C^μ of step functions is a real vector space and a lattice, and therefore a class of elementary functions (Definition 12.1.1), and that the integral [2] is linear, homogeneous and monotonic. Only the continuity of [2] remains to be proved in order to assert that [2] is a Daniell integral on C^μ, after which it may be extended to Γ^μ and L_1^μ. The verification of the continuity property for integrals of step functions requires first that it be proved that the μ-integral [2] can be written as an *iterated integral*:

$$\int \left(\cdots \int \left(\int \sigma \, d\mu_1 \right) d\mu_2 \cdots \right) d\mu_n. \tag{3}$$

[3] is interpreted in the usual way: Let x_i, $i \geq 2$ denote fixed but arbitrary points of S_i. Then

$$\sigma(x) = \sigma_{2,\ldots,n}(x_1) = \sum_{j=1}^{m} \alpha_j \chi_{S_1^j}(x_1)[\chi_{S_2^j}(x_2)\cdots\chi_{S_n^j}(x_n)]$$

is a generalized step function on Ω_1 (Definition 12.5.12), since the quantity in brackets is either 1 or 0 depending upon whether or not $x_i \in S_i^j$, $i = 2, 3, \ldots, n$, and $j = 1, 2, \ldots, m$. The integration of $\sigma(x) = \sigma_{2,\ldots,n}(x_1)$ with respect to μ_1 yields

[4] $$\int \sigma \, d\mu_1 = \sum_{j=1}^{m} \alpha_j \mu_1(S_1^j)[\chi_{S_2^j}(x_2)\cdots\chi_{S_n^j}(x_n)].$$

The quantity on the right-hand side is a function of the $n - 1$ variables, $x_2, \ldots, x_n$. In fact, it is a step function on $\Omega_2 \otimes \cdots \otimes \Omega_n$, so that if $x_3, \ldots, x_n$ are held fixed, a step function on Ω_2 is obtained which may be integrated with respect to μ_2. This procedure may be repeated until the integration with respect to μ_n is carried out. In order to show that [2] and [3] are equal, it suffices to verify the equality for the characteristic function of a (finite) rectangular set. The linearity of the integral yields the desired result for all step functions.

The proofs will be carried out for $n = 2$. This is simply a notational convenience, and the general case is not conceptually different.

4 Proposition. Let $S = S_1 \otimes S_2$ be a finite rectangular set in $\Omega = \Omega_1 \otimes \Omega_2$. Then

[5] $$\int \chi_S \, d\mu = \int \left(\int \chi_S \, d\mu_1 \right) d\mu_2 = \int \left(\int \chi_S \, d\mu_2 \right) d\mu_1.$$

Proof: The definition of a characteristic function yields

$$\chi_S(x_1, x_2) = \begin{cases} \chi_{S_1}(x_1) & \text{if } x_2 \in S_2 \\ 0 & \text{if } x_2 \in \Omega_2 - S_2, \end{cases}$$

and it follows that

$$\int \chi_S \, d\mu_1 = \begin{cases} \mu_1(S_1) & \text{if } x_2 \in S_2 \\ 0 & \text{if } x_2 \in \Omega_2 - S_2. \end{cases}$$

Thus, the first iterated integral of [5] is equal to

$$\int \left(\int \chi_S \, d\mu_1 \right) d\mu_2 = \mu_1(S_1) \int \chi_{S_2} \, d\mu_2 = \mu_1(S_1) \cdot \mu_2(S_2),$$

which is precisely the μ-integral of χ_S (see [2]). Reversing the order of integration, the second equality is obtained. ∎

The linearity of the integral implies the following more general result:

5 Corollary (Fubini's Theorem for step functions). If σ is a step function defined on $\Omega = \Omega_1 \otimes \Omega_2$, then the μ-integral of σ and the two iterated integrals exist, and

$$\int \sigma \, d\mu = \int \left(\int \sigma \, d\mu_1 \right) d\mu_2 = \int \left(\int \sigma \, d\mu_2 \right) d\mu_1 .$$

We show next that the continuity of the μ-integral is a simple consequence of the continuity of the μ_i-integrals.

6 Proposition. If $\sigma_n = \sigma_n(x_1,x_2)$ is a sequence of nonnegative step functions tnat converge monotonically to zero, then

$$\int \sigma_n \, d\mu \downarrow 0.$$

Proof: Corollary 5 yields $\int \sigma_n \, d\mu = \int \left(\int \sigma_n \, d\mu_1 \right) d\mu_2$. Moreover, $\int \sigma_n \, d\mu_1 = \tau_n(x_2)$ is a step function defined on Ω_2, and so the monotonicity and the continuity of the μ_1-integral imply that $\tau_n \downarrow 0$ on Ω_2. Integrating again, and using this time the continuity of the μ_2-integral, we obtain

$$\int \sigma_n \, d\mu = \int \tau_n \, d\mu_2 \downarrow 0. \quad \blacksquare$$

This completes the proof that [2] is a Daniell integral defined on the class C^μ of step functions. The next step is to extend C^μ, first to Γ^μ by taking monotone limits, and then to L_1^μ, which may be thought of as the space of functions that can be written as the difference of functions in Γ^μ. The Daniell integral is extended in the manner described in Chapter 12. Measurable functions and sets are defined in the usual way, the latter being those sets whose characteristic functions are measurable. If S is finitely measurable, then $\mu(S) = \int \chi_S \, d\mu$. μ is called the *product measure*, and indeed if S is a finite rectangular set, then $\mu(S)$ is the product $\mu_1(S_1) \cdot \mu_2(S_2) \cdot \cdots \cdot \mu_n(S_n)$ (Proposition 4). The class of measurable sets is denoted by $\mathfrak{M}^\mu$.

In formulating and proving Fubini's theorem for step functions, the existence of the μ_1-integral (the first iterate) $\int \sigma \, d\mu_1$, was a consequence of the simplicity of the function σ. In general, if $f \in L_1^\mu$, then the function of x_1 which is obtained from $f(x_1,x_2)$ by fixing x_2, may not be μ_1-integrable for all values of x_2, so that the argument given in Proposition 4 is not generally applicable. The following simple example illustrates this difficulty:

7 Example. Let $\Omega_i = R$, $\mu_i = m$, $i = 1,2$. Then $\Omega = R^2$ and μ is the two-dimensional Lebesgue measure. If $g(x)$ is an nonmeasurable function defined on [0,1] ($g(x)$ may be taken as the characteristic function of a nonmeasurable set (see Sec. 4 of Chapter 7), and if $S = [0,1] \otimes [0,1]$, then

$$f(x_1,x_2) = \begin{cases} 1 & \text{if } x_1 \neq 1/2 \text{ and } (x_1,x_2) \in S \\ g(x_2) & \text{if } x_1 = 1/2 \text{ and } (x_1,x_2) \in S \\ 0 & \text{if } (x_1,x_2) \in R^2 - S \end{cases}$$

is integrable over R^2. Indeed, $f(x_1,x_2) = \chi_S(x_1,x_2)$ a.e. (μ) since $T = \{(x_1,x_2): x_1 = 1/2 \text{ and } 0 \leq x_2 \leq 1\}$ has μ-measure equal to zero. (T is nonmeasurable only when considered as a one-dimensional set, that is as a subset of R.) The iterated integral $\int\left(\int f\, d\mu_1\right) d\mu_2$ exists and is equal to $\int f\, d\mu$. However, $f(1/2,x_2) = g(x_2)$ is not μ_2-integrable, so that the iterated integral taken in the reversed order does not exist, at least not in the narrow sense described above. In the statement of the general theorem, we shall see that whenever f is μ-integrable, then the first iterate, $\int f\, d\mu_1 = p(x_2)$ exists for "almost all values of x_2," that is, except on a μ_2-null set.

The μ-null set T described in Example 7 is a special case of a *cross section* of a given set:

8 Definition. If S is a subset of the product space $\Omega = \Omega_1 \otimes \Omega_2$, and $a \in \Omega_1$, then the *cross section* of S determined by a is the following subset of Ω_2:

$$S_2(a) = \{x_2 \in \Omega_2 : (a,x_2) \in S\}$$

It is easily checked that $\{a\} \otimes S_2(a) \subset S$.

A similar definition may be given for a cross section determined by a point in Ω_2 (see Example 9).

9 Example. If $\Omega_i = R$, and $S = \{(x,y): 0 \leq x,\ y \leq 1\} \cup \{(x,y): x \geq 2\}$, then $S_1(1/2) = [0,1] \cup [2,\infty)$, $S_1(3) = [2,\infty)$, $S_2(4) = \Omega_2 = R$, and $S_2(3/2) = \phi$. (See Figure 35. The shaded area represents S and $S_1(1/2)$ is shown by the heavy line.)

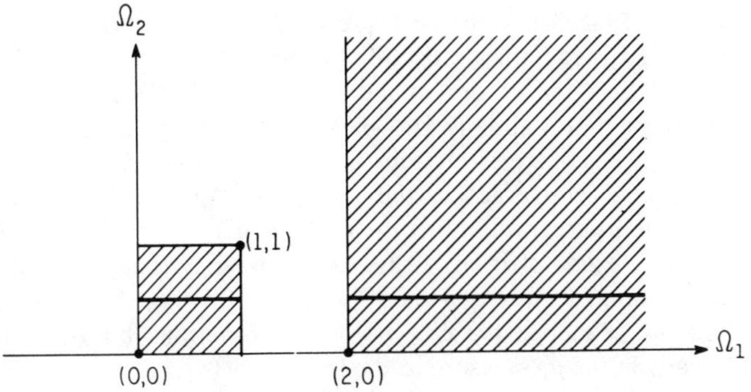

FIGURE 35 •

10 Theorem (Fubini). Let $(\Omega_i,\mathcal{C}_i,\mu_i)$, $i = 1,2$, be complete measure spaces, and let μ be the product measure defined on $\Omega_1 \otimes \Omega_2 = \Omega$. If f is a function in L_1^μ, then for almost all values of x_2, that is, except on a

•See special note on page 447.

μ_2-null set E_2, $f(x_1,x_2)$ is μ_1-integrable, the function

$$p(x_2) = \begin{cases} 0 & \text{if } x_2 \in E_2 \\ \int f\,d\mu_1 & \text{if } x_2 \in \Omega_2 - E_2 \end{cases}$$

is μ_2-integrable, and

[6] $$\int p\,d\mu_2 = \int\left(\int f\,d\mu_1\right)d\mu_2 = \int f\,d\mu.$$

A similar statement can be made for the other iterated integral.

Proof: We may assume that f is in Γ^μ. The linearity of the integral permits the extension to the general case. If $\{\sigma_n\}$ is a nondecreasing sequence of step functions which converge monotonically to f a.e. (μ), then

[7] $$\int f\,d\mu = \lim_{n\to\infty} \int \sigma_n\,d\mu.$$

Proposition 4 may be applied to the right side of [7] to obtain

[8] $$\int f\,d\mu = \lim_{n\to\infty} \int\left(\int \sigma_n\,d\mu_1\right)d\mu_2.$$

We must show that this limit may be taken under the integral, for this will imply that [8] is equal to

[9] $$\int\left(\int \lim_{n\to\infty} \sigma_n\,d\mu_1\right)d\mu_2 = \int\left(\int f\,d\mu_1\right)d\mu_2.$$

The equality of [8] and [9] is simply [6]. The proof is carried out in three steps:

Step (i):

We prove first that there is a μ_2-null set E_2 with the following property: If $x_2 \in \Omega_2 - E_2$, then $\sigma_n(x_1,x_2) \uparrow f(x_1,x_2)$ a.e. (μ_1). (The μ_1-integrability of $f(x_1,x_2)$ for almost all x_2 is proved in step (iii).)

Remark. Lest the reader mistakenly think that we are trying to prove something that is already known, we wish to emphasize that $\{\sigma_n\}$ is a monotone sequence of step functions defined on the *product space* Ω, and that the sequence converges in the *μ-measure* (that is, in the product measure) to the μ-integrable function f. In order to prove that the first iterate, $\int f\,d\mu_1$, exists, it is necessary to show first that $f(x_1,x_2)$ is μ_1-integrable for almost all x_2. This is done by showing that if x_2 is held fixed, then the sequence $\{\sigma_n(x_1,x_2)\}$ of step functions on Ω_1 converges monotonically in the μ_1-measure to the function $f(x_1,x_2)$.

Step (ii):

We prove next that the nondecreasing sequence of μ_2-step functions,

$$\tau_n(x_2) = \int \sigma_n(x_1,x_2)\,d\mu_1,$$

converges monotonically to a μ_2-integrable function $p(x_2)$, and that

[10] $$\int p\,d\mu_2 = \lim_{n\to\infty} \int \tau_n\,d\mu_2 = \lim_{n\to\infty} \int \left(\int \sigma_n\,d\mu_1\right) d\mu_2,$$

which is the right-hand side of [8].

Step (iii):

The μ_2-integrability of $p(x_2)$ will imply that the sequence of integrals, $\int \sigma_n(x_1,x_2)\,d\mu_1$, is uniformly bounded for almost all x_2. The theorem on monotone convergence yields the μ_1-integrability of $f(x_1,x_2)$ for almost all x_2, and

[11] $$\int f(x_1,x_2)\,d\mu_1 = \lim_{n\to\infty} \int \sigma_n(x_1,x_2)\,d\mu_1 = p(x_2) \text{ a.e. } (\mu_2).$$

This, however, implies that the right-hand side of [10] is the iterated integral of f, thus proving that [8] and [9] are equal.

The existence of the μ_2-null set E_2 of step (i) is guaranteed by the following lemma:

11 Lemma. Suppose that E is a μ-null set in the product space Ω, and that

[12] $$E_2 = \{x_2 \in \Omega_2 : E_1(x_2) \text{ is not } \mu_1\text{-null}\}.$$

($E_1(x_2)$ is a cross-section. See Definition 8.) Then $\mu_2(E_2) = 0$.

Proof: $\mu(E) = 0$ implies that there is a nondecreasing sequence $\{\gamma_n\}$ of step functions (on Ω) satisfying

(a) $\gamma_n(x_1,x_2) \uparrow \infty$ if $(x_1,x_2) \in E$;

(b) If A is any positive number, then a sequence can be found which satisfies (a) and whose integrals are uniformly bounded by A.

By integrating once, a nondecreasing sequence

$$\varphi_n(x_2) = \int \gamma_n(x_1,x_2)\,d\mu_1$$

of generalized step functions on Ω_2 is obtained, and Corollary 5 implies that

[13] $$\int \varphi_n\,d\mu_2 = \int \gamma_n\,d\mu < A.$$

Since the Ω_i are complete measure spaces, a set $E_i \in \mathfrak{C}_i$ has μ_i-measure zero if and only if E_i is a μ_i-null set arising from the Daniell integral defined on the space of generalized step functions (Lemma 12.5.17). Thus to prove that $\mu_2(E_2) = 0$, it suffices to show that the sequence $\{\varphi_n\}$, of generalized step functions on Ω_2, diverges on E_2, for we already know that the sequence of integrals is uniformly bounded [13]. If $x_2 \in \Omega_2$, then

$$\{x_2\} \otimes E_1(x_2) \subset E.$$

Therefore, if $x_1 \in E_1(x_2)$, then $\gamma_n(x_1,x_2) \uparrow \infty$. (This follows from (a).) However, the set $E_1(x_2)$ has positive μ_1-measure whenever $x_2 \in E_2$ (see [2]). It follows that if $x_2 \in E_2$,

$$\varphi_n(x_2) = \int \gamma_n(x_1,x_2)\, d\mu_1 \uparrow \infty,$$

which completes the proof of the lemma.

Let us return to Step (i). It is given that $\sigma_n \uparrow f$ on $\Omega - E$, where E is a μ-null set. Fixing x_2, it follows from Lemma 11 that unless x_2 is in the μ_2-null set E_2, the cross section of E determined by x_2,

$$E_1(x_2) = \{x_1 \in \Omega_1 : (x_1,x_2) \in E\},$$

is a μ_1-null set. Thus, except for $x_2 \in E_2$,

$$\sigma_n(x_1,x_2) \uparrow f(x_1,x_2) \text{ a.e. } (\mu_1),$$

which completes Step (i).

The sequence $\{\tau_n(x_2)\}$ of step functions on Ω_2 [see Step (ii)] is nondecreasing, and

$$\int \tau_n\, d\mu_2 = \int \Big(\int \sigma_n(x_1,x_2)\, d\mu_1 \Big)\, d\mu_2 = \int \sigma_n\, d\mu < A.$$

Hence the theorem on monotone convergence implies that except on a μ_2-null set, $\{\tau_n\}$ converges monotonically to a μ_2-integrable function $p(x_2)$, and [10] follows, which is Step (ii).

The μ_2-integrability of $p(x_2)$ implies that except on a μ_2-null set,

$$\tau_n(x_2) = \int \sigma_n(x_1,x_2)\, d\mu_1 < p(x_2) < \infty.$$

Again, the theorem on monotone convergence yields that for almost all x_2, the limit function $f(x_1,x_2)$ of the sequence $\{\sigma_n(x_1,x_2)\}$ is μ_1-integrable and [11] follows, which completes the proof of the theorem. ▌

Exercises

***1.** Instead of defining the product measure in terms of a Daniell integral, it is possible to define first a "measure function" on the semialgebra $\mathcal{S}$ of rectangular subsets of Ω, and then to extend this function in the manner described in Sec. 1.

The reader is asked first to show that the family $\mathcal{S}$ of rectangular subsets (finite and infinite) of Ω is a semialgebra. A measure function μ is defined on $\mathcal{S}$ as follows: Let $S = S_1 \otimes S_2 \otimes \cdots \otimes S_n$ be a rectangular set. If the S_i are finitely measurable, then set $\mu(S) = \mu_1(S_1) \cdot \mu_2(S_2) \cdot \cdots \cdot \mu_n(S_n)$. If at least one of the S_i has infinite measure, then put $\mu(S) = \infty$. Show that μ satisfies the conditions of Exercise 13.1.4, and that therefore it may be extended to a complete measure on a σ-algebra of sets which contains $\mathcal{S}$. Why is this measure the same as the one generated by the Daniell integral? (See Sec. 5 of Chapter 12.)

2. Let $f_i(x_i)$ be μ_i-integrable functions on the complete measure spaces $(\Omega_i,\mathcal{C}_i,\mu_i)$, $i = 1, 2$. Prove that the function $f(x_1, x_2) = f_1(x_1) f_2(x_2)$ is integrable over the product space, and that

$$\int f\, d\mu = \Big(\int f_1\, d\mu_1\Big) \cdot \Big(\int f_2\, d\mu_2\Big).$$

3. (Tonelli's theorem.) Prove that if f is a nonnegative, measurable function defined on the product space Ω of two complete, σ-finite, measure spaces $(\Omega_i, \mathcal{C}_i, \mu_i)$, $i = 1, 2$, then f is integrable over Ω if either of its iterated integrals exists. (Fubini's theorem implies further that if the conditions of Tonelli's theorem are satisfied, then both the iterated integrals exist and are equal to the integral over the product space.)
4. Let Z_+ denote the set of positive integers, and let $\mathcal{C}$ be the collection of all subsets of Z_+.
   (a) Prove that the set function on $\mathcal{C}$ defined by

$$\mu(S) = \begin{cases} \text{number of elements in } S \text{ if } S \text{ is finite} \\ \infty \quad \text{if } S \text{ is infinite} \end{cases}$$

   is a complete measure, but is not σ-finite.
   (b) Describe the integrable and measurable functions and the generalized step functions.
   (c) Form the product of this measure space with itself. Describe the measurable sets, measurable and integral functions, and step functions.
   (d) Interpret Fubini's theorem for this product space.

*Exercises which request the proof of a statement made in the text, as well as those referred to in the text, are starred.

Bibliography

E. Asplund and L. Bungart, [1] *A First Course In Integration.* 3d ed. New York: Macmillan, 1965.

L. Bers, [1] *Topology* (Lecture Notes). New York: New York University, 1957.

G. Birkhoff and S. Maclane, [1] *A Survey Of Modern Algebra.* 3d ed. New York: Macmillan, 1965.

G. Birkhoff and G. Rota, [1] *Ordinary Differential Equations.* Boston: Ginn, 1962.

R. P. Boas, [1] *A Primer of Real Functions*, Carus Mathematical Monograph No. 13, Mathematical Association of America, 1960.

C. W. Burrill and J. R. Knudsen, [1] *Real Variables.* New York: Holt, 1969.

W. G. Chinn and N. E. Steenrod, [1] *First Concepts of Topology*, S.M.S.G. Monograph Project. New York: Random House and L. W. Singer, 1966.

R. Courant and D. Hilbert, [1] *Methods of Mathematical Physics* (English ed.), Vol. I. New York: Interscience, 1954.

P. J. Daniell, [1] "A General Form Of The Integral," *Annals of Mathematics* (2), Vol. 19 (1917–1918).

[2] "Further Properties of the General Integral," *Annals of Mathematics* (2), Vol. 22 (1919–1920).

B. R. Gelbaum and J. M. H. Olmstead, [1] *Counterexamples in Analysis.* San Francisco: Holden-Day, 1964.

C. Goffman, [1] *Real Functions*, New York: Holt, 1953.

L. M. Graves, [1] *The Theory of Functions of Real Variables.* New York: McGraw-Hill, 1956.

D. W. Hall and G. L. Spencer, [1] *Elementary Topology.* New York: Wiley, 1955.

P. R. Halmos, [1] *Naive Set Theory.* Princeton: Van Nostrand, 1960.

[2] *Measure Theory.* Princeton: Van Nostrand, 1950.

I. N. Herstein, [1] *Topics in Algebra.* New York: Blaisdell, 1964.

E. Hewitt and K. Stromberg, [1] *Real and Abstract Analysis.* New York: Springer-Verlag, 1965.

E. Hille, [1] *Analytic Function Theory* (Vol. I). Boston: Ginn, 1962.

J. G. Hocking and G. S. Young, [1] *Topology*, Reading Mass.: Addison-Wesley, 1961.

G. Isaacs, [1] *Real Numbers.* New York: McGraw-Hill, 1968.

E. Kamke, [1] *Theory of Sets.* New York: Dover, 1950.

N. D. Kazarinoff, [1] *Analytic Inequalities.* New York: Holt, 1961.

J. L. Kelley, [1] *General Topology.* Princeton: Van Nostrand, 1955.

H. Kestelman, [1] *Modern Theories of Integration.* London: Oxford U. P., 1937.

A. N. Kolmogorov and S. V. Fomin, [1] *Measure, Lebesgue Integrals, and Hilbert Space.* New York: Academic, 1961.

R. G. Kuller, [1] *Topics in Modern Analysis.* Englewood Cliffs, N.J.: Prentice-Hall, 1969.

E. Landau, [1] *Foundations of Analysis.* New York: Chelsea, 1951.

H. Lebesgue, [1] *Measure And The Integral.* San Francisco: Holden-Day, 1966.

S. Lefschetz, [1] *Introduction to Topology.* Princeton: Princeton U. P., 1949.

L. H. Loomis, [1] *An Introduction to Abstract Harmonic Analysis.* Princeton: Van Nostrand, 1953.

N. H. McCoy, [1] *Introduction to Modern Algebra.* Boston: Allyn and Bacon, 1960.

E. J. McShane and T. A. Botts, [1] *Real Analysis.* Princeton: Van Nostrand, 1959.

B. Mendelson, [1] *Introduction to Topology.* Boston: Allyn and Bacon, 1962.

M. E. Munroe, [1] *Introduction to Measure and Integration.* Cambridge, Mass.: Addison-Wesley, 1953.

I. P. Natanson, [1] *Theory of Functions of a Real Variable.* 2 vols. New York: Ungar (Vol. I, 1955; Vol. II, 1960).

Z. Nehari, [1] *Conformal Mapping.* New York: McGraw-Hill, 1952.

M. H. A. Newman, [1] *Elements of the Topology of Plane Sets.* London: Cambridge U. P., 1964.

I. G. Petrovski, *Ordinary Differential Equations.* Englewood Cliffs, N.J.: Prentice-Hall, 1966.

F. Riesz and B. Nagy, [1] *Functional Analysis* (English ed.). New York: Ungar, 1956.

H. L. Royden, [1] *Real Analysis.* New York: Macmillan, 1963.

W. Rudin, [1] *Principles of Mathematical Analysis.* New York: McGraw-Hill, 1964.

S. Saks, [1] *Theory of The Integral.* 2d ed. Warsaw: Monografje Matematyczne, 1937.

G. E. Shilov and B. L. Gurevich, [1] *Integral, Measure and Derivative: A Unified Approach* (English ed.). Englewood Cliffs, N.J.: Prentice-Hall, 1966.

M. Shinbrot, [1] "Fixed Point Theorems," *Mathematics in the Modern World: Readings from Scientific American.* San Francisco: Freeman, 1968.

G. F. Simmons, [1] *Introduction to Topology and Modern Analysis.* New York: McGraw-Hill, 1963.

M. H. Stone, [1] "The Generalized Weierstrass Approximation Theorem," *Mathematics Magazine*, Vol. 21 (1948).
[2] "A Generalized Weierstrass Approximation Theorem," *Studies in Mathematics.* Vol. I, *Studies in Analysis.* Englewood Cliffs, N.J.: Mathematical Association of America and Prentice-Hall, 1962.

R. Solovay, [1] "The Measure Problem," *American Mathematical Society Notices*, 12 (1965).

P. C. Suppes, [1] *Axiomatic Set Theory.* Princeton: Van Nostrand, 1960.

A. E. Taylor, [1] *General Theory of Functions and Integration*, Waltham, Mass.: Blaisdell, 1965.
[2] *Introduction to Functional Analysis.* New York: Wiley, 1958.

E. C. Titchmarsh, [1] *The Theory of Functions.* 2d ed. London: Oxford U. P., 1939.

A. J. White, [1] *Real Analysis: An Introduction.* Reading, Mass.: Addison-Wesley, 1968.

J. H. Williamson, [1] *Lebesgue Integration.* New York: Holt, 1962.

Symbols

The following is a list of symbols and the numbers of the pages on which they either first appear or on which they appear with a new or more general meaning.

*Ordered sets and ordered fields are denoted in the same way. In each case the context will determine the meaning of the symbol.

Notes to the Dover Edition

Page

14. It is true that Property (i) implies that an ordered field is *not* reflexive—and for that reason it cannot be partially ordered by the relation $<$. Moreover, for any elements x and y, it cannot be the case that both $x < y$ and $y > x$. Hence an ordered field is *vacuously* antisymmetric.

57. The step function does not necessarily vanish at the endpoints of the intervals. However, this does not affect the argument, as noted on page 58.

365. In the proof of Proposition 3(d) it should be added that, since $\sigma \in C_0$, it follows that $\sigma^+ = \max(\sigma, \bar{0})$ and $\sigma^- = \min(\sigma, \bar{0})$ are also in C_0 (Definition 1 (ii)).

368. In Exercise 6(a), note that each function in C_0 vanishes outside *some* interval. In 6(b) there is a single interval, $[a, b]$, outside of which *all* the functions vanish.

435. In Figure 35, since $S_1(1/2) \subset \Omega_1$, the heavy line shows the set $S_1(1/2) \otimes \{1/2\}$, not $S_1(1/2)$, which should be indicated by a heavy line on the horizontal axis.

Index